From Quantum Paraelectric/Ferroelectric Perovskite Oxides to High Temperature Superconducting Copper Oxides

From Quantum Paraelectric/Ferroelectric Perovskite Oxides to High Temperature Superconducting Copper Oxides

In Honor of Professor K.A. Müller for His Lifework

Editors
Annette Bussmann-Holder
Hugo Keller
Antonio Bianconi

MDPI • Basel • Beijing • Wuhan • Barcelona • Belgrade • Manchester • Tokyo • Cluj • Tianjin

Editors

Annette Bussmann-Holder
Max-Planck-Institut für
Festkörperforschung
Germany

Hugo Keller
Physik-Institut der Universität
Zürich
Switzerland

Antonio Bianconi
Rome International Center for
Materials Science Superstripes
(RICMASS)
Italy

Editorial Office
MDPI
St. Alban-Anlage 66
4052 Basel, Switzerland

This is a reprint of articles from the Special Issue published online in the open access journal *Condensed Matter* (ISSN 2410-3896) (available at: https://www.mdpi.com/journal/condensedmatter/special_issues/KAMUller).

For citation purposes, cite each article independently as indicated on the article page online and as indicated below:

LastName, A.A.; LastName, B.B.; LastName, C.C. Article Title. *Journal Name* **Year**, *Volume Number*, Page Range.

ISBN 978-3-0365-0474-2 (Hbk)
ISBN 978-3-0365-0475-9 (PDF)

Cover image courtesy of Hugo Keller.

© 2021 by the authors. Articles in this book are Open Access and distributed under the Creative Commons Attribution (CC BY) license, which allows users to download, copy and build upon published articles, as long as the author and publisher are properly credited, which ensures maximum dissemination and a wider impact of our publications.

The book as a whole is distributed by MDPI under the terms and conditions of the Creative Commons license CC BY-NC-ND.

Contents

About the Editors vii

Preface to "From Quantum Paraelectric/Ferroelectric Perovskite Oxides to High Temperature Superconducting Copper Oxides" ix

Annette Bussmann-Holder and Hugo Keller
From $SrTiO_3$ to Cuprates and Back to $SrTiO_3$: A Way Along Alex Müller's Scientific Career
Reprinted from: *Condens. Matter* **2021**, 6, 2, doi:10.3390/condmat6010002 1

Giorgio Benedek, Joseph R. Manson, Salvador Miret-Artés, Adrian Ruckhofer, Wolfgang E. Ernst, Anton Tamtögl and Jan Peter Toennies
Measuring the Electron–Phonon Interaction in Two-Dimensional Superconductors with He-Atom Scattering
Reprinted from: *Condens. Matter* **2020**, 5, 79, doi:10.3390/condmat5040079 29

Isaac B. Bersuker and Victor Polinger
Perovskite Crystals: Unique Pseudo-Jahn–Teller Origin of Ferroelectricity, Multiferroicity, Permittivity, Flexoelectricity, and Polar Nanoregions
Reprinted from: *Condens. Matter* **2020**, 5, 68, doi:10.3390/condmat5040068 51

Andreas Bill, Vladimir Hizhnyakov, Reinhard K. Kremer, Götz Seibold, Aleksander Shelkan, and Alexei Sherman
Phase Separation and Pairing Fluctuations in Oxide Materials
Reprinted from: *Condens. Matter* **2020**, 5, 65, doi:10.3390/condmat5040065 79

Hans Bill
Color Centers and Jahn-Teller Effect in Ionic Crystals—My Scientific Encounters with Alex Müller
Reprinted from: *Condens. Matter* **2020**, 5, 59, doi:10.3390/condmat5040059 103

Alan R. Bishop
A Lattice Litany for Transition Metal Oxides
Reprinted from: *Condens. Matter* **2020**, 5, 46, doi:10.3390/condmat5030046 113

Kazimierz Conder, Albert Furrer and Ekaterina Pomjakushina
A Retrospective of Materials Synthesis at the Paul Scherrer Institut (PSI)
Reprinted from: *Condens. Matter* **2020**, 5, 55, doi:10.3390/condmat5040055 135

Guy Deutscher
The Role of the Short Coherence Length in Unconventional Superconductors
Reprinted from: *Condens. Matter* **2020**, 5, 77, doi:10.3390/condmat5040077 145

Carlo Di Castro
Revival of Charge Density Waves and Charge Density Fluctuations in Cuprate High-Temperature Superconductors
Reprinted from: *Condens. Matter* **2020**, 5, 70, doi:10.3390/condmat5040070 149

Zurab Guguchia
Unconventional Magnetism in Layered Transition Metal Dichalcogenides
Reprinted from: *Condens. Matter* **2020**, 5, 42, doi:10.3390/condmat5020042 165

Hiroshi Kamimura, Masaaki Araidai, Kunio Ishida, Shunichi Matsuno, Hideaki Sakata, Kenji Shiraishi, Osamu Sugino and Jaw-Shen Tsai
First-Principles Calculation of Copper Oxide Superconductors That Supports the Kamimura-Suwa Model
Reprinted from: *Condens. Matter* **2020**, 5, 69, doi:10.3390/condmat5040069 181

Rustem Khasanov, Alexander Shengelaya, Roland Brütsch and Hugo Keller
Suppression of the *s*-Wave Order Parameter Near the Surface of the Infinite-Layer Electron-Doped Cuprate Superconductor $Sr_{0.9}La_{0.1}CuO_2$
Reprinted from: *Condens. Matter* **2020**, 5, 50, doi:10.3390/condmat5030050 193

Wolfgang Kleemann, Jan Dec, Alexander Tkach and Paula M. Vilarinho
$SrTiO_3$—Glimpses of an Inexhaustible Source of Novel Solid State Phenomena
Reprinted from: *Condens. Matter* **2020**, 5, 58, doi:10.3390/condmat5040058 205

Boris I. Kochelaev
Detection of Two Phenomena Opposite to the Expected Ones
Reprinted from: *Condens. Matter* **2020**, 5, 56, doi:10.3390/condmat5040056 219

Thomas W. Kool
Meetings with a Remarkable Man, Alex Müller—The Professor of SrTiO3
Reprinted from: *Condens. Matter* **2020**, 5, 44, doi:10.3390/condmat5030044 231

Reinhard K. Kremer, Annette Bussmann-Holder, Hugo Keller and Robin Haunschild
The Crucial Things in Science often Happen Quite Unexpectedly—Das Entscheidende in der Wissenschaft Geschieht oft Ganz Unerwartet (K. Alex Müller)
Reprinted from: *Condens. Matter* **2020**, 5, 43, doi:10.3390/condmat5030043 243

Jakob Nachtigal, Marija Avramovska, Andreas Erb, Danica Pavićević, Robin Guehne and Jürgen Haase
Temperature-Independent Cuprate Pseudogap from Planar Oxygen NMR
Reprinted from: *Condens. Matter* **2020**, 5, 66, doi:10.3390/condmat5040066 253

Ekhard K. H. Salje
Polaronic States and Superconductivity in WO_{3-x}
Reprinted from: *Condens. Matter* **2020**, 5, 32, doi:10.3390/condmat5020032 273

Gernot Scheerer, Margherita Boselli, Dorota Pulmannova, Carl Willem Rischau, Adrien Waelchli, Stefano Gariglio, Enrico Giannini, Dirk van der Marel and Jean-Marc Triscone
Ferroelectricity, Superconductivity, and $SrTiO_3$—Passions of K.A. Müller
Reprinted from: *Condens. Matter* **2020**, 5, 60, doi:10.3390/condmat5040060 281

Alexander Shengelaya, Fabio La Mattina and Kazimierz Conder
Unconventional Transport Properties of Reduced Tungsten Oxide $WO_{2.9}$
Reprinted from: *Condens. Matter* **2020**, 5, 63, doi:10.3390/condmat5040063 291

About the Editors

Annette Bussmann-Holder
Professor of Physics
Emphasis on solid state physics: phase transitions, ferro- and antiferroelectricity, perovskite oxides, nonlinear phenomena and excitations, magnetism, superconductivity, high-temperature superconductivity, multiferroic systems.

Methods: analytical approaches, self-consistent phonon approximation, Bogoliubov transformation, Lang–Firsov transformation.

Hugo Keller
Professor of Experimental Physics
Emphasis on solid state physics: magnetism, superconductivity, high-temperature superconductivity, phase transitions and critical phenomena, multiferroic systems, muonium physics.

Experimental Methods: Mössbauer spectroscopy, muon spin rotation (μSR), electron paramagnetic resonance (EPR), nuclear magnetic resonance (NMR), SQUID and torque magnetometry.

Antonio Bianconi
Prof. Dr., European Academy of Science Member
Emphasis on Quantum Complex Matter: high-temperature superconductors, multiple scattering theory, multigap superconductors, Fano resonances, spin-orbit interaction, quantum materials for quantum computers, multiscale correlated disorder, polarons, phase separation, polymorphism, physics of life and quantum biology, metalloproteins, intrinsic disordered proteins, myelin fluctuations, epidemic spreading.

Experimental Methods: X-ray spectroscopy, X-ray absorption near edge structure (XANES), scanning micro X-ray diffraction, photo-induced effects, scanning micro-XANES, EXAFS.

Preface to "From Quantum Paraelectric/Ferroelectric Perovskite Oxides to High Temperature Superconducting Copper Oxides"

This Special Issue of the journal is dedicated to the lifework of Professor Dr. Dr. h.c. mult. K. Alex Müller. It comprises multiple contributions to solid state physics starting early on in the field of structural phase transitions in perovskite oxides with emphasis on their investigation using electron paramagnetic resonance (EPR). As a result, essential knowledge was gained on the order parameter of phase transition and the driving mechanism. He coined the term quantum paraelectricity to refer to the suppression of the expected transition to a polar state in $SrTiO_3$, which has raised enormous interest in the research community. Further EPR studies by Müller are related to the resonance of ions corresponding to impurities in diverse perovskites and related crystals and, thereby, the discovery of experimentally negative U-centers that were later on established theoretically. The Jahn–Teller effect attracted his attention early on and played a key role in the discovery of high-temperature superconductivity in cuprates, where the Jahn–Teller polaron is especially at the heart of this phenomenon. The polaron has also been shown to be vital in Fermi glasses, $LaBaNiO_4$, and doped polythiophene. The concept of the Jahn–Teller polaron in connection with perovskite oxides inspired Müller to search for superconductivity in these compounds which—as everybody knows—was successful. In 1986, together with Georg Bednorz, he discovered high-temperature superconductivity in LaBaCuO, with the highest ever observed transition temperatures at ambient pressure—a discovery which was awarded with the Nobel prize only one year later. This discovery caused an enormous worldwide breakthrough in basic research as well as in possible applications. In order to explain these findings, Müller suggested the polaron concept and bipolaron condensation. In order to verify this concept, he proposed searching for unconventional isotope effects which have indeed been observed in cuprates.

The above topics have been reviewed in detail in the first contribution to this Special Issue by two of the Guest Editors, A. Bussmann-Holder and H. Keller.

The contributions of all other authors are either in direct relation to K. A. Müller or are indirectly inspired by his work. The chosen topics range from perovskite oxides to cuprates and beyond, and address other systems and/or different aspects and viewpoints connected in one or the other way to K. A. Müller.

Annette Bussmann-Holder, Hugo Keller, Antonio Bianconi
Editors

Review

From SrTiO₃ to Cuprates and Back to SrTiO₃: A Way Along Alex Müller's Scientific Career

Annette Bussmann-Holder [1,*] and Hugo Keller [2,*]

1 Max-Planck-Institute for Solid State Research, Heisenbergstr. 1, D-70569 Stuttgart, Germany
2 Physik-Institut der Universität Zürich, Winterthurerstr. 190, CH-8057 Zürich, Switzerland
* Correspondence: a.bussmann-holder@web.de or a.bussmann-holder@fkf.mpg.de (A.B.-H.); keller@physik.uzh.ch (H.K.)

Received: 25 November 2020; Accepted: 27 December 2020; Published: 31 December 2020

Abstract: K.A. Müller took a long route in science leaving many traces and imprints, which have been and are still today initiations for further research activities. We "walk" along this outstanding path but are certainly not able to provide a complete picture of it, since the way was not always straight, often marked by unintended detours, which had novel impact on the international research society.

Keywords: perovskite oxides; phase transitions; high-temperature cuprate superconductors

---

1. Introduction

After World War II, science started a new blooming period, especially in solid state physics, which used to be the poor cousin in physics. In this early period, a number of novel systems were discovered, mostly in the need of applications to support the reconstruction of Europe. A lot of attention was devoted then to perovskite oxides, which, because of their ferroelectric properties, were excellent ultrasound transducers, piezoelectric converters, and pyroelectric devices. The most prominent members of this material class are $PbTiO_3$ and its mixed ceramics with $PbZrO_3$, still today installed in many applications, and being necessary ingredient for many techniques. $BaTiO_3$ is similarly important, together with $SrTiO_3$. Many other related compounds originated from this time, which all turned out to be of high scientific interest and relevant for new technologies (for a comprehensive review, see, Reference [1]). It was this exciting period when K.A. Müller started his research in experimental solid state physics.

2. Polar and Rotational Instabilities in Perovskite Oxides

In the early 1950s, the investigations of perovskites were enormously intense, since their ferroelectric properties were relevant for novel industrial applications. Understanding the background for the appearance of these phenomena intensified the experimental and theoretical activities and lead to new tools, like neutron scattering, nuclear and electron (para)magnetic resonance (NMR, EPR), and low-temperature techniques. These techniques evidenced that the ferroelectric perovskites have unique properties related to their dynamics, namely a distinct soft mode behavior of a long wave length transverse optic mode, where the related frozen-in ionic displacement pattern determines the low temperature structure and the magnitude of the polarization [2–4]. Thus, a unique characterization of ferroelectrics was possible through the identification of a soft optic mode. Besides the polar properties, further structural phase transitions were discovered, where, especially, $SrTiO_3$ [5], $LaAlO_3$ [6], and $BaTiO_3$ were of uttermost interest. In the former two compounds, phase transitions from cubic to tetragonal, cubic to $R\bar{3}c$, were observed and believed to be characteristic for oxide perovskites [5]. The origin of these transitions remained unknown, however, for a decade, as well as their precise transition temperatures. This issue was clarified by K.A. Müller et al. in 1968 [7], who showed that the rotation angle φ of the oxygen ion octahedra is the order parameter of these phase transitions (see Figure 1, where the perovskite structure and the rotation of the octahedra are schematically illustrated).

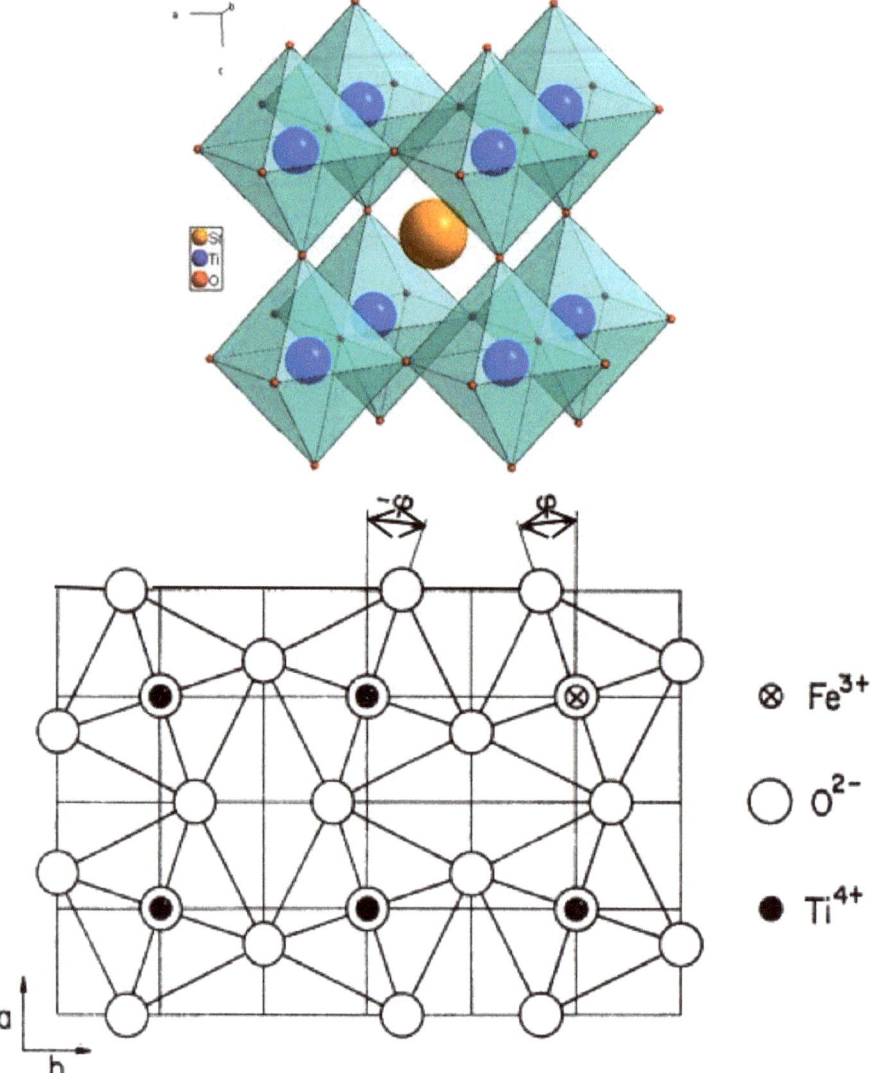

Figure 1. (**Top**) Perovskite structure of SrTiO$_3$. (**Bottom**) Oxygen octahedra rotated around the tetragonal c-axis in SrTiO$_3$ below the transition temperature T_s. The rotation angle φ represents the order parameter of the structural phase transition below T_s. (from References [5,8]).

He identified this assignment via electron paramagnetic resonance (EPR) experiments where tiny amounts of a magnetic ion (e.g., Fe as indicated in the lower part of Figure 1) were implanted in the respective samples. By comparing the normalized rotation angles of different perovskite oxides, he concluded that, indeed, a unique similarity between them was present, since the temperature dependence of the rotation angle follows one single law independent of the corresponding compound.

Since the rotation angle continuously goes to zero upon approaching the phase transition temperature, the transition has been classified as second order type.

Prior to this well-known work, K.A. Müller was interested in the Jahn-Teller effect, which continued to remain in his focus throughout his career, also and importantly, in the search for superconductivity. Already, in 1937, Jahn and Teller showed that, for a polyatomic molecule, "All nonlinear nuclear configurations are unstable for an orbitally degenerate ground state". Using EPR to detect d^7 impurity ions in Al_2O_3 and $SrTiO_3$, he showed that line broadening due to resonance relaxation occurs already above room temperature. By determining the level splitting between the ground and excited levels, he concluded that phonons are sufficient to overcome this splitting, thereby arriving at the concept of the dynamic Jahn-Teller effect [9], which later turned out to be at the heart for the discovery of high-temperature superconductivity.

The previous observations of a soft transverse optic mode at $q = 0$ gave rise to speculations that a coupling to an acoustic mode initiates the phase transition. This could rapidly be excluded, since the transition in $LaAlO_3$ has been related to the softening of a transverse acoustic mode at $(1/2, 1/2, 1/2)$ with no further evidence of any other soft mode being present [10]. On the opposite side, in $SrTiO_3$, there is indeed a soft transverse optic mode at $q = 0$. This observation led K.A. Müller et al. to propose that a second zone boundary soft acoustic mode must exist in $SrTiO_3$. This idea was directly proven to be true by Raman scattering experiments and, shortly afterwards, additionally verified by inelastic neutron scattering [11].

An interesting discovery was made by K.A. Müller, as early as 1969, where he experimentally demonstrated the existence of a negative U-center in Ni-doped $SrTiO_3$ [12]. At that time, this discovery did not draw much attention. However, six years later, P.W. Anderson [13], and, simultaneously and independently, R.A. Street and N.F. Mott [14], introduced theoretically the negative U centers to account for the fact that most glasses and amorphous semiconductors are diamagnetic. Obviously, they were unaware that this had already been discovered before experimentally [12]. There are important consequences from these centers for superconductivity, namely that there is an analogy between the existence of an effective attraction between electrons in a superconductor and the negative U centers in an insulating glass. Since the origin of this electron-electron attractive interaction stems from the lattice, it also bears a close resemblance to bipolaron formation, later on proposed by Chakraverty [15], to give rise to superconductivity.

The increasing interest in $SrTiO_3$ and other perovskite titanates successfully encouraged crystal growers to optimize their growth conditions, and it became possible to obtain almost domain-free samples of $SrTiO_3$ and $LaAlO_3$ in their low-temperature tetragonal, $R\bar{3}c$ phase. These were the starting point to investigate in deep detail, and with much improved resolution, the temperature dependence of the order parameter, the rotation angle $\varphi(T)$, in the vicinity of the phase transition temperature (see Figure 1) [16]. While it was long taken for given that Landau theory universally describes the behavior of φ by the law: $\varphi \propto \varepsilon^\beta, \varepsilon = (T_s - T)/T_s$, T_s being the phase transition temperature and $\beta = 1/2$, the better sample quality enabled to measure β with higher accuracy, especially in the vicinity of T_s. The new data evidenced deviations from this law close to T_s, thus questioning the global validity of Landau theory ($\beta = 1/2$). From this work, it was concluded that the temperature dependence of φ in $SrTiO_3$ and $LaAlO_3$, below the second-order phase transition temperatures T_s, is described by an exponent $\beta = 0.33(2)$ (critical exponent of the order parameter) in the temperature region $0.9 \leq t = T/T_s \leq 0.96$ (critical region). For $t < 0.9$, there occurs a change to Landau behavior (mean-field behavior), as indicated by the solid line in Figure 2. This observation of static critical exponents near displacive phase transitions confirmed the notion of universality classes in this field [16].

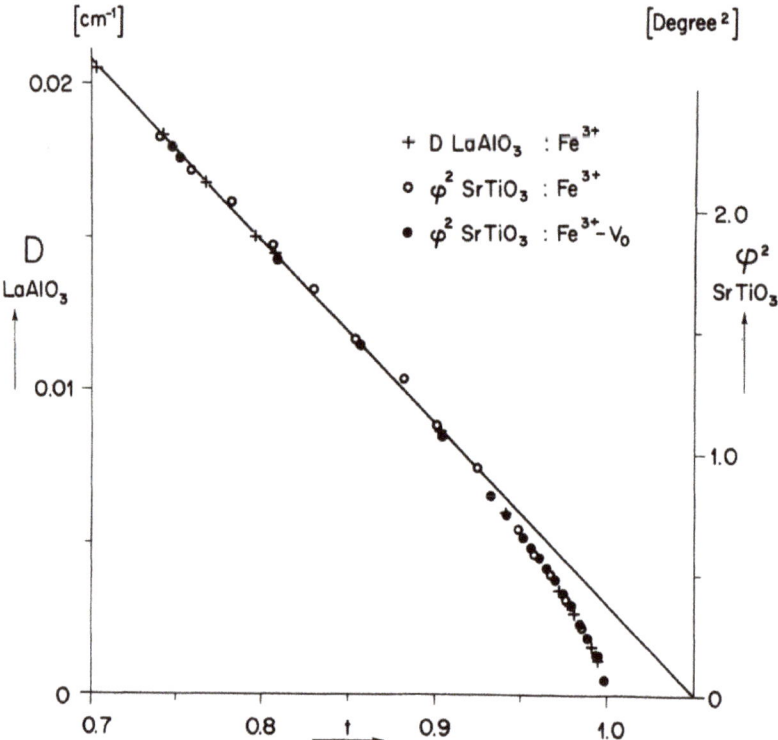

Figure 2. $\varphi^2(t)$ of SrTiO$_3$ and crystal field parameter D of LaAlO$_3$ vs reduced temperature $t = T/T_s$ between 0.7 and 1, showing the changeover from Landau to critical behavior at $t \simeq 0.9$ (from Reference [16]).

The above comprehensive results on SrTiO$_3$ set off more detailed investigations on the long wave length soft zone center transverse optic mode, which was speculated to be related to a ferroelectric phase transition in analogy to other titanites. However, a real transition had never been reported, but the extrapolation of this mode to zero led to a transition temperature around 20–30 K. Since the transverse optic soft mode driven phase transitions are related to a temperature dependent dielectric constant $\varepsilon(T)$ via the Lyddane-Sachs-Teller relation, measurements of $\varepsilon(T)$ can uniquely clarify the occurrence of a true phase transition. These were performed by Müller and Burkard [17] in 1979 and had the unexpected outcome that $\varepsilon(T)$ saturates below 4 K exhibiting extremely high values there (Figure 3). The results have been interpreted in terms of quantum fluctuations which suppress a true phase transition and coined the name "quantum paraelectricity" [17]. After the Nobel Prize awarded paper, it is the most frequently cited paper of K.A. Müller, as noted in a bibliographic study by Kremer et al. in Reference [18].

In the search for superconductivity in perovskites, Binnig and coauthors went back to SrTiO$_3$, which was already known to become superconducting at low temperature ($T_c \simeq 0.5$K) upon changing the oxygen stoichiometry [19–21]. Instead of manipulating the oxygen content, they doped the compound with small amounts of Nb [22] in order to enhance the superconducting transition temperature T_c. Even though this methodology did not lead to a significant enhancement of T_c, they discovered that their compound has two superconducting gaps and verified, thereby, the long known prediction that superconductors can exhibit more than one gap [23,24]. The most amazing thing about Nb-doped SrTiO$_3$ is, however, that it has an extremely low carrier concentration, such that its Fermi energy is smaller or comparable to the Debye energy. This "strange" behavior has very recently been taken up again and gave rise to new theoretical [25–27] and experimental interest, results, concepts, and speculations [28] (also see Scheerer et al. in Reference [29]). These will be discussed in more detail at the end of this paper.

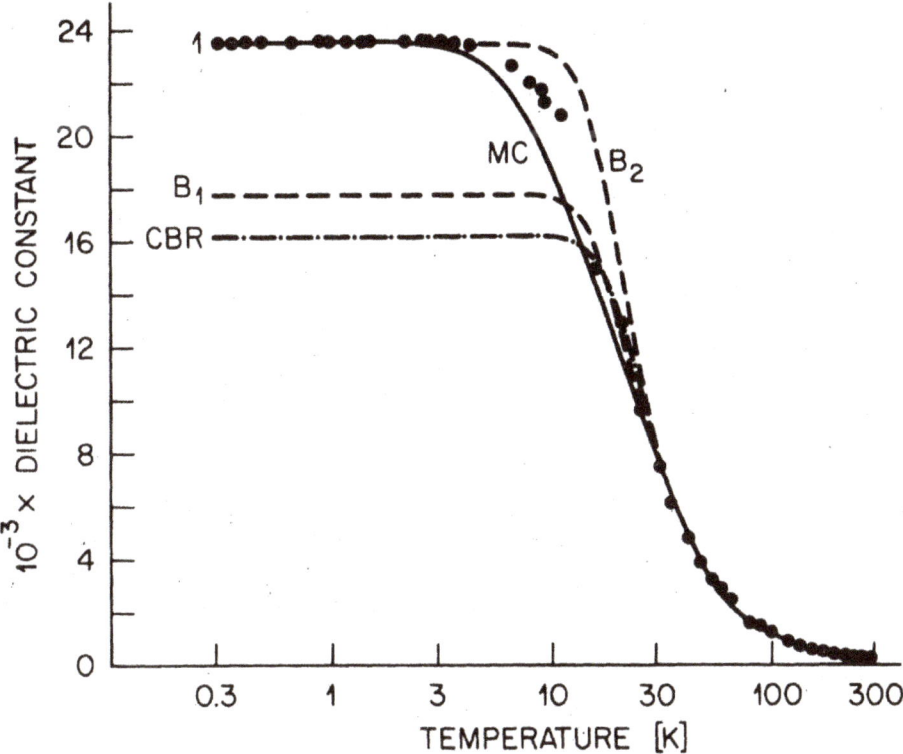

Figure 3. Comparison of the temperature dependent dielectric constant ε of SrTiO$_3$ in the quantum dominated temperature region with the mean-field theory; B1 adjusted to fit the crossover below 150 K; B2 adjusted to fit ε at $T \to 0$; CBR: effective acoustic mode coupling; MC dipolar mode coupling with biquadratic electron-phonon interaction (from Reference [17]).

For many years, the proximity of SrTiO$_3$ to a polar instability has challenged researchers to induce this order by different means. A first success was achieved by electric field effects, where polar domains could be induced [30]. Pressure on the contrary proved to be counter-productive, since the soft mode was observed to harden [31–33]. Similar effects were observed with doping, where an increase in the carrier density reduces the gap but also enhances the soft mode frequency [34]. It was up to J. G. Bednorz and K.A. Müller to achieve a breakthrough by replacing tiny amounts of Sr with Ca in SrTiO$_3$ [34]. Since Ca is smaller than Sr, a spontaneous polarization appears in the basal planes and not along the c-direction and is switchable between the two equivalent a-axes. This behavior was named "XY quantum ferroelectric" and shown to vanish rapidly with increasing Ca content [34]. An alternative approach has been undertaken by Itoh et al. by replacing O^{16} by its isotope O^{18} [35]. Indeed, ferroelectricity was observed there, but soon shown to be incomplete through theoretical modeling [36] and Raman spectroscopy [37]. By "incomplete", it is meant that ferroelectricity is not of long-range order, but appears locally only. Kleemann et al., in Reference [38], focus exactly on the possibility to achieve a true polar ground state in SrTiO$_3$ but also strive through doping-dependent novel phenomena achievable in this compound.

Even though the discovery of high-temperature superconductivity in the La-Ba-Cu-O perovskite slowed down the work of K.A. Müller with respect to oxide perovskites, he continued beforehand existing collaborations and produced interesting new results. Here, it is important to mention his work on BaTiO$_3$, which exhibits three phase transitions with decreasing temperature: the first from cubic to tetragonal at 408 K induces ferroelectricity, the following at 278 K lowers the symmetry to orthorhombic and is accompanied by a change in the polarization direction; at 183 K, the symmetry becomes rhombohedral and again the polarization reorients. The mechanisms of all transitions remained controversial for a long time, since experimental data testing different time and length scales suggested different mechanisms to be at play. Local probe experiments were in accordance with order-disorder type transitions [39,40], whereas long range testing experiments were in accordance with purely displacive ones [41]. The puzzles could be resolved again by EPR, where Mn^{4+} ions were doped in BaTiO$_3$ to randomly substitute for the isovalent Ti^{4+} ion. These experiments clearly showed that dynamic precursor clusters are formed above the respective transition temperature, which signify a coexistence of order/disorder and displacive features [42]. Depending on the testing tool, either of both appears dominant and yields a corresponding interpretation. Note that, in the paper by Bishop in Reference [43], similar aspects and beyond are addressed, and more detailed consequences are given.

A representative and comprehensive collection of the broad work of K.A. Müller on the properties of perovskites and other oxides (except cuprates) is presented in Reference [8] and supplemented in the work by Kool in Reference [44].

The above described results have an important consequence, since they suggest that perovskites are locally inhomogeneous with different coexisting time and length scales. As will be shown below, this is also true for cuprate superconductors, where distinctive properties stem from inherent inhomogeneity [45].

3. High-Temperature Superconductivity in Cuprates

3.1. Discovery of Superconductivity in the Cuprates

While K.A. Müller was already famous for his work on oxide perovskites in the eighties, he became inspired by the work on granular Al to move to superconductivity [46], a field he had never been interested in before. Granular Al has an extremely short coherence length, untypical for BCS superconductors, which later was shown to be also realized in cuprate superconductors [47]. For more details on the physical meaning of a short coherence length in unconventional superconductors, see the contribution of Deutscher in Reference [48].

K.A. Müller almost immediately turned to his favorite materials, the oxide perovskites, since the observation of a low carrier density and a large Debye energy yielded astonishingly high superconducting transition temperatures in oxides, incompatible with BCS theory. The above described discovery of two-band superconductivity in Nb-doped SrTiO$_3$ was the first result in that direction and actually was considered as very disappointing from the perspective of J.G. Bednorz. However, he, together with K.A. Müller, did not give up, and, finally, they succeeded in their world famous discovery of the La-Ba-Cu-O compound, which was honored with the Nobel prize only a year later, the fastest Nobel prize [49–52].

This discovery was inspired, as mentioned above, by the observation of the combination of low carrier density and a rather large superconducting transition temperature and the knowledge of Jahn-Teller polarons [53]. From this observation, and theoretical results of Höck et al. [53], K.A. Müller concluded that the coupling between the ions and the electrons/holes must be extraordinarily strong and unconventional, as suggested by Chakravarty based on a (bi-) polaronic concept [15]. Indeed, the cuprates also have a smaller carrier density than conventional superconductors, which is why they were named "bad" metals. Already, in his Nobel prize award speech, K.A. Müller depicted this scenario with alternating CuO pyramids, being charge rich and charge poor (Figure 4) [52], which was shortly afterwards verified experimentally by the group of A. Bianconi [54], suggesting a stripe-like patterning consisting of alternating distorted (D) and undistorted (U) CuO_6 octahedra with widths W and L (Figure 5). Having such a scenario in mind, it is obvious from the very beginning that cuprates, being non-stoichiometric, cannot be homogeneous. This is discussed in the next section.

Jahn-Teller polarons (1983)

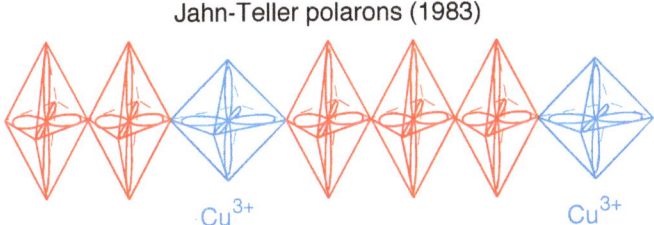

Figure 4. Schematic representation of the Jahn-Teller polaron in a linear chain substitution of trivalent La by a divalent alkaline-earth element would lead to a symmetric change in the oxygen polyhedral in the presence of Cu^{3+} (after Reference [52]).

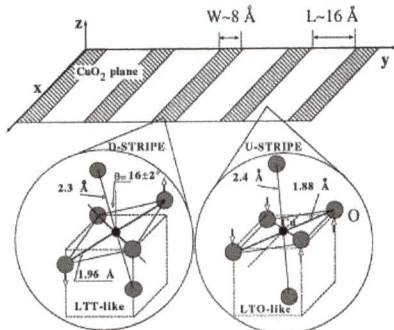

Figure 5. Pictorial view of the distorted CuO_6 octahedra, left side, of the "LTT (low-temperature tetragonal) type" assigned to the distorted stripe (D stripes) of width $W \simeq 8$ Å and of the undistorted octahedra, right side, of the "LTO (low-temperature orthorhombic) type" assigned to the undistorted stripes (U stripes) of width $L \simeq 16$ Å. The superlattice of quantum stripes of wavelength $L + W$ is shown in the upper part (from Reference [54]).

As mentioned above, the discovery of high-temperature superconductivity in the cuprates was not accidental but based on the idea that an unconventionally strong electron-lattice interaction may lead to superconductivity at high T_c. A possible way to achieve such a strong coupling is the formation of Jahn-Teller (JT) polarons (bipolarons) in doped perovskite oxides, as proposed by Bednorz and Müller [49,50]. In order to test and support the polaron concept, K.A. Müller made, from the beginning, several suggestions for key experiments (see, e.g., References [55–57]). Some of his main proposals for experiments—dealing with inhomogeneity, mixed oder parameters, and unconventional isotope effects in cuprate superconductors—are discussed in the following sections.

3.2. Essential Heterogeneities and Mixed Order Parameters in Cuprate Superconductors

In spite of many efforts in understanding the electron (hole) pairing mechanism in these materials, K.A. Müller pursued his ideas which culminated - driven by not understood experimental results—in his suggestion that two coexisting order parameters are characteristic for these compounds [58,59]. In view of the majority line that cuprates have a single d-wave order parameter (see, e.g., References [60–62]), he intensely pointed out to a combination of $s + d$ wave order parameters [58], thereby supporting his viewpoint of intrinsic inhomogeneity [45] characterized by local and global features which are beyond lattice periodicity. These ideas have subsequently been supported by detailed muon-spin rotation (μSR) experiments on various families of cuprate superconductors by the University of Zurich group [63–67]. The temperature dependence of the magnetic penetration depth λ, which can be extracted from the μSR relaxation rate σ_{sc} according to the relation $\sigma_{sc} \propto \lambda^{-2}$ [68], is sensitive to the gap symmetry.

As an example, Figure 6a shows the temperature dependence of the μSR relaxation rate σ_{sc} for single-crystal La$_{1.83}$Sr$_{0.17}$CuO$_4$ measured in various external magnetic fields perpendicular to the CuO$_2$ plane [63]. In this case, $\sigma_{sc}(T) \propto \lambda_{ab}^{-2}(T)$, where λ_{ab} is the in-plane magnetic penetration depth. Note that $\sigma_{sc}(T)$ shows an inflection point at low temperature which is most pronounced in the lowest magnetic field of $\mu_0 H = 0.02$ T. By analyzing the μSR data within a two-component model with coupled $s + d$-wave order parameters, a good agreement between experiment and theory was achieved [63,67], as demonstrated in Figure 6b. In particular, it is evident from the analysis that the s-wave gap contributes less to the superfluid density (32%) as compared to the d-wave gap (68%), but that its contribution is essential in order to describe the μSR data consistently. The fact that various cuprate families exhibit essentially the same behavior of the μSR relaxation rate $\sigma_{sc}(T)$ [63–65] suggests that this behavior is generic to all cuprate superconductors and reflects also the intrinsic inhomogeneity of these materials [45,69].

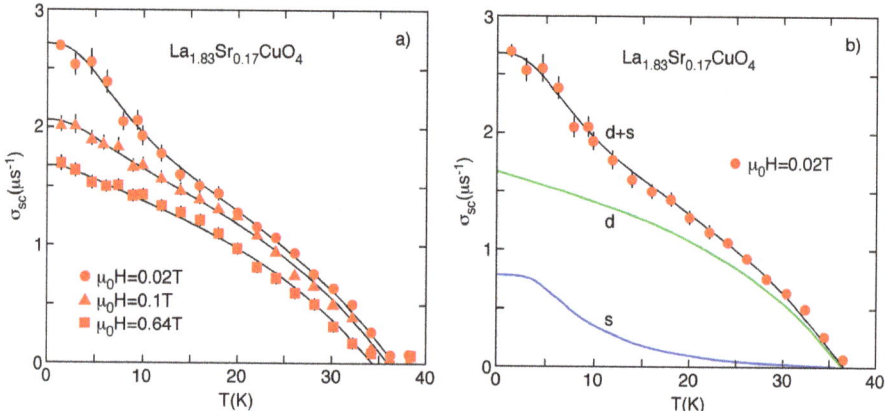

Figure 6. (**a**) Temperature dependence of the muon-spin rotation (μSR) relaxation rate $\sigma_{sc} \propto \lambda_{ab}^{-2}$ of single-crystal La$_{1.83}$Sr$_{0.17}$CuO$_4$ measured at various magnetic fields, as indicated in the figure. The solid lines correspond to results from a $s + d$ two-gap model. (**b**) The green and blue lines represent the contributions of d- and s-wave gap (after Reference [63]).

Testing the gap symmetry experimentally crucially depends on the experimental technique involved. For example, ARPES (angle-resolved photoemission spectroscopy) or phase-sensitive tunneling experiments on single-crystal samples probe the gap symmetry near the surface, whereas μSR experiments probe the gap symmetry in the bulk. Inconsistent results concerning the gap symmetry determined by different experimental techniques [59], as well as group theoretical considerations inspired K.A. Müller, to conclude that the gap symmetry may change from purely d-wave near the surface to more s-wave, like in the bulk of the superconductor [69,70]. We investigated the temperature dependencies of the in-plane (λ_{ab}) and out-of-plane (λ_c) magnetic field penetration depths near the surface and in the bulk of the electron-doped superconductor Sr$_{0.9}$La$_{0.1}$CuO$_2$ by means of AC magnetization and μSR measurements [71]. These studies provide evidence that the scenario of coexisting mixed $s + d$ order parameters is indeed also realized in this electron-doped cuprate superconductor, as proposed by K.A. Müller [58,69]. The temperature dependence of λ_{ab}^{-2} was found to be consistent with a dominant d-wave component ($\approx$96%) and only a tiny s-wave component near the surface of the sample, whereas $\lambda_c(T)$ is well described by a mixed $s + d$ order parameter with an s-wave component of more than 50% in the bulk. A comparison of the magnetization (surface probe) and the μSR (bulk probe) data for λ_{ab}^{-2} reveals that the s-wave component of the superfluid density is strongly suppressed by more than a factor of two near the surface of the sample, in accordance with the proposal of K.A. Müller [69]. For more details of this study, we refer to the contribution of Khasanov et al. in Reference [71].

Since the involvement of the lattice is essential in this model of mixed $s + d$ order parameters, K.A. Müller proposed to investigate isotope effects in the cuprate superconductors in order to highlight the important role played by the lattice. The most prominent isotope effects observed in cuprate superconductors are summarized in the following.

3.3. Unconventional Isotope Effects in Cuprate Superconductors

The observation of an isotope effect (IE) on the superconducting transition temperature T_c in mercury in 1950 played a key role in the development of the weak-coupling microscopic BCS theory of superconductivity, where the electron-phonon interaction leads to the formation of Cooper pairs (two electrons with opposite spin and momentum) at T_c. The isotope shift on T_c is expressed by the relation

$$T_c \propto M^{-\alpha}, \alpha = -d \ln T_c / d \ln M, \qquad (1)$$

where M is the isotope mass, and α is the IE exponent. Weak-coupling BCS theory predicts $T_c \propto M^{-1/2}$ with $\alpha_{BCS} \simeq 0.5$. Indeed, for many low-temperature conventional superconductors, values of $\alpha \approx 0.5$ were found. The observation of an IE in conventional superconductors is thus consistent with the BCS theory of superconductivity, where the electron-phonon interaction is the pairing mechanism.

The discovery of high-temperature superconductivity in the cuprates by Bednorz and Müller [49] with transitions temperatures well above those found in conventional superconductors raised the question of the electron/hole pairing nature of superconductivity in these novel systems. The observation of only a tiny oxygen (^{16}O/^{18}O)-isotope effect (OIE) in optimally doped $YBa_2Cu_3O_{7-\delta}$ [72] led many theoreticians to conclude that the electron-phonon interaction, or, more generally, lattice effects, cannot be the pairing glue in high-temperature superconductivity. Alternatively, several other models of purely electronic origin were proposed, and a majority of scientists working in the field ignored the role of the lattice in the cuprates. However, it is important to note that all cuprates show a finite OIE on T_c at all doping levels, increasing substantially with reduced doping [73,74].

From the very beginning, K.A Müller was convinced that the IE plays a crucial role to understand the microscopic pairing mechanism in the cuprates. The concept of the Jahn-Teller (JT) polaron, which originally was his basic motivation to search for superconductivity in perovskite oxides, involves local lattice deformations. Consequently, strong lattice interactions and unconventional isotope effects are expected to be present in the cuprates. In an early paper, K.A. Müller reviews the importance of IEs in the cuprates [75]. In 1990, K.A. Müller, therefore, initiated a new project, *Isotope Effects in Cupate Superconductors*, at the University of Zurich (see, e.g., References [76–78]), in close collaboration with material scientists from the ETH Zurich and the Paul Scherrer Institute (PSI) (see the contribution of Conder et al. in Reference [79]). The main goal of this project was to investigate which oxygen atoms (plane (p), apical (a), or chain (c) oxygens) in the crystal lattice of $YBa_2Cu_3O_{7-\delta}$ contribute mainly to the OIE on T_c. It was expected that, within the JT concept, the apical oxygen ions should play a crucial role and give rise to a pronounced OIE. Experimentally, this can be tested by the so-called site-selective oxygen-isotope effect (SOIE) study on T_c in optimally doped $YBa_2Cu_3O_{7-\delta}$ [80]. For this purpose, fully ^{16}O and ^{18}O exchanged, as well as site-selective exchanged samples, were prepared. In the site-selective samples, the planar sites were substituted by ^{18}O (^{16}O) and the apical and chain sites by ^{16}O (^{18}O). Surprisingly, it was found that the planar oxygen atoms contribute dominantly (≥80%) to the total OIE shift on T_c. This finding was later confirmed for $Y_{1-x}Pr_xBa_2Cu_3O_{7-\delta}$ for various dopings ($x = 0, 0.3, 0.4$) [81,82]. All these SOIE results on T_c for $Y_{1-x}Pr_xBa_2Cu_3O_{7-\delta}$ are summarized in Figure 7. It is evident that the main contribution to the OIE on T_c for all dopings originates from the oxygen atoms in the CuO_2 planes, opposite to what was expected by K.A. Müller in his original proposal, where he assumed that the apical oxygen ion displacement is subject to strong anharmonicity analogous to perovskite ferroelectrics [75,83]. Apparently, these assumptions did not meet the experimental results, which he immediately accepted.

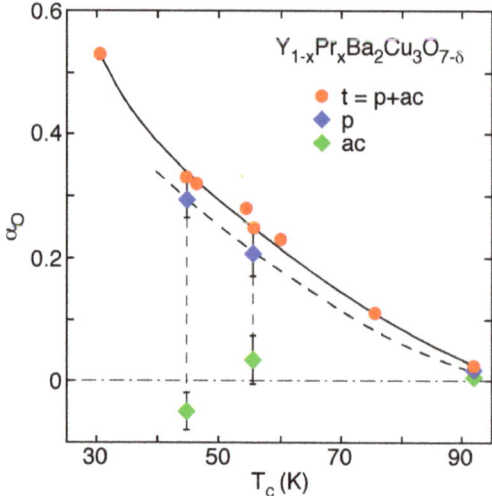

Figure 7. Total (t) and partial (p, ac) oxygen-isotope exponent α_O as a function of T_c for $Y_{1-x}Pr_xBa_2Cu_3O_{7-\delta}$ (t = total: all oxygen sites, p: planar oxygen sites, ac: apex and chain oxygen sites). Solid and dashed lines are guides to the eye (after Reference [78]).

In subsequent years, doping dependent unconventional OIEs on several physical quantities (besides T_c) were observed, including, for instance, the antiferromagnetic transition temperature T_N, the spin-glass transition temperature T_g, the spin-stripe ordering tempertaure T_{so}, the magnetic penetration depth λ, and the pseudogap temperature T^* [67,76–78,84,85]. In the following, we will only focus on the OIEs on λ and T^*.

The BCS theory is based on the Migdal adiabatic approximation, in which the effective mass m^* of the supercarriers is not sensitive to the mass M of the lattice atoms. However, if the supercarriers interact strongly with the lattice as in a polaronic concept (JT representation), the adiabatic approximation is no longer valid, and m^* depends on M (see, e.g., Reference [86]). A direct way to test this notion is to search for a possible IE on the magnetic penetration depth λ. In the simplest case of a spherical or ellipsoidal Fermi surface, the zero-tempertaure in-plane magnetic penetration depth λ_{ab} may be expressed by the London formula [78]:

$$\lambda_{ab}(0) = \sqrt{\frac{1}{\mu_0 e^2} \frac{m^*_{ab}}{n_s}}, \qquad (2)$$

where n_s is the superconducting carrier density, and m^*_{ab} is the in-plane effective mass of the carriers. The in-plane superfluid density ρ_s is then given by the relation:

$$\rho_s \propto 1/\lambda^2_{ab}(0) \propto n_s/m^*_{ab}. \qquad (3)$$

This means that a possible OIE shift of the superfluid density ρ_s arises from an OIE shift of n_s and/or m^*_{ab} according to the expression:

$$\Delta\lambda^{-2}_{ab}(0)/\lambda^{-2}_{ab}(0) = \Delta n_s/n_s - \Delta m^*_{ab}/m^*_{ab}. \qquad (4)$$

The first OIE experiments on the magnetic penetration depth in polycrystalline $YBa_2Cu_3O_{6.94}$ were performed by means of magnetization measurements [87], and, later on, in fine-grained samples of $La_{2-x}Sr_xCuO_4$ ($0.06 \leq x \leq 0.15$) [76,88]. For $La_{2-x}Sr_xCuO_4$, the observed oxygen-isotope shift of $\lambda_{ab}^{-2}(0)$ was found to decrease substantially with increasing doping x [76,88]. For $x = 0.105$, an oxygen-isotope shift of $\Delta\lambda_{ab}^{-2}(0)/\lambda_{ab}^{-2}(0) = -9(1)\%$ was found [88].

In addition, we performed OIE studies on microcrystals of underdoped $La_{2-x}Sr_xCuO_4$ with $x = 0.080$ and $x = 0.086$ (volume $\approx 150 \times 150 \times 50$ μm^3) by means of high-sensitive torque magnetometry [89]. A substantial OIE on $\lambda_{ab}^{-2}(0)$ was observed, namely $\Delta\lambda_{ab}^{-2}(0)/\lambda_{ab}^{-2}(0) = -10(2)\%$ for $x = 0.080$ and $-8(1)\%$ for $x = 0.086$, respectively, in agreement with the results obtained for fine-grained powder samples [76,88].

The University of Zurich group carried out a μSR OIE study of λ_{ab} in fine-grained powder samples of underdoped $Y_{1-x}Pr_xBa_2Cu_3O_{7-\delta}$ ($x = 0.3$ and 0.4) [90], which were in line with those of magnetic torque data for underdoped $La_{2-x}Sr_xCuO_4$ [89]. In order to investigate which oxygen atoms in the lattice mainly contribute to the OIE on λ_{ab}, we initiated a site-selective OIE μSR study of λ_{ab} in underdoped $Y_{0.6}Pr_{0.4}Ba_2Cu_3O_{7-\delta}$ [82]. A substantial total OIE of T_c, as well as of λ_{ab}, is observed. Moreover, the SOIE experiments clearly indicates that the planar oxygens account within experimental error for 100% to both OIE shifts, yielding $\Delta T_c/T_c = -3.7(4)\%$ and $\Delta\lambda_{ab}^{-2}(0)/\lambda_{ab}^{-2}(0) = -6.2(1.0)\%$, consistent with the SOIE results on T_c presented above.

In a further OIE study, the low-energy μSR (LEμSR) technique was applied, which allows a *direct measurement* of the magnetic penetration depth by measuring the magnetic field profile $B(z)$ inside a superconductor in the Meissner state just below a distance z from the surface [91,92]. From the measured $B(z)$, the magnetic penetration depth λ is then extracted directly. This method was used to detect the OIE on the in-plane penetration depth λ_{ab} in a nearly optimally doped $YBa_2Cu_3O_{7-\delta}$ thin film (600 nm thick) [92]. The analysis of the LEμSR data for the ^{16}O and ^{18}O substituted thin films yielded $^{16}\lambda_{ab}(4K) = 151.8(1.1)$ nm and $^{18}\lambda_{ab}(4K) = 155.8(1.0)$ nm. Correcting for the incomplete ^{18}O exchange of 95%, one obtains for the relative OIE shift $\Delta\lambda_{ab}/\lambda_{ab} = (^{18}\lambda_{ab} - ^{16}\lambda_{ab})/^{16}\lambda_{ab} = 2.8(1.0)\%$ at 4 K. This value is in good correspondence with the values reported for optimally doped $YBa_2Cu_3O_{7-\delta}$, $La_{1.85}Sr_{0.15}CuO_4$, and $Bi_{1.6}Pb_{0.4}Sr_2Ca_2Cu_3O_{10+\delta}$ evaluated *indirectly* from magnetization measurements [78].

The OIE results on T_c and λ_{ab} presented in this chapter are summarized in Figure 8, where the OIE shifts $\Delta\lambda_{ab}(0)/\lambda_{ab}(0)$ versus the OIE shifts $-\Delta T_c/T_c$ are shown. In order to discuss these findings, we use the OIE shift $\Delta\lambda_{ab}(0)/\lambda_{ab}(0)$, instead of $\Delta\lambda_{ab}^{-2}(0)/\lambda_{ab}^{-2}(0) = -2 \Delta\lambda_{ab}(0)/\lambda_{ab}(0)$. Figure 8 clearly demonstrates that there is a *correlation* between the OIE on T_c and $\lambda_{ab}(0)$. At optimal doping, both isotope shifts are small, but finite with $\Delta\lambda_{ab}(0)/\lambda_{ab}(0) \approx 10 |\Delta T_c/T_c|$. With decreasing doping (decreasing T_c), $\Delta\lambda_{ab}(0)/\lambda_{ab}(0)$ remains almost constant as underlined by the blue dashed line, then starts to increase, and, in the underdoped regime, the two relative oxygen isotope shifts are almost equal ($\Delta\lambda_{ab}(0)/\lambda_{ab}(0) \approx |\Delta T_c/T_c|$), as indicated by the red solid line. This behavior is apparently *generic* for different families of cuprates, and, at first glance, one may speculate that this is a direct consequence of the empirical Uemura relation [93,94]. In the underdoped regime, the simple relation $\lambda_{ab}^{-2}(0) \simeq C T_c^\alpha$ with $\alpha \simeq 1$ holds, and C is a "universal constant". With this relation, one readily gets for the isotope shift on $\lambda_{ab}(0)$:

$$\Delta\lambda_{ab}(0)/\lambda_{ab}(0) = -1/2\, \alpha\, \Delta T_c/T_c. \qquad (5)$$

For $\alpha \simeq 1$ (the Uemura relation), one obtains $\Delta\lambda_{ab}(0)/\lambda_{ab}(0) \simeq -1/2\, \Delta T_c/T_c$, as marked by the dashed red line in Figure 8, which deviates from the experimental data. In the underdoped regime, these are much better described by the red solid line with $\alpha \simeq 2$ ($\Delta\lambda_{ab}(0)/\lambda_{ab}(0) \simeq -\Delta T_c/T_c$). However, the physical meaning of this *factor 2* remains unclear.

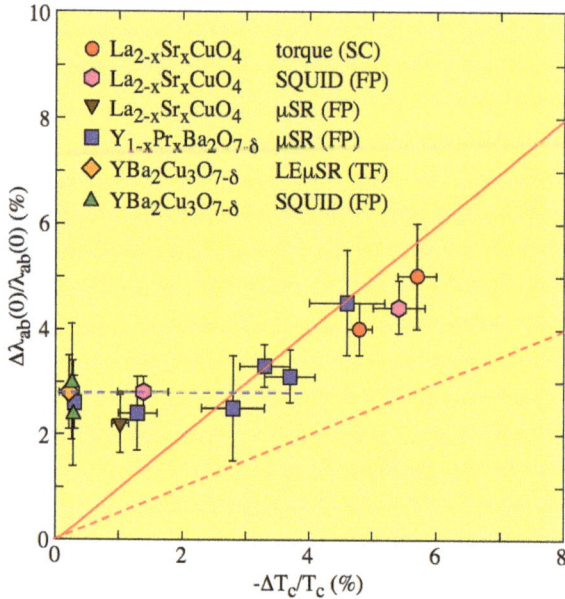

Figure 8. Plot of the oxygen (^{16}O/^{18}O)-isotope effect (OIE) shift $\Delta\lambda_{ab}(0)/\lambda_{ab}(0)$ versus the OIE shift $-\Delta T_c/T_c$ for La$_{1-x}$Sr$_x$CuO$_4$ and Y$_{1-x}$Pr$_x$Ba$_2$Cu$_3$O$_{7-\delta}$ using different experimental techniques as described in the text and various types of samples (SC: single crystal, FP: fine powder, TF: thin film). The meaning of the dashed lines is explained in the text (after Reference [78]).

S. Weyeneth and K.A. Müller [95] analyzed the doping dependence of the OIE on T_c for various cuprate systems in terms of a "polaronic model" proposed by Kresin and Wolf [96,97], which was derived for polarons forming perpendicular to the superconducting CuO$_2$ planes. It is important to note that, in References [96,97], polarons are not mentioned. However, S. Weyeneth and K. A. Müller [95] reinterpreted the model by them in terms of a polaronic one. The doping dependence of the measured OIE exponent follows well the predicted behavior, in agreement with previous results based on a purely empirical model [98]. In addition, the OIE exponent of the pseudogap temperature T^* with reversed sign compared to that of T_c is also described by this polaronic model. These findings suggest that superconductivity in the cuprates is driven by polaron, or rather bipolaron, formation in the CuO$_2$ planes [95]. In this model, it is assumed that the isotope effect on T_c is only determined by the isotope effect on the supercarrier density n_s. The same argument holds for T^*. Since $dT_c/dn_s > 0$, but $dT^*/dn_s < 0$, the two isotope effects are sign reversed (see Figure 9). However, this concept underestimates the observed OIE on λ_{ab} in the underdoped regime by almost a factor 2, as shown by the dashed red line ($\alpha \simeq 1$) in Figure 8, and even closer to optimal doping where $\Delta T_c/T_c \simeq 0$. This suggests the conclusion that the OIE on $\lambda_{ab}(0)$ must arise from an OIE on n_s and m^*_{ab}, according to Equation (4).

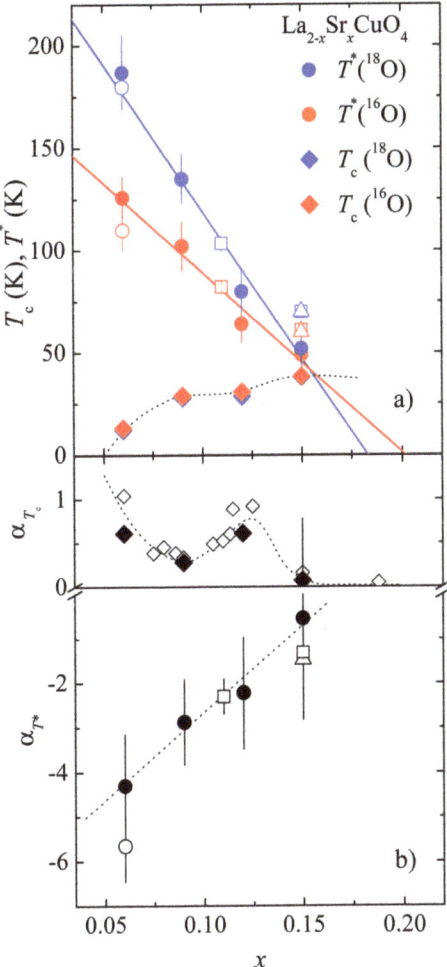

Figure 9. (a) The superconducting transition temperature T_c and the pseudogap temperature T^* of $La_{2-x}Sr_xCuO_4$ as a function of doping x for ^{16}O (red symbols) and ^{18}O (blue symbols). The solid lines are obtained from a linear fitting. The dashed line is a guide to the eye. The data are results from X-ray absorption near edge structure (XANES) and neutron crystal field spectroscopy (NCFS) experiments (see Reference [99]). (b) Doping dependence of the isotope effect exponent α_{T_c} and α_{T^*} for $La_{2-x}Sr_xCuO_4$. The data of α_{T^*} are obtained by various XANES and NCFS experiments and data of α_{T_c} are from magnetization measurements (see Reference [99]). The dashed lines are a guide to the eye (from Reference [99]).

In another project proposed by K.A. Müller, we explored a possible OIE on the pseudogap (charge-stripe ordering temperature) T^* in cuprate superconductors. It is generally accepted that T^* plays a fundamental role in understanding the complex phase diagram und the underlying physics of cuprates. However, from an experimental point of view, the quantity T^* is ill defined, depending on the experimental technique and the corresponding time and length scales involved (see, e.g., Reference [99]). In addition, the meaning of T^* and its origin are controversely discussed. Here, we define T^* as the temperature where local deviations from the average structure occur (also named charge ordering temperature) and where no direct magnetic effects are present. Thereby, it is evident that OIE studies of T^* are crucial to test lattice/polaron effects in the cuprates. Up to now, only a few such studies have been carried out [99–105]. The first OIE on T^* was performed in underdoped $La_{1.94}Sr_{0.06}CuO_4$ by means of Cu K-edge X-ray absorption near edge structure (XANES) studies [100], yielding a huge and sign reversed OIE exponent of $\alpha_{T^*} \simeq -5$. Later on, similar XANES experiments on $La_{2-x}Sr_xCuO_4$ as a function of doping x ($0.06 \leq x \leq 0.15$) completed this result [99]. All XANES data, together with additional outcome from neutron crystal field spectroscopy (NCFS) studies of $La_{1.96-x}Ho_{0.04}Sr_xCuO_4$ [104,105], are summarized Figure 9. The observed OIE exponent α_{T^*} changes almost linearly from $\alpha_{T^*} \simeq -0.6$ for $x = 0.15$ to $\alpha_{T^*} \simeq -5$ for $x = 0.06$. This is in contrast to the doping dependence of the OIE exponent α_{T_c} which exhibits a characteristic anomaly at $x \simeq 1/8$ (Figure 9), which is absent in $\alpha_{T^*}(x)$, implying that there is no simple correlation between $\alpha_{T^*}(x)$ and $\alpha_{T_c}(x)$, at least for $La_{2-x}Sr_xCuO_4$. Furthermore, NCFS experiments on slightly underdoped $HoBa_2Cu_4O_8$ [101] revealed $\alpha_{T^*} = -2.2(6)$, in agreement with values for $La_{2-x}Sr_xCuO_4$. Additional NCFS studies of the $^{63}Cu/^{65}Cu$ isotope effect show a large negative isotope shift of T^* for $HoBa_2Cu_4O_8$ [103], which was not observed for optimally doped $La_{1.81}Ho_{0.04}Sr_{0.15}CuO_4$ [104]. A consistent explanation of this observation was given by K.A. Müller [104]. In both compounds, single-layer $La_{1.81}Ho_{0.04}Sr_{0.15}CuO_4$ and double-layer $HoBa_2Cu_4O_8$, local oxygen and copper lattice JT-type modes have to be considered. Whereas oxygen and copper modes are relevant for the bilayer compound, the umbrella-type copper modes are absent in the single-layer compound, thus explaining the observed oxygen and copper isotope effects on T^* [104]. As shown in Figure 9a, the pseudogap temperature T^* decreases linearly with increasing doping x with pronounced different slopes for ^{16}O and ^{18}O. The limit $T^* = 0$ K has frequently been interpreted as quantum critical point (see, e.g., References [106,107]). However, since, in this limit, an appreciable isotope effect is also observed, [$\alpha_{x_c} = 0.84(22)$] [99], a purely electronic model can be excluded as origin of this point.

A theoretical explanation of the above described isotope effects has been given in References [67,108], where it was shown that the original idea as introduced by K.A. Müller, namely Jahn-Teller (bi-)polarons [50,52], indeed provide a consistent explanation of them. Polaron formation occurs in the case of strong local electron-lattice coupling where the individual particles lose their meaning to form a new quasi-particle. In such a case, the electrons are renormalized by the lattice, and the lattice, in turn, is renormalized by the electrons. This introduces an exponential slowing down of the electron (hole) hopping, i.e., their kinetic energy, whereas the phonons are rigidly displaced. Upon adopting a three band picture to capture the essential physics of cuprates, namely nearest t_1 and next-nearest neighbor hopping t_2, together with an inter-planar hopping term t_4, their polaronic coupling consequences have been investigated. Interestingly, only the second and inter-planar couplings are relevant for the isotope effects and describe the correct trends as observed experimentally for all those compounds mentioned above. This observation admits to draw conclusions on the local displacement involved in the dynamics, namely a Q_2-type phonon mode, as shown in Figure 10.

On the other hand, the lattice renormalization caused by the polaron formation plays a decisive role to understand experimental XAFS (X-ray absorption fine structure) data of $La_{2-x}Sr_xCuO_4$ [109]. These show two anomalies as a function of temperature, namely an upturn at T^* and a second anomaly at T_c. Both are consistently explained by the polaronic approach [110], emerging in a picture where (bi)polarons become coherent at T^*, being gaseous above T^*, to adopt a stripe-type dynamical pattern below T^*, as illustrated in Figure 11. At T_c, the superconducting pairing condensate symmetry becomes relevant, since a divergence, as observed by XAFS, takes place only if the symmetry is s-wave-like or has at least a substantial s-wave contribution. For a purely d-wave order parameter, no divergence takes place. This substantiates the original suggestion of K.A. Müller that cuprates must have coexisting $s + d$ order parameters [58].

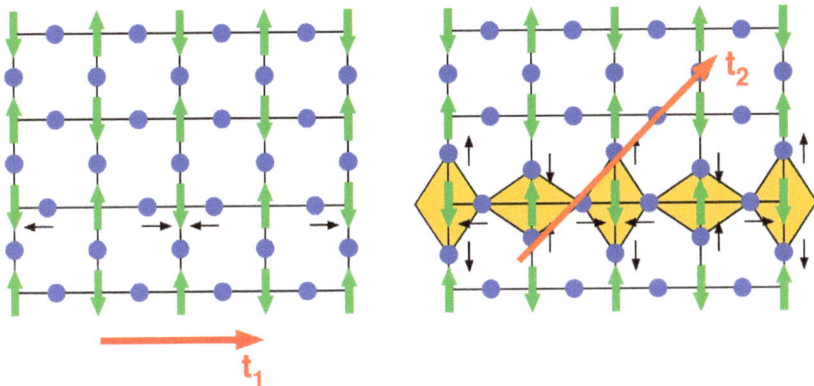

Figure 10. The relevant ionic displacements in the CuO_2 plane dominated by the nearest (t_1) and or second nearest (t_2) hopping integrals, giving rise to a Q_2-type phonon mode visualized by the yellow areas (from Reference [67]).

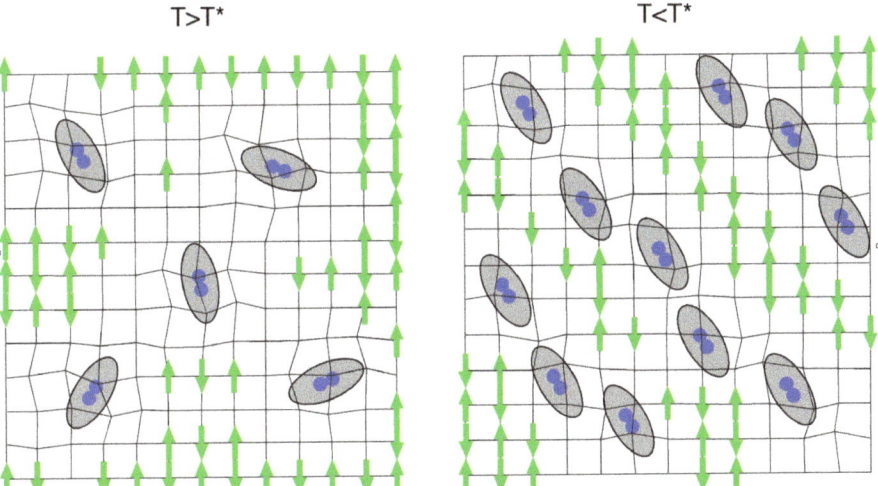

Figure 11. Disordered bipolaron gas above T^* (**left**) and ordered bipolaron liquid below T^* (**right**) (from Reference [67]).

With these clear conclusions, we move back to $SrTiO_3$ and one of the first famous papers of K.A. Müller, namely the introduction of quantum paraelectricity in this compound [17].

4. $SrTiO_3$—An Exotic Low Carrier Density Superconductor

One of the most frequently cited papers of K.A. Müller is his work on the low-temperature behavior of the dielectric constant in $SrTiO_3$ [17]. The saturation of the dielectric constant ε in this temperature regime has been termed "quantum paraelectricity" (see Figure 3) and gave rise to numerous speculations of its role played for superconductivity in doped or oxygen reduced $SrTiO_3$. When, in the 1960s, superconductivity in this compound was first reported [19,20], the pairing mechanism was interpreted in terms of intervalley scattering or soft mode induced. A typical BCS pairing scheme was excluded, since the low carrier density does not support this. For a rather long period, the work on superconductivity in $SrTiO_3$ slowed down, essentially until Binnig et al. [22] discovered multiband superconductivity in Nb-doped SrTiO$_3$, inspired by K.A. Müller. With the discovery of high-temperature superconductivity in the cuprates, the majority interest was devoted to these compounds, and, only after many years of hype, novel or formerly discussed systems reemerged.

During the high-temperature superconductivity peak activities, SrTiO$_3$ also gained renewed interest, since its lattice parameters match extraordinarily well with those of many cuprates, and, correspondingly, it was used as substrate material for the film growth of these superconductors. Scientifically, focused interest appeared in 2004 with the discovery of the two-dimensional electron (2DEG) or hole gas (2DHG) interface conductivity. In particular, the discovery of 2DEG at the interface between two oxide insulators LaAlO$_3$ and SrTiO$_3$ provided new opportunities for research and applications [111]. Especially, superconductivity [112] and magnetism [113] have been observed at these interface 2DEGs, which are not found in typical semiconductor interfaces. By using strain as an additional tuning parameter, a variety of new research areas have been reemerged [114–116].

The low carrier density of doped SrTiO$_3$ has been in the focus during the last 5 years, since it was found that, even smaller than first thought, carrier densities support superconductivity [26,117]. These new findings gave rise to novel theories concentrating often on the proximity of SrTiO$_3$ to quantum criticality and its incipient polar properties [25,118]. Experimentally, the carrier density has been determined from macroscopic measurements, namely conductivity, resistivity, and Hall measurements [117], thereby overlooking that all perovskite oxides have an inherent tendency to inhomogeneity, meaning that the doped carriers are not homogeneously distributed in the respective sample. This is already reflected in undoped perovskite oxides where finite size precursors occur, signaling locally distorted regions in an intact matrix [42,119]. In addition, extended X-ray absorption fine structure (EXAFS) [120,121], electron paramagnetic resonance (EPR) [8], atomic pair distribution function (PDF) (for a recent review, see, e.g., Reference [122]), and similar data are strong indications that the real space properties differ substantially from the momentum space derived ones. For this reason, novel experiments have been carried out in order to arrive at a clue for the understanding of the insulator/metal (I/M) and/or insulator/superconducting (I/S) transition. Both of these are achieved by either introducing oxygen vacancies or by doping the A or B site in ABO$_3$ SrTiO$_3$ by aliovalent ions. The data are interpreted by theoretical modeling based on the polarizability model [123–125].

From the latter approach, the dynamics of SrTiO$_3$ can be studied in detail when concentrating on local anomalies in the momentum dependence of the two lowest transverse optic and acoustic modes, which represent the ferroelectric soft mode and the soft acoustic mode responsible for the antiferroelectric rotational instability. Anomalies in these two branches have already been analyzed previously and shown to be signatures for local polar and piezoelectric nano-regions [126]. By calculating the phonon group velocities for the two considered branches, the above described anomalies become very apparent and define the momentum at which local, spatially confined soft modes occur. The decisive momentum value is the one where the scattering between optic and acoustic mode is strongest. The corresponding squared local frequency $\omega_{TO}^2(q)$ is shown in Figure 12 as a function of carrier concentration, temperature, and momentum q.

As is obvious from the figure, this local mode softens with decreasing temperature and simultaneously moves to lower momentum values but never reaches the long wave length $q = 0$ limit. In addition, the softening slows down with increasing carrier concentration, and its momentum space spread increases, highlighting the growing spatial confinement of these polar nano-domains. In contrast to a long wave length "true" soft mode, it is not linearly dependent on temperature but substantially nonlinear below ≈ 150 K. Since $\omega_{TO}^2(q) \approx 1/\varepsilon_0$, this mode is directly linked to the dielectric permittivity ε_0, which is reduced by approximately 40% as compared to the long wave length limit, but still exhibits an appreciable temperature dependence which is typical for an almost ferroelectric compound [127]. From the calculated thermal average of the displacement-displacement correlation function, a local dipole moment has been derived, which decreases with decreasing temperature and increasing carrier concentration (Figure 13), where the inset shows the zero temperature limit of it in dependence of the carrier concentration.

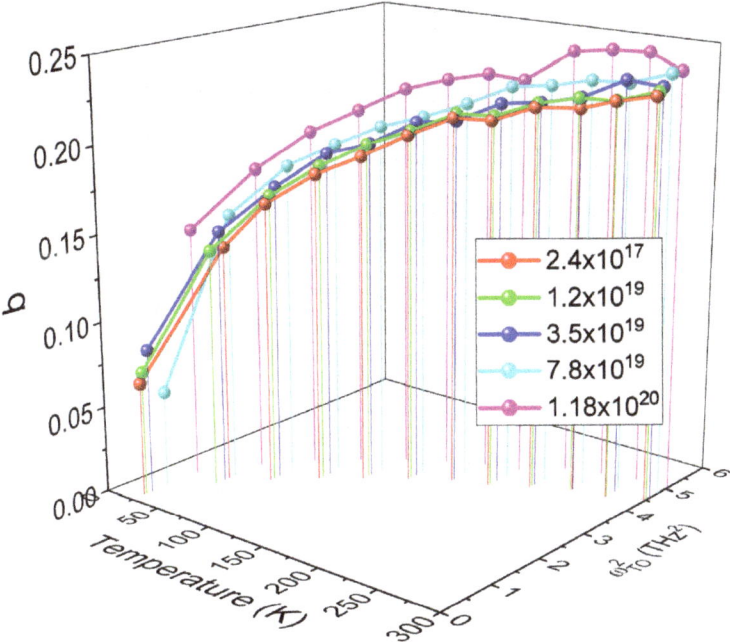

Figure 12. The squared transverse local optic mode frequency ω_{TO}^2 as a function of temperature, momentum, and carrier concentration.

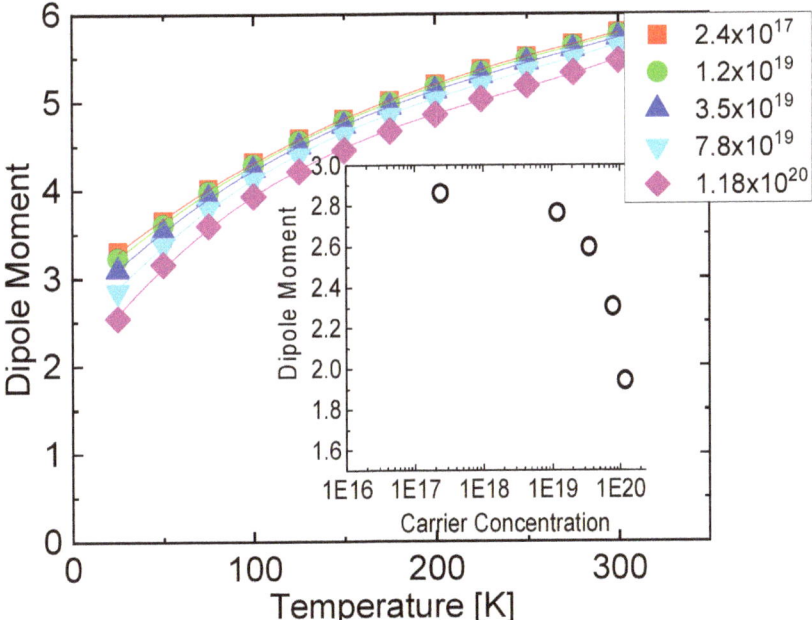

Figure 13. Temperature dependence of the local dipole moment for various carrier concentrations. The inset shows the local dipole moment in the zero temperature limit as a function of the carrier concentration.

Experimentally, the electrical properties of doped SrTiO$_3$ depend on the volume concentration of carriers which are, however, inhomogeneously distributed in the samples. Self-doped samples exhibit different electronic, chemical, and structural properties in highly doped regions as compared to the matrix, which is almost unaffected by doping [128]. According to References [129,130], the removal of oxygen preferentially takes place along extended defects, thus giving rise to a local reduction, which cannot be detected by average macroscopic property detecting methods, like, e.g., mobility and Hall measurements. This argument is supported by local conductivity atomic force microscopy (LC-AFM) data [131,132], which demonstrate that very low concentrations of oxygen deficiency assemble along these dislocations. In order to prove that a random distribution of "defects" is absent, a decisive experiment has been carried out, which underlines the important role of dislocations and filament formation on the electrical transport properties of reduced SrTiO$_3$. As outlined above, even with increasing carrier density, mode softening persists on a dynamic time scale of ps and length scale of nm. This can be confirmed experimentally on the nano-scale by using, e.g., time-resolved infrared spectroscopy or scanning near-field optical microscopy (SNOM). Alternatively, piezoelectric force microscopy (PFM) offers the possibility of locally detecting a piezoelectric response which is an indirect probe of mode softening through the creation of induced dipole moments. This has been shown in Reference [133], where, in the vicinity of the core of edge dislocations in a SrTiO$_3$ bicrystal, a polarization of the order of 20 µC/cm^2 has been detected. This is shown in Figure 14, where, along a sharp step of a plastically deformed SrTiO$_3$ crystal, corresponding to a broad band of dislocations, piezoelectric responses are observed.

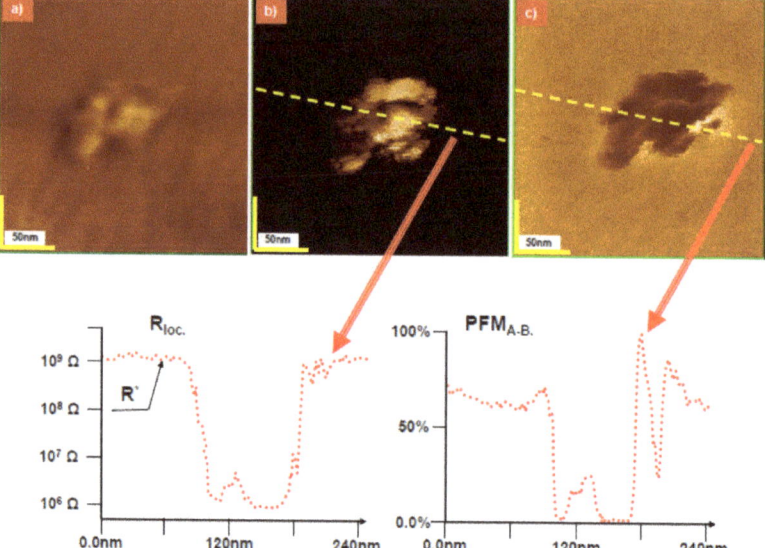

Figure 14. Topographic image of the exit of a dislocation bundle in thermally reduced SrTiO$_3$: (**a**) the topography variation is about 0–4.8 nm. In the area of the bundle (local conductivity atomic force microscopy (LC-AFM)) (**b**) with metallic properties, the piezo response (**c**) is absent. This can be observed at the cross sections of LC-AFM (**b**) and piezoelectric force microscopy (PFM) (**c**) responses. The distribution of the resistance along the cross section (lower part of the figure) shows an anti-correlation with the distribution of the piezo activity on the cross section of the PFM. The resistance outside the bundle is much higher than on the cross section, which is related to the finite resolution of the analog-to-digital converters (here, 4 decades), so that, using additional measurements (only in this area), the resistance is higher than 10$^{12-13}$ Ω (from Reference [127]).

As expected, this rapidly decreases with increasing carrier concentration and approaches a constant value for small densities, thus clearly supporting the polar character of the matrix and the inherent inhomogeneity of doped SrTiO$_3$. Correspondingly, in the metallic and superconducting low carrier density limit of SrTiO$_3$, metallicity is not globally present but appears only in filaments which coexist with elastically and polar distorted domains where the latter shrink in size with increasing carrier density and have completely vanished beyond a critical carrier density n_c to give space for a homogeneous metal where superconductivity is absent. In terms of the electronic band structure, highly localized polaronic bands can be attributed to the matrix, whereas Luttinger liquid-type behavior must be present in the filaments. With increasing carrier density, the localized band adopts dispersion from the itinerant one, which corresponds to a steep band/flat band scenario, where superconductivity is a consequence of interband interactions [134].

In conclusion, the dynamical properties of SrTiO$_3$ have been calculated as a function of carrier concentration. Up to a critical concentration n_c, the lattice potential is of double-well character, and strong transverse optic mode softening takes place [127]. This is linked to the formation of polar nano-regions, which grow in size with decreasing temperature, implying substantial sample inhomogeneity. The "insulating" nano-domains coexist with the filamentary conductivity; hence, possible links to superconductivity are apparent. As long as this coexistence persists, metallicity/superconductivity survives. Thus, its essential origin must be associated with the polar character of the matrix and the two-component properties. Beyond n_c, almost "normal" dynamics are observed with negligible mode softening and nano-domain formation. Typical metallicity is expected there, as well as homogeneous sample properties. The theoretical results are in agreement with experiments which demonstrate the local metallic character and the insulating behavior of the matrix, which are schematically summarized in Figure 15, where the change in the character of the potential along a core of dislocations and the entry into the matrix is depicted.

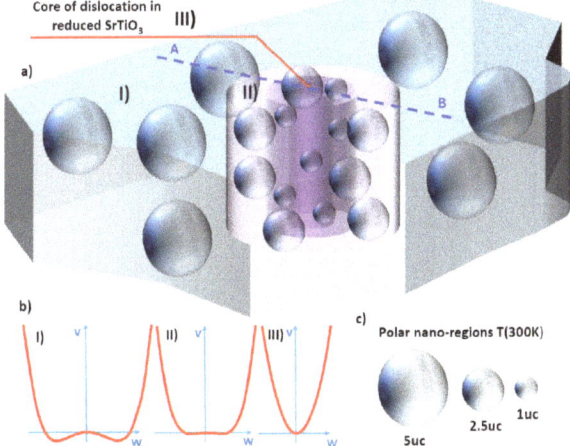

Figure 15. (a) Schematic representation of the structural and chemical development along a pathway from A to B crossing a core of dislocations in SrTiO$_3$. (b) The change of the local potential along the same pathway where in the region I) an intact matrix is present, while, in II), a crossover region exists with flat potential, which changes over to single well in III), where the metallic properties of the dislocation core dominate. (c) Sizes of the polar nano-regions (uc: unit cell) (from Reference [127]).

5. Another Rather Mysterious Perovskite: WO$_3$

WO$_3$ is a perovskite where the A-site ion is missing or only in part occupied by M=Na, K, Rb, or Cs [135]. The dominating physics in these materials stem from the valence fluctuations of W, which dominantly adopts a 3$^+$ or 5$^+$ valency. Structurally, depending on the dopant atom M, carrier-doped tungsten oxides (M_xWO_3) are classified into two different categories: perovskite or hexagonal. Both structures consist of WO_6 octahedra that form a corner-sharing network, though, in many cases, each octahedron exhibits a significant distortion [136–138]. Early on, it was discovered that these tungsten bronzes become superconductors when small amounts x of M are added. Typical superconducting transition temperatures for the three compounds are shown in Figure 16.

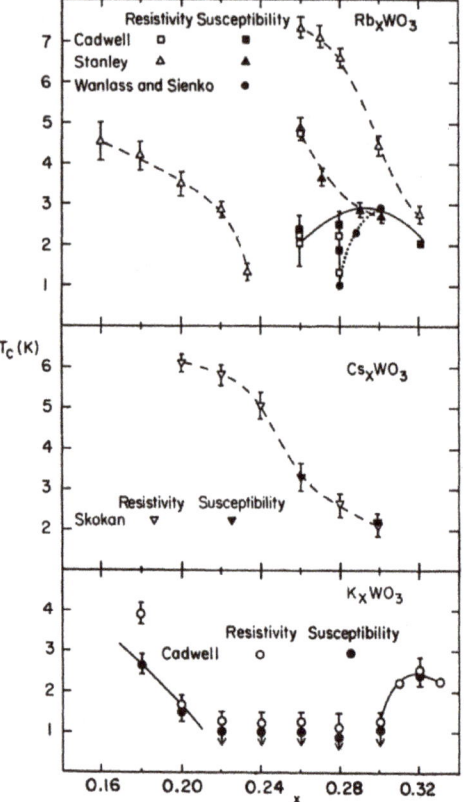

Figure 16. Superconducting transition temperature T_c as a function of x for Rb_xWO_3, Cs_xWO_3, and K_xWO_3 (from Reference [135]).

As is obvious from the figure, there are two superconducting regions in the K- and Rb-doped compounds which adopt a dome or half-dome, like x-dependence. The highest T_cs are observed for the Rb-doped system reaching more than 7 K. What is of particular interest in WO_3 is the fact that charge accumulations occur around the W^{5+} ion with the tendency to localization, i.e., polaron formation [139]. Early on, detailed experiments by Salje and Schirmer convincingly showed that these polarons pair to form bipolarons [140,141], which yielded one of the first realizations for obtaining bipolaronic superconductivity, the original concept behind the discovery of cuprates [49]. This connection was taken up by Reich and Tsabba [142] to get back to WO_3 and search for superconductivity in a WO_3 sample with surface composition of $Na_{0.05}WO_3$. These crystals showed a sharp diamagnetic step in their magnetization at 91 K, and a magnetic hysteresis below this temperature. Transport measurements below 100 K show a sharp metal to insulator transition, which is followed by a rapid decrease in the resistivity when the temperature is lowered to about 90 K. These results point to a possible nucleation of a superconducting phase on the surface of these crystals. The results of Ref. [142] initialized an

EPR and magnetic susceptibility study of Na-doped WO_3 samples, suggesting traces of possible superconductivity [143]. In a later investigation of a related compound, namely lithium intercalated $WO_{2.9}$, a small superconducting fraction was observed [144]. The absence of a clear transition in resistivity measurements indicated, however, that the superconductivity is localized in small regions which do not percolate. EPR experiments showed the presence of W^{5+} - W^{5+} electron bipolarons in reduced tungsten oxide samples, thus rendering the compound as a promising candidate for high-temperature superconductivity. More details with respect to recent progress on superconductivity in WO_3-type compounds can be found in the contributions of Shengelaya et al. [145] and Salje [146] in this issue.

6. Concluding Remarks

In our contribution to the special issue honoring K.A. Müller's life work, we tried to walk along his scientific path, which is a long one and covers almost 7 decades. Since his productivity and scientific activity was enormous, we certainly cannot follow all the detours he made during his career, but we concentrated on the two uttermost important systems: $SrTiO_3$, his scientific favorite child, and the cuprates. In addition, we added an outlook for the future research in the former where, again, superconductivity is in the focus with emphasis on its inherent heterogeneous filamentary character, parallelizing his viewpoints for cuprates. We both are deeply indebted to our friend, K. Alex Müller, who not only scientifically supported and inspired both of us, and never stopped encouraging us in following even controversial scientific topics but also for his open minded attitudes and his honest interest in our private life. His black humor has promoted us throughout our collaborations and beyond and today also adds spice to our life.

- Thank you very much, Alex! -

In order to come back to the starting ideas leading to high-temperature superconductivity, we finish our conclusions with the final Figure 17, the Jahn-Teller bipolaron [55,147], which is the essential ingredient and motivation for the discovery of high-temperature superconductivity [49–52].

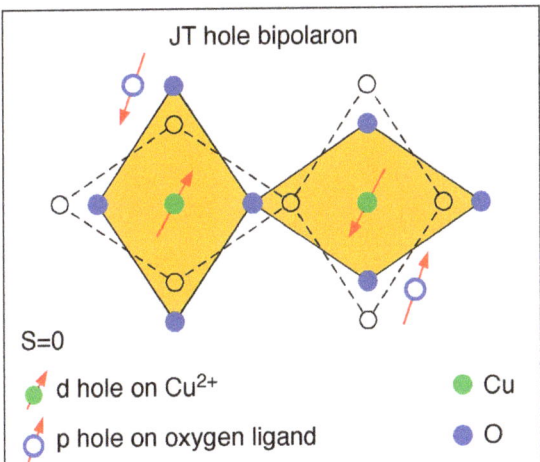

Figure 17. Schematic sketch of the inter-site Jahn-Teller (JT) bipolaron (from Reference [67]).

Author Contributions: Both authors (A.B.-H. and H.K.) have contributed equally to this work. Both authors have read and agreed to the published version of the manuscript.

Funding: This research received no external funding.

Acknowledgments: We would like to thank K.A. Müller for the fruitful and encouraging collaboration, his great support, and his warm friendship over many years.

Conflicts of Interest: The authors declare no conflict of interest.

References

1. Lines, M.E.; Glass, A.M. *Principles and Applications of Ferroelectrics and Related Materials*; Clarendon Press: Oxford, UK, 1977.
2. Cochran, W. Crystal stability and the theory of ferroelectricity. *Adv. Phys.* **1960**, *9*, 387–423. [CrossRef]
3. Anderson, P.W. *Concepts in Solids*; Benjamin: New York, NY, USA, 1963.
4. Cowley, R.A. Lattice dynamics and phase transitions of strontium titanate. *Phys. Rev.* **1964**, *134*, A981–A997. [CrossRef]
5. Müller, K.A. Paramagnetic Resonance of Fe^{3+} in $SrTiO_3$ single crystals. *Helv. Phys. Acta* **1958**, *31*, 173–204.
6. Geller, S.; Bala, V.B. Crystallographic studies of perovskite-like compounds. II. Rare earth alluminates. *Acta Cryst.* **1957**, *9*, 1019–1025. [CrossRef]
7. Müller, K.A.; Berlinger, W.; Waldner, F. Characteristic Structural Phase Transitions in Perovskite-Type Compounds. *Phys. Rev. Lett.* **1968**, *21*, 814–817. [CrossRef]
8. Müller, K.A.; Kool, T.W. *Properties of Perovskites and Other Oxides*; World Scientific Publishing: Singapore, 2010.
9. Höchli, U.; Müller, K.A. Observation of the Jahn-Teller splitting of three-valent d^7 ions via Orbach relaxation. *Phys. Rev. Lett.* **1964**, *12*, 730–733. [CrossRef]
10. Cochran, W.; Zia, A. Structure and dynamics of perovskite-type crystals. *Phys. Stat. Sol.* **1968**, *25*, 273–283. [CrossRef]
11. Fleury, P.A.; Scott, J.F.; Worlock, J.M. Soft Phonon Modes and the 110 K Phase Transition in $SrTiO_3$. *Phys. Rev. Lett.* **1968**, *21*, 16–19. [CrossRef]
12. Müller, K.A.; Berlinger, W.; Rubins, R.S. Observation of Two Charged States of a Nickel-Oxygen Vacancy Pair in $SrTiO_3$ by Paramagnetic Resonance. *Phys. Rev.* **1969**, *186*, 361–371. [CrossRef]
13. Anderson, P.W. Model for the electronic structure of amorphous semiconductors. *Phys. Rev. Lett.* **1975**, *34*, 953–956. [CrossRef]
14. Street, R.A.; Mott, N.F. States in the Gap in Glassy Semiconductors. *Phys. Rev. Lett.* **1975**, *35*, 1293–1296. [CrossRef]
15. Chakraverty, B.K. Possibility of insulator to superconductor phase transition. *J. Phys. Lett.* **1979**, *40*, 99–100. [CrossRef]
16. Müller, K.A.; Berlinger, W. Static critical exponents at structural phase transitions. *Phys. Rev. Lett.* **1971**, *26*, 13–16. [CrossRef]
17. Müller, K.A.; Burkard, H. $SrTiO_3$: An intrinsic quantum paraelectric below 4 K. *Phys. Rev. B* **1979**, *19*, 3593–3602. [CrossRef]
18. Kremer, R.K.; Bussmann-Holder, A.; Keller, H.; Haunschild, R. The Crucial Things in Science often Happen Quite Unexpectedly - Das Entscheidende in der Wissenschaft Geschieht of Ganz Unerwartet (K. Alex Müller). *Condens. Matter* **2020**, *5*, 43. [CrossRef]
19. Schooley, F.; Hosler, W.R.; Cohen, M.L. Superconductivity in semiconducting $SrTiO_3$. *Phys. Rev. Lett.* **1964**, *12*, 474–475. [CrossRef]
20. Appel, J. Soft-mode superconductivity in $SrTiO_{3-x}$. *Phys. Rev.* **1969**, *180*, 508–516. [CrossRef]
21. Cohen, M.L. Superconductivity in many-valley semiconductors and in semimetals. *Phys. Rev.* **1964**, *134*, A511–A521. [CrossRef]
22. Binnig, G.; Baratoff, A.; Hoenig, H.E.; Bednorz, J.G. Two-band superconductivity in Nb-doped $SrTiO_3$. *Phys. Rev. Lett.* **1980**, *45*, 1352–1355. [CrossRef]
23. Suhl, H.; Matthias, B.T.; Walker, L.R. Bardeen-Cooper-Schrieffer Theory of Superconductivity in the Case of Overlapping Bands. *Phys. Rev. Lett.* **1959**, *3*, 552–555. [CrossRef]
24. Moskalenko, V. Superconductivity in metals with overlapped energy bands. *Fiz. Metal. Metalloved.* **1959**, *8*, 503–513.
25. Gor'kov, L.P. Phonon mechanism in the most dilute superconductor n-type $SrTiO_3$. *Proc. Natl. Acad. Sci. USA* **2016**, *113*, 4646–4651. [CrossRef]

26. Thiemann, M.; Beutel, M.H.; Dressel, M.; Lee-Hone, N.R.; Broun, D.M.; Fillis-Tserakis, E.; Boschker, H.; Mannhart, J.; Scheffler, M. Single-Gap Superconductivity and Dome of Superfluid Density in Nb-Doped SrTiO$_3$. *Phys. Rev. Lett.* **2018**, *120*, 37002. [CrossRef] [PubMed]
27. Collignon, C.; Lin, X.; Rischau, C.W.; Fauqué, B.; Behnia, K. Metallicity and Superconductivity in Doped Strontium Titanate. *Annual Rev. Cond. Mat. Phys.* **2019**, *10*, 25–44. [CrossRef]
28. For a recent review see: Gastiasoro, M.N.; Ruhman, J.; Fernandes, R.M. Superconductivity in dilute SrTiO$_3$: A review. *Ann. Phys.* **2020**, *417*, 168107. [CrossRef]
29. Scheerer, G.; Boselli, M.; Pulmannova, D.; Rischau, C.W.; Waelchli, A.; Gariglio, S.; Giannini, E.; van der Marel, D.; Triscone, J.M. Ferroelectricity, Superconductivity, and SrTiO$_3$—Passions of K.A. Müller. *Condens. Matter* **2020**, *5*, 60. [CrossRef]
30. Hemberger, J.; Lunkenheimer, P.; Viana, R.; Böhmer, R.; Loidl, A. Electric-field-dependent dielectric constant and nonlinear susceptibility in SrTiO$_3$. *Phys. Rev. B* **1995**, *52*, 13159–13162. [CrossRef]
31. Samara, G.A. Pressure and Temperature Dependences of the Dielectric Properties of the Perovskites BaTiO$_3$ and SrTiO$_3$. *Phys. Rev.* **1966**, *151*, 378–386. [CrossRef]
32. Berre, B.; Fossheim, K.; Müller, K.A. Critical attenuation of the soft mode in SrTiO$_3$. *Phys. Rev. Lett.* **1969**, *23*, 589–591. [CrossRef]
33. Bussmann-Holder, A.; Bilz, H.; Bäuerle, D.; Wagner, D. A polarizability model for the ferroelectric mode in semiconducting SrTiO$_3$. *Z. Physik B* **1981**, *41*, 353–355. [CrossRef]
34. Bednorz, J.G.; Müller, K.A. Sr$_{1-x}$Ca$_x$TiO$_3$: An XY Quantum Ferroelectric with Transition to Randomness. *Phys. Rev. Lett.* **1984**, *52*, 2289–2292. [CrossRef]
35. Itoh, M.; Wang, R.; Inaguma, Y.; Yamaguchi, T.; Shan, Y.-J.; Nakamura, T. Ferroelectricity Induced by Oxygen Isotope Exchange in Strontium Titanate Perovskite. *Phys. Rev. Lett.* **1999**, *82*, 3540–3543. [CrossRef]
36. Bussmann-Holder, A.; Bishop, A.R. Incomplete ferroelectricity in SrTi^{18}O$_3$. *Eur. Phys. J. B* **2006**, *53*, 279–282. [CrossRef]
37. Shigenari, T.; Abe, K. Raman Spectra of Soft Modes of SrTiO$_3$. *Ferroelectrics* **2010**, *369*, 117–126. [CrossRef]
38. Kleemann, W.; Dec, J.; Tkach, A.; Vilarinho, P.M. SrTiO$_3$—Glimpses of an Inexhaustible Source of Novel Solid State Phenomena. *Condens. Matter* **2020**, *5*, 58. [CrossRef]
39. Ravel, B.; Stern, E.A.; Verdinskii, R.I.; Kraizman, V. Local structure and the phase transitions of BaTiO$_3$. *Ferroelectrics* **1998**, *206–207*, 407–430. [CrossRef]
40. Zalar, B.; Laguta, V.V.; Blinc, R. NMR Evidence for the Coexistence of Order-Disorder and Displacive Components in Barium Titanate. *Phys. Rev. Lett.* **2003**, *90*, 037601. [CrossRef]
41. Harada, J.; Axe, J.D.; Shirane, G. Neutron-Scattering Study of Soft Modes in Cubic BaTiO$_3$. *Phys. Rev. B* **1971**, *4*, 155–162. [CrossRef]
42. Völkel, G.; Müller, K.A. Order-disorder in the low-temperature phase of BaTiO$_3$. *Phys. Rev. B* **2007**, *76*, 094105. [CrossRef]
43. Bishop, A.R. A Lattice Litany for Transition Metal Oxides. *Condens. Matter* **2020**, *5*, 46. [CrossRef]
44. Kool, T.W. Meetings with a Remarkable Man, Alex Müller—The Professor of SrTiO$_3$. *Condens. Matter* **2020**, *5*, 44. [CrossRef]
45. Müller, K.A. Essential Heterogeneities in Hole-Doped Cuprate Superconductors. In *Superconductivity in Complex Systems*; Müller, K.A., Bussman-Holder, A., Eds.; Structure and Bonding 114; Springer: Berlin/Heidelberg, Germany, 2005; pp. 1–11.
46. Deutscher, G.; Fenichel, H.; Gershenson, M.; Grünbaum, E.; Ovadyahu, Z. Transition to zero dimensionality in granular aluminum superconducting films. *J. Low Temp. Phys.* **1973**, *10*, 231–243. [CrossRef]
47. Deutscher, G.; Müller, K.A. Origin of superconductive glassy state and extrinsic critical currents in high-T_c oxide. *Phys. Rev. Lett.* **1987**, *59*, 1745–1748. [CrossRef] [PubMed]
48. Deutscher, G. The role of the Short Coherence Length in Unconventional Superconductors. *Condens. Matter* **2020**, *5*, 77. [CrossRef]
49. Bednorz, J.G.; Müller, K.A. Possible High T_c Superconductivity in the Ba-La-Cu-O Sysyem. *Z. Phys. B* **1986**, *64*, 189–193. [CrossRef]
50. Bednorz, J.G.K. Alexander Müller Nobel Lecture. Available online: https://www.nobelprize.org/prizes/physics/1987/muller/lecture/ (accessed on 29 May 2020).
51. Müller, K.A.; Bednorz, J.G. The Discovery of a Class of High-Temperature Superconductors. *Science* **1987**, *237*, 1133–1139. [CrossRef] [PubMed]

52. Bednorz, J.G.; Müller, K.A. Perovskite-type oxides—The new approach to high-T_c superconductivity. *Rev. Mod. Phys.* **1988**, *60*, 585–600. [CrossRef]
53. Höck, K.-H.; Nickisch, H.; Thomas, H. Jahn-Teller effect in itinerant electron systems: The Jahn-Teller polaron. *Helv. Phys. Acta* **1983**, *56*, 237–243.
54. Bianconi, A.; Saini, N.L.; Lanzara, A.; Missori, M.; Rosselli, T.; Oyanagi, H.; Yamaguchi, H.; Oka, K.; Ito, T. Determination of the Local Lattice Distortions in the CuO_2 Plane of $La_{1.85}Sr_{0.15}CuO_4$. *Phys. Rev. Lett.* **1996**, *76*, 3412–3415. [CrossRef]
55. Müller, K.A. On the superconductivity in hole doped cuprates. *J. Phys. Condens. Matter* **2007**, *19*, 251002. [CrossRef]
56. Müller, K.A. The Unique Properties of Superconductivity in Cuprates. *J. Supercond. Nov. Magn.* **2014**, *27*, 2163–2179. [CrossRef]
57. Müller, K.A. The Polaronic Basis for High-Temperature Superconductivity. *J. Supercond. Nov. Magn.* **2017**, *30*, 3007–3018. [CrossRef]
58. Müller, K.A. Possible coexistence of s- and d-wave condensates in copper oxide superconductors. *Nature* **1995**, *377*, 133–135. [CrossRef]
59. Müller, K.A.; Keller, H. *s* and *d* Wave Symmetry Components in High-Temperature Cuprate Superconductors. In *High-T_c Superconductivity 1996: Ten Years after the Discovery*; Kaldis, E., Liarokapis, E., Müller, K.A., Eds.; Kluwer Academic Publishers: Dordrecht, The Netherlands, 1997; pp. 7–29.
60. Wollman, D.A.; Van Harlingen, D.J.; Lee, W.C.; Ginsberg, D.M.; Leggett, A.J. Experimental determination of the superconducting pairing state in YBCO from the phase coherence of YBCO-Pb dc SQUIDs. *Phys. Rev. Lett.* **1993**, *71*, 2134–2137. [CrossRef]
61. Tsuei, C.C.; Kirtley, J.R.; Chi, C.C.; Yu-Jahnes, L.S.; Gupta, A.; Shaw, T.; Sun, J.Z.; Ketchen, M.B. Pairing Symmetry and Flux Quantization in a Tricrystal Superconducting Ring of $YBa_2Cu_3O_{7-\delta}$. *Phys. Rev. Lett.* **1994**, *73*, 593–596. [CrossRef]
62. Brawner, D.A.; Ott, H.R. Evidence for an unconventional superconducting order parameter in $YBa_2Cu_3O_{6.9}$. *Phys. Rev. B* **1994**, *50*, 6530(R)–6533(R). [CrossRef]
63. Khasanov, R.; Shengelaya, A.; Maisuradze, A.; La Mattina, F.; Bussmann-Holder, A.; Keller, H.; Müller, K.A. Experimental evidence for two gaps in the high-temperature $La_{1.83}Sr_{0.17}CuO_4$ superconductor. *Phys. Rev. Lett.* **2007**, *98*, 057007. [CrossRef]
64. Khasanov, R.; Strässle, S.; Di Castro, D.; Masui, T.; Miyasaka, S.; Tajima, S.; Bussmann-Holder, A.; Keller, H. Multiple gap symmetries for the order parameter of cuprate superconductors from penetration depth measurements. *Phys. Rev. Lett.* **2007**, *99*, 237601. [CrossRef]
65. Khasanov, R.; Shengelaya, A.; Karpinski, J.; Bussmann-Holder, A.; Keller, H.; Müller, K.A. s-wave symmetry along the c-axis and s+d in-plane superconductivity in bulk $YBa_2Cu_4O_8$. *J. Supercond. Nov. Magn.* **2008**, *21*, 81–85. [CrossRef]
66. Bussmann-Holder, A.; Khasanov, R.; Shengelaya, A.; Maisuradze, A.; La Mattina, F.; Keller, H.; Müller, K.A. Mixed order parameter symmetries in cuprate superconductors. *Europhys. Lett.* **2007**, *77*, 27002. [CrossRef]
67. Keller, H.; Bussmann-Holder, A.; Müller, K.A. Jahn-Teller physics and high-T_c superconductivity. *Mater. Today* **2008**, *11*, 38–46. [CrossRef]
68. Zimmermann, P.; Keller, H.; Lee, S.L.; Savić, I.M.; Warden, M.; Zech, D.; Cubitt, R.; Forgan, E.M.; Kaldis, E.; Karpinski, J.; et al. Muon-spin rotation studies of the temperature dependence of the magnetic penetration depth in the $YBa_2Cu_3O_x$ family and related compounds. *Phys. Rev. B* **1995**, *52*, 541–552. [CrossRef] [PubMed]
69. Müller, K.A. On the macroscopic s- and d-wave symmetry in cuprate superconductors. *Philos. Mag. Lett.* **2002**, *82*, 279–288. [CrossRef]
70. Iachello, F. A model of cuprate superconductors based on the analogy with atomic nuclei. *Philos. Mag. Lett.* **2002**, *82*, 289–295. [CrossRef]
71. Khasanov, R.; Shengelaya, A.; Brütsch, R.; Keller, H. Suppression of the s-wave Order Parameter Near the Surface of the Infinite-Layer Electron-Doped Cuprate Superconductor $Sr_{0.9}La_{0.1}Cu_2$. *Condens. Matter* **2020**, *5*, 50. [CrossRef]
72. Batlogg, B.; Cava, R.J.; Jayaraman, A.; van Dover, R.B.; Kourouklis, G.A.; Sunshine, S.; Murphy, D.W.; Rupp, L.W.; Chen, H.S.; White, A.; et al. Isotope Effect in the High-Tc Superconductors $Ba_2YCu_3O_7$ and $Ba_2EuCu_3O_7$. *Phys. Rev. Lett.* **1987**, *58*, 2333–2336. [CrossRef]

73. Franck, J.P.; Jung, J.; Mohamed, A.K.; Gygax, S.; Sproule, G.I. Observation of an oxygen isotope effect in superconducting $(Y_{1-x}Pr_x)Ba_2Cu_3O_{7-\delta}$. *Phys. Rev. B* **1991**, *44*, 5318–5321. [CrossRef] [PubMed]
74. Franck, J.P. Experimental studies of the isotope effect in high temperature superconductors. In *Physical Properties of High Temperature Superconductors IV*; Ginsberg, D.M., Ed.; World Scientific: Singapore, 1994; pp. 189–293.
75. Müller, K.A. On the oxygen isotope effect and apex anharmonicity in high-T_c cuprates. *Z. Phys. B Condens. Matter* **1990**, *80*, 193–201. [CrossRef]
76. Zhao, G.M.; Conder, K.; Keller, H.; Müller, K.A. Oxygen isotope effects in $La_{2-x}Sr_xCuO_4$: Evidence for polaronic charge carriers and their condensation. *J. Phys. Condens. Matter* **1998**, *10*, 9055–9066. [CrossRef]
77. Zhao, G.M.; Keller, H.; Conder, K. Unconventional isotope effects in the high-temperature cuprate superconductors. *J. Phys. Condens. Matter* **2001**, *13*, R569–R587. [CrossRef]
78. Keller, H. Unconventional Isotope Effects in Cuprate Superconductors. In *Superconductivity in Complex Systems*; Müller, K.A., Bussmann-Holder, A., Eds.; Springer: Berlin/Heidelberg, Germany, 2005; pp. 143–169.
79. Conder, K.; Furrer, A.; Pomjakushina, E. A Retrospective of Materials Synthesis at the Paul Scherrer Institut (PSI). *Condens. Matter* **2020**, *5*, 55. [CrossRef]
80. Zech, D.; Keller, H.; Conder, K.; Kaldis, E.; Liarokapis, E.; Poulakis, N.; Müller, K.A. Site-selective oxygen isotope effect in optimally doped $YBa_2Cu_3O_{6+x}$. *Nature* **1994**, *371*, 681–683. [CrossRef]
81. Zhao, G.; Ager, J.W., III; Morris, D.E. Site dependence of large oxygen isotope effect in $Y_{0.7}Pr_{0.3}Ba_2Cu_3O_{6.97}$. *Phys. Rev. B* **1996**, *54*, 14982–14985. [CrossRef] [PubMed]
82. Khasanov, R.; Shengelaya, A.; Morenzoni, E.; Angst, M.; Conder, K.; Savić, I.M.; Lampakis, D.; Liarokapis, E.; Tatsi, A.; Keller, H. Site-selective oxygen isotope effect on the magnetic field penetration depth in underdoped $Y_{0.6}Pr_{0.4}Ba_2Cu_3O_{7-\delta}$. *Phys. Rev. B* **2003**, *68*, 220506(R). [CrossRef]
83. Bussmann-Holder, A.; Genzel, L.; Bishop, A.R.; Simon, A. The role of apical oxygen in superconducting cuprates. *Philos. Mag. B* **1997**, *75*, 463–469. [CrossRef]
84. Khasanov, R.; Shengelaya, A.; Di Castro, D.; Morenzoni, E.; Maisuradze, A.; Savić, I.M.; Conder, K.; Pomjakushina, E.; Bussmann-Holder, A.; Keller, H. Oxygen isotope effect on the superconducting transition and magnetic states within the phase diagram of $Y_{1-x}Pr_xBa_2Cu_3O_{7-\delta}$. *Phys. Rev. Lett.* **2008**, *101*, 077001. [CrossRef] [PubMed]
85. Guguchia, Z.; Khasanov, R.; Bendele, M.; Pomjakushina, E.; Conder, K.; Shengelaya, A.; Keller, H. Negative Oxygen Isotope Effect on the Static Spin Stripe Order in Superconducting $La_{2-x}Ba_xCuO_4$ ($x=1/8$) Observed by Muon-Spin Rotation. *Phys. Rev. Lett.* **2014**, *113*, 057002. [CrossRef]
86. Alexandrov, A.S.; Mott, N.F. Spin and charge bipolaron kinetics of high T_c superconductors. *Int. J. Mod. Phys. B* **1994**, *8*, 2075–2109. [CrossRef]
87. Zhao, G.; Morris, D.E. Observation of a possible oxygen isotope effect on the effective mass of carriers in $YBa_2Cu_3O_{6.94}$. *Phys. Rev. B* **1995**, *51*, 16487–16490. [CrossRef]
88. Zhao, G.M.; Hunt, M.B.; Keller, H.; Müller, K.A. Evidence for polaronic supercarriers in the copper oxide superconductors $La_{2-x}Sr_xCuO_4$. *Nature* **1997**, *385*, 236–239. [CrossRef]
89. Hofer, J.; Conder, K.; Sasagawa, T.; Zhao, G.M.; Willemin, M.; Keller, H.; Kishio, K. Oxygen-isotope effect on the in-plane penetration depth in underdoped $La_{2-x}Sr_xCuO_4$ single crystals. *Phys. Rev. Lett.* **2000**, *84*, 4192–4195. [CrossRef] [PubMed]
90. Khasonov, R.; Shengelaya, A.; Conder, K.; Morenzoni, E.; Savić, I.M.; Keller, H. The oxygen-isotope effect on the in-plane penetration depth in underdoped $Y_{1-x}Pr_xBa_2Cu_3O_{7-\delta}$ as revealed by muon-spin rotation. *J. Phys. Condens. Matter* **2003**, *15*, L17–L23. [CrossRef]
91. Jackson, T.J.; Riseman, T.M.; Forgan, E.M.; Glückler, H.; Prokscha, T.; Morenzoni, E.; Pleines, M.; Niedermayer, C.; Schatz, G.; Luetkens, H.; et al. Depth-Resolved Profile of the Magnetic Field beneath the Surface of a Superconductor with a Few nm Resolution. *Phys. Rev. Lett.* **2000**, *84*, 4958–4961. [CrossRef] [PubMed]
92. Khasanov, R.; Eshchenko, D.G.; Luetkens, H.; Morenzoni, E.; Prokscha, T.; Suter, A.; Garifinov, N.; Mali, M.; Roos, J.; Conder, K.; et al. Direct observation of the oxygen isotope effect on the in-plane magnetic field penetration depth in optimally doped $YBa_2Cu_3O_{7-\delta}$. *Phys. Rev. Lett.* **2004**, *92*, 057602. [CrossRef] [PubMed]
93. Uemura, Y.J.; Luke, G.M.; Sternlieb, B.J.; Brewer, J.H.; Carolan, J.F.; Hardy, W.N.; Kadono, R.; Kempton, J.R.; Kiefl, R.F.; Kreitzman, S.R.; et al. Universal Correlations between T_c and n_s/m^* (Carrier Density over Effective Mass) in High-T_c Cuprate Superconductors. *Phys. Rev. Lett.* **1989**, *62*, 2317–2320. [CrossRef]

94. Uemura, Y.J.; Le, L.P.; Luke, G.M.; Sternlieb, B.J.; Wu, W.D.; Brewer, J.H.; Riseman, T.M.; Seaman, C.L.; Maple, M.B.; Ishikawa, M.; et al. Basic similarities among cuprate, bismuthate, organic, Chevrel-phase, and heavy-fermion superconductors shown by penetration-depth measurements. *Phy. Rev. Lett.* **1991**, *66*, 2665–2668. [CrossRef]
95. Weyeneth, S.; Müller, K.A. Oxygen Isotope Effect in Cuprates Results from Polaron-Induced Superconductivity. *J. Supercond. Nov. Magn.* **2011**, *24*, 1235–1239. [CrossRef]
96. Kresin, V.Z.; Wolf, S.A. Microscopic model for the isotope effect in high-T_c oxides. *Phys. Rev. B* **1994**, *49*, 3652–3654. [CrossRef]
97. Bill, A.; Kresin, V.Z.; Wolf, S.A. Isotope effect for the penetration depth in superconductors. *Phys. Rev. B* **1998**, *57*, 10814–10824. [CrossRef]
98. Schneider, T.; Keller, H. Universal trends in extreme type-II superconductors. *Phys. Rev. Lett.* **1992**, *69*, 3374–3377. [CrossRef]
99. Bendele, M.; von Rohr, F.; Gugchia, Z.; Pomjakushina, E.; Conder, K.; Bianconi, A.; Simon, A.; Bussmann-Holder, A.; Keller, H. Evidence for strong lattice effects as revealed from huge unconventional oxygen isotope effects on the pseudogap temperature in $La_{2-x}Sr_xCuO_4$. *Phys. Rev. B* **2017**, *95*, 014514. [CrossRef]
100. Lanzara, A.; Zhao, G.; Saini, N.L.; Bianconi, A.; Conder, K.; Keller, H.; Müller, K.A. Oxygen-isotope effect of the charge-stripe ordering temperature in $La_{2-x}Sr_xCuO_4$ from X-ray absorption spectroscopy. *J. Phys. Condens. Matter* **1999**, *11*, L541–L546. [CrossRef]
101. Rubio Temprano, D.; Mesot, J.; Janssen, S.; Conder, K.; Furrer, A.; Mutka, H.; Müller, K.A. Large Isotope Effect on the Pseudogap in the High-Temperature Superconductor $HoBa_2Cu_4O_8$. *Phys. Rev. Lett.* **2000**, *84*, 1990–1993. [CrossRef] [PubMed]
102. Rubio Temprano, D.; Furrer, A.; Conder, K.; Mutka, H. A neutron crystal-field study of the pseudogap in the underdoped high T_c superconductor $HoBa_2Cu_4{}^{18}O_8$. *Physica B* **2000**, *276–278*, 762–763. [CrossRef]
103. Rubio Temprano, D.; Mesot, J.; Janssen, S.; Conder, K.; Furrer, A.; Sokolov, A.; Trounov, V.; Kazakov, S.M.; Karpinski, J.; Müller, K.A. Large copper isotope effect on the pseudogap in the high-temperature superconductor $HoBa_2Cu_4O_8$. *Eur. Phys. J. B* **2001**, *19*, 5–8. [CrossRef]
104. Rubio Temprano, D.; Conder, K.; Furrer, A.; Mutka, H.; Trounov, V.; Müller, K.A. Oxygen and copper isotope effects on the pseudogap in the high-temperature superconductor $La_{1.81}Ho_{0.04}Sr_{0.15}CuO_4$ studied by neutron crystal-field spectroscopy. *Phys. Rev. B* **2002**, *66*, 184506. [CrossRef]
105. Häfliger, P.S.; Podlesnyak, A.; Conder, K.; Pomjakushina, E.; Furrer, A. Pseudogap of the high-temperature superconductor $La_{1.96-x}Sr_xHo_{0.04}CuO_4$ as observed by neutron crystal-field spectroscopy. *Phys. Rev. B* **2006**, *74*, 184520. [CrossRef]
106. Varma, C.M. Non-Fermi-liquid states and pairing instability of a general model of copper oxide metals. *Phys. Rev. B* **1997**, *55*, 14554–14580. [CrossRef]
107. Li, Y.; Balédent, V.; Barisić, N.; Cho, Y.; Fauqué, B.; Sidis, Y.; Yu, G.; Zhao, X.; Bourges, P.; Greven, M. Unusual magnetic order in the pseudogap region of the superconductor $HgBa_2CuO_{4+\delta}$. *Nature* **2008**, *455*, 372–375. [CrossRef]
108. Bussmann-Holder, A.; Keller, H. Polaron Effects in High-Temperature Cuprate Superconductors. In *Polarons in Advanced Materials*; Alexandrov, S.A., Ed.; Springer: Dordrecht, The Netherlands, 2007; pp. 599–621.
109. Zhang, C.J.; Oyanagi, H. Local lattice instability and superconductivity in $La_{1.85}Sr_{0.15}Cu_{1-x}M_xO_4$ (M=Mn, Ni, and Co). *Phys. Rev. B* **2009**, *79*, 064521. [CrossRef]
110. Bussmann-Holder, A.; Simon, A.; Keller, H.; Bishop, A.R. Polaron signatures in the phonon dispersion of high-temperature superconducting copper oxides. *Eur. Phys. Lett.* **2013**, *101*, 47004. [CrossRef]
111. Ohtomo, A.; Hwang, H.Y. A high-mobility electron gas at the $LaAlO_3/SrTiO_3$ heterointerface. *Nature* **2004**, *427*, 423–426. [CrossRef] [PubMed]
112. Reyren, N.; Thiel, S.; Caviglia, A.D.; Hammerl, G.; Richter, C.; Schneider, C.W.; Kopp, T.; Rüetschi, A.-S.; Jaccard, D.; Gabay, M.; et al. Superconducting interfaces between insulating oxides. *Science* **2007**, *317*, 1196–1199. [CrossRef] [PubMed]
113. Brinkman, A.; Huijben, M.; van Zalk, M.; Huijben, J.; Zeitler, U.; Maan, J.C.; van der Wiel, W.; Rijnders, D.; Blank, D.H.A.; Hilgenkamp, H. Magnetic effects at the interface between non-magnetic oxides. *Nat. Mater.* **2007**, *6*, 493–496. [CrossRef] [PubMed]

114. Mannhart, J.; Schlom, D.G. Oxide interfaces—An opportunity for electronics. *Science* **2010**, *327*, 1607–1611. [CrossRef]
115. Zubko, P.; Gariglio, S.; Gabay, M.; Ghosez, P.; Triscone, J.-M. Interface physics in complex oxide heterostructures. *Ann. Rev. Cond. Mat. Phys.* **2011**, *2*, 141–156. [CrossRef]
116. Hwang, H.Y.; Iwasa, Y.; Kawasaki, M.; Keimer, B.; Nagaosa, N.; Tokura, Y. Emergent phenomena at oxide interfaces. *Nat. Mater.* **2012**, *11*, 103–113. [CrossRef]
117. Lin, X.; Zhu, Z.; Fauqué, B.; Behnia, K. Fermi surface of the most dilute superconductor. *Phys. Rev. X* **2013**, *3*, 021002. [CrossRef]
118. Edge, J.M.; Kedem, Y.; Aschauer, U.; Spaldin, N.A.; Balatsky, A.V. Quantum critical origin of the superconducting dome in $SrTiO_3$. *Phys. Rev. Lett.* **2015**, *115*, 247002. [CrossRef]
119. Bussmann-Holder, A.; Beige, H.; Völkel, G. Precursor effects, broken local symmetry, and coexistence of order-disorder and displacive dynamics in perovskite ferroelectrics. *Phys. Rev. B* **2009**, *79*, 18411. [CrossRef]
120. Egami, T.; Billinge, S.J.L. Underneath the Bragg Peaks. In *Structural Analysis of Complex Materials*; Pergamon Press: New York, NY, USA, 2012.
121. Sato, K.; Miyanaga, T.; Ikeda, S.; Diop, D. XAFS Study of Local Structure Change in Perovskite Titanates. *Physica Scripta* **2005**, *T115*, 359–361. [CrossRef]
122. Hou, D.; Zhao, C.; Paterson, A.R.; Li, S.; Jones, J.L. Local structures of perovskite dielectrics and ferroelectrics via pair distribution function analyses. *J. Eur. Ceramic Soc.* **2017**, *38*, 971–987. [CrossRef]
123. Bilz, H.; Benedek, G.; Bussmann-Holder, A. Theory of ferroelectricity: The polarizability model. *Phys. Rev. B* **1987**, *35*, 4840–4848. [CrossRef] [PubMed]
124. Bussmann-Holder, A.; Büttner, H. Ferroelectricity in oxides. *Nature* **1992**, *360*, 541. [CrossRef]
125. Bussmann-Holder, A. The polarizability model for ferroelectricity in perovskite oxides. *J. Phys. Condens. Matter* **2012**, *24*, 273202. [CrossRef]
126. Bussmann-Holder, A.; Roleder, K.; Ko, J.-H. What makes the difference in perovskite titanates? *J. Phys. Chem. Solids* **2018**, *117*, 148–157. [CrossRef]
127. Bussmann-Holder, A.; Keller, H.; Simon, A.; Bihlmayer, G.; Roleder, K.; Szot, K. Unconventional Co-Existence of Insulating Nano-Regions and Conducting Filaments in Reduced $SrTiO_3$: Mode Softening, Local Piezoelectricity, and Metallicity. *Crystals* **2020**, *10*, 437. [CrossRef]
128. Calvani, P.; Capizzi, M.; Donato, F.; Lupi, S.; Maselli, P.; Peschiaroli, D. Observation of a midinfrared band in $SrTiO_{3-y}$. *Phys. Rev. B* **1993**, *47*, 8917–8922. [CrossRef]
129. Waser, R.; Dittmann, R.; Staikov, G.; Szot, K. Redox-Based Resistive Switching Memories—Nanoionic Mechanisms, Prospects, and Challenges. *Adv. Mater.* **2009**, *21*, 2632–2663. [CrossRef]
130. Wrana, D.; Rodenbücher, C.; Bełza, W.; Szot, K.; Krok, F. In situ study of redox processes on the surface of $SrTiO_3$ single crystals. *Appl. Surf. Sci.* **2018**, *432*, 46–52. [CrossRef]
131. Rodenbücher, C.; Menzel, S.; Wrana, D.; Gensch, T.; Korte, C.; Krok, F.; Szot, K. Current channeling along extended defects during electroreduction of $SrTiO_3$. *arXiv* **2019**, arXiv:1910.02748. [CrossRef] [PubMed]
132. Szot, K.; Speier, W.; Bihlmayer, G.; Waser, R. Switching the electrical resistance of individual dislocations in single-crystalline $SrTiO_3$. *Nat. Mater.* **2006**, *5*, 312–320. [CrossRef] [PubMed]
133. Gao, P.; Yang, S.; Ishikawa, R.; Li, N.; Feng, B.; Kumamoto, A.; Shibata, N.; Yu, P.; Ikuhara, Y. Atomic-Scale Measurement of Flexoelectric Polarization at $SrTiO_3$ Dislocations. *Phys. Rev. Lett.* **2018**, *120*, 267601. [CrossRef] [PubMed]
134. Bussmann-Holder, A.; Keller, H.; Simon, A.; Bianconi, A. Multi-Band Superconductivity and the Steep Band/Flat Band Scenario. *Condens. Matter* **2019**, *4*, 91. [CrossRef]
135. Cadwell, L.H.; Morris, R.C.; Moulton, W.G. Normal and superconducting properties of K_xWO_3. *Phys. Rev. B* **1981**, *23*, 2219–2223. [CrossRef]
136. Wiseman, P.J.; Dickens, P.G. Neutron diffraction studies of cubic tungsten bronzes. *J. Solid State Chem.* **1976**, *17*, 91–100. [CrossRef]
137. Brusetti, R.; Bordet, P.; Bossy, J.; Schober, H.; Eibl, S. Superconductivity in the tungsten bronze Rb_xWO_3 ($0.20 \leq x \leq 0.33$) in connection with its structure, electronic density of states, and phonon density of states. *Phys. Rev. B* **2007**, *76*, 174511. [CrossRef]
138. Lee, K.S.; Seo, D.K.; Whangbo, M.H. Electronic Band Structure Study of the Anomalous Electrical and Superconducting Properties of Hexagonal Alkali Tungsten Bronzes A_xWO_3 (A = K, Rb, Cs). *J. Am. Chem. Soc.* **1997**, *119*, 4043–4049. [CrossRef]

139. Bousquet, E.; Hamdi, H.; Aguado-Puente, P.; Salje, E.K.H.; Artacho, E.; Ghosez, P. First-principles characterization of single-electron polaron in WO$_3$. *Phys. Rev. Res.* **2020**, *2*, 012052. [CrossRef]
140. Schirmer, O.F.; Salje, E. Conducting bi-polarons in low-temperature crystalline WO$_{3-x}$. *J. Phys. C* **1980**, *13*, 1067–1072. [CrossRef]
141. Schirmer, O.F.; Salje, E. The W^{5+} polaron in crystalline low temperature WO$_3$ ESR and optical absorption. *Solid State Commun.* **1980**, *33*, 333–336. [CrossRef]
142. Reich, S.; Tsabba, Y. Possible nucleation of a 2D superconducting phase on WO$_3$ single crystals surface doped with Na$^+$. *Eur. Phys. J. B* **1999**, *1*, 1–4. [CrossRef]
143. Shengelaya, A.; Reich, S.; Tsabba, Y.; Müller, K.A. Electron spin resonance and magnetic susceptibility suggest superconductivity in Na doped WO$_3$ samples. *Eur. Phys. J. B* **1999**, *12*, 13–15. [CrossRef]
144. Shengelaya, A.; Conder, K.; Müller, K.A. Signatures of filamentary superconductivity up to 94 K in tungsten oxide WO$_{2.9}$. *J. Supercond. Nov. Magn.* **2020**, *33*, 301–306. [CrossRef]
145. Shengelaya, A.; Mattina, F.L.; Conder, K. Unconventional Transport Properties of Reduced Tungsten Oxide WO$_{2.9}$. *Condens. Matter* **2020**, *5*, 63. [CrossRef]
146. Salje, E.K.H. Polaronic States and Superconductivity in WO$_{3-x}$. *Condens. Matter* **2020**, *5*, 32. [CrossRef]
147. Kochelaev, B.I.; Safina, A.M.; Shengelaya, A.; Keller, H.; Müller, K.A.; Conder, K. Three-Spin-Polarons and Their Elastic Interaction in Cuprates. *Mod. Phys. Lett. B* **2003**, *17*, 415–421. [CrossRef]

Publisher's Note: MDPI stays neutral with regard to jurisdictional claims in published maps and institutional affiliations.

© 2020 by the authors. Licensee MDPI, Basel, Switzerland. This article is an open access article distributed under the terms and conditions of the Creative Commons Attribution (CC BY) license (http://creativecommons.org/licenses/by/4.0/).

Article

Measuring the Electron–Phonon Interaction in Two-Dimensional Superconductors with He-Atom Scattering

Giorgio Benedek [1,2], Joseph R. Manson [2,3], Salvador Miret-Artés [2,4], Adrian Ruckhofer [5], Wolfgang E. Ernst [5], Anton Tamtögl [5,*] and Jan Peter Toennies [6]

1. Dipartimento di Scienza dei Materiali, Università di Milano-Bicocca, Via R. Cozzi 55, 20185 Milano, Italy; giorgio.benedek@unimib.it
2. Donostia International Physics Center (DIPC), Paseo M. de Lardizabal 4, 20018 Donostia/San Sebastián, Basque Country, Spain; jmanson@clemson.edu (J.R.M.); s.miret@iff.csic.es (S.M.-A.)
3. Department of Physics and Astronomy, Clemson University, Clemson, SC 29630, USA
4. Institute of Fundamental Physics, Spanish Research Council, 28006 Madrid, Spain
5. Institute of Experimental Physics, Graz University of Technology, 8010 Graz, Austria; ruckhofer@tugraz.at (A.R.); wolfgang.ernst@tugraz.at (W.E.E.)
6. Max-Planck-Institut für Dynamik und Selbstorganisation, Am Fassberg 17, 37077 Göttingen, Germany; jtoenni@gwdg.de
* Correspondence: tamtoegl@tugraz.at

Received: 27 October 2020; Accepted: 27 November 2020; Published: 3 December 2020

Abstract: Helium-atom scattering (HAS) spectroscopy from conducting surfaces has been shown to provide direct information on the electron–phonon interaction, more specifically the mass-enhancement factor λ from the temperature dependence of the Debye–Waller exponent, and the mode-selected electron–phonon coupling constants $\lambda_{\mathbf{Q}\nu}$ from the inelastic HAS intensities from individual surface phonons. The recent applications of the method to superconducting ultra-thin films, quasi-1D high-index surfaces, and layered transition-metal and topological pnictogen chalcogenides are briefly reviewed.

Keywords: electron-phonon interaction; superconductivity; topological insulator; topological materials; transition metal dichalcogenide; charge density wave; helium atom scattering

1. Introduction

Helium-atom scattering (HAS) from a conducting surface can exchange energy and momentum with the lattice vibrations of the surface exclusively via the surface charge-density oscillations produced by the atomic motion, i.e., via electron–phonon (e–ph) interaction. Although this mechanism has been suggested since the early days of HAS spectroscopy [1], as a consequence of the discovery by HAS of the ubiquitous anomalous longitudinal surface resonance at metal surfaces [2–5], only more recent theoretical studies based on density functional perturbation theory (DFPT) [6,7] proved that the inelastic HAS intensities from surface phonons are directly proportional to their specific e–ph coupling constants $\lambda_{\mathbf{Q}\nu}$, inaugurating what has been termed as *mode-lambda spectroscopy*.

An interesting aspect of the e–ph mechanism is that subsurface phonons that produce a modulation of the surface electron density can also be detected by HAS, the detection depth being equal to the range of the e–ph interaction. Thus, a surface probe such as He atoms with incident energies in the range of tens of meV that only tickle the surface where the electron density is ~10^{-4} a.u. [8], can even measure the dispersion curves of phonons propagating at several atomic planes beneath the surface (*quantum sonar effect*), e.g., at the interface of ultra-thin metal films with the substrate, provided the e–ph coupling is sufficiently strong [7,9].

A natural consequence of this mechanism is that the thermal mean-square distortion of the surface charge-density profile, providing the Debye–Waller attenuation of the specular intensity with increasing temperature, is proportional to the mean-square phonon displacement via the total e–ph coupling, represented by the mass-enhancement factor λ. This permits the direct derivation of λ from the temperature dependence of HAS reflectivity for any conducting surface, as demonstrated in a recent series of papers devoted to metal surfaces [10], ultra-thin metal films [11], layered transition-metal chalcogenides [12,13], topological semimetals [14,15] and one-dimensional metals [16], multidimensional materials [17] and graphene [18]. The very high surface sensitivity of HAS for surface dynamic corrugations of the order of 10^{-2} a.u. also permits the detection of surface charge-density waves, undetectable by other current surface probes, together with the associated charge density wave (CDW) excitations like phasons and amplitons [14,19,20].

This ability of HAS spectroscopy, including its comparatively high resolution in energy and parallel momentum (see Chap. 9 of Ref. [1]), makes it a choice tool to investigate various aspects of the e–ph interaction in two-dimensional (2D) superconductors. After a theoretical summary (Section 2), with the relevant equations whose detailed derivation is found in refs. [7,10,11,15,17], a few examples for different classes of 2D superconducting materials are discussed in the following sections.

2. The Electron–Phonon Theory of He-Atom Scattering from Conducting Surfaces

The repulsive part of the interaction between a He atom and a conducting surface is described to a good approximation by the Esbjerg–Nørskov (EN) potential [21]

$$V(\mathbf{r}) = An(\mathbf{r}), \quad (1)$$

where $n(\mathbf{r})$ is the surface electron density, $A = 364$ eV a_0^3 the EN constant, and a_0 the Bohr's radius [22]. Consider in Figure 1 an inelastic scattering process of a He atom from an initial state $|i\rangle$ of energy E_i and wave vector $\mathbf{k}_i = (\mathbf{K}_i, k_{iz})$ into a final state $\langle f|$ of energy E_f and wave vector $\mathbf{k}_f = (\mathbf{K}_f, k_{fz})$, where $\mathbf{K}_i$ and $\mathbf{K}_f$ are the components parallel to the surface and k_{iz}, k_{fz} the respective normal components. The inelastic scattering probability $P(\mathbf{k}_i, \mathbf{k}_f)$ for one-phonon creation processes in the standard distorted wave Born approximation takes the form (up to a constant factor) [23]:

$$P(\mathbf{k}_i, \mathbf{k}_f) \propto f(\Delta E) N(E_F) \sum_{\mathbf{Q}\nu} \lambda_{\mathbf{Q}\nu} \delta(\Delta E - \varepsilon_{\mathbf{Q}\nu}), \quad (2)$$

where $N(E_F)$ is the electron density of states (DOS) at the Fermi level, $\Delta E = E_f - E_i$ the energy gain, $\varepsilon_{\mathbf{Q}\nu}$ the phonon energy of wave vector $\mathbf{Q} = \mathbf{K}_f - \mathbf{K}_i$ and branch index ν, and $\lambda_{\mathbf{Q}\nu}$ the mode-selected e–ph coupling constants (mode lambdas) [24,25]. The coefficient

$$f(\Delta E) \equiv \frac{k_f}{|k_{iz}|} \Delta E[1 + n_{BE}(\Delta E)] A^2 \left| \langle f | \psi^*_{\mathbf{K}n}(\mathbf{r}) \psi_{\mathbf{K}+\mathbf{Q}n'}(\mathbf{r}) | i \rangle \right|^2, \quad (3)$$

with $n_{BE}(\Delta E)$ the Bose-Einstein phonon occupation number, is proportional to the square matrix element between final and initial He-atom wave functions of the electron density matrix, connecting two Fermi level electronic states of bands n and n' and wave vectors $\mathbf{K}$ and $\mathbf{K} + \mathbf{Q}$, respectively. Please note that the He-atom wave-functions decay very rapidly as they penetrate into the surface electron density near the classical turning point, far away from the first surface atomic plane, where the wave function $\psi_{\mathbf{K}n}(\mathbf{r})$ decays asymptotically as $e^{-\kappa z}$. In the Wentzel-Kramers-Brillouin (WKB) approxi-mation, $\kappa = (2m^*\phi)^{1/2}\hbar$ where ϕ is the surface work function and m^* the effective mass of the surface free electrons at the Fermi level. Thus, the square matrix element in Equation (3) can be taken as independent of the electronic band indices and of ΔE, the latter being much smaller than ϕ. Since at sufficiently high temperature $\Delta E[1 + n_{BE}(\Delta E)] \cong k_B T$, the coefficient $f(\Delta E)$ is also approximately constant, and the inelastic HAS intensity for a phonon $(\mathbf{Q}, \nu)$ is just proportional to the mode-selected e–ph coupling constant $\lambda_{\mathbf{Q}\nu}$. This conclusion enables inelastic HAS spectra to indicate which phonons are strongly coupled to electrons, and which are less so. As compared to other methods which permit an extraction of the Eliashberg function, the information obtained from HAS spectroscopy is particularly relevant for 2D Bardeen-Cooper-Schrieffer superconductors for providing, besides the frequency, the parallel momentum **Q** and branch index ν of phonons that are mostly involved in pairing.

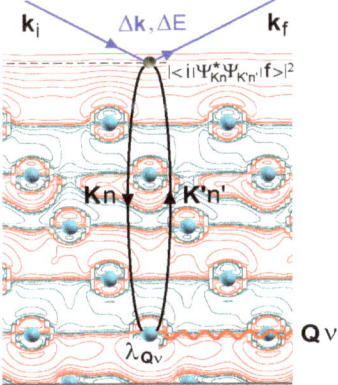

Figure 1. Diagram representing the inelastic scattering of a He atom of incident wavevector $\mathbf{k}_i$ and final wavevector $\mathbf{k}_f$ at the surface of a metal film, where a virtual electron-hole pair is created at the scattering turning point well above the first atomic layer, and recombines into a phonon of wavevector **Q** and branch index ν. The phonon displacement field produces an oscillation of the charge density via the mode-selected e–ph coupling constant $\lambda_{\mathbf{Q}\nu}$ (red/blue contour lines correspond to positive/negative charge oscillations). The fairly long range of the e–ph interaction determines the depth at which phonons can be created/annihilated by inelastic He-atom scattering (quantum sonar effect). The inelastic amplitude is proportional to $\lambda_{\mathbf{Q}\nu}$ (mode-lambda spectroscopy) [7].

Another important piece of information is obtained from the temperature dependence of HAS reflectivity. Reflectivity, expressed by the ratio $I(T)/I_i$ of the specular peak intensity to the intensity of the incident beam, is attenuated with increasing surface temperature T due to lattice vibrations that modulate the profile of the surface electron density. At conducting surfaces this modulation also contains the contribution of thermal elementary excitations of the electronic Fermi sea, which is however negligible at the usual surface temperatures of HAS experiments. Thus, only phonon excitations, in a first approximation, are responsible for reflectivity attenuation, and this occurs via e–ph interaction. The attenuation of the specular peak intensity is expressed by the Debye–Waller (DW) factor

$$I(T) = I_0 e^{-2W(\mathbf{k}_f, \mathbf{k}_i; T)}, \qquad (4)$$

where I_0 is the rigid surface intensity, and the so-called DW exponent for HAS from insulators, as well as for other probes directly scattered by the atom cores like X-rays and neutrons, is simply related to the mean-square displacement of surface atoms by

$$2W(\mathbf{k}_f, \mathbf{k}_i; T) = -\ln[I(T)/I_0] = <[(\mathbf{k}_f - \mathbf{k}_i) \cdot \mathbf{u}]^2>_T . \qquad (5)$$

For a conducting surface the vibrational displacement **u** must be replaced by the corresponding modulation of the surface profile at the He-surface classical turning point, proportional to the mean-square phonon displacement. Since the latter includes the contribution of all phonons weighted by the respective mode lambdas, it was shown that at sufficiently high temperature, where $<u^2>_T$ grows linearly with T [10,11] according to

$$2W(\mathbf{k}_f, \mathbf{k}_i, T) \cong N(E_F) \frac{\hbar^2 (\Delta k_z)^2}{2m^* \phi} k_B T \lambda. \tag{6}$$

For a two-dimensional electron gas (2DEG) the Fermi level DOS referred to the surface unit cell of area a_c, is given by $N(E_F) = m^* a_c / \pi \hbar^2$. A multilayer system as considered in the following example is conveniently viewed as a stack of n_s 2DEGs, provided n_s does not exceed the number n_{sat} of layers encompassed by the e–ph interaction range. With these definitions, the mass-enhancement factor λ (also denoted λ_{HAS} when obtained from HAS DW data) can be written as

$$\lambda = \frac{\pi}{2n_s} \alpha, \ \alpha \equiv -\frac{4\phi}{a_c (\Delta k_z)^2} \frac{\partial \ln I(T)}{k_B \partial T}, \ n_s \leq n_{sat}. \tag{7}$$

3. Mode-Selected Electron–Phonon Coupling from HAS: Lead Ultra-thin Films

The early HAS studies of surface phonons in Pb(111) ultra-thin films (3–8, 10 and 50 monolayers (ML)) grown on a Cu(111) substrate [26–28] showed something surprising: the number of dispersion curves detected by HAS increases with the number of layers, well beyond the few dispersion curves observed in the semi-infinite crystal. Due to the large acoustic mismatch between Pb and Cu (acoustic impedance ratio $Z_{Pb}/Z_{Cu} = 0.56$), the observed phonons are confined within the film. Actually the number of modes resolved by HAS reaches a maximum at about 8 ML, then it decreases to the four typical surface phonon branches of the face-centered-cubic (111) metal surface localized on the surface atomic bilayer, with prevalent shear-vertical (SV) or longitudinal (L) polarization. This suggests $n_{sat} \cong 8$ for Pb(111). Figure 2a,b illustrates the HAS dispersion curves for 5 ML- and 7 ML-Pb(111)/Cu(111) (filled circles) as compared to the theoretical dispersion curves calculated with DFPT calculations [7]: substantially all branches with prevalent SV polarization are detected. The calculated eigenvectors indicate that the highest mode (ε_2) is localized on the surface bilayer, whose interlayer distance is quite contracted, whereas the second highest mode (ε_1) is localized near the film–substrate interface, a few layers beneath the surface. This offers a clear demonstration of the quantum sonar effect.

The inelastic HAS spectra measured in the 90° planar scattering geometry along the scan curves at incident angles of 37° and 39.5° for 5 ML (Figure 2a), and 35.5° and 37.5° for 7 ML (Figure 2b) are displayed in Figure 2c,e, respectively. The spectra of the mode-lambdas λ_{Qv} calculated with DFPT for all the theoretical quasi-SV modes intersected by the scan curves (Figure 2d,f) compare very well with the experimental inelastic spectra, thus demonstrating the approximate proportionality predicted by Equation (2).

The calculated λ_{Qv} for the different modes ($v = \alpha_1, \alpha_2, \ldots, \varepsilon_1, \varepsilon_2$) are plotted as functions of $\mathbf{Q}$ in Figure 3 for both 5 ML and 7 ML. It is rather instructive that the maxima of λ_{Qv} occur for the $\mathbf{Q} = 0$ acoustic modes α_1. These modes have a small but finite frequency at $\mathbf{Q} = 0$ (Figure 2a,b), because of the film–substrate interaction, and consist of an almost rigid translation of the whole film against the substrate. Since the wave penetration into the substrate is quite small, due to the large acoustic mismatch, the displacement field gradient, measuring the local stress, is concentrated at the interface, and therefore a large contribution to the e–ph coupling comes in this case from the interface electrons.

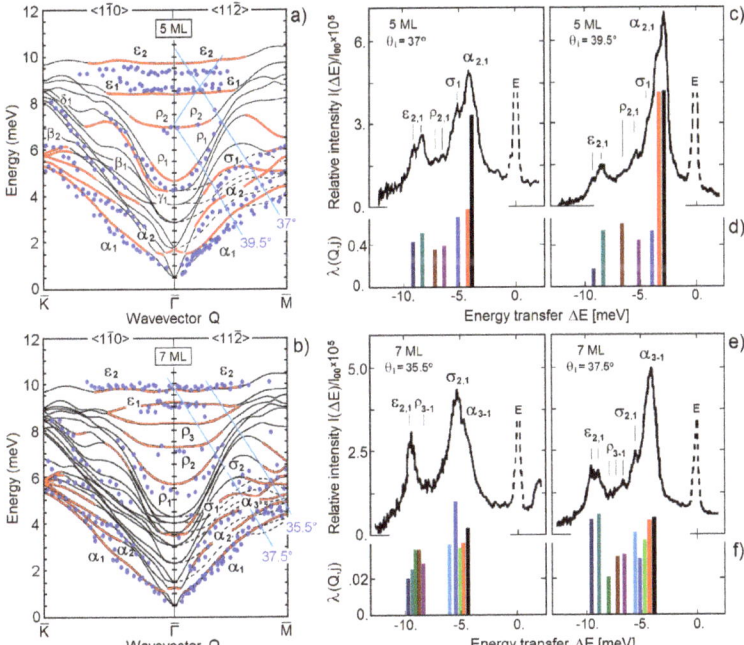

Figure 2. The experimental HAS phonon dispersion curves (•) for (**a**) 5 ML and (**b**) 7 ML of Pb grown on a Cu(111) substrate compared with the theoretical DFPT dispersion curves. The heavier (red) lines refer to quasi-shear-vertical modes, thin lines are for quasi-longitudinal modes, broken lines for shear-horizontal modes, not observable in the planar HAS configuration. The intersections of the two HAS scan curves with the dispersion curves for incident angles of 37° and 39.5° (**a**) and of 35.5° and 37.5° (**b**) provide the energy and momentum of the observed phonons. The corresponding HAS energy-gain spectra (**c,e**) show some prominent peaks, approximately proportional to the respective mode lambdas $\lambda_{Q\nu}$ calculated with DFPT (**d,f**); the colors correspond to those of the $\lambda_{Q\nu}$ plots in Figure 3. E labels the diffuse elastic peaks (broken lines). a,b reproduced with permission from [7], copyright 2011 by the American Physical Society. c-f from [9], with permission from The Royal Society of Chemistry.

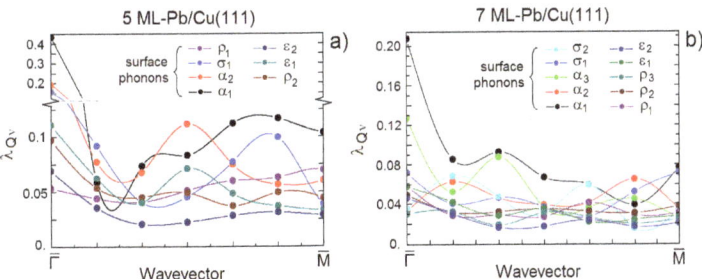

Figure 3. Plots of the mode-selected e–ph coupling constants $\lambda_{Q\nu}$ for different branch indices ν as functions of the wavevectors **Q** in the $\overline{\Gamma M}$ direction for 5ML-Pb/Cu(111) (**a**) and 7ML-Pb/Cu(111) (**b**), calculated with DFPT [7] (reproduced with permission, copyright 2011 by the American Physical Society). Indexed Greek letters label the same modes of Figure 2a,b; colors correspond to those of the $\lambda_{Q\nu}$ bins in Figure 2d,f.

Measurements of the temperature-dependent DW exponent during the layer-by-layer deposition of Pb films on Cu(111) permit the obtaining of λ as a function of the number of layers via Equation (7) (Figure 4a) [27]. Please note that due to oscillating surface relaxation where the interlayer distance oscillates so as to form a sequence of atomic bilayers near the surface, the reflectivity peaks actually occur at $n = 1$ (wetting layer), 3, 4.5, 6.5, 8 ... ML. As appears in Figure 4a, $\lambda < 0.8$ for a single Pb wetting layer and for 3 ML, then rises to values above 1 for $n \geq 4$ ML, exhibiting quantum-size oscillations [26]. This behavior corresponds quite well to what has been reported from scanning-tunneling microscopy/spectroscopy (STM/STS) for the superconducting T_c of Pb ultra-thin films grown on semiconductor surfaces (Figure 4b) [29–34].

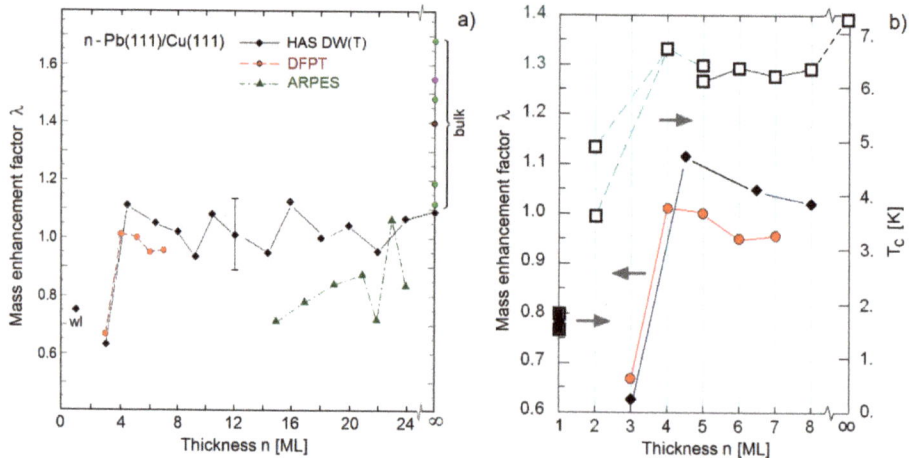

Figure 4. (a) Values of the mass-enhancement factor λ extracted from the temperature dependence of the HAS Debye–Waller exponent, measured for a He-beam incident energy of 5.97 meV (♦) [11,26,27], (reproduced with permission, copyright 2014 by the American Chemical Society) are compared to values from ARPES measurements (▲) [35] and from density functional perturbation theory (•) [7]. The HAS value at $n = 1$ ML refers to the Pb wetting layer (wl). A collection of values for bulk Pb taken from literature [24,25] (filled circles at $n = \infty$) are shown for comparison. (b) The experimental HAS (♦) and theoretical (•) values of λ shown in (a) for $n \leq 8$ ML are compared with the experimental superconducting T_c of ultra-thin Pb films on Si(111) [9] (with permission from The Royal Society of Chemistry) for two different equilibrium configurations of the 2 ML and for $n = 4$, 5 ML (open squares) [31] and of the wetting layer, $n = 1$ ML (filled squares) [33].

For 3 to 8 ML (Figure 4b) the HAS data reproduce quite well the results of DFPT calculations [7]. For thicknesses of 15 to 24 ML, the HAS data in Figure 4a show quantum-size oscillations that look different from those of the corresponding angle-resolved photoemission spectroscopy (ARPES) data [35]. It should be noted that ARPES values for λ are derived from the deformation of an electronic band induced by the e–ph coupling near the Fermi surface, and therefore refer to that specific electronic band, whereas the HAS values from the DW exponent are averages over all possible virtual phonon-induced electronic transitions around the Fermi level, as considered in the definition of λ.

4. A Quasi-1D Metal: Bi(114)

Although bulk crystalline bismuth at normal pressure is not superconducting above 0.53 mK [36,37], nanostructured [38] and amorphous bismuth [37] are reported to become superconductors with critical temperatures above 1 K. It is, therefore, interesting to investigate the e–ph interaction in Bi nanostructures by means of HAS. Of particular interest is the (114) surface of bismuth (Figure 5a).

Bi(114) exhibits, in its stable reconstructed (1 × 2) form, resulting from a missing-row transition, the features of a one-dimensional (1D) topological metal consisting of conducting atomic rows 28.4 Å apart [15,38]. Below a critical temperature of ≈240 K, a CDW arises along the rows with a critical exponent of 1/3 due to a Peierls dimerization along the chains [16], leading to a (2 × 2) reconstruction with an oblique unit cell in real (Figure 5a) and reciprocal space (Figure 5b). The electronic structure in the (1 × 2) phase, calculated by Wells et al. [39], is characterized at the Fermi level by carrier pockets at the zone center $\bar{\Gamma}$ and the zone-boundary points $\bar{X}$ (Figure 5c). The $\bar{\Gamma}$-$\bar{X}$ intervalley e–ph coupling explains the (2 × 2) reconstruction.

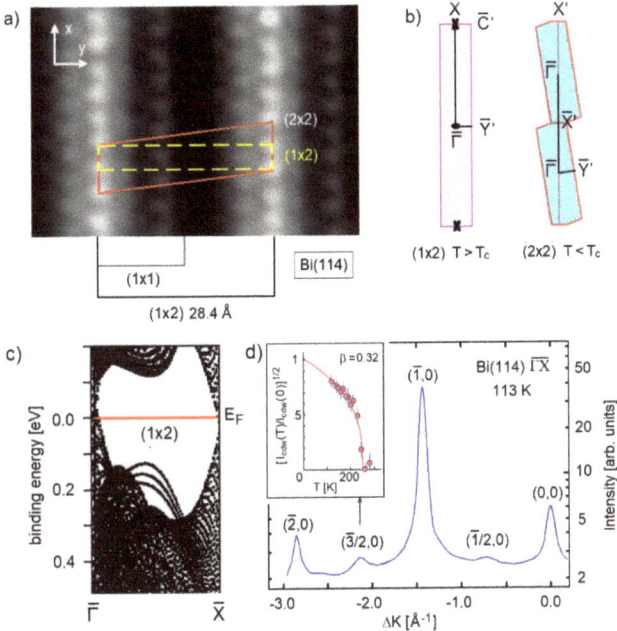

Figure 5. (a) STM image of the (1 × 2) Bi(114) surface resulting from a missing-row reconstruction, leading to parallel atomic rows in the x-direction, 28.4 Å apart, that confers a quasi-1D metallic character to the surface [39]. Below ≈140 K the atomic rows undergo a Peierls dimerization with an oblique (2 × 2) unit cell in real (a) and reciprocal space (b). Dimerization is due to intervalley nesting between Fermi level pockets at $\bar{\Gamma}$ and $\bar{X}$ points of the (2 × 1) Brillouin zone (c). The transition is monitored by the intensity of HAS diffraction from semi-integer CDW G-vectors (d) as a function of temperature. The inset of panel (d) shows that the order parameter grows with a critical exponent $\beta \cong 1/3$ [15] published 2020 by the American Chemical Society).

The HAS diffraction spectra of Bi(114) along the row direction, reported in [15,16], show the temperature evolution of the CDW peaks, the most prominent being that at (±3/2,0); an example, measured at 113 K, is shown in Figure 5d. The best fit of the order parameter, proportional to $[I_{CDW}(T)/I_{CDW}(0)]^{1/2}$ for the CDW peak (inset of Figure 5d), yields a CDW critical temperature of 242 ± 7 K and a critical exponent $\beta = 0.32 \pm 0.03$, consistent with the universal exponent of 1/3 as predicted in the presence of fluctuations [40,41], and in agreement with the critical exponent for other CDWs in layered chalcogenides and quasi-1D systems [17,42–45].

The e–ph coupling constant λ can be derived from the temperature dependence of the DW exponent, Equation (7), with the experimental parameters taken from Ref. [16]: $a_c = 129$ Å^2, $(\Delta k_z)^2 = 54.24$ Å^{-2}, $\phi = 4.23$ eV (for Bi), $-\partial \ln I(T)/k_B \partial T = 104.3$ eV^{-1} and $n_s = 1$ (assuming that only the surface atomic rows contribute). This gives $\lambda = 0.40$, to be compared with the HAS value 0.57 for Bi(111) [10]. A slightly larger value for Bi(114), $\lambda = 0.45$, is obtained when treated as a 1D free-electron gas system, as discussed in Ref. [17]. This value is probably more reliable, in consideration of the quasi-1D metallic nature of Bi(114)(1 × 2), although the difference is well within the uncertainty of the present method.

5. Superconducting Layered Chalcogenides

The large family of layered chalcogenides includes several superconductors. Although their superconductivity generally occurs at fairly low temperatures and under special conditions (proximity, pressure, high doping, low dimensionality), their great interest is related to the frequent coexistence/competition with CDWs and the interplay of electron-electron and e–ph interactions.

The high sensitivity and resolution of HAS permit detection of CDWs where other conventional surface probes fail, and measurement of the dispersion curves of their collective excitations [14,16,19]. The additional possibility to directly estimate the e–ph interaction makes HAS an excellent tool to investigate superconducting (SC) layered chalcogenides.

5.1. 1T-TaS$_2$(001)

Tantalum di-chalcogenides as superconductors exemplify the special conditions mentioned above. The SC critical temperature of 1T-TaS$_2$(001), originally reported by Wilson et al. to be about 0.5 K [46], was later determined to be about 1.5 K at normal pressure [47]. Pressure was shown by Sipos et al. [48] to produce a large increase of T_c up to 5 K at 5 GPa and then to keep it about constant at larger values up to 25 GPa. A raise of T_c can also be obtained, as recently demonstrated [49,50], by reducing the sample thickness to a few layers, and even in a monolayer an increase of T_c up to 3.61 K is induced by the formation of structural defects [51]. An alternate stacking of the 1T and 2H layers of TaS$_2$ to form the so-called 4Hb polytype was also shown to raise T_c to 2.7 K [52]. The coexistence of the CDW with superconductivity in 1T-TaS$_2$ has been investigated by Sipos et al. [48] and more recently by Ritschel et al. [53]. Wagner et al. [54] reported on the interesting observation that a Cu doping of 2H-TaS$_2$ raises the SC T_c from 0.5 K of the undoped material to a maximum of 4.5 K for a 4% doping, while dramatically reducing the CDW coherence length by a factor ten.

The complex CDW structure of 1T-TaS$_2$, as described by Wilson et al. [46,55], Coleman et al. [56], and more recently by Yu et al. [57], has been studied with HAS by Cantini et al. [58] and by Heimlich et al. [59,60]. The slopes of the HAS specular intensity as a function of temperature, shown in Figure 6a [59] for increasing temperature in the low-T commensurate CDW (C-CDW), intermediate-T nearly commensurate CDW (NC-CDW) and high-T incommensurate CDW (I-CDW) regions, permit to obtain from Equation 7 the respective e–ph coupling constant. With the input data $k_{iz}^2 = 14.58$ Å^{-2} [60], $\phi = 5.2$ eV [61], $a_c = 9.43$ Å^2 [62] and $n_s = 2$, from a Thomas-Fermi screening length not exceeding 1 nm [57], the observed specular intensity slopes give λ(C-CDW) = 0.61 ± 0.06, λ(NC-CDW) = 0.91 ± 0.09, and λ(I-CDW) = 0.61 ± 0.10 for the three phases. The data for decreasing temperature, presented in [52], give for the I-CDW and NC-CDW comparable values within the experimental error. For the low temperature C-CDW phase, the McMillan formula yields, with $\mu^* = 0.20$ [63] and a Debye temperature Θ_D of 249 K [46,64,65], a superconducting critical temperature $T_c = 1.2$ K, in fair agreement with the value at normal pressure of 1.5 K reported by Nagata et al. [47].

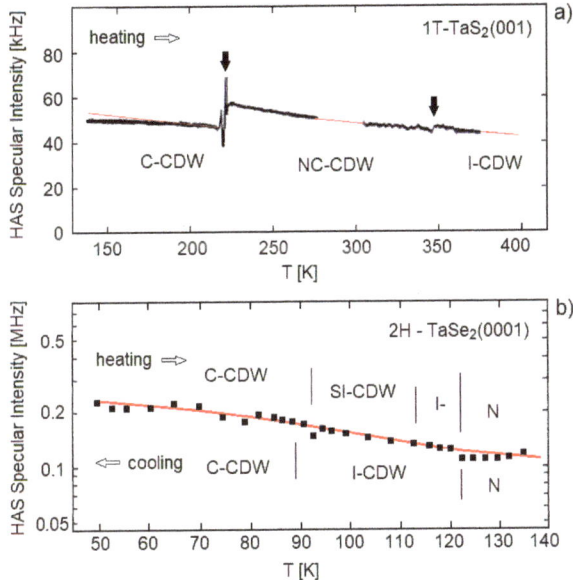

Figure 6. Thermal attenuation (Debye–Waller factor) data for scattering of He atoms from two chalcogenide surfaces. The specular intensity is plotted as a function of temperature. (**a**) 1T-TaS$_2$(001) plotted on a linear scale, and (**b**) 2H-TaSe$_2$(0001) plotted on a logarithmic scale. In (**a**), the crystal was initially at low temperature and then heated, and the two vertical arrows indicate the charge density wave transitions. The transition from incommensurate charge density wave (I-CDW) to non-commensurate (NC-CDW) occurs at about 350 K, while the transition from the NC-CDW to the commensurate (C-CDW) phase occurs at about 220 K. The three red segments indicate the slopes in the C-CDW phase near the transition, and the average slopes in the NC-CDW and I-CDW phases. In (**b**) the experimental data are also shown for increasing temperature from the C-CDW through the transitions to the striped incommensurate (SI-CDW) at 92 K, the I-CDW at 113 K, and finally the normal phase at 122 K, as marked by the vertical lines. Also marked are the transition temperatures for decreasing T, with relevant hysteresis effects: no SI-CDW is observed and the I-CDW to C-CDW occurs at 88 K [59] (adapted from [17]). The red line is a smooth interpolation of the data points.

For the high-pressure (5 GPa) NC-CDW phase, Rossnagel derives from the McMillan formula and the input data $T_c = 5$ K, $\mu^* = 0.20$ and $\Theta_D = 249$ K, an e–ph interaction constant $\lambda = 0.85$ [63], also in agreement with the present value of λ for the NC-CDW phase, although HAS experiments are performed in vacuum conditions. Density functional theory calculations by Amy Liu [66] also predict a dramatic effect of pressure, with λ varying from 2.09 at 5 GPa ($T_c = 16$ K) to 0.69 at 30 GPa ($T_c = 5.9$ K). Please note that the Debye temperature, here corresponding to a phonon energy of 21.6 meV, is adopted in the McMillan formula as an average energy of phonons contributing to λ, and may have little to do with that derived from the specific heat, according to Debye theory. Moreover, this theory has been conceived for monoatomic crystals with only three acoustic phonon branches, whereas for the additional flat optical branches of polyatomic crystals the Einstein model would be more appropriate. The simple fit of the Debye theory to polyatomic crystals with high-energy optical phonon branches leads to an important temperature dependence of Θ_D. Since in general $T_c \ll \Theta_D$, it is implicit that Θ_D should be conveniently taken in its low-T limit. In the present three-atomic case, the Debye phonon energy $\varepsilon_D = k_B \Theta_D = 21.6$ meV falls just above the acoustic region but well below the optical phonon region (27–40 meV), according to pseudo-charge model calculations for the C-CDW phase in its ($\sqrt{13} \times \sqrt{13}$)R13.9° superstructure (1T-TaS$_2$) [67]. This is indicative of a relevant contribution of

5.2. 2H-TaSe$_2$(0001)

HAS reflectivity data for 2H-TaSe$_2$ (Figure 6b) show a decrease for increasing temperature similar to that of 1T-TaS$_2$, with imperceptible discontinuities at the CDW transition temperatures. The HAS intensities of the CDW diffraction peaks [59,60] show instead clear features at the transition temperatures, from the C-CDW to the striped-incommensurate (SI-CDW) phase (92 K), from the SI-CDW to the I-CDW phase (113 K) and from the I-CDW to the normal phase (122 K), in agreement with previous McWhan et al. X-ray data [69]. Similar agreement is found for the transition temperatures during cooling, with hysteresis at the I-CDW → C-CDW transition (88 K). The decay slopes of the specular intensity for increasing temperature (Figure 6b) in the C-CDW linear region (70–90 K) and in the (SI + I) – CDW region (90–122 K) are about the same, within the experimental uncertainty. With the input data $k_{iz}^2 = 68.45$ Å^{-2} [59,60], $\phi = 5.5$ eV [70], $a_c = 10.22$ Å^2 [71] and $n_s = 2$ (from a spherically averaged coherence length of ≈14 Å [72], vs. the unit cell thickness including two Ta layers of 12.70 Å), a single value for $\lambda = 0.57 \pm 0.08$ is obtained. The data for decreasing temperature displayed in [59] show a smaller slope in the I-CDW region (122–88 K), giving $\lambda = 0.30 \pm 0.06$, but approximately the same slope in the C-CDW region (<88 K). The factor of two difference in λ found in the incommensurate phase between heating and cooling data may reflect the structural dependence on the thermal history, as suggested by the fact that the SI phase is observed when heating and not when cooling the sample. In any case the superconducting temperature resulting from McMillan's formula, with a Debye temperature of 140 K [73], $\mu^* = 0.20$, and $\lambda = 0.57$ for the C-CDW phase is equal to 410 mK. The C-CDW values of λ and the corresponding T_c are larger than those found in the literature, e.g., $\lambda = 0.49$ reported by Rossnagel [64], 0.39 by Bhoi et al. [73], 0.4 by Luo et al. [74], and correspondingly rather small values of T_c, such as 200 mK by Kumakura et al. [75] and 130 mK by Yokota et al. [76].

As in the case of 1T-TaS$_2$ the interplay between CDW and superconductivity is an interesting issue, as discussed in recent works by Lian et al. [77], Wu et al. [78], Liu et al. (for 1T-TaSe$_{2-x}$Te$_x$) [79], etc. Similarly, a large rise of T_c up to above 2 K has been demonstrated as an effect of doping [80] or ion gating (intercalation) [78], as well as of film thickness reduction down to 1 ML. For example, Lian et al. have shown that a value of $\lambda = 0.4$, valid for bulk 2H-TaSe$_2$ and giving a T_c of 100 mK, rises to $\lambda = 0.74$ in a single monolayer, with a $T_c = 2.2$ K [78]. Thus, the larger values of λ observed with HAS clearly refer to the surface, more precisely to the surface bilayer ($n_s = 2$).

The conjecture, originally formulated for graphite, that in layered crystals with weak interlayer van der Waals forces the surface phonon dispersion curves should be almost coincident with those of the bulk, proved invalid for layered chalcogenides (Figure 7) [1]. A specific role of e–ph interaction at the surface of 2H-TaSe$_2$, and in general of transition-metal chalcogenides, has been surmised since the early HAS measurements of the surface phonon dispersion curves [43,81,82]. An intriguing HAS observation in 2H-TaSe$_2$ was a temperature-dependent Kohn anomaly in the Rayleigh wave (RW ≡ S$_1$) dispersion curve at about $\frac{1}{2}\overline{\Gamma M}$ [82], whereas the bulk longitudinal acoustic (Σ_1) mode, as measured with inelastic neutron spectroscopy [70], shows a T-dependent anomaly at $^2/_3\overline{\Gamma M}$(Figure 7a). Moreover, the maximum softening observed in the bulk Σ_1 mode occurs at the normal-to-CDW transition (≈120 K), whereas the deepest S$_1$ Kohn anomaly occurs at ≈110 K. Actually the HAS intensities of the CDW diffraction peaks [43] exhibit hysteresis loops around the I↔SI transition temperature (113 K), which would associate in some way the SI phase to the surface region, within the penetration depth of the RW at $\frac{1}{2}\overline{\Gamma M}$. As mentioned above, HAS data on the temperature dependence of the CDW diffraction peaks also provided the order-parameter critical exponent for the CDW transition at 122 K equal to the universal $\beta = 1/3$ [42], as generally found for CDW transitions driven by e–ph interaction in this class of materials.

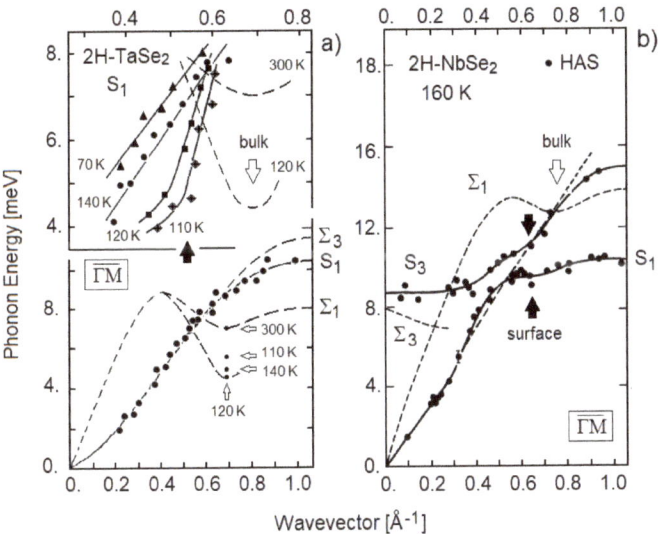

Figure 7. (a) The Rayleigh wave dispersion curve (S_1) in 2H-TaSe$_2$ measured with HAS (larger dots) [81,82] at 140 K along the $\overline{\Gamma M}$ direction (bottom) and at different temperatures around $\frac{1}{2}\overline{\Gamma M}$ (top), where a Kohn anomaly develops with a minimum at 110 K (upward filled arrow), close to the I-CDW → SI-CDW transition. For comparison the deep anomaly in the bulk longitudinal acoustic mode (Σ_1) (broken lines) is located at $2/3\overline{\Gamma M}$ and reaches its minimum at about the I-CDW → N transition temperature of ~120 K (open downward arrow) [71]. (b) A similar situation is found in HAS data for 2H-NbSe$_2$ measured at 160 K [83], where the shift from the bulk anomaly at about $3/4\overline{\Gamma M}$ [71] (open downward arrow) to the surface anomalies at $2/3\overline{\Gamma M}$ in both the Rayleigh wave (S_1) and the optical S_3 branches (filled black arrows) is smaller than in 2H-TaSe$_2$, although the anomalous S_1 mode softens at lower temperature down to zero [84,85].

The shift of the surface phonon anomaly to a wave vector different from that for the bulk phonon anomaly may be ascribed to a different nesting of the Fermi level surface electronic states with respect to that for the bulk states. A similar behavior is found with HAS in 2H-NbSe$_2$ (Figure 7b), where the dip in the dispersion curve of the bulk Σ_1 phonon also falls at ≈0.76 Å^{-1} while the dips of both S_1 and S_3 surface phonons branches, as measured by HAS, are shifted to a smaller wave vector, ≈0.65 Å^{-1}. It is unfortunate that this HAS data, originally measured in 1986 by Brusdeylins and Toennies and reproduced in [83], have not been extended to lower temperatures, where, as found with inelastic X-ray spectroscopy (IXS) [84,85], the S_1 frequency at ≈0.65 Å^{-1} vanishes at about 33 K. The Weber et al. analysis based on state-of-the-art *ab-initio* calculations indicate the wave vector dependence of the e–ph coupling as the driving mechanism for CDW formation in 2H-NbSe$_2$ [85]. Considering the superior resolution of inelastic HAS spectroscopy in the meV range with respect to IXS, and its ability to obtain direct information on e–ph coupling constants, revisiting with HAS the CDW phases and transitions in this class of 2D superconductors would be highly desirable. Recent examples in this direction are the HAS measurements of λ in 2H-MoS$_2$ [12], 1T-PtTe$_2$ [13] and 1T-PdTe$_2$ [86].

5.3. Pnictogen Chalcogenides

Layered pnictogen chalcogenides, with a bulk electronic band gap and a cone of topological Dirac states localized at the surface and crossing the Fermi level, are topological insulators (TI) unsuitable to be superconductors at any temperature. They can however become so by intercalating donor atoms [87–89], doping [90], by interfacing with conventional superconductors via the proximity effect [91–96], or under pressure [97–99]. Although critical temperatures of induced superconductivity are in the best cases in the range of a few Kelvin, the involvement of surface topologically protected states confers a particular stability against external disturbances. This would make pnictogen chalcogenides and their heterostructures particularly suitable for device applications. Moreover, TI-superconductor heterostructures have provided a quantum playground for some fundamental advances, e.g., the evidence of Majorana fermions in condensed matter with promising applications to quantum computing (for brief accounts see, e.g., [100–102]).

A successful option was looking for Majorana fermions in the vortices of a 2D chiral topological superconductor [103–109]. Xu et al. [110], were able to detect with STM/STS a Majorana mode in a heterostructure made of the topological insulator Bi_2Te_3 and the superconductor $NbSe_2$. The spin-selected Andreev reflection effect [111] was exploited by Sun et al. [112] in a similar 2D heterostructure where five quintuple layers of the topological insulator Bi_2Se_3 are interfaced to a $NbSe_2$ superconductor, which enabled unveiling Majorana zero modes by probing the vortex core states with spin-polarized-STM/STS. An important issue is the coexistence of superconductivity with the topological order [113–115]. The recent HAS observation of diffraction features attributed to long-period CDWs in Bi_2Se_3 [19] raises another coexistence issue, as mentioned above for layered transition-metal chalcogenides, with the corresponding question about the origin and role of e–ph interaction in topological pnictogen chalcogenides.

HAS reflectivity measurements on $Bi_2Se_3(111)$, $Bi_2Te_3(111)$, $Bi_2Te_2Se(111)$ [15,116,117] and $Sb_2Te_3(111)$ [118] have permitted obtaining λ from the temperature dependence of the DW exponent and Equation (7) (λ_{HAS} in Figure 8b). Their respective values are 0.23, 0.19, 0.14 and 0.08, all with an experimental uncertainty of around 10%. The large range of values as compared to the similarity of these compounds can be understood by considering their electronic structure at the Fermi level, as resulting from the ARPES data for the bismuth chalcogenides used in HAS experiments [15,19,116] and from *ab-initio* calculations for $Sb_2Te_3(111)$ [118] (Figure 8a,c). The substantial differences in the sets of topological states concurring to the e–ph interaction can qualitatively account for the values of λ and indicate which Fermi level transitions contribute most.

In $Bi_2Se_3(111)$ the Fermi level cuts the Dirac cone arising from the Dirac point, as well as pairs of topological quantum-well states and the bottom of the conduction band; in $Bi_2Te_3(111)$ the Dirac cone and apparently only the conduction-band bottom; in $Bi_2Te_2Se(111)$ only the Dirac states, in $Sb_2Te_3(111)$, where the Dirac point is above the Fermi level and there are altogether six hole pockets (three spin-up and three spin-down, as indicated by arrows in Figure 8c), there is an intervalley e–ph coupling, in addition to the contribution from Dirac holes. It appears that the Dirac states alone contribute little to the total e–ph interaction, and that larger contributions come from the surface quantum-well states and the bottom of the conduction band at the surface. The observation that λ_{HAS} increases when quantum-well and conduction-band states are involved agrees with the result of DFPT calculations of λ as a function of the Fermi level position recently reported by Heid et al. [119]. The above values of λ_{HAS} are also in agreement with the corresponding values from other sources: for $Bi_2Se_3(111)$ $\lambda = 0.17$ [120], 0.25 [121], 0.26 [122]; for $Bi_2Te_3(111)$ $\lambda = 0.19$ [120], for $Bi_2Te_2Se(111)$ $\lambda = 0.12$ [15]. No other value of λ to compare with λ_{HAS} is available, to our knowledge, for $Sb_2Te_3(111)$, though it is noted in [123] that in Sb_2Te_3 the e–ph interaction from *ab-initio* calculations is weaker than in Bi_2Te_3, as actually found here.

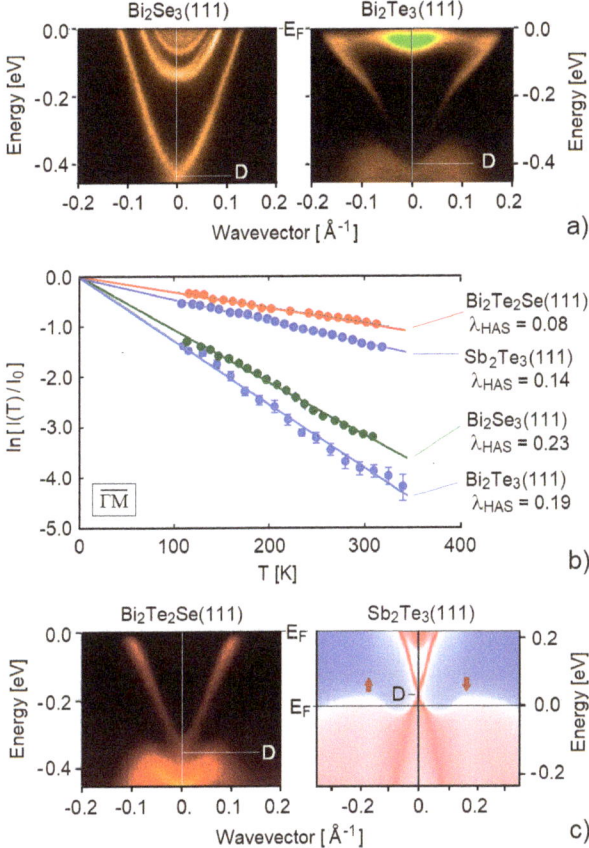

Figure 8. ARPES data for the (111) surface of n-type pnictogen chalcogenides: (a) Bi_2Se_3 and Bi_2Te_3 and (c) Bi_2Te_2Se, with decreasing binding energy of the Dirac point (D) and of the surface conduction-band minimum (from 0.15 eV in Bi_2Se_3 to 0.08 eV in Bi_2Te_3 and ≈0 in Bi_2Te_2Se (adapted from [15] published 2020 by the American Chemical Society)). For comparison *ab-initio* calculations of the surface band structure of Sb_2Te_3 (111) [118], where in addition to the Dirac cone, hole pockets of opposite spin are also found (upward and downward arrows). (b) The DW exponent slopes obtained from HAS specular intensity measured as functions of temperature with the scattering plane in the $\overline{\Gamma M}$ direction for the four samples. The corresponding e–ph coupling constants $\lambda \equiv \lambda_{HAS}$ decrease from Bi_2Se_3(111) to Bi_2Te_2Se(111), suggesting a dominant role in the e–ph interaction of the conduction-band quantum-well electronic states over the Dirac electrons and the intervalley transitions.

6. Conclusions

In this brief review HAS from conducting surfaces has been shown to provide quantitative information on the electron–phonon (e–ph) coupling constants in some classes of 2D superconductors. Although the temperature dependence of HAS reflectivity, expressed via the DW exponent, allows for the determination of the global e–ph interaction through the mass-enhancement factor λ, inelastic HAS intensities from single phonons turn out to be proportional to the mode-selected e–ph coupling constants λ_{Qv} (*mode-lambda spectroscopy*). The latter possibility has been illustrated by the study of Pb ultra-thin films of 5 and 7 monolayers, while the derivation of λ from the temperature dependence of the DW exponent has been discussed for Pb films 1 to 24 ML thick, as well as for a high-index, quasi-1D bismuth surface and for superconducting layered chalcogenides. Concerning transition-metal dichalcogenides, an important aspect that HAS spectroscopy can help elucidate is whether and where the coexistence of CDWs and superconductivity is competitive or cooperative. Pnictogen chalcogenides, where topological superconductivity can be induced by pressure, intercalation, or proximity to superconductors, offer possible scenarios of coexistence with CDWs, after the recent observation of CDWs in some of these materials. The values of λ obtained by HAS in n-type Bi chalcogenides permit the appreciation of the role of surface quantum well and conduction-band states with respect to that of Dirac states.

In view of these potentialities it would be desirable to investigate by HAS various classes of high-T_c superconductors, in particular cuprates. No HAS study addressing e–ph interaction has been carried out so far in these materials. E–ph interaction in perovskites has been shown in the 1980s by Heinz Bilz and his school to receive an important contribution from the non-linear polarizability of oxygen ions [124], and to be therefore responsible for ferroelectricity [125–129] and other structural phase transitions [130,131], with possible implications for high-T_c superconductivity in perovskites [131]. Given the layered structure of these materials and the good quality of their surfaces, high-resolution HAS studies of their complex phonon structure and of mode-selected e–ph coupling constants would certainly help to further advance the understanding of high-T_c superconductivity. The simultaneous information on dynamics and structure of the surface electron density that can be obtained with HAS would also be instrumental in investigating certain aspects of high-T_c materials, e.g., phase separation [132], heterogeneity [133,134] and the superstripes landscape [135–137] made of multiscale puddles of CDWs from 3 nm to hundreds of nm [138] controlled by doping and elastic strain [139]. The nanoscale CDW texturing has been observed in 1T-TiSe$_2$ [140] where by Ti self-doping the Fermi level is tuned near a Lifshitz transition in correlated multi-band systems as was predicted theoretically [141]. Moreover, it has been confirmed that the CDW texture in Sr doped Bi$_2$Se$_3$ [142] can be manipulated by uniaxial strain to control superconductivity.

Inelastic HAS spectroscopy, besides measuring single surface phonon frequencies and the related λ_{Qv}, allows for the observation of low-energy collective electronic excitations like phasons, amplitons and acoustic surface plasmons (ASP) [1,19,20,143] due to the direct "mechanical" interaction (Pauli repulsion) of He atoms with the surface electron density. When the ASP dispersion curve (linear in the long-wavelength limit [144]) enters the surface-projected phonon density, there are avoided crossings with the surface phonon dispersion curves and a robust renormalization of ASP phase velocity, all driven by e–ph interaction. Actually, a strong ASP-phonon interaction is at the basis of a recent prediction by Shvonski et al. [145] of polaron plasmons as possible hybrid excitations at topological metal surfaces, reviving the old concept by Lemmens and Devreese [146] of collective excitations of the polaron gas. In view of the importance that has been attributed to polaron pairing (bipolarons) in the early days of high-T_c superconductivity [147–150], the strong e–ph coupling in cuprates has been confirmed experimentally by the anomalous doping dependent isotope effect: (i) on the superconductivity critical temperature [151], which shows the proximity of the chemical potential to a Lifshitz transition at 1/8 doping [152], (ii) on the polaronic CDW ordering temperature T^* [153] and (iii) the kink in the electronic dispersion from angle-resolved photoemission spectra [154]. Therefore, the study of exotic collective polaron excitations in topological superconductors by HAS could have many surprises in store.

Author Contributions: G.B., J.R.M., S.M.-A. and J.P.T. developed the physical interpretation of the data. A.T. and A.R. performed the experimental measurements in terms of the topological insulator and semimetal surfaces together with most of the data analysis concerned with these samples. G.B. wrote the paper, and all authors discussed the results and contributed to the preparation of the manuscript. All authors have read and agreed to the published version of the manuscript.

Funding: A.R., W.E.E., and A.T. acknowledge financial support provided by the FWF (Austrian Science Fund) within the projects J3479-N20 and P29641-N36.

Acknowledgments: This work is dedicated to Alex Müller with profound admiration for having opened new horizons in modern physics with his discovery, with Georg Bednorz, of high-temperature superconductivity. GB is particular grateful to Alex for the privilege of having been a witness of the early days of high-Tc superconductivity at the Ettore Majorana Foundation and Centre for Scientific Culture (EMFCSC) in Erice (Sicily), and for the stimulus to start, with his active participation and the collaboration of the present Festschrift editors, a series of periodic courses on superconductivity (now continued by Antonio Bianconi) within the International School of Solid State Physics at EMFCSC [155]. We thank Martin Bremholm, Philip Hofmann and co-workers for the synthesis of the topological insulator samples and providing us with the Bi(114) sample.

Conflicts of Interest: The authors declare no conflict of interest.

References

1. Benedek, G.; Toennies, J.P. *2018 Atomic-Scale Dynamics at Surfaces—Theory and Experimental Studies with Helium Atom. Scattering*; Springer: Berlin, Germany, 2018.
2. Doak, R.B.; Harten, U.; Toennies, J.P. Anomalous Surface Phonon Dispersion Relations for Ag(111) Measured by Inelastic Scattering of He. *Atoms Phys. Rev. Lett.* **1983**, *51*, 578–581. [CrossRef]
3. Harten, U.; Toennies, J.P.; Wöll, C. Helium time-of-flight spectroscopy of surface-phonon dispersion curves of the noble metals. *Faraday Discuss. Chem. Soc.* **1985**, *80*, 137–149. [CrossRef]
4. Jayanthi, C.S.; Bilz, H.; Kress, W.; Benedek, G. Nature of surface-phonon anomalies in noble metals. *Phys. Rev. Lett.* **1987**, *59*, 795–798. [CrossRef] [PubMed]
5. Kaden, C.; Ruggerone, P.; Toennies, J.P.; Zhang, G.; Benedek, G. Electronic pseudocharge model for the Cu(111) longitudinal-surface-phonon anomaly observed by helium-atom scattering. *Phys. Rev. B* **1992**, *46*, 13509–13525. [CrossRef] [PubMed]
6. Benedek, G.; Bernasconi, M.; Chis, V.; Chulkov, E.; Echenique, P.M.; Hellsing, B.; Toennies, J.P. Theory of surface phonons at metal surfaces: Recent advances. *J. Phys. Condens. Matter* **2010**, *22*, 084020. [CrossRef] [PubMed]
7. Sklyadneva, I.Y.; Benedek, G.; Chulkov, E.V.; Echenique, P.M.; Heid, R.; Bohnen, K.-P.; Toennies, J.P. Mode-Selected Electron-Phonon Coupling in Superconducting Pb Nanofilms Determined from He Atom Scattering. *Phys. Rev. Lett.* **2011**, *107*, 095502. [CrossRef]
8. Tamtögl, A.; Ruckhofer, A.; Campi, D.; Allison, W.; Ernst, W.E. Atom-surface van der Waals Potentials of Topological Insulators and Semimetals from Scattering Measurements. *Phys. Chem. Chem. Phys.* **2020**, in press.
9. Benedek, G.; Bernasconi, M.; Bohnen, K.-P.; Campi, D.; Chulkov, E.V.; Echenique, P.M.; Heid, R.; Sklyadneva, I.Y.; Toennies, J.P. Unveiling mode-selected electron–phonon interactions in metal films by helium atom scattering. *Phys. Chem. Chem. Phys.* **2014**, *16*, 7159–7172. [CrossRef]
10. Manson, J.R.; Benedek, G.; Miret-Artés, S. Electron–Phonon Coupling Strength at Metal Surfaces Directly Determined from the Helium Atom Scattering Debye–Waller Factor. *J. Phys. Chem. Lett.* **2016**, *7*, 1016–1021. [CrossRef]
11. Benedek, G.; Miret-Artés, S.; Toennies, J.P.; Manson, J.R. Electron–Phonon Coupling Constant of Metallic Overlayers from Specular He Atom Scattering. *J. Phys. Chem. Lett.* **2018**, *9*, 76–83. [CrossRef]
12. Anemone, G.; Taleb, A.A.; Benedek, G.; Castellanos-Gomez, A.; Farías, D. Electron–Phonon Coupling Constant of 2H-MoS_2(0001) from Helium-Atom Scattering. *J. Phys. Chem. C* **2019**, *123*, 3682–3686. [CrossRef]
13. Anemone, G.; Garnica, M.; Zappia, M.; Aguilar, P.C.; Taleb, A.A.; Kuo, C.-N.; Lue, C.S.; Politano, A.; Benedek, G.; Parga, A.L.V.; et al. Experimental determination of surface thermal expansion and electron–phonon coupling constant of 1T-$PtTe_2$. *2D Mater.* **2020**, *7*, 025007.
14. Tamtögl, A.; Kraus, P.; Mayrhofer-Reinhartshuber, M.; Benedek, G.; Bernasconi, M.; Dragoni, D.; Campi, D.; Ernst, W.E. Statics and dynamics of multivalley charge density waves in Sb(111). *npj Quantum Mater.* **2019**, *4*, 28. [CrossRef]
15. Benedek, G.; Miret-Artés, S.; Manson, J.R.; Ruckhofer, A.; Ernst, W.E.; Tamtögl, A. Origin of the Electron–Phonon Interaction of Topological Semimetal Surfaces Measured with Helium Atom Scattering. *J. Phys. Chem. Lett.* **2020**, *11*, 1927–1933. [CrossRef]
16. Hofmann, P.; Ugeda, M.M.; Tamtögl, A.; Ruckhofer, A.; Ernst, W.E.; Benedek, G.; Martínez-Galera, A.J.; Stróżecka, A.; Gómez-Rodríguez, J.M.; Rienks, E.; et al. Strong-coupling charge density wave in a one-dimensional topological metal. *Phys. Rev. B* **2019**, *99*, 035438. [CrossRef]

17. Benedek, G.; Manson, J.R.; Miret-Artés, S. The Electron–Phonon Interaction of Low-Dimensional and Multi-Dimensional Materials from He Atom Scattering. *Adv. Mater.* **2020**, *32*, 2002072. [CrossRef]
18. Benedek, G.; Manson, J.R.; Miret-Artés, S. The electron–phonon coupling constant for single-layer graphene on metal substrates determined from He atom scattering. *Phys. Chem. Chem. Phys.* **2020**, in press. [CrossRef]
19. Ruckhofer, A.; Campi, D.; Bremholm, M.; Hofmann, P.; Benedek, G.; Bernasconi, M.; Ernst, W.E.; Tamtögl, A. Terahertz surface modes and electron-phonon coupling on Bi_2Se_3(111). *Phys. Rev. Res.* **2020**, *2*, 023186. [CrossRef]
20. Ruckhofer, A.; Halbritter, S.E.; Lund, H.; Holt, A.J.U.; Bianchi, M.; Bremholm, M.; Benedek, G.; Hofmann, P.; Ernst, W.E.; Tamtögl, A. Inelastic helium atom scattering from Sb_2Te_3(111): Phonon dispersion, focusing effects and surfing. *Phys. Chem. Chem. Phys.* **2020**, in press. [CrossRef]
21. Esbjerg, N.; Nørskov, J.K. Dependence of the He-Scattering Potential at Surfaces on the Surface-Electron-Density Profile. *Phys. Rev. Lett.* **1980**, *45*, 807–810. [CrossRef]
22. Cole, M.W.; Toigo, F. Energy of immersing a He, Ne, or Ar atom or H_2 molecule into a low-density electron gas. *Phys. Rev. B* **1985**, *31*, 727–729. [CrossRef] [PubMed]
23. Manson, R.; Celli, V. Inelastic surface scattering of non-penetrating particles. *Surf. Sci.* **1971**, *24*, 495–514. [CrossRef]
24. Grimvall, G. The Electron-Phonon Interaction in Normal. *Metals Phys. Scr.* **1976**, *14*, 63. [CrossRef]
25. Allen, P.B. The Electron-Phonon Coupling Constant λ. In *Handbook of Superconductivity*; Poole, C.P., Jr., Ed.; Academic Press: New York, NY, USA, 1999; pp. 478–483.
26. Hinch, B.J.; Koziol, C.; Toennies, J.P.; Zhang, G. Evidence for Quantum Size Effects Observed by Helium Atom Scattering during the Growth of Pb on Cu(111). *Europhys. Lett. EPL* **1989**, *10*, 341–346. [CrossRef]
27. Zhang, G. Max-Planck-Institut für Strömungsforschung. Ph.D. Thesis, University of Göttingen, Göttingen, Germany, 1991.
28. Braun, J.; Ruggerone, P.; Zhang, G.; Toennies, J.P.; Benedek, G. Surface phonon dispersion curves of thin Pb films on Cu(111). *Phys. Rev. B* **2009**, *79*, 205423. [CrossRef]
29. Guo, Y.; Zhang, Y.-F.; Bao, X.-Y.; Han, T.-Z.; Tang, Z.; Zhang, L.-X.; Zhu, W.-G.; Wang, E.G.; Niu, Q.; Qiu, Z.Q.; et al. Superconductivity Modulated by Quantum Size Effects. *Science* **2004**, *306*, 1915–1917. [CrossRef]
30. Eom, D.; Qin, S.; Chou, M.-Y.; Shih, C.K. Persistent Superconductivity in Ultrathin Pb Films: A Scanning Tunneling Spectroscopy Study. *Phys. Rev. Lett.* **2006**, *96*, 027005. [CrossRef]
31. Özer, M.M.; Jia, Y.; Zhang, Z.; Thompson, J.R.; Weitering, H.H. Tuning the Quantum Stability and Superconductivity of Ultrathin Metal Alloys. *Science* **2007**, *316*, 1594–1597. [CrossRef]
32. Qin, S.; Kim, J.; Niu, Q.; Shih, C.-K. Superconductivity at the Two-Dimensional Limit. *Science* **2009**, *324*, 1314–1317. [CrossRef]
33. Brun, C.; Hong, I.-P.; Patthey, F.; Sklyadneva, I.Y.; Heid, R.; Echenique, P.M.; Bohnen, K.P.; Chulkov, E.V.; Schneider, W.-D. Reduction of the Superconducting Gap of Ultrathin Pb Islands Grown on Si(111). *Phys. Rev. Lett.* **2009**, *102*. [CrossRef]
34. Zhang, T.; Cheng, P.; Li, W.-J.; Sun, Y.-J.; Wang, G.; Zhu, X.-G.; He, K.; Wang, L.; Ma, X.; Chen, X.; et al. Superconductivity in one-atomic-layer metal films grown on Si(111). *Nat. Phys.* **2010**, *6*, 104–108. [CrossRef]
35. Zhang, Y.-F.; Jia, J.-F.; Han, T.-Z.; Tang, Z.; Shen, Q.-T.; Guo, Y.; Qiu, Z.Q.; Xue, Q.-K. Band Structure and Oscillatory Electron-Phonon Coupling of Pb Thin Films Determined by Atomic-Layer-Resolved Quantum-Well States. *Phys. Rev. Lett.* **2005**, *95*, 096802. [CrossRef] [PubMed]
36. Prakash, O.; Kumar, A.; Thamizhavel, A.; Ramakrishnan, S. Evidence for bulk superconductivity in pure bismuth single crystals at ambient pressure. *Science* **2017**, *355*, 52–55. [CrossRef] [PubMed]
37. Mata-Pinzón, Z.; Valladares, A.A.; Valladares, R.M.; Valladares, A. Superconductivity in Bismuth. A New Look at an Old Problem. *PLoS ONE* **2016**, *11*, e0147645.
38. Tian, M.; Wang, J.; Ning, W.; Mallouk, T.E.; Chan, M.H.W. Surface Superconductivity in Thin Cylindrical Bi Nanowire. *Nano Lett.* **2015**, *15*, 1487–1492. [CrossRef]
39. Wells, J.W.; Dil, J.H.; Meier, F.; Lobo-Checa, J.; Petrov, V.N.; Osterwalder, J.; Ugeda, M.M.; Fernandez-Torrente, I.; Pascual, J.I.; Rienks, E.D.L.; et al. Nondegenerate Metallic States on Bi(114): A One-Dimensional Topological Metal. *Phys. Rev. Lett.* **2009**, *102*, 096802. [CrossRef]
40. Ma, T.; Wang, S. *Phase Transition Dynamics*; Springer: New York, NY, USA, 2013.
41. Liu, R.; Ma, T.; Wang, S.; Yang, J. *Dynamic Theory of Fluctuations and Critical Exponents of Thermodynamic Phase Transitions*. 2019. Available online: https://hal.archives-ouvertes.fr/hal-01674269 (accessed on 2 December 2020).

42. Brusdeylins, G.; Heimlich, C.; Skofronick, J.G.; Toennies, J.P.; Vollmer, R.; Benedek, G. Determination of the Critical Exponent for a Charge Density Wave Transition in 2H-TaSe$_2$ by Helium Atom Scattering. *Europhys. Lett.* **1989**, *9*, 563–568. [CrossRef]
43. Brusdeylins, G.; Heimlich, C.; Skofronick, J.G.; Toennies, J.P.; Vollmer, R.; Benedek, G.; Miglio, L. He-atom scattering study of the temperature-dependent charge-density-wave surface structure and lattice dynamics of 2H-TaSe$_2$(001). *Phys. Rev. B* **1990**, *41*, 5707–5716. [CrossRef]
44. Requardt, H.; Kalning, M.; Burandt, B.; Press, W.; Currat, R. Critical x-ray scattering at the Peierls transition in the quasi-one-dimensional system. *J. Phys. Condens. Matter* **1996**, *8*, 2327–2336. [CrossRef]
45. Lorenzo, J.E.; Currat, R.; Monceau, P.; Hennion, B.; Berger, H.; Levy, F. A neutron scattering study of the quasi-one-dimensional conductor. *J. Phys. Condens. Matter* **1998**, *10*, 5039–5068. [CrossRef]
46. Wilson, J.A.; Salvo, F.J.D.; Mahajan, S. Charge-density waves and superlattices in the metallic layered transition metal dichalcogenides. *Adv. Phys.* **1975**, *24*, 117–201. [CrossRef]
47. Nagata, S.; Aochi, T.; Abe, T.; Ebisu, S.; Hagino, T.; Seki, Y.; Tsutsumi, K. Superconductivity in the layered compound 2H-TaS$_2$. *J. Phys. Chem. Solids* **1992**, *53*, 1259–1263. [CrossRef]
48. Sipos, B.; Kusmartseva, A.F.; Akrap, A.; Berger, H.; Forró, L.; Tutiš, E. From Mott state to superconductivity in 1T-TaS$_2$. *Nat. Mater.* **2008**, *7*, 960–965. [CrossRef] [PubMed]
49. Navarro-Moratalla, E.; Island, J.O.; Mañas-Valero, S.; Pinilla-Cienfuegos, E.; Castellanos-Gomez, A.; Quereda, J.; Rubio-Bollinger, G.; Chirolli, L.; Silva-Guillén, J.A.; Agraït, N.; et al. Enhanced superconductivity in atomically thin TaS$_2$. *Nat. Commun.* **2016**, *7*, 11043. [CrossRef]
50. Yang, Y.; Fang, S.; Fatemi, V.; Ruhman, J.; Navarro-Moratalla, E.; Watanabe, K.; Taniguchi, T.; Kaxiras, E.; Jarillo-Herrero, P. Enhanced superconductivity upon weakening of charge density wave transport in 2H-TaS$_2$ in the two-dimensional limit. *Phys. Rev. B* **2018**, *98*, 035203. [CrossRef]
51. Peng, J.; Yu, Z.; Wu, J.; Zhou, Y.; Guo, Y.; Li, Z.; Zhao, J.; Wu, C.; Xie, Y. Disorder Enhanced Superconductivity toward TaS$_2$ Monolayer. *ACS Nano* **2018**, *12*, 9461–9466. [CrossRef]
52. Ribak, A.; Skiff, R.M.; Mograbi, M.; Rout, P.K.; Fischer, M.H.; Ruhman, J.; Chashka, K.; Dagan, Y.; Kanigel, A. Chiral superconductivity in the alternate stacking compound 4Hb-TaS$_2$. *Sci. Adv.* **2020**, *6*, eaax9480. [CrossRef]
53. Ritschel, T.; Trinckauf, J.; Garbarino, G.; Hanfland, M.; Zimmermann, M.V.; Berger, H.; Büchner, B.; Geck, J. Pressure dependence of the charge density wave in 1T-TaS$_2$ and its relation to superconductivity. *Phys. Rev. B* **2013**, *87*, 125135. [CrossRef]
54. Wagner, K.E.; Morosan, E.; Hor, Y.S.; Tao, J.; Zhu, Y.; Sanders, T.; McQueen, T.M.; Zandbergen, H.W.; Williams, A.J.; West, D.V.; et al. Tuning the charge density wave and superconductivity in Cu$_x$TaS$_2$. *Phys. Rev. B* **2008**, *78*, 104520. [CrossRef]
55. Wilson, J.A.; Salvo, F.J.D.; Mahajan, S. Charge-density waves and superlattices in the metallic layered transition metal dichalcogenides. *Adv. Phys.* **2001**, *50*, 1171–1248. [CrossRef]
56. Coleman, R.V.; McNairy, W.W.; Slough, C.G. Amplitude modulation of charge-density-wave domains in 1T-TaS$_2$ at 300 K. *Phys. Rev. B* **1992**, *45*, 1428–1431. [CrossRef] [PubMed]
57. Yu, Y.; Yang, F.; Lu, X.F.; Yan, Y.J.; Cho, Y.-H.; Ma, L.; Niu, X.; Kim, S.; Son, Y.-W.; Feng, D.; et al. Gate-tunable phase transitions in thin flakes of 1T-TaS$_2$. *Nanotechnology* **2015**, *10*, 270–276. [CrossRef] [PubMed]
58. Cantini, P.; Boato, G.; Colella, R. Surface charge density waves observed by atomic beam diffraction. *Physica B+C* **1980**, *99*, 59–63. [CrossRef]
59. Heimlich, C. Max-Planck Inst. für Strömungsforschung. Ph.D. Thesis, University of Göttingen, Göttingen, Germany, October 1987.
60. Brusdeylins, G.; Heimlich, C.; Toennies, J.P. Helium scattering from the layered compound single crystal surface of 1T-TaS$_2$ in the temperature region of charge density wave reconstruction. *Surf. Sci.* **1989**, *211–212*, 98–105. [CrossRef]
61. Shimada, T.; Ohuchi, F.S.; Parkinson, B.A. Work Function and Photothreshold of Layered Metal Dichalcogenides. *Jpn. J. Appl. Phys.* **1994**, *33*. [CrossRef]
62. Zhao, R.; Grisafe, B.; Ghosh, R.K.; Holoviak, S.; Wang, B.; Wang, K.; Briggs, N.; Haque, A.; Datta, S.; Robinson, J. Two-dimensional tantalum disulfide: Controlling structure and properties via synthesis. *2D Materials* **2018**, *5*, 025001. [CrossRef]
63. Rossnagel, K. On the origin of charge-density waves in select layered transition-metal dichalcogenides. *J. Phys. Condens. Matter.* **2011**, *23*, 213001. [CrossRef]

64. Shiino, O.; Watanabe, T.; Endo, T.; Hanaguri, T.; Kitazawa, K.; Nohara, M.; Takagi, H.; Murayama, C.; Takeshita, N.; Môri, N.; et al. Metal–insulator transition in 1T-TaS$_{2-x}$Se$_x$. *Phys. B Condens. Matter* **2000**, *284–288*, 1673–1674. [CrossRef]
65. Balaguru-Rayappan, J.B.; Raj, S.A.C.; Lawrence, N. Thermal properties of 1T-TaS$_2$ at the onset of charge density wave states. *Phys. B Condens. Matter.* **2010**, *405*, 3172–3175. [CrossRef]
66. Liu, A.Y. Electron-phonon coupling in compressed 1T-TaS$_2$: Stability and superconductivity from first principles. *Phys. Rev. B* **2009**, *79*, 220515. [CrossRef]
67. Benedek, G.; Hofmann, F.; Ruggerone, P.; Onida, G.; Miglio, L. Surface phonons in layered crystals: Theoretical aspects. *Surf. Sci. Rep.* **1994**, *20*, 1–43. [CrossRef]
68. Benedek, G.; Brusdeylins, G.; Hofmann, F.; Ruggerone, P.; Toennies, J.P.; Vollmer, R.; Skofronick, J.G. Strong coupling of Rayleigh phonons to charge density waves in 1T-TaS$_2$. *Surf. Sci.* **1994**, *304*, 185–190. [CrossRef]
69. McWhan, D.B.; Axe, J.D.; Youngblood, R. Pressure dependence of the striped-to-hexagonal charge-density-wave transition in 2H-TaSe2. *Phys. Rev. B* **1981**, *24*, 5391–5393. [CrossRef]
70. Tsoutsou, D.; Aretouli, K.E.; Tsipas, P.; Marquez-Velasco, J.; Xenogiannopoulou, E.; Kelaidis, N.; Aminalragia-Giamini, S.; Dimoulas, A. Epitaxial 2D MoSe$_2$ (HfSe$_2$) Semiconductor/2D TaSe$_2$ Metal van der Waals Heterostructures. *ACS Appl. Mater. Interfaces* **2016**, *8*, 1836–1841. [CrossRef]
71. Moncton, D.E.; Axe, J.D.; DiSalvo, F.J. Neutron scattering study of the charge-density wave transitions in 2H–TaSe$_2$ and 2H–NbSe$_2$. *Phys. Rev. B* **1977**, *16*, 801–819. [CrossRef]
72. Craven, R.A.; Meyer, S.F. Specific heat and resistivity near the charge-density-wave phase transitions in 2H–TaSe$_2$ and 2H–TaS$_2$. *Phys. Rev. B* **1977**, *16*, 4583–4593. [CrossRef]
73. Bhoi, D.; Khim, S.; Nam, W.; Lee, B.S.; Kim, C.; Jeon, B.-G.; Min, B.H.; Park, S.; Kim, K.H. Interplay of charge density wave and multiband superconductivity in 2H-Pd$_x$TaSe$_2$. *Sci. Rep.* **2016**, *6*, 24068. [CrossRef]
74. Luo, H.; Xie, W.; Tao, J.; Inoue, H.; Gyenis, A.; Krizan, J.W.; Yazdani, A.; Zhu, Y.; Cava, R.J. Polytypism, polymorphism, and superconductivity in TaSe$_{2-x}$Te$_x$. *Proc. Natl. Acad. Sci. USA* **2015**, *112*, E1174–E1180. [CrossRef]
75. Kumakura, T.; Tan, H.; Handa, T.; Morishita, M.; Fukuyama, H. Charge density waves and superconductivity in 2H-TaSe$_2$. *Czechoslov. J. Phys.* **1996**, *46*, 2611–2612. [CrossRef]
76. Yokota, K.; Kurata, G.; Matsui, T.; Fukuyama, H. Superconductivity in the quasi-two-dimensional conductor 2H-TaSe$_2$. *Phys. B Condens. Matter.* **2000**, *284–288*, 551–552. [CrossRef]
77. Lian, C.-S.; Heil, C.; Liu, X.; Si, C.; Giustino, F.; Duan, W. Coexistence of Superconductivity with Enhanced Charge Density Wave Order in the Two-Dimensional Limit of TaSe$_2$. *J. Phys. Chem. Lett.* **2019**, *10*, 4076–4081. [CrossRef] [PubMed]
78. Wu, Y.; He, J.; Liu, J.; Xing, H.; Mao, Z.; Liu, Y. Dimensional reduction and ionic gating induced enhancement of superconductivity in atomically thin crystals of 2H-TaSe$_2$. *Nanotechnology* **2018**, *30*, 035702. [CrossRef] [PubMed]
79. Liu, Y.; Shao, D.F.; Li, L.J.; Lu, W.J.; Zhu, X.D.; Tong, P.; Xiao, R.C.; Ling, L.S.; Xi, C.Y.; Pi, L.; et al. Nature of charge density waves and superconductivity in 1T–TaSe$_{2-x}$Te$_x$. *Phys. Rev. B* **2016**, *94*, 045131. [CrossRef]
80. Li, X.C.; Zhou, M.H.; Yang, L.H.; Dong, C. Significant enhancement of superconductivity in copper-doped 2H-TaSe$_2$. *Supercond. Sci. Technol.* **2017**, *30*, 125001. [CrossRef]
81. Benedek, G. Summary Abstract: Surface phonon dynamics of 2H–TaSe$_2$(001). *J. Vac. Sci. Technol. A* **1987**, *5*, 1093–1094. [CrossRef]
82. Benedek, G.; Brusdeylins, G.; Heimlich, C.; Miglio, L.; Skofronick, J.G.; Toennies, J.P.; Vollmer, R. Shifted surface-phonon anomaly in 2H-TaSe$_2$. *Phys. Rev. Lett.* **1988**, *60*, 1037–1040. [CrossRef] [PubMed]
83. Hulpke, E. *Helium Atom Scattering from Surfaces*; Springer: Berlin, Germany, 1992.
84. Murphy, B.M.; Requardt, H.; Stettner, J.; Serrano, J.; Krisch, M.; Müller, M.; Press, W. Phonon Modes at the 2H–NbSe$_2$ Surface Observed by Grazing Incidence Inelastic X-Ray Scattering. *Phys. Rev. Lett.* **2005**, *95*, 256104. [CrossRef]
85. Weber, F.; Rosenkranz, S.; Castellan, J.-P.; Osborn, R.; Hott, R.; Heid, R.; Bohnen, K.-P.; Egami, T.; Said, A.H.; Reznik, D. Extended Phonon Collapse and the Origin of the Charge-Density Wave in 2H–NbSe$_2$. *Phys. Rev. Lett.* **2011**, *107*, 107403. [CrossRef]
86. Anemone, G.; Casado Aguilar, M.P.; Garnica, A.; Al Taleb, C.-N.; Kuo, C.; Shan-Lue, A.; Politano, A.L.; Vàzquez de Parga, G.; Benedek, D.; Farìas, R.M. Electron-Phonon Coupling in Superconducting 1T-PdTe2. To be publish.

87. Hor, Y.S.; Williams, A.J.; Checkelsky, J.G.; Roushan, P.; Seo, J.; Xu, Q.; Zandbergen, H.W.; Yazdani, A.; Ong, N.P.; Cava, R.J. Superconductivity in $Cu_xBi_2Se_3$ and its Implications for Pairing in the Undoped Topological Insulator. *Phys. Rev. Lett.* **2010**, *104*, 057001. [CrossRef]
88. Hor, Y.S.; Checkelsky, J.G.; Qu, D.; Ong, N.P.; Cava, R.J. Superconductivity and non-metallicity induced by doping the topological insulators Bi_2Se_3 and Bi_2Te_3. *J. Phys. Chem. Solids* **2011**, *72*, 572–576. [CrossRef]
89. Bray-Ali, N. Anon How to turn a topological insulator into a superconductor. *Physics* **2010**, *3*, 11. [CrossRef]
90. Yonezawa, S. Nematic Superconductivity in Doped Bi_2Se_3 Topological Superconductors. *Condens. Matter.* **2019**, *4*, 2. [CrossRef]
91. Zhang, H.; Li, H.; He, H.; Wang, J. Enhanced superconductivity in Bi_2Se_3/Nb heterostructures. *Appl. Phys. Lett.* **2019**, *115*, 113101. [CrossRef]
92. Matsubayashi, K.; Terai, T.; Zhou, J.S.; Uwatoko, Y. Superconductivity in the topological insulator Bi_2Te_3 under hydrostatic pressure. *Phys. Rev. B* **2014**, *90*, 125126. [CrossRef]
93. Yano, R.; Hirose, H.T.; Tsumura, K.; Yamamoto, S.; Koyanagi, M.; Kanou, M.; Kashiwaya, H.; Sasagawa, T.; Kashiwaya, S. Proximity-Induced Superconducting States of Magnetically Doped 3D Topological Insulators with High Bulk Insulation. *Condens. Matter.* **2019**, *4*, 9. [CrossRef]
94. Koren, G.; Kirzhner, T.; Lahoud, E.; Chashka, K.B.; Kanigel, A. Proximity-induced superconductivity in topological Bi_2Te_2Se and Bi_2Se_3 films: Robust zero-energy bound state possibly due to Majorana fermions. *Phys. Rev. B* **2011**, *84*, 224521. [CrossRef]
95. Qin, H.; Guo, B.; Wang, L.; Zhang, M.; Xu, B.; Shi, K.; Pan, T.; Zhou, L.; Chen, J.; Qiu, Y.; et al. Superconductivity in Single-Quintuple-Layer Bi_2Te_3 Grown on Epitaxial FeTe. *Nano Lett.* **2020**, *20*, 3160–3168. [CrossRef]
96. Charpentier, S.; Galletti, L.; Kunakova, G.; Arpaia, R.; Song, Y.; Baghdadi, R.; Wang, S.M.; Kalaboukhov, A.; Olsson, E.; Tafuri, F.; et al. Induced unconventional superconductivity on the surface states of Bi_2Te_3 topological insulator. *Nat. Commun.* **2017**, *8*, 2019. [CrossRef]
97. Kong, P.P.; Zhang, J.L.; Zhang, S.J.; Zhu, J.; Liu, Q.Q.; Yu, R.C.; Fang, Z.; Jin, C.Q.; Yang, W.G.; Yu, X.H.; et al. Superconductivity of the topological insulator Bi_2Se_3 at high pressure. *J. Phys. Condens. Matter* **2013**, *25*, 362204. [CrossRef]
98. Cai, S.; Kushwaha, S.K.; Guo, J.; Sidorov, V.A.; Le, C.; Zhou, Y.; Wang, H.; Lin, G.; Li, X.; Li, Y.; et al. Universal superconductivity phase diagram for pressurized tetradymite topological insulators. *Phys. Rev. Mater.* **2018**, *2*, 114203. [CrossRef]
99. Zhang, J.; Zhang, S.; Kong, P.; Yang, L.; Jin, C.; Liu, Q.; Wang, X.; Yu, J. Pressure induced electronic phase transitions and superconductivity in n-type Bi_2Te_3. *J. Appl. Phys.* **2018**, *123*, 125901. [CrossRef]
100. Zhang, S. *Discovery of the Chiral Majorana Fermion and Its Application to Topological Quantum Computing*; APS March Meeting: Nashville, TN, USA, 2018.
101. Kauffman, L.H.; Lomonaco, S.J. Braiding, Majorana fermions, Fibonacci particles and topological quantum computing. *Quantum Inf. Process.* **2018**, *17*, 201. [CrossRef]
102. Benedek, G. *Majorana Fermions in Condensed Matter Scientific Papers of Ettore Majorana: A New Expanded Edition*; Cifarelli, L., Ed.; Springer International Publishing: Cham, Switzerland, 2020; pp. 159–168.
103. Volovik, G.E. Fermion zero modes on vortices in chiral superconductors. *J. Exp. Theor. Phys. Lett.* **1999**, *70*, 609–614. [CrossRef]
104. Read, N.; Green, D. Paired states of fermions in two dimensions with breaking of parity and time-reversal symmetries and the fractional quantum Hall effect. *Phys. Rev. B* **2000**, *61*, 10267–10297. [CrossRef]
105. Kitaev, A. Anyons in an exactly solved model and beyond. *Ann. Phys.* **2006**, *321*, 2–111. [CrossRef]
106. Fu, L.; Kane, C.L. Superconducting Proximity Effect and Majorana Fermions at the Surface of a Topological Insulator. *Phys. Rev. Lett.* **2008**, *100*, 096407. [CrossRef]
107. Akhmerov, A.R.; Nilsson, J.; Beenakker, C.W.J. Electrically Detected Interferometry of Majorana Fermions in a Topological Insulator. *Phys. Rev. Lett.* **2009**, *102*, 216404. [CrossRef]
108. Alicea, J. New directions in the pursuit of Majorana fermions in solid state systems. *Rep. Prog. Phys.* **2012**, *75*, 076501. [CrossRef]
109. He, Q.L.; Pan, L.; Stern, A.L.; Burks, E.C.; Che, X.; Yin, G.; Wang, J.; Lian, B.; Zhou, Q.; Choi, E.S.; et al. Chiral Majorana fermion modes in a quantum anomalous Hall insulator–superconductor structure. *Science* **2017**, *357*, 294–299. [CrossRef]

110. Xu, J.-P.; Wang, M.-X.; Liu, Z.L.; Ge, J.-F.; Yang, X.; Liu, C.; Xu, Z.A.; Guan, D.; Gao, C.L.; Qian, D.; et al. Experimental Detection of a Majorana Mode in the core of a Magnetic Vortex inside a Topological Insulator-Superconductor Bi_2Te_3/NbSe2 Heterostructure. *Phys. Rev. Lett.* **2015**, *114*, 017001. [CrossRef]

111. Lee, E.J.H.; Jiang, X.; Houzet, M.; Aguado, R.; Lieber, C.M.; De Franceschi, S. Spin-resolved Andreev levels and parity crossings in hybrid superconductor–semiconductor nanostructures. *Nat. Nanotechnol.* **2014**, *9*, 79–84. [CrossRef] [PubMed]

112. Sun, H.-H.; Zhang, K.-W.; Hu, L.-H.; Li, C.; Wang, G.-Y.; Ma, H.-Y.; Xu, Z.-A.; Gao, C.-L.; Guan, D.-D.; Li, Y.-Y.; et al. Majorana Zero Mode Detected with Spin Selective Andreev Reflection in the Vortex of a Topological Superconductor. *Phys. Rev. Lett.* **2016**, *116*, 257003. [CrossRef] [PubMed]

113. Kitaev, A. Periodic table for topological insulators and superconductors. *AIP Conf. Proc.* **2009**, *1134*, 22–30.

114. Qi, X.-L.; Zhang, S.-C. Topological insulators and superconductors. *Rev. Mod. Phys.* **2011**, *83*, 1057–1110. [CrossRef]

115. Wang, M.-X.; Liu, C.; Xu, J.-P.; Yang, F.; Miao, L.; Yao, M.-Y.; Gao, C.L.; Shen, C.; Ma, X.; Chen, X.; et al. The Coexistence of Superconductivity and Topological Order in the Bi_2Se_3. *Thin Films Sci.* **2012**, *336*, 52–55.

116. Tamtögl, A.; Campi, D.; Bremholm, M.; Hedegaard, E.M.J.; Iversen, B.B.; Bianchi, M.; Hofmann, P.; Marzari, N.; Benedek, G.; Ellis, J.; et al. Nanoscale surface dynamics of Bi_2Te_3(111): Observation of a prominent surface acoustic wave and the role of van der Waals interactions. *Nanoscale* **2018**, *10*, 14627–14636. [CrossRef]

117. Tamtögl, A.; Kraus, P.; Avidor, N.; Bremholm, M.; Hedegaard, E.M.J.; Iversen, B.B.; Bianchi, M.; Hofmann, P.; Ellis, J.; Allison, W.; et al. Electron-phonon coupling and surface Debye temperature of B_2Te_3(111) from helium atom scattering. *Phys. Rev. B* **2017**, *95*, 195401. [CrossRef]

118. Lund, H.E.; Campi, D.; Ruckhofer, A.; Halbritter, S.; Holt, A.J.U.; Bianchi, M.; Bremholm, M.; Benedek, G.; Hofmann, P.; Ernst, W.E.; et al. Electron-Phonon Coupling and (Bulk) Electronic Structure of Sb_2Te_3 with intrinsic doping. 2021; To be publish.

119. Heid, R.; Sklyadneva, I.Y.; Chulkov, E.V. Electron-phonon coupling in topological surface states: The role of polar optical modes. *Sci. Rep.* **2017**, *7*, 1095. [CrossRef]

120. Chen, C.; Xie, Z.; Feng, Y.; Yi, H.; Liang, A.; He, S.; Mou, D.; He, J.; Peng, Y.; Liu, X.; et al. Tunable Dirac Fermion Dynamics in Topological Insulators. *Sci. Rep.* **2013**, *3*, 2411. [CrossRef]

121. Hatch, R.C.; Bianchi, M.; Guan, D.; Bao, S.; Mi, J.; Iversen, B.B.; Nilsson, L.; Hornekær, L.; Hofmann, P. Stability of the Bi_2Se_3(111) topological state: Electron-phonon and electron-defect scattering. *Phys. Rev. B* **2011**, *83*, 241303. [CrossRef]

122. Zeljkovic, I.; Scipioni, K.L.; Walkup, D.; Okada, Y.; Zhou, W.; Sankar, R.; Chang, G.; Wang, Y.J.; Lin, H.; Bansil, A.; et al. Nanoscale determination of the mass enhancement factor in the lightly doped bulk insulator lead selenide. *Nat. Commun.* **2015**, *6*, 6559. [CrossRef] [PubMed]

123. Campi, D.; Bernasconi, M.; Benedek, G. Ab-initio calculation of surface phonons at the Sb_2Te_3(111) surface. *Surf. Sci.* **2018**, *678*, 46–51. [CrossRef]

124. Bussmann, A.; Bilz, H.; Roenspiess, R.; Schwarz, K. Oxygen polarizability in ferroelectric phase transitions. *Ferroelectrics* **1980**, *25*, 343–346. [CrossRef]

125. Bussmann-Holder, A.; Benedek, G.; Bilz, H.; Mokross, B. Microscopic polarizability model of ferroelectric soft modes. *J. Phys. Colloq.* **1981**, *42*, C6-409–C6-411. [CrossRef]

126. Bilz, H.; Bussmann-Holder, A.; Benedek, G. Ferroelectricity in ternary compounds. *Nuovo Cimento D* **1983**, *2*, 1957–1963. [CrossRef]

127. Bilz, H.; Benedek, G.; Bussmann-Holder, A. Theory of ferroelectricity: The polarizability model. *Phys. Rev. B* **1987**, *35*, 4840–4849. [CrossRef]

128. Benedek, G.; Bussmann-Holder, A.; Bilz, H. Nonlinear travelling waves in ferroelectrics. *Phys. Rev. B* **1987**, *36*, 630–638. [CrossRef]

129. Bussmann-Holder, A.; Bilz, H.; Benedek, G. Applications of the polarizability model to various displacive-type ferroelectric systems. *Phys. Rev. B* **1989**, *39*, 9214–9223. [CrossRef]

130. Bilz, H.; Büttner, H.; Bussmann-Holder, A.; Kress, W.; Schröder, U. Nonlinear Lattice Dynamics of Crystals with Structural Phase Transitions. *Phys. Rev. Lett.* **1982**, *48*, 264–267. [CrossRef]

131. Bilz, H.; Büttner, H.; Bussmann-Holder, A.; Vogl, P. Phonon anomalies in ferroelectrics and superconductors. *Ferroelectrics* **1987**, *73*, 493–500. [CrossRef]

132. Sigmund, E.; Müller, K.A. (Eds.) Phase Separation in Cuprate Superconductors. In *Proceedings of the Second International Workshop on Phase Separation in Cuprate Superconductors, Cottbus, Germany, 4–10 September 1993*; Springer Science & Business Media: Berlin, Germany, 2012.
133. *Symmetry and Heterogeneity in High Temperature Superconductors*; Bianconi, A. (Ed.) Springer Science & Business Media: Berlin, Germany, 2006.
134. Shengelaya, A.; Müller, K.A. The intrinsic heterogeneity of superconductivity in the cuprates. *Europhys. Lett.* **2014**, *109*, 27001. [CrossRef]
135. Bianconi, A. Resonances and Complexity: From Stripes to Superstripes. *J. Supercond. Nov. Magn.* **2011**, *24*, 1117–1121. [CrossRef]
136. Bianconi, A.; Innocenti, D.; Campi, G. Superstripes and Superconductivity in Complex Granular Matter. *J. Supercond. Nov. Magn.* **2013**, *26*, 2585–2588. [CrossRef]
137. Bianconi, A. Superstripes in the Low Energy Physics of Complex Quantum Matter at the Mesoscale. *J. Supercond. Nov. Magn.* **2015**, *28*, 1227–1229. [CrossRef]
138. Campi, G.; Bianconi, A.; Poccia, N.; Bianconi, G.; Barba, L.; Arrighetti, G.; Innocenti, D.; Karpinski, J.; Zhigadlo, N.D.; Kazakov, S.M.; et al. Inhomogeneity of charge-density-wave order and quenched disorder in a high-T_c superconductor. *Nature* **2015**, *525*, 359–362. [CrossRef]
139. Agrestini, S.; Saini, N.L.; Bianconi, G.; Bianconi, A. The strain of CuO_2 lattice: The second variable for the phase diagram of cuprate perovskites. *J. Phys. Math. Gen.* **2003**, *36*, 9133–9142. [CrossRef]
140. Jaouen, T.; Hildebrand, B.; Mottas, M.-L.; Di Giovannantonio, M.; Ruffieux, P.; Rumo, M.; Nicholson, C.W.; Razzoli, E.; Barreteau, C.; Ubaldini, A.; et al. Phase separation in the vicinity of Fermi surface hot spots. *Phys. Rev. B* **2019**, *100*, 075152. [CrossRef]
141. Bianconi, A.; Poccia, N.; Sboychakov, A.O.; Rakhmanov, A.L.; Kugel, K.I. Intrinsic arrested nanoscale phase separation near a topological Lifshitz transition in strongly correlated two-band metals. *Supercond. Sci. Technol.* **2015**, *28*, 024005. [CrossRef]
142. Kostylev, I.; Yonezawa, S.; Wang, Z.; Ando, Y.; Maeno, Y. Uniaxial-strain control of nematic superconductivity in $Sr_xBi_2Se_3$. *Nat. Commun.* **2020**, *11*, 4152. [CrossRef]
143. Benedek, G.M.; Bernasconi, D.; Campi, I.V.; Silkin, I.P.; Chernov, V.M.; Silkin, E.V.; Chulkov, P.M.; Echenique, J.P.; Toennies, G.; Anemone, A.; et al. Evidence for Acoustic Surface Plasmons from Inelastic Atom Scattering. 2020; Unpublished work.
144. Silkin, V.M.; García-Lekue, A.; Pitarke, J.M.; Chulkov, E.V.; Zaremba, E.; Echenique, P.M. Novel low-energy collective excitation at metal surfaces. *Europhys. Lett.* **2004**, *66*, 260. [CrossRef]
145. Shvonski, A.; Kong, J.; Kempa, K. Plasmon-polaron of the topological metallic surface states. *Phys. Rev. B* **2019**, *99*, 125148. [CrossRef]
146. Lemmens, L.F.; Devreese, J.T. Collective excitations of the polaron-gas. *Solid State Commun.* **1974**, *14*, 1339–1341. [CrossRef]
147. Mott, N.F. The bipolaron theory of high-temperature superconductors. *Phys. Stat. Mech. Appl.* **1993**, *200*, 127–135. [CrossRef]
148. Müller, K.A. Large, Small, and Especially Jahn–Teller Polarons. *J. Supercond.* **1999**, *12*, 3–7. [CrossRef]
149. Müller, K.A. The unique properties of superconductivity in cuprates. *J. Supercond Nov. Magn.* **2014**, *27*, 2163–2179. [CrossRef]
150. Müller, K.A. *Essential Heterogeneities in Hole-Doped Cuprate Superconductors Superconductivity in Complex. Systems: Structure and Bonding*; Müller, K.A., Bussmann-Holder, A., Eds.; Springer: Berlin, Germany, 2015; pp. 1–11.
151. Zhao, G.; Conder, K.; Keller, H.; Müller, K.A. Oxygen isotope effects in: Evidence for polaronic charge carriers and their condensation. *J. Phys. Condens. Matter* **1998**, *10*, 9055–9066. [CrossRef]
152. Perali, A.; Innocenti, D.; Valletta, A.; Bianconi, A. Anomalous isotope effect near a 2.5 Lifshitz transition in a multi-band multi-condensate superconductor made of a superlattice of stripes. *Supercond. Sci. Technol.* **2012**, *25*, 124002. [CrossRef]
153. Lanzara, A.; Zhao, G.; Saini, N.L.; Bianconi, A.; Conder, K.; Keller, H.; Müller, K.A. Oxygen-isotope shift of the charge-stripe ordering temperature in $La_{2-x}Sr_xCuO_4$ from x-ray absorption spectroscopy. *J. Phys. Condens. Matter.* **1998**, *11*, L541. [CrossRef]

154. Lanzara, A.; Bogdanov, P.V.; Zhou, X.J.; Kellar, S.A.; Feng, D.L.; Lu, E.D.; Yoshida, T.; Eisaki, H.; Fujimori, A.; Kishio, K.; et al. Evidence for ubiquitous strong electron–phonon coupling in high-temperature superconductors. *Nature* **2011**, *412*, 510–514. [CrossRef]
155. Benedek, G. The Erice Legacy. *Il Nuovo Saggiatore* **2019**, *35*, 47.

Publisher's Note: MDPI stays neutral with regard to jurisdictional claims in published maps and institutional affiliations.

 © 2020 by the authors. Licensee MDPI, Basel, Switzerland. This article is an open access article distributed under the terms and conditions of the Creative Commons Attribution (CC BY) license (http://creativecommons.org/licenses/by/4.0/).

Article

Perovskite Crystals: Unique Pseudo-Jahn–Teller Origin of Ferroelectricity, Multiferroicity, Permittivity, Flexoelectricity, and Polar Nanoregions

Isaac B. Bersuker [1,*] and Victor Polinger [2]

1 Department of Chemistry, University of Texas at Austin, Austin, TX 78712-1229, USA
2 Department of Chemistry, University of Washington, Seattle, WA 98195-1700, USA; vpolinger@msn.com
* Correspondence: bersuker@cm.utexas.edu

Received: 23 September 2020; Accepted: 27 October 2020; Published: 2 November 2020

Abstract: In a semi-review paper, we show that the local pseudo-Jahn–Teller effect (PJTE) in transition metal B ion center of ABO_3 perovskite crystals, notably $BaTiO_3$, is the basis of all their main properties. The vibronic coupling between the ground and excited electronic states of the local BO_6 center results in dipolar distortions, leading to an eight-well adiabatic potential energy surface with local tunneling or over-the-barrier transitions between them. The intercenter interaction between these dipolar dynamic units results in the formation of the temperature-dependent three ferroelectric and one paraelectric phases with order–disorder phase transitions. The local PJTE dipolar distortion is subject to the presence of sufficiently close in energy local electronic states with opposite parity but the same spin multiplicity, thus limiting the electronic structure and spin of the $B(d^n)$ ions that can trigger ferroelectricity. This allowed us to formulate the necessary conditions for the transition metal perovskites to possess both ferroelectric and magnetic (multiferroic) properties simultaneously. It clarifies the role of spin in the spontaneous polarization. We also show that the interaction between the independently rotating dipoles in the paraelectric phase may lead to a self-assembly process resulting in polar nanoregions and relaxor properties. Exploring interactions of PJTE ferroelectrics with external perturbations, we revealed a completely novel property—*orientational polarization in solids*—a phenomenon first noticed by P. Debye in 1912 as a possibility, which was never found till now. The hindered rotation of the local dipole moments and their ordering along an external field is qualitatively similar to the behavior of polar molecules in liquids, thus adding a new dimension to the properties of solids—notably, the perovskite ferroelectrics. We estimated the contribution of the orientational polarization to the permittivity and flexoelectricity of perovskite crystals in different limiting conditions.

Keywords: perovskite crystals; Pseudo-Jahn-Teller effect; ferroelectricity; multiferroicity; permittivity; flexoelectricity; polar nanoregions; orientational polarization

1. Introduction

There are innumerable publications devoted to the study of crystals with perovskite ABO_3 structure, reflecting the rich variety of their properties with wide-ranging applications in physics, chemistry, and materials science. In this paper, we show that there is a fundamental structural feature of such crystals related to the high, cubic-octahedral symmetry of their metallic B center, which defines their main properties. This feature is the pseudo-Jahn–Teller effect (PJTE). Its role in triggering the ferroelectricity of perovskite crystals was revealed first more than half a century ago [1] but gained a more developed form in recent decades [2–7]. Similar to the Jahn–Teller effect (JTE), the PJTE, under certain conditions, makes a high-symmetry configuration unstable with respect to lower symmetry distortions, but distinguished from the JTE, it does not require electronic degeneracy.

Instead, the two or more electronic states that mix under the nuclear displacements (vibronic coupling) may have any energy gap, provided that the other relevant parameters of the system are appropriate to obey the PJTE condition of "pseudodegeneracy" [8–11]. Most importantly, the JTE in degenerate states and the PJTE in nondegenerate states are the only source of spontaneous symmetry breaking (SSB) in polyatomic systems [8–10], while degeneracy and pseudodegeneracy are the only source of SSB in the whole spectrum of transformations from elementary particles to nucleus, to atoms, molecules, crystals, and phase transitions [12]. Emerging from the first principles, the JTE and PJTE compensate for the violation of the adiabatic approximation in defining polyatomic space configurations by the high-symmetry positions of the atoms.

The metallic B center in perovskites is in ideal conditions to exhibit a local PJTE. Indeed, as illustrated in the next section, the crystal is centrosymmetric, and for some of the electronic configurations of the B ion, it has a nondegenerate ground state with a relatively low-lying excited state of opposite parity (but the same spin multiplicity), which mix under vibronic coupling to produce the PJTE dipolar distortions (note that the JTE does not produce dipolar distortions in centrosymmetric crystals). This B center dipolar instability plays a key role in all the main properties of the appropriate perovskites (in some perovskites, the A center may play a similar role, not fully explored so far).

Below we illustrate these statements using the most studied perovskite, $BaTiO_3$, as a basic example. In Section 2, we demonstrate how the PJTE operates in the B center of the perovskite crystal, with numerical estimates of the parameters for this crystal. Section 3 shows how the local polar distortions induced by the PJTE cooperate in the crystal to produce its paraelectric and ferroelectric phases. We discuss also the relation between the predicted by the vibronic theory order–disorder nature of the phase transitions in ferroelectric perovskites and the (apparently) observed partial displacive components in some experimental observations. The presence of sufficiently close in energy local ground and excited states that produce the dipolar distortion strongly depends on the electronic structure of the transition metal B ion in the BO_6 cluster. It takes place only for a limited number of cases with unpaired electrons, thus limiting the necessary conditions of multiferroicity and revealing explicitly the role of spin in spontaneous polarization (Section 4). Section 5 is devoted to a novel, unique property of the perovskites with the PJTE, namely the orientational polarization. The almost free or hindered rotation of the local dipole moments results in strong orientational effects in interaction with external perturbations (similar to polar liquids), which enhances permittivity, flexoelectricity, electrostriction, etc., by orders of magnitude. The possible existence of orientational polarizable solids was indicated more than a century ago, but not found till the present work. Section 6 demonstrates how the PJTE-induced local dipolar distortions explain the formation of polar nanoregions and relaxor properties of ferroelectrics in a Gibbs free energy controlled self-assembly process.

In entirety, the origin of a bundle of properties (practically, all main properties) of perovskite ferroelectrics, notably $BaTiO_3$, is thus explained based on one fundamental feature, the local PJTE, and new properties are predicted and confirmed experimentally. This result, as a pattern and together with a variety of other similar generalizations [8–14], strongly supports the role of the PJTE as a unique tool in exploring molecular and solid-state properties. It is remarkable that the same JTE and PJTE ideas served also as a basis in the understanding of the formation of local polarons, bipolarons, and bipolaron-strips in perovskite cuprates that led to the discovery and explanation of the origin of high-temperature superconductivity [15–26]. Note also that the treatment of the ferroelectric and superconductivity phenomena based on the local vibronic coupling significantly deviates from the traditional band theory of crystals, because the disordered local (JTE and PJTE) distortions violate the translational symmetry.

2. The Pseudo-Jahn–Teller Effect in ABO_3 Perovskite Crystals: B-Center Instability, Mueller's ESR Experiments

As outlined in the Introduction, the high symmetry of perovskite ABO_3 crystals makes them outstanding with respect to a variety of important properties. Among them, properties related to electronic degeneracy and pseudodegeneracy, which are directly controlled by symmetry, seem to be most important. The cases of electronic degeneracy leading to the Jahn–Teller effect (JTE) in the local states and their cooperative interactions in crystals are relatively easily recognized and well-studied [9,10]. Less attention has been paid until recently to the pseudo-JTE (PJTE), which is not seen directly from the initial structural and electronic data of the crystal. Meanwhile, as shown in a series of works [1–7], the main properties of some important perovskite ABO_3 crystals are controlled by the PJTE.

Consider the local PJTE in the paraelectric phase of $BaTiO_3$. Very briefly, in cluster language for the octahedral unit $[TiO_6]^{8-}$ (Figure 1), the local PJTE emerges from the vibronic mixing of the ground state A_{1g}, formed by the fully occupied six HOMO ((highest occupied (HO) molecular orbitals (MO)) t_{1u} and t_{2u} (mostly oxygen $2p$ orbitals) and three LUMO (lowest unoccupied, LU) t_{2g}, mostly titanium $3d$ orbitals (a total of nine MOs with the one-electron energy levels shown in Figure 2), by the polar t_{1u} type normal coordinates Q_x, Q_y, and Q_z. This PJTE $(A_{1g} + T_{1u}) \otimes t_{1u}$ problem [9,10] results in a 9×9 secular equation. Its solution yields the following adiabatic potential energy surface (APES) of the $TiO_6{}^{8-}$ cluster, obtained already in the first paper on the subject [1]:

$$U(Q) = \frac{1}{2}K_0Q^2 - 2\left[\sqrt{\Delta^2 + 2F^2(Q^2 - Q_x^2)} + \sqrt{\Delta^2 + 2F^2(Q^2 - Q_y^2)} + \sqrt{\Delta^2 + 2F^2(Q^2 - Q_z^2)}\right] \quad (1)$$

where $Q^2 = Q_x{}^2 + Q_y{}^2 + Q_z{}^2$, 2Δ is the energy gap between the mixing electronic states, K_0 is the primary force constant for the Q displacements (stiffness of the crystal without the vibronic coupling), and F is the vibronic coupling constant (H is the Hamiltonian),

$$F = \langle 2p_z(O) | \left(\frac{\partial H}{\partial Q_x}\right)_0 | 3d_{xz}(Ti) \rangle \quad (2)$$

In a recent, more rigorous treatment by the Green's functions approach [5], the local PJTE in the cluster unit was appended with the bulk crystal properties by taking into account the interaction of the Ti ion with the whole crystal via its electronic and vibrational bands. It improved the results by yielding appropriately band-averaged parameters instead of the local cluster ones.

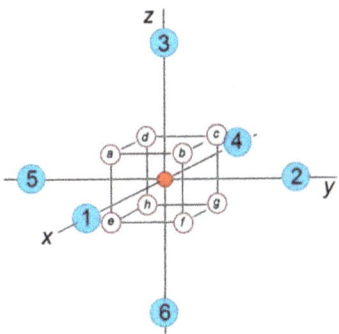

Figure 1. The octahedral fragment of the perovskite crystal structure ABO_3 with the transition metal atom B at the center (red) and six oxygen atoms at the apexes of the octahedron (numbered, blue). The letters a, b, c, etc., denote the induced by the PJTE eight equivalent off-center positions of the atom B in the eight wells of the APES (reprinted from [4]).

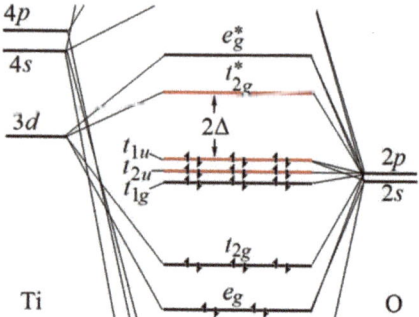

Figure 2. The energy-level correlation diagram (not to scale) for the octahedral cluster [TiO$_6$]. HOMO t_{1u} and t_{2u} (mostly oxygen) and LUMO t_{2g} (mostly titanium) are one-electron energy levels of the orbitals that are mixed in the PJTE. The shown electron population corresponds to the particular case of [TiO$_6$]$^{8-}$ [4,10].

The three-dimensional APES (1) has a specific form (Figure 3). Under the condition

$$\Delta < 8F^2/K_0 \quad (3)$$

the surface (1) has a maximum (meaning instability) when the Ti ion is in the center of the octahedron, eight equivalent minima placed along the four trigonal axes, in which the Ti ion is displaced toward three oxygen ions (away from the other three); higher-in-energy 12 equivalent saddle points along the six C_{2v} axes, at which the Ti ion is displaced toward two oxygen ions (at the top of the lowest barrier between two near-neighbor minima); and next six higher-in-energy equivalent saddle points, at which the Ti ion is displaced to one of the oxygen ions along the fourfold axes [1,2,5].

With an additional two experimentally determined structural constants, the band-averaged energy gap $2\Delta = 2.8$ eV, and the vibrational frequency at the bottom of the trigonal minimum $\hbar\omega_E = 193$ cm^{-1}, all the main parameters of this APES, shown in Table 1, were estimated [5], including K_0, F, the positions of the minima $Q_x = Q_y = Q_z = Q_0$ and first saddle points $Q_x = Q_y = q_0$, $Q_z = 0$, their PJTE stabilization energies, and the tunneling splitting δ_0. The latter is a characteristic measure of the energy barrier between the near-neighbor minima of the APES. Its order of magnitude was first estimated by K. A. Muller and coworkers [27–29] in ESR experiments with probing ions by substituting Ti^{4+} with the Mn^{4+} ion, which produces a very similar APES (Mn^{4+} is a "ferroelectric" ion, see below, Section 5), yielding approximate lifetimes in the minima (in sec.): $10^{-9} > \tau > 10^{-10}$. It is also seen in the NMR [30,31] and EXAFS [32,33] experiments: in the former, the characteristic "time unit" $\tau' \sim 10^{-15}$ and the Ti ion are seen in the trigonal minima in all the phases of BaTiO$_3$, while in NMR experiments, $\tau' \sim 10^{-8}$, and only an averaged picture is revealed (see the discussion in Section 3).

Table 1. Numerical values of the PJTE vibronic coupling and APES parameters of the Ti active centers in the BaTiO$_3$ crystal. E_{PJT} [111] is the PJTE stabilization energy at the trigonal minima, E_{PJT} [110] is the stabilization energy at the maximum of the barrier between two near-neighbor minima [5].

K_0 (eV/Å^2)	2Δ (eV)	F (eV/Å)	$\hbar\omega_E$ (cm^{-1})	Q_0 (Å)	q_0 (Å)	E_{PJT} [111] (cm^{-1})	E_{PJT} [110] (cm^{-1})	δ_0 (cm^{-1})
55	2.8	3.42	193	0.14	0.16	−1250	−1130	35

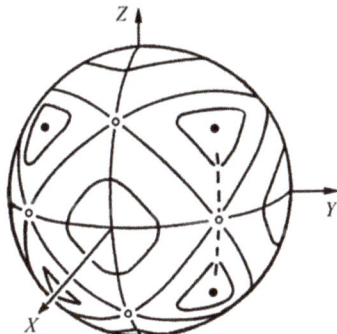

Figure 3. Contour map of the ground-state APES $U(Q_x,Q_y,Q_z)$ for the octahedral cluster $[BO_6]^{8-}$ due to the PJTE coupling of its oxygen HOMOs to the central-atom's LUMOs. Equipotential level curves are projected on the surface of a sphere. Solid dots represent trigonal (rhombohedral) minima, open dots are orthorhombic saddle points. Broken curve is the path of steepest descent from one of the saddle points to the closest wells [6].

The four phases of BaTiO$_3$ emerge directly from this APES by gradually populating the states in the minima with temperature and stepwise overcoming its barriers [1,5,6] (see Section 3). Of particular interest is the prediction of disorder in the orthorhombic and tetragonal ferroelectric phases and in the cubic paraelectric phase, and order–disorder phase transitions between them, discussed below in more detail.

3. Ferroelectricity in ABO$_3$ Perovskite Crystals–Order–Disorder Phase Transitions

Due to the mentioned above high symmetry, perovskite crystals like BaTiO$_3$ have an inversion center and no local dipole moments. For such systems, the previous dominant theories of ferroelectricity assumed that the ferroelectric distortion occurs as a result of the compensation of the local repulsions (resisting dipolar displacements) by long-range dipole–dipole attractions. In such "displacive" theories, it is assumed that the local charge-separating displacements are induced by a self-consistent interaction with displacements in the bulk.

On the other hand, it has been proven that the JTE and PJTE are the only source of spontaneous distortions of high-symmetry configurations of polyatomic systems (at $T = 0$). As shown in the previous section, in perovskites with an inversion center, spontaneous local polar distortions are possible due to the PJTE. This triggered the idea that the spontaneous polarization in ferroelectric crystals is due to the temperature-dependent ordering of considered above PJTE-induced local dipole moments [1], presently developed in the vibronic PJTE theory of ferroelectricity (see the review [2] and references therein). According to this theory, the local dipole moments should be present in the crystal before its spontaneous polarization (meaning in its paraelectric phase), in contrast to the displacive theories, where the dipole moments occur as a result of the spontaneous polarization.

At the time of the first paper on the vibronic theory [1], there were no experiments to test the drastic differences between the two theories. Later on, the number of experimental observations that contradicted the displacive theories of ferroelectricity gradually increased (continuing), including X-ray scattering and diffraction data [34], Raman and optical reflective experiments [35–38], the mentioned above ESR with probing ions [27–29], EXAFS measurements [32,33], NMR experiments [30,31], neutron scattering [39], and later in a variety of experimental studies (see more details and references in [2]; the authors of [34] did not cite the first prediction of the order–disorder phase transitions in ferroelectric perovskites [1] but acknowledged it in a personal letter; see [40]). In these experiments, several "peculiar" (for displacive theories) features in the ferroelectric properties were revealed in some ABO$_3$ perovskites, notably BaTiO$_3$. Among them, we emphasize here the instant trigonal displacement of the Ti ion in BaTiO$_3$ in all its four phases, ferroelectric and paraelectric [32,33], in strong disagreement with displacive theories, in which the metal off-center displacement occurs as a result of the phase transition to the polarized phase. This fundamental property of perovskite ferroelectrics is confirmed by a variety of experimental data, including the above-mentioned ones, and actually by any observations that implicate the Ti ion. Not only is this ion instantly displaced along one of the [111]-type directions in the paraelectric phase, where the averaged symmetry is cubic, but it is as well displaced in this trigonal direction in the tetragonal phase, where the crystal symmetry and the macroscopic polarization are tetragonal [39]. "The most striking example is the off-center displacement of the Ti atom in barium titanate observed in the paraelectric phase way above the Curie temperature of the tetragonal-to-cubic phase transition. As accurate measurements indicate [32,33], the Ti ion remains instantly displaced closely along [111] directions *throughout all the four BaTiO$_3$ phases*, and the magnitude of the off-center displacement decreases monotonically by only 13% when heating from 35 to 590 K, showing *no steps at the phase transitions.*" [32,33].

All these experimental observations and other empirical data (see [2]), as a pattern and in details, confirm the predicted earlier picture of the special, induced by the PJTE local dipolar distortions, which in cooperative interactions produce all observed ferroelectric properties. In a more recent work [6], using the numerical estimates of Table 1, the interaction between the PJTE-induced local dipole moments in BaTiO$_3$ was taken into account in a mean-field approximation, in which both the local off-center displacement and the mean-field of the environment are interdependent in a self-consistent way (see below). It yields the experimentally observed phase transitions in BaTiO$_3$ with reasonable values of Curie temperatures [6].

The picture of the B center of the ABO$_3$ crystal with the PJTE-induced eight-minimum APES in Equation (1) and local dipolar displacement in the minima, and their changes under the influence of the external mean-field **E** of the environment, is illustrated in Figure 4.

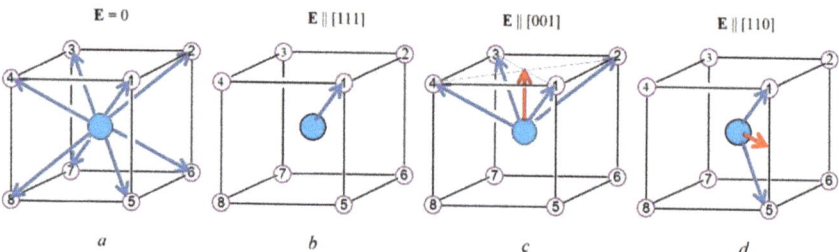

Figure 4. Polar displacements (blue arrows) of the central atom (blue) in the octahedron [BO$_6$]. White circles correspond to off-center equilibrium positions of the central ion B (the blue ball) in eight trigonal wells. (**a**) At no polar perturbation, the trigonal wells are symmetry-equivalent and therefore equally populated. The ion B has an equal probability to be off-center shifted (blue arrows) to each well and, on average, its off-center shift is zero. (**b**) Under a trigonal field **E** ∥ [111], one well is deeper than the others, and at low temperature, it is the only one temperature populated. The vector of average displacement (the blue arrow) is along the field parallel to the axis [111]. (**c**) In the tetragonal field **E** ∥ [001], the wells 1, 2, 3, and 4 remain symmetry-equivalent to one another; they are the lowest in energy and hence most populated. The mean displacement (red vector) points are along the axis [001]. (**d**) With the field E along the second-order axis, **E** ∥ [110], the wells 1 and 5 are the lowest and the only ones thermally populated. Therefore, the mean displacement (red vector) is along the electric field, **E** ∥ [110] [6].

Qualitatively, the phenomenon of ferroelectric ordering of the PJTE-induced local dipolar distortions is similar to magnetization in ferromagnetic crystals. The Hamiltonian of the system is $H = \sum_m H_m + \sum_{mn} H_{mn}$, where m and n label different unit cells of the lattice. Similar to Section 2, the one-cell term $H_m = H_0(m)$ includes the PJTE in the octahedral unit [BO$_6$]. As for the inter-cell coupling term H_{mn}, we employ the simplified version of dipole–dipole coupling, which follows from the physical nature of ferroelectric ordering. In the mean-field **E** of all other dipoles in the crystal, the energy of a selected dipole $\mathbf{p}_m$ equals $-\mathbf{p}_m \cdot \mathbf{E}$. At each site, the local dipole is influenced by the mean-field produced by the ordering of all other dipoles. At every other site, the induced electric dipole $\mathbf{p}_n$ is proportional to the applied electric field **E**. Therefore, $H_{mn} = -\frac{1}{2} A_{mn} \mathbf{p}_m \cdot \mathbf{p}_n$ with $m \neq n$. Due to the symmetry of the crystal lattice, the inter-cell coupling parameters A_{mn} are symmetric to swapping the indices, $A_{mn} = A_{nm}$. In what follows, the inter-cell constants A_{mn} are combined into one correlation constant, which is determined by fitting to experimental data.

To decouple H_{mn}, we apply the mean-field approximation. Based on the smallness of the fluctuation $\Delta \mathbf{p}_m = \mathbf{p}_m - \langle \mathbf{p}_m \rangle$ from the average value $\langle \mathbf{p}_m \rangle$, it provides reasonable answers far from the phase transition when $|\Delta \mathbf{p}_m| \ll p_m$ but does not apply close to the Curie temperature T_C. In the ferroelectric phase with a uniform polarization, the average $\langle \mathbf{p}_m \rangle$ is the same at different sites, so $\langle \mathbf{p}_m \rangle = \langle \mathbf{p} \rangle$. We insert $\mathbf{p}_m = \langle \mathbf{p} \rangle + \Delta \mathbf{p}_m$ into H_{mn} and neglect the terms quadratic in $\Delta \mathbf{p}_m$. Omitting the constant terms and plugging the resultant approximated expression into H_{mn} yields the decoupled Hamiltonian of the whole crystal as a sum of independent (mean-field) Hamiltonians, $H = \sum_m H_{MF}(m)$. In each of its ferroelectric phases, the perovskite crystal has translation symmetry. The one-site Hamiltonian is the same for different unit cells, $H_{MF}(m) = H_{MF} = H_0 - \mathbf{p} \cdot \mathbf{E}$. Here, H_0 is the one-site vibronic Hamiltonian yielding the PJTE in Section 2, and **E** is the mean-field, $\mathbf{E} = A \langle \mathbf{p} \rangle$, with the inter-cell correlation parameter $A = A(m) = \sum_n A_{mn}$. Due to the translational symmetry, **E** is the same for all sites, so is $A = A(0) = \sum_m A_{m0}$.

In the paraelectric phase with zero mean-field, $\mathbf{p} \cdot \mathbf{E} = 0$, and there is no intercenter dipolar coupling. The one-site nuclear dipolar displacements are independent of the other sites. It reduces to uncorrelated hindered rotations of the polar distortions $\mathbf{Q}_m$ at different sites (via tunneling or over-the-barrier transitions between the equivalent minima of the APES), yielding the bulk average $\langle \mathbf{Q} \rangle = 0$, meaning a disorder of local dipole moments. Lowering the temperature below T_C creates conditions for spontaneous polarization. Even a weak mean-field **E** can violate the equivalence between the wells and quench the tunneling; by locking the system in the lowest-in-energy potential wells and orienting the local dipoles in the direction **E** (Figure 4), it polarizes the crystal.

The effect of the mean-field depends on the magnitude of the PJTE induced dipole moment **p**. The latter is a combination of nuclear and electronic contributions, $\mathbf{p} = \mathbf{p}_{nucl} + \mathbf{p}_{el}$. It is easy to prove that due to the vibronic coupling, both $\mathbf{p}_{nucl}$ and $\mathbf{p}_{el}$ are proportional to the polar distortion **Q** induced by the applied electric field. Therefore, in what follows, we assume $\mathbf{p} = Z_B e \mathbf{Q}$, where Z_B is the so-called Born charge, and e is the elementary charge. The average value $\langle \mathbf{p} \rangle$ of the local dipole moment is proportional to the off-center nuclear displacement, $\langle \mathbf{p} \rangle = Z_B e \langle \mathbf{Q} \rangle$. Induced by the mean-field, it depends on $\langle \mathbf{Q} \rangle$ in a self-consistent way (see below). Therefore, we present the mean-field Hamiltonian as $H_{MF}(m) = H_{MF} = H_0 - \lambda \mathbf{Q} \cdot \langle \mathbf{Q} \rangle$ with $\lambda = Z_B^2 e^2 A$.

In the lowest-temperature rhombohedral phase at $T = 0$ K, the strength of the mean-field is maximal, and the octahedral unit cell [BO_6] is locked in one of the trigonal wells at the bottom of the trough with the radius Q_0. Therefore, at $T = 0$ K, the magnitude of $\langle \mathbf{Q} \rangle$ approaches its maximum value Q_0. At a higher temperature $T \neq 0$ K, the nuclear motion delocalizes over different wells and the vector $\langle \mathbf{Q} \rangle$ decreases in magnitude and changes its direction. Therefore, for the sequence of the corresponding ferroelectric phase transitions, we take $\langle \mathbf{Q} \rangle / Q_0$ as the order parameter.

In a certain unit cell, the delocalization changes with temperature. Nuclear motion depends on the radius Q_0 of the trough and the height of potential barriers between trigonal wells. Both are due to the one-site PJTE (see Table 1 for $BaTiO_3$), whereas the mean-field constant A is due to the inter-cell interaction, $A = \sum_m A_{m0}$. Generally, in cubic perovskites, there is a wide range of coupling parameters. In what follows, we consider two limiting cases: (a) deep trigonal wells where the tunneling model applies, and (b) shallow wells along the warped trough of the APES with the hindered rotation of the nuclei. The two cases correspond to significantly different mechanisms of delocalization and, correspondingly, of the ferroelectric phase transitions.

a   The tunneling model for deep trigonal wells. Most of the time, the nuclear motion is localized in the trigonal wells with low frequency tunneling through the barriers from one well into another. In a typical perovskite with PJTE, the tunneling energy gaps are of the order of several meV (see the estimates for BaTiO$_3$ in Table 1). The excited states of hindered rotations along the trough are a few dozen meV higher in energy. In such cases, the tunneling model applies, in which we can (approximately) consider just the eight tunneling states and neglect any quantum entanglement of all other excited states.

At $T > T_C$, in the paraelectric phase with zero mean-field, $E = 0$, self consistently, the order parameter is zero, $\langle \mathbf{Q} \rangle = 0$. The one-site Hamiltonian $H_{MFA} = H_0$ has cubic symmetry, and all trigonal wells are symmetry-equivalent. With no tunneling (assuming that the potential barriers are infinitely high), the ground state corresponds to eight Born–Oppenheimer states localized in the wells $|a\rangle$, $|b\rangle$, $|c\rangle$, ..., $|h\rangle$, labeled in Figure 1. Tunneling results in an even spread of the nuclear motion over the potential trough, lifting the eight-fold degeneracy of the ground state. The ground term splits into two triplets and two singlets, $A_{1g} + T_{1u} + A_{2u} + T_{2g}$ [41]. The tunneling splitting depends on the three overlap integrals, $\langle a|b \rangle$, $\langle a|c \rangle$, and $\langle a|g \rangle$. Evidently, for neighboring wells, $\langle a|b \rangle$ dominates. Due to the greater distance, in a good approximation, the other two, $\langle a|c \rangle$ and $\langle a|g \rangle$, can be neglected. Measured from the ground-state energy level E_0 in the infinite-deep wells, the tunneling energy levels are equidistant, $E(A_{1g}) \approx -3\Gamma$, $E(T_{1u}) \approx -\Gamma$, $E(T_{2g}) \approx \Gamma$, and $E(A_{2u}) \approx 3\Gamma$, where Γ is the tunneling parameter,

$$\Gamma \approx \frac{\hbar \omega_E}{2\pi} \exp\left(-\frac{1}{\hbar} \int_v^w \sqrt{2m[U(s) - E_0]}\, ds\right) \quad (4)$$

$\hbar \omega_E$ being the energy quantum of the transversal mode in a trigonal well, the one perpendicular to the corresponding trigonal axis, $U(s)$ is the potential energy (the APES) in Equation (1), and v and w are the classical turning points where the system hits the barrier wall. The integral is taken over the arc length s (broken curve in Figure 3) along the path of steepest descent from the orthorhombic saddle point to the closest trigonal minimum points of the APES. For BaTiO$_3$, the numeric estimate is $\Gamma \approx 35$ cm^{-1} = 4.3 meV [5].

Within the basis set of the eight tunneling states, the one-site Hamiltonian $H = H_0 - \lambda \mathbf{Q} \cdot \langle \mathbf{Q} \rangle$ of the mean-field approximation is represented by the following matrix [41],

$$\mathbf{H}(\gamma) = \Gamma \begin{pmatrix} 3 & -\gamma_x & -\gamma_y & -\gamma_z & 0 & 0 & 0 & 0 \\ -\gamma_x & 1 & 0 & 0 & 0 & -\gamma_z & -\gamma_y & 0 \\ -\gamma_y & 0 & 1 & 0 & -\gamma_z & 0 & -\gamma_x & 0 \\ -\gamma_z & 0 & 0 & 1 & -\gamma_y & -\gamma_x & 0 & 0 \\ 0 & 0 & -\gamma_z & -\gamma_y & -1 & 0 & 0 & -\gamma_x \\ 0 & -\gamma_z & 0 & -\gamma_x & 0 & -1 & 0 & -\gamma_y \\ 0 & -\gamma_y & -\gamma_x & 0 & 0 & 0 & -1 & -\gamma_z \\ 0 & 0 & 0 & 0 & -\gamma_x & -\gamma_y & -\gamma_z & -3 \end{pmatrix} \quad (5)$$

with basis $|A_{2u}\rangle$, $|T_{2g}\xi\rangle$, $|T_{2g}\eta\rangle$, $|T_{2g}\zeta\rangle$, $|T_{1u}x\rangle$, $|T_{1u}y\rangle$, $|T_{1u}z\rangle$, $|A_{1g}\rangle$.

Here, $\gamma_j = \lambda Q_0 \langle Q_j \rangle / (\Gamma \sqrt{3})$ and $j = x, y, z$ is the dimensionless mean-field order parameter in terms of the tunneling constant Γ. The three order parameters γ_j are solutions of the following system of coupled transcendental equations, $Z_\gamma \langle Q_j \rangle_\gamma = \mathrm{Tr}(e^{-\beta H} Q_j)$ and $Z_\gamma = \mathrm{Tr}(e^{-\beta H})$ with $j = x, y$, and z. Here, $\beta = 1/kT$, and the index γ means self-consistent dependence on the components γ_j included in the mean-field Hamiltonian (5). In the lowest-temperature rhombohedral phase, all atoms B are coherently shifted into the same trigonal well, say, in the direction [111]. In this case, $\gamma_x = \gamma_y = \gamma_z = \gamma$, and there is just one transcendental equation to solve. The mean-field $\mathbf{E}$ deepens the respective trigonal well. If $\mathbf{E}$ is strong enough, with the ordering energy $\mathbf{p} \cdot \mathbf{E}$ comparable to Γ, it quenches the tunneling and locks the system in this well. Rising temperature weakens the mean-field, reducing its locking effect.

For different values of the correlation parameter $j = p_0 A/\Gamma$, the calculated temperature dependence of the order parameter γ is shown in Figure 5. Plugging the resultant value of γ into the expression for the Helmholtz free energy, $\Phi = -\partial (\ln Z_a)/\partial \beta$ with $\beta = 1/kT$, we find the free energy Φ_{111} of the rhombohedral phase. Figure 6 shows the temperature dependence of Φ for different ferroelectric phases where we assume $j = p_0 A/\Gamma = 5$.

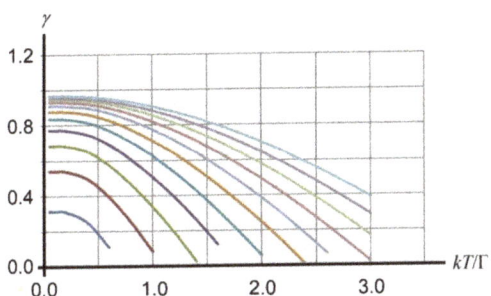

Figure 5. Temperature dependence of the order parameter γ in the rhombohedral phase when the tunneling mechanism of disorder dominates. The correlation parameter, $j = p_0 A/\Gamma$, varies from $j = 1$ for the left-bottom graph to $j = 3$ for the graph at the right top [6].

In the orthorhombic phase, all atoms B are off-center shifted along the two-fold symmetry axis [110]. As there is no minimum on the APES in this direction, we treat this "displacement" as resulting from an averaged motion evenly spread over two near-neighbor trigonal wells, the well b and the well f in Figure 1, separated by the orthorhombic potential barrier. In this case, $\gamma_x = \gamma_y = \gamma$ and $\gamma_z = 0$, and again we have just one equation to solve. Plugging the evaluated values of $\gamma_x = \gamma_y = \gamma$ and $\gamma_z = 0$ into the Helmholtz free energy, we find Φ_{110} in the orthorhombic phase (Figure 6). Solving the equation $\Phi_{111} = \Phi_{110}$ for temperature, we find T_C for the rhombohedral-to-orthorhombic phase transition. At around $kT/\Gamma = 3.8$, the graph of Φ_{111} intersects with that of Φ_{110}. Above $T_C = 3.8\Gamma/k$, the free energy Φ_{111} of the rhombohedral phase is above the orthorhombic phase Φ_{110}. The latter becomes energy-advantageous, and therefore, at $T_C = 3.8\Gamma/k$, the rhombohedral-to-orthorhombic phase transition takes place. In this model, its dependence on the inter-cell coupling parameter A is close to linear, $kT_C \approx 0.8 p_0 A - 0.3\Gamma$ [6].

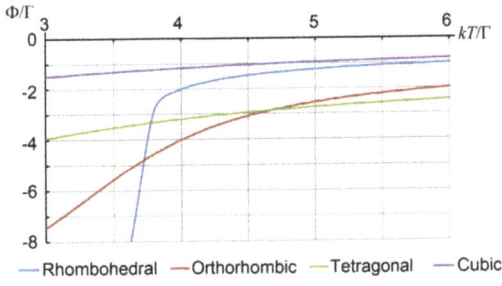

Figure 6. Temperature dependence of the Helmholtz free energy Φ (in units of Γ) for the four phases, rhombohedral, orthorhombic, tetragonal, and the paraelectric, at $j = 5$. At $kT < 3.8\Gamma$, the free energy of the rhombohedral phase is the lowest. With temperature, at $kT > 3.8\Gamma$, the free energy of the orthorhombic phase becomes the lowest, causing the rhombohedral-to-orthorhombic phase transition. Above $kT \approx 4.6\Gamma$, the tetragonal phase becomes most energy-advantageous [6].

In the tetragonal phase, all atoms B are off-center "shifted" to a tetragonal position, parallel to one of the four-fold symmetry axes, say [001]. Due to tunneling through four potential barriers, similar to the orthorhombic phase, this apparent shift is the average over four close-neighbor trigonal wells, a, b, c, and d in Figure 4, and the respective mean-field order parameter can be set as $\gamma_x = \gamma_y = 0$ and $\gamma_z = \gamma$. Plugging it into the expression of Helmholtz free energy Φ_{001} and solving the equation $\Phi_{110} = \Phi_{001}$, we find the temperature of the orthorhombic-to-tetragonal phase transition. For $j = 5$, it takes place at $kT_C \approx 4.7\Gamma$. Similar to the previous case, in this phase transition, the dependence of T_C on the coupling parameter A is linear, $kT_C \approx p_0 A - 0.3\Gamma$ [6].

For BaTiO$_3$, the experimental values of the two low-temperature phase transitions, rhombohedral-to-orthorhombic $T_C(I)$ and orthorhombic-to-tetragonal $T_C(II)$, are $T_C(I) = 178$ K and $T_C(II) = 278$ K. Repeated with different values of the correlation parameter A, the tunneling model calculated values of $T_C(I) = 204$ K and $T_C(II) = 256$ K are closest to the experimental values when $j = 5$. The percentage errors are 15% and 8%, respectively. A lower value of the mean-field parameter A better describes the lower-temperature phase transition than the higher-temperature one. This result is quite understandable because of the larger amplitudes of vibration in the wells at higher temperatures and hence the stronger inter-cell interaction of the distortions.

Among the essential results of the tunneling model is the first-order type of ferroelectric phase transition. The equations which we used to evaluate Helmholtz free energy follow from the first-order perturbation approach to the eight-fold degenerate ground state. In this theory, the small parameter is the tunneling overlap Γ. In the case of barium titanate, as one can see from Figure 6, the tunneling model fails to provide a reasonable description of the high-temperature phase transition, tetragonal-to-cubic. It could happen by averaging over all eight trigonal wells. Such an averaging implies coherent tunneling through all twelve orthorhombic barriers. Obviously, due to the relatively high entropy factor, the respective probability is too low. An alternative tunneling path is through the high-symmetry point of the APES, which in the oxygen octahedron [BO$_6$] corresponds to the on-center position of the atom B. This potential barrier, of the order of 0.2 eV = 1600 cm^{-1}, is too high for tunneling.

b. *Shallow trigonal wells, low potential barriers.* The warping of the two-dimensional trough on the APES may happen to be relatively weak, which results in moderately low potential barriers between trigonal wells, but the high-symmetry point, like in barium titanate, may be too high in energy, so the first-order tunneling does not hold. In this case, the high-temperature disorder is achieved by second-order processes through a temperature population of the excited states that are close in energy to the top of the corresponding potential barriers between near-neighbor wells. Such over-the-barrier hopping is similar to the Arrhenius-type activation in chemical reactions, and it resembles the Orbach relaxation in magnetic resonance.

The temperature of the ferroelectric phase transition T_C is assumed high enough to populate the close in energy excited states, $kT_C \geq \Delta E_{JT}$. Then, the hindered rotation in the warped trough can be treated in terms of classical physics. The temperature averaging contains the exponent with the potential energy $U = U_0 - \lambda \mathbf{Q} \cdot \langle \mathbf{Q} \rangle$ of the mean-field Hamiltonian $H = H_0 - \lambda \mathbf{Q} \cdot \langle \mathbf{Q} \rangle$. Since it includes $\langle \mathbf{Q} \rangle$ as a parameter, we come to the following system of coupled transcendental equations:

$$Z\langle Q_j \rangle = \int Q_j e^{-U(\mathbf{Q})/kT} d^3 Q, \qquad Z = \int e^{-U(\mathbf{Q})/kT} d^3 Q \qquad (6)$$

In the tetragonal phase, the atom B is off-center shifted in one of the three tetragonal directions, say, along the axis [001]. In this case, $\langle Q_x \rangle = \langle Q_y \rangle = 0$ and $\langle Q_z \rangle = \langle Q \rangle \neq 0$. Out of the three equations (6), we have just one to solve. With the same parameter values as used above in the tunneling model, the numerical solution of this equation is shown in Figure 7. With temperature, the order parameter $\langle Q \rangle / Q_0$ decreases smoothly to zero at $kT_C \approx 0.1236 E_{JT}$. In the case of barium titanate with $F_0^2/K_0 \approx 0.25$ eV, this corresponds to $kT_C \approx 250$ cm^{-1} = 359 K, reasonably close to the experimental value of 373 K.

Thus, distinguished from the two low-temperature phase transitions, trigonal-to-rhombohedral and rhombohedral-to-tetragonal, the transition to the cubic phase is of second order.

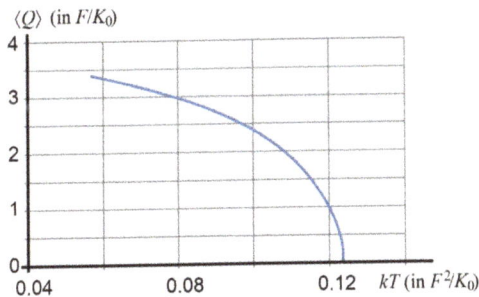

Figure 7. Temperature dependence of the order parameter $\langle Q \rangle$ (in units of F/K_0) in the tetragonal phase versus kT (in units of F_0^2/K_0). At $kT \approx 0.123\, F_0^2/K_0$, the off-center displacement of the atom B smoothly drops to zero, manifesting the second-order phase transition [6].

A fundamental conclusion, which was outlined already in the first paper of the vibronic PJTE theory [1], is that the ferroelectric phase transitions in such perovskite crystals are of order–disorder nature. This was first confirmed for $BaTiO_3$ in the experiments with diffuse X-ray scattering [32,33,40]. However, some more recent experimental findings—in particular, the low-frequency (soft) mode at the Curie temperature, observed in neutron scattering [42], IR absorption [43], hyper-Raman scattering [44], etc.—were interpreted as demonstrating displacive features of the phase transition (see also [45]).

As we noted earlier [2,6], there is no contradiction between these experimental observations and the vibronic PJTE theory. To begin with, the latter does not exclude the possibility of a low-frequency mode near the phase transition: it follows directly from the temperature dependence of the Helmholtz' free energy, Φ, with respect to the limiting phonon TO mode, $[\omega_{eff}(T)]^2 = (\partial^2 \Phi/\partial q^2)_0$. The consideration of the phase transition above in this section is based on the mean-field approximation which does not include any details about phonon dispersion. It follows from the EXAFS experiments [32,33] that in $BaTiO_3$, the magnitude of the local instant dipolar distortions does not change significantly as a result of the phase transition, meaning that the polarization process is similar to the magnetic ordering of local spins, which is of pure order–disorder type.

An attempt to include the phonon dispersion in a similar scheme involving the PJTE was undertaken in several papers [46–50], in which the PJTE is applied in a two-band approach, instead of our local two-state (in fact, several state) approach, outlined in Sections 2 and 3. The major difference between the two approaches is in the way the vibronic Hamiltonian is treated. In our paper, the mean-field approximation is used before any Fourier transformation to crystal waves is applied. It addresses the orientation degrees of freedom of the "pseudo spin" at each unit cell. Therefore, obviously, it provides a better description of the properties of local origin and the order–disorder phase transitions. In the two-band theory, the mean-field approximation is applied after the Jahn–Teller Hamiltonian is presented in crystal plane-wave form. It operates with crystal lattice modes that are uniform over the whole crystal, showing that the growing population of the low-energy portion of the electron energy band with lowering temperature strengthens the pseudo Jahn–Teller effect and reduces the respective curvature, $(\partial^2 \Phi/\partial q^2)_0$. Evidently, this approach is adjusted to describe displacive transitions, but it fails to explain the variety of the experimental data of local origin [25–39]—notably, the instant trigonal displacements of the Ti ion in all the four phases, irrelevant to the phase transition [32,33].

The difference between the two approaches to the solid-state problem with the PJTE, local versus bulk (cooperative), is of a more fundamental nature than it may seem at first sight. Indeed, both the JTE and PJTE are of local origin: they are defined by off-diagonal matrix elements of the vibronic coupling terms in the Hamiltonian, which are significant only for the non-zero overlap of the electronic wavefunctions of near-neighbor atoms. Therefore, if the interaction between the JTE or PJTE centers are not strong enough, meaning before their ordering (before the structural phase transition), their local JT dynamics are not correlated, the translation symmetry of the crystal is obeyed in average, but not in each moment of time, and the traditional presentation of the crystal structure by electron and phonon bands is inadequate. To our knowledge, there are no worked out general methods (or computer programs) to handle such systems beyond our approach, described above based on the cited papers.

In view of this significantly local origin of the PJTE, another important issue emerges with regard to the interpretation of the mentioned above experimental data [42–44] as showing elements of displacive phase transitions in BaTiO$_3$: it did not take into account the relativity to the means of observation (see [2,6,12]). Indeed, in the eight-minimum APES, the B ion under consideration has a characteristic "lifetime" $\tau \approx \hbar/\Gamma$, where Γ is the introduced above tunneling parameter. In cubic perovskites, Γ ranges from 0.01 to 50 cm^{-1}. Respectively, the characteristic time is from $\tau \approx 10^{-9}$ s = 1 ns to $\tau \approx 10^{-13}$ s = 0.1 ps. Similarly, since (before the ordering) the transitions between the minima at different centers are not correlated, there is a related characteristic dimension l, which is the size of one elementary cell. For barium titanate, it is of the order of $l \approx 4$ Å.

On the other hand, the experimental methods of observation have their own characteristic "time of measurement" τ_{exp}, which is directly related to their frequencies. For example, the NMR technique measures the nuclear quadruple transitions and motional average dynamic displacements with a characteristic $\tau_{exp} = 10^{-8}$ s, whereas in EXAFS experiments, $\tau_{exp} = 10^{-15}$. All the other widely used experimental techniques lie in between these limits. Similarly, they have characteristic length limitations l_{exp} mostly determined by the wavelengths. Except for the EXAFS measurements, l_{exp} for all the other widely used experimental methods are several orders of magnitude larger than the unit cell dimension where the disorder begins. Obviously, the experimental results in [42–44] reflect the crystal processes averaged over many unit cells, both in space and time, and hence they cannot be used as an indication of displacive versus order–disorder phase transition.

It follows from this discussion that the most appropriate experimental method to reveal the microscopic origin of ferroelectric properties of perovskite crystals is EXAFS, and the already performed experiments with this method fully support the conclusions from the vibronic (PJTE) theory.

4. PJTE-Induced Orientational Polarization in Solids—Application to Dielectric Susceptibility and Flexoelectricity

One of the most important novel properties of ferroelectric ABO$_3$ perovskites, revealed by the vibronic (PJTE) theory, is their strongly enhanced interaction with external perturbations (electric fields, pressure, strain, etc.). As outlined in Section 3, in the absence of external influence, the PJTE-induced local dipolar distortions in [BO$_6$] units are of dynamic nature, moving between the eight equivalent minima of its APES by tunneling or over-the-barrier transitions. In fact, the orthorhombic barriers between neighboring trigonal wells are relatively low, so the dynamics of the dipolar distortions may be regarded as some hindered rotations along a three-dimensional trough. In this respect, the crystal acquires properties of polar liquids, with all the consequences for the observable properties, including orientational polarization of crystals, first revealed in the vibronic PJTE theory [51–53]. Remarkably, the possible existence of dielectric crystals with randomly oriented dipole moments (similar to ferromagnetics), which may behave like a polar liquid above its freezing point, was first suggested by P. Debye [54] in 1912 (long before the experimental discovery of ferroelectricity), but not discovered till the present work. Similar to polar molecules in liquids, the orientational polarizability of solids with dynamic, PJTE-induced dipolar displacements is expected to be larger than displacive polarizability by orders of magnitude [51–53].

Different crystals vary with respect to the barrier height and the radius of the trough Q_0 (see Section 2). At high temperatures, $T > T_C$, in the paraelectric phase with zero mean-field, the trigonal wells are symmetry-equivalent, the tunneling evenly distributes the nuclear motion over the trough, and for the ion B, on average, its off-center shift $\langle Q \rangle$ is zero (Figure 4a). The tunneling splitting of localized states in eight trigonal wells is proportional to the tunneling parameter Γ. The necessary condition of tunneling is the resonance of the states, localized in the minima. In a typical ferroelectric perovskite with the PJTE, the energy gaps 2Γ between tunneling energy levels are of the order of magnitude of several meV. Higher in energy with a few dozen meV are excited states of hindered rotations along the trough (for $BaTiO_3$, the estimates are in Table 1). Any polar perturbation W, even not very strong, lowering one or several of the eight wells and/or raising the other ones violates the resonance, quenches the tunneling, and locks the system in the lowest well(s) (Figure 4b–d). The perturbation W causes a non-zero off-center shift $\langle Q \rangle$, which produces anisotropy in the dielectric and elastic properties. A remarkably small magnitude of $W \sim \Gamma$, of the order of, or even less than, meV, diminishes the tunneling and locks the system in the deepest minimum. A similar sensitivity, though to a lesser extent, is in the excited states of the hindered rotations. *This unique feature of the PJTE in cubic perovskites determines their significant response to polar perturbations resulting in orientational polarization* [51–53].

There are two types of polar perturbations: (a) electromagnetic fields targeting the electronic subsystem, and (b) applied strain distorting the crystal lattice. In what follows, we discuss both effects in ferroelectric perovskites. In both cases, we consider the paraelectric phase at $T > T_C$ with no mean-field and a relatively strong PJTE, when the tunneling approximation applies to the ground state. This allows us to reduce the problem to just eight tunneling states. For all the remaining excited states, we assume that their thermal population is small, and hence their contribution to the corresponding response function is negligible. The effect of locking the PJTE dipolar distortions is shown in Figure 8 [53], where we use an applied electric field $\mathbf{E}$ as an example of the polar perturbation, $W = -\mathbf{p}\cdot\mathbf{E}$, and the corresponding Hamiltonian is $H = H_0 - \mathbf{p}\cdot\mathbf{E}$, with H_0 as the Jahn–Teller Hamiltonian (Section 3). The magnitude of the induced dipole moment $\mathbf{p}$ is a combination of nuclear and electronic contributions, $\mathbf{p} = \mathbf{p}_n + \mathbf{p}_e$. As shown in [53], in the case of static applied field, the nuclear contribution dominates, $|\mathbf{p}_n| \gg |\mathbf{p}|$. Therefore, as mentioned above, $\mathbf{p} \approx \mathbf{p}_n = Z_B e \mathbf{Q}$, where Z_B is the so-called Born charge, and e is the elementary charge constant. The induced off-center nuclear displacement $\langle Q \rangle$ is expected to approach its maximum value Q_0 under a trigonal field of infinite strength, when $E \to \infty$ and the octahedral unit cell $[BO_6]$ is entirely locked in one of the trigonal wells (Figure 4b). Accordingly, for the induced dipole moment, its asymptotic value is $p_0 \approx Z_B e Q_0$. When $\mathbf{E} \parallel [110]$, the perpendicular component averages out, $\langle Q_z \rangle = 0$, and the asymptotic value is $\langle Q \rangle \to (Q_0\sqrt{6})/3 \approx 0.8 Q_0$. Similarly, under the tetragonal field $\mathbf{E} \parallel [001]$, the perpendicular components average out, $\langle Q_x \rangle = \langle Q_y \rangle = 0$, with the asymptote value $Q_0/\sqrt{3} \approx 0.6 Q_0$. The general outcome for $\langle Q \rangle$ (in units of Q_0) as a function of E (in units of Γ/p_0), shown in Figure 8, was obtained by diagonalizing the Hamiltonian matrix 8×8 with the subsequent thermal population of the resultant vibronic states [53].

4.1. Dielectric Susceptibility

In a cubic perovskite, in its paraelectric phase with zero mean-field and no applied field, $E = 0$, the crystal is not polarized. Under an applied electric field, a non-zero dipole moment $\mathbf{p}$ is induced in each unit cell, and the crystal becomes polarized. By definition, the vector of polarization $\mathbf{P}$ is the average dipole moment per unit volume, $\mathbf{P} = \langle \mathbf{p} \rangle / a^3$. Here, a is the lattice constant, $\langle \mathbf{p} \rangle = \text{Tr}(p_i e^{-\beta H})/\text{Tr}(e^{-\beta H})$, with the averaging over both the temperature and (statistical) over different unit cells, $H = H_0 - \mathbf{p}\cdot\mathbf{E}$ is the vibronic Hamiltonian of the octahedral site $[BO_6]$ in the applied electric field, and $\beta = 1/kT$. Regularly, there is a linear relationship between the generalized "displacement" $\mathbf{P}$ and the "force" $\mathbf{E}$, namely $P_i = \varepsilon_0 \Sigma_j \alpha_{ij} E_j$. It corresponds to the first terms in the power expansion of $\mathbf{P}$ in terms of $\mathbf{E}$. The linear factor α_{ij} is the rank-two tensor of dielectric susceptibility or polarizability.

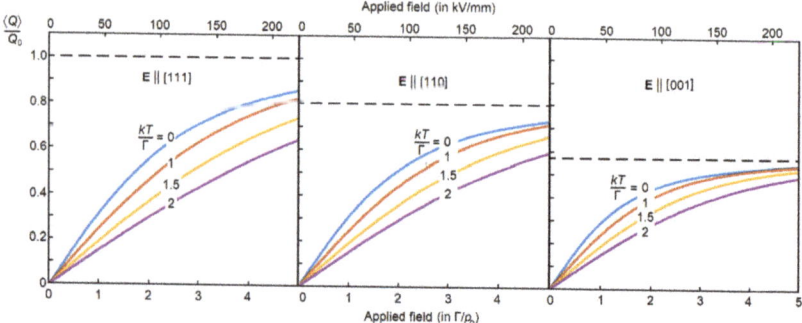

Figure 8. Approximate values of $\langle Q \rangle / Q_0$ versus the applied electric field E (in Γ/p_0) resulting from the tunneling Hamiltonian at three different directions of the applied electric field, (a) **E** ∥ [111], (b) **E** ∥ [110], and (b) **E** ∥ [001], and at four different temperatures, $kT = 0$, Γ, 1.5Γ, and 2Γ. The broken lines are horizontal asymptotes when $E \to \infty$. Shown above is the scale corresponding to the case of barium titanate with $\Gamma = 35$ cm^{-1} [53].

In cubic crystals with no PJTE, the three principal values of the tensor α_{ij} are equivalent, and the polarizability is the same in all directions. The dielectric properties of such a crystal are isotropic, and the principal axes have no special direction. As follows from the numeric results in Figure 8, to be specified below, in cubic perovskites, the PJTE brings about two important properties: (1) the dielectric susceptibility has cubic anisotropy; the principal axes are "tied" to the symmetry axes [100], [010], and [001], and (2) the induced dielectric susceptibility is non-linearly dependent on the applied electric field. The polarizability, and, correspondingly, the dielectric permeability, is not a constant but depends on the strength of the applied field.

a. *Low potential barriers and/or relatively weak electric fields, $E \ll \Gamma/p_0$.* This limiting case of $E \to 0$ corresponds to the left side of Figure 8, close to its vertical axis where the field-dependence on $\langle Q \rangle$, and hence on P, is linear. The consideration can be limited to just one localized state in each trigonal well (the potential trough is two-dimensional; even in a very shallow well, there is at least one localized quantum state). Under the applied field, the wells are not exactly symmetry-equivalent. However, since the electric field is relatively weak, the field-induced symmetry breaking is not significant. The localized states in trigonal wells are close to the resonance, with an even distribution of the nuclear motion over the trigonal wells, and the unit-cell octahedron [BO$_6$] carries frequent tunneling transitions or/and hindered rotations close to the bottom of the trough. Accordingly, assuming $P_i = \varepsilon_0 \Sigma_j \alpha_{ij} E_j$, we have $(\partial P_i / \partial E_j)_0 = \varepsilon_0 \alpha_{ij}$, where the subscript "0" means the derivative at $E = 0$. Plugging $\langle p_i \rangle = \text{Tr}(p_i e^{-\beta H})/\text{Tr}(e^{-\beta H})$ into the definition $\mathbf{P} = \langle \mathbf{p} \rangle / a^3$, we come to $\alpha_{ij} = \alpha \delta_{ij}$ with

$$\alpha \approx \frac{p_0^2}{3\varepsilon_0 a^3 \Gamma} \frac{\sinh(3\Gamma\beta) + \sinh(\Gamma\beta)}{\cosh(3\Gamma\beta) + 3\cosh(\Gamma\beta)} = \frac{Z_B^2 e^2 Q_0^2}{3\varepsilon_0 a^3 \Gamma} \frac{\sinh(3\Gamma\beta) + \sinh(\Gamma\beta)}{\cosh(3\Gamma\beta) + 3\cosh(\Gamma\beta)} \quad (7)$$

Hence, in a weak electric field, the polarizability is isotropic and the principal axes of the tensor α_{ij} have no specific direction. At relatively low temperatures and/or large Γ, when $kT \ll \Gamma$ (meaning $\Gamma\beta \to \infty$), the temperature factor in Equation (7) approaches unity, and $\alpha \approx p_0^2/(3\varepsilon_0 a^3 \Gamma)$. At high temperatures and/or relatively narrow tunneling gap Γ, when $kT \gg \Gamma$ (meaning $\Gamma\beta \to 0$), the temperature factor in Equation (7) approaches $\Gamma/(kT)$, and $\alpha \approx p_0^2/(3\varepsilon_0 a^3 kT)$. To be expected, this result coincides with the classic Langevin–Debye equation for orientational polarizability of dipolar molecules [55]. For BaTiO$_3$, assuming $Z_B \approx 7.8$ and $Q_0 \approx 0.19$ Å, we have $p_0 \approx 7$ D. Therefore, at $a \approx 4$ Å and $T \approx 500$ K, the Equation (7) gives $\alpha \approx 33$, close to the polarizability of some polar liquids.

b. *Strong electric fields and/or deep potential wells,* $E \geq \Gamma/p_0$ (this case apparently does not apply to BaTiO$_3$, with $\Gamma \approx 35$ cm^{-1} and $p_0 \approx 5$ D, $E = \Gamma/p_0 = 46$ kV/mm, which is above its dielectric breakdown). In this case, as follows from Figure 8, the field-dependence of $\langle Q \rangle$ and, hence, of $\langle p \rangle$ is non-linear. Therefore, for the dielectric susceptibility, instead of the derivative $(\partial P_i/\partial E_j)_0$, we can use the average $\alpha_{ij} = (\varepsilon_0)^{-1}(\Delta P_i/\Delta E_j)$. Since $\mathbf{P}(0)= 0$, we find $\Delta P_i/\Delta E_j = (P_i - 0)/(E_j - 0) = P_i/E_j$. For the applied electric field pointing along the trigonal axis, $\mathbf{E} \parallel [111]$, we have $E_x = E_x = E_x = E/\sqrt{3}$. According to Figure 4b, it lowers the trigonal well 1 in Figure 1, lifts the opposite well 7, and keeps the remaining potential wells unchanged. At low temperatures, the dipole moment $\langle Q_x \rangle = \langle Q_y \rangle = \langle Q_z \rangle \approx Q_0/\sqrt{3}$, so that $\langle Q \rangle = Q_0$. In this case, the polarization $P \approx \langle p \rangle/a^3 = p_0/a^3$. Similar results can be obtained for the cases when the applied electric field points along the symmetry axes [110] or [001]. The only difference is in the corresponding magnitude of $\langle p \rangle$. For $\mathbf{E} \parallel [110]$, we find $\langle p \rangle \approx 2p_0/\sqrt{6}$ whereas for a tetragonal field, $\mathbf{E} \parallel [001]$, we have $\langle p \rangle \approx p_0/\sqrt{3}$. Thus, for the case of a strong electric field, we come to

$$\alpha[111] \approx \frac{Z_B e Q_0}{\varepsilon_0 a^3 E}, \alpha[001] = \frac{\alpha[111]}{\sqrt{3}} \approx \frac{Z_B e Q_0}{\varepsilon_0 a^3 E \sqrt{3}}, \alpha[110] = \frac{2\alpha[111]}{\sqrt{6}} \approx \frac{2 Z_B e Q_0}{\varepsilon_0 a^3 E \sqrt{6}} \tag{8}$$

This angular dependence can be approximated by the cubic invariant,

$$\alpha(E, \varphi, \theta) \approx \frac{12 Z_B e Q_0}{\varepsilon_0 a^3 E}[0.8 - 0.5 Y_4(\varphi, \theta) + 2.8 Y_6(\varphi, \theta)] \tag{9}$$

where $Y_4(\phi, \theta)$ and $Y_6(\phi, \theta)$ are the spherical harmonics, $Y_4(\theta, \varphi) = (x^4 + y^4 + z^4 - \frac{3}{5}r^4)/r^4$ and $Y_6(\theta, \varphi) = x^2 y^2 z^2/r^6$. As follows from Figure 8, if the applied electric field is strong enough, the system is locked at the bottom of the lowest wells of the APES, and the polarization approaches its asymptotic value shown by the dotted line in Figure 8. Therefore, in all three formulas of Equation (8), the polarizability is inversely proportional to E, approaching zero at $E \to \infty$.

c. *High temperature and/or deep potential wells,* $kT \gg \Gamma$. This case is of special interest. In most ferroelectric perovskites, the Curie temperature T_C is of the order of a few dozen meV, while Γ is only a few meV. Therefore, for most cubic perovskites in the paraelectric phase, the condition $kT \gg \Gamma$ applies. Tunneling energy gaps are of the order of 2Γ. Therefore, in the high-temperature case when $\beta \to 0$, the Boltzmann exponent $\exp(-\beta H)$ can be approximated as $1-\beta H$, and the temperature average $\langle \mathbf{p} \rangle = \text{Tr}(\mathbf{p}e^{-\beta H})/\text{Tr}(e^{-\beta H})$ simplifies $\langle \mathbf{p} \rangle \approx \text{Tr}[\mathbf{p}(1 - \beta H)]/\text{Tr}(1 - \beta H) \approx (-\beta/8)\text{Tr}(\mathbf{p}H)$. At this point, one can take advantage of the invariance of the trace of a matrix to any unitary transformations of the basis set. In particular, we can use the basis of the eight tunneling states when $\mathbf{E} = 0$ with the well-known matrices 8 × 8 for the dipole moment $\mathbf{p}$ and the Hamiltonian H. Therefore, under any applied field, $\text{Tr}(\mathbf{Q}) = 0$, and $\text{Tr}(\mathbf{p}H) = -8p_0E/3$. Hence, at relatively high temperatures in the applied field of any strength, we have $\langle \mathbf{p} \rangle \approx p_0^2/(3kT)$, resulting in the Langevin–Debye equation $\alpha \approx p_0^2/(3\varepsilon_0 a^3 kT)$. The dielectric susceptibility is isotropic with the arbitrary direction of the principal axes. The orientational contribution to the dielectric susceptibility α caused by the PJTE-induced dipole moment rotations in the trough in the limits of high temperatures and deep potential wells is thus inversely proportional to temperature. However, the experimentally measured value of α also includes the significant contribution of the PNR (see Equation (16) in Section 6).

4.2. Flexoelectricity

Similarly, in cubic perovskites, the PJTE produces an enhanced response to another kind of dipolar perturbation, the strain induced by applied forces, which acts directly upon the crystal lattice. Caused by the stress, the corresponding strain is a rank-two symmetric tensor u with components $u_{mn} = \frac{1}{2}(\partial U_m/\partial x_n + \partial U_n/\partial x_m)$, where $\mathbf{U}$ is the vector of deformation, excluding the rigid-body motions. We follow the traditional notation x_n with the index values $n = 1, 2$, or 3 corresponding to rectangular coordinates x, y, and z. In centrosymmetric crystals, the parity of u is even. Therefore, in the paraelectric phase, a uniform strain does not cause any polarization. A non-uniform strain with a non-zero gradient $\partial u_{mn}/\partial x_j \neq 0$ is odd, and hence it can polarize the crystal. Like in the electric-field case, any non-uniform strain, even a weak one, breaks the resonance between the tunneling states, quenches the tunneling, and locks the system in lowest well(s) (Figure 4). The polarization **P** induced by the strain gradient is called the flexoelectric effect [56–58]. The stronger the perturbation $\partial u_{mn}/\partial x_j$, the greater the induced polarization. As shown in Figure 8, if strong enough, it can induce a non-linear polarization with a cubic anisotropy. In what follows, we limit the consideration with a relatively weak strain gradient $\partial u_{mn}/\partial x_j$ and its linear response.

Expanding polarization **P** in terms of $\partial u_{mn}/\partial x_j$, we can keep just the first non-zero terms,

$$P_i = \sum_{k,m,n} f_{ikmn} \frac{\partial u_{mn}}{\partial x_k} \tag{10}$$

Similar to the influence of an external electric field, due to the break of the local inversion symmetry induced by the PJTE, a relatively small strain gradient can induce a measurable polarization. The rate of change of P_i with the strain gradient $\partial u_{mn}/\partial x_k$ may be called flexoelectric susceptibility. The strength of the flexoelectric coupling is determined by the four-rank tensor f with components f_{ijmn}, called flexotensor. The elements $\partial u_{mn}/\partial x_j$ of the strain gradient can be treated as components of the rank-three tensor $u' = \nabla \otimes u$.

In the cubic symmetry group O_h, the components of the vector operator ∇ form the basis of the irreducible representation T_{1u}, while the components of the symmetric tensor u transform as the symmetric square $[T_{1u}^2] = A_{1g} + E_g + T_{2g}$. Therefore, $u' = \nabla \otimes u$ transforms as the product $T_{1u} \times (A_{1g} + E_g + T_{2g}) = A_{2u} + E_u + 3T_{1u} + 2T_{2u}$. The components of the rank-three tensor u' form the basis of a reducible representation. To find the irreducible combinations $u'_{\Lambda\lambda}(G)$, we apply the so-called Clebsch–Gordan decomposition of the group theory. Here, Λ is the resultant irreducible representations A_{2u}, E_u, $3T_{1u}$, and $2T_{2u}$, λ being their rows, and G, similar to the quantum number seniority in atomic spectra, labels the original term in the product, $T_{1u} \times A_{1g}$, $T_{1u} \times E_g$, or $T_{1u} \times T_{2g}$.

The left side of Equation (10) is a vector transforming as T_{1u}. Hence, of the seven representations A_{2u}, E_u, $3T_{1u}$, and $2T_{2u}$, we keep only the three vector representations T_{1u}, originating from the above products. We present components of the vectors $u'(A_{1g})$, $u'(E_g)$ and $u'(T_{2g})$ by the expressions $u'_j(\Lambda) = \sum_{k,\lambda}(\partial u_{\Lambda\lambda}/\partial x_k)T_{1u}k\Lambda\lambda|T_{1u}j\rangle$, where $\langle \Lambda_1\gamma_1\Lambda_2\gamma_2|\Lambda\lambda\rangle$ are the Clebsch–Gordan coefficients. In these terms, Equation (10) takes the form similar to the one with susceptibility,

$$P_i = \sum_j \left[f_{ij}(A_{1g})u'_j(A_{1g}) + f_{ij}(E_g)u'_j(E_g) + f_{ij}(T_{2g})u'_j(T_{2g}) \right],$$

$$P_i = \sum_\Lambda \sum_j f_{ij}(\Lambda)u'_j(\Lambda), \text{ with } f_{ij}(\Lambda) = \sum_{k,m,n,\lambda} f_{ikmn}\langle\Lambda\lambda|T_{1u}mT_{1u}n\rangle\langle T_{1u}j|T_{1u}k\Lambda\lambda\rangle \tag{11}$$

Instead of just one tensor $\varepsilon_0\chi_{ij}$, it includes three tensors, $f_{ij}(A_{1g})$, $f_{ij}(E_g)$, and $f_{ij}(T_{2g})$.

The applied non-uniform strain induces a local electric field **E'** polarizing the crystal, so that the perturbation becomes $W = -p_0(E'_x C_x + E'_y C_y + E'_z C_z)$ with $\mathbf{E'} = V_A \mathbf{u'}(A_{1g}) + V_E \mathbf{u'}(E_g) + V_T \mathbf{u'}(T_{2g})$. Here, V_Λ (with $\Lambda = A_{1g}$, E_g, and T_{2g}) are the corresponding coupling constants. Assuming that the strain gradient is relatively weak, and involving the relations $\partial P_i/\partial u_j' = (\partial P_i/\partial E_j')(\partial E_j'/\partial u_j') = V_\Gamma(\partial P_i/\partial E_j')$, we get $f_{ij}(\Lambda) = f(\Lambda)\delta_{ij}$ with Λ of the products $T_{1u} \times A_{1g}$, $T_{1u} \times E_g$ and $T_{1u} \times T_{2g}$. Quite similar to the procedure that leads to Equation (7), Equation (11) yields the following:

$$f(\Lambda) \approx \frac{p_0^2 V_\Lambda}{3a^3\Gamma} \frac{\sinh(3\Gamma\beta) + \sinh(\Gamma\beta)}{\cosh(3\Gamma\beta) + 3\cosh(\Gamma\beta)} \tag{12}$$

Thus, in cubic perovskites, among the 81 components of f_{ijkl}, only three reduced matrix elements $f(\Lambda)$ with $\Lambda = A_{1g}$, E_g, and T_{2g} remain independent. In addition to the cubic symmetry-related constraints, the flexotensor is invariant to index transpositions [59] $f_{ijkl} = f_{jikl}$ and $f_{ijkl} = f_{ilkj}$. Therefore, the three reduced matrix elements can be expressed in terms of just two components, $f(A_{1g}) = f_{1111}\sqrt{3}, f(E_g) = (f_{1111} - f_{1122})\sqrt{2}$, and $f(T_{2g}) = 2f_{1122}\sqrt{3}$ [60]. Accordingly, just two measurements are required to determine $f(\Lambda)$ experimentally. For example, to find the tetragonal component $f(E_g) = (2P_z\sqrt{2})(\partial u_{zz}/\partial z)^{-1}$, it is sufficient to measure the polarization that occurs under a non-uniform strain applied along the crystal axis [001], with the strain gradient $\partial u_{zz}/\partial z \neq 0$. Similarly, if the strain gradient is along the trigonal direction [111], one can find $f(T_{2g}) = P/u'(T_{2g})$.

The applied strain shifts the equilibrium positions of all ions in the unit-cell octahedron [BO$_6$]. Therefore, in Equation (12), the coupling constants V_Γ are related to the vibronic coupling constant F in Sections 2 and 3. For example, consider a tetragonal uniaxial strain along [001] with $u_q = u_{zz}$ and $u_e = (\sqrt{3}/2)(u_{xx} - u_{yy}) = 0$ or, equivalently, $u_{xx} = u_{yy} = -\frac{1}{3}u_q$ and $u_{zz} = \frac{2}{3}u_q$. It shifts the equilibrium position $\mathbf{R}_{nj}^{(0)} = \langle X_{nj}^{(0)}, Y_{nj}^{(0)}, Z_{nj}^{(0)} \rangle$ of the jth ion in the nth unit cell to $\mathbf{R}'_{nj} = \langle X_{nj}^{(0)}(1 - \frac{1}{3}u_{n\theta}), Y_{nj}^{(0)}(1 - \frac{1}{3}u_{n\theta}), Z_{nj}^{(0)}(1 + \frac{2}{3}u_{n\theta})\rangle$, where $j = 1, 2, \ldots, 7$ labels the seven ions in the octahedron [BO$_6$]. We need the site index $\mathbf{n}$ in the symmetry-adapted strain $u_{n\theta}$ to include its non-uniform nature. In the perovskite crystal lattice, the adjacent octahedrons [BO$_6$] share a bridge oxygen atom. In the next-neighbor octahedron [BO$_6$]$_{n+1}$ shifted in the direction [001] by one lattice constant a, due to the non-zero strain gradient, the symmetry-adapted coordinates are slightly different. Comparing the tetragonal deformation of the adjacent unit cells, let $u'_\theta = \Delta u_q/\Delta z = (u_{n+1,q} - u_{nq})/a$ where, as above, a is the lattice constant. It follows that the non-uniform tetragonal deformation creates a dipolar distortion $Q_z \rightarrow Q_z + 0.1a^2 u'_\theta$ or, in other words, $\Delta Q_z = 0.1a^2 u'_\theta$. This gives $V_E = 0.1Fa^2 = 0.1Z_B ea^2 u'_\theta$. The corresponding energy increment $F\Delta Q_z = V_E u'_\theta = 0.1Fa^2 u'_\theta$ can be treated as being due to a local electric field E' induced by the non-zero gradient u'_θ. Then, $0.1Fa^2 u'_\theta = p_0 E'$ and, therefore, $E' = 0.1Fa^2 u'_\theta/p_0$. Correspondingly, the induced polarization is $P_z = \varepsilon_0 \alpha E' = 0.1\varepsilon_0 \alpha Fa^2 u'_\theta/p_0$.

The resultant component of the flexotensor, $f_{1111} = \partial P_z/\partial u'_\theta = 0.1\varepsilon_0 \alpha Fa^2/p_0$, is proportional to the dielectric susceptibility α. For BaTiO$_3$, with the experimental value of α of the order of 2×10^4, we find $f_{1111} = 0.1\varepsilon_0 \alpha Fa^2/p_0 \approx 0.66$ μC/m, which is around 2600 times greater than was found in [58] and much nearer to the experimental value. Besides, with the orientational polarizability, as it follows from the outlined here PJTE theory, the flexoelectric factor is positive (in accordance with experimental data), whereas without the latter, it emerges as negative [58]. For other limiting conditions, rough estimates [51–53] yield even higher orientational contributions to the flexoelectric coefficients.

5. Multiferroicity in ABO$_3$ Perovskites with B(d^n) Configurations

The vibronic theory of ferroelectricity in ABO$_3$ perovskite crystals based on PJTE-induced local dipolar distortions in the B centers of the perovskite crystal was extended to formulate the necessary condition of coexisting magnetic and ferroelectric (multiferroicity) properties [3,4]. "Multiferroicity implies that the ferroelectric crystal, which is a dielectric, has also a nonzero magnetic moment, meaning unpaired electrons. In the ferroelectric BaTiO$_3$ the d^0 configuration of the Ti^{4+} ion has no unpaired electrons, and attempts to obtain ferroelectricity in perovskites with d^n, $n>0$, configurations of the transition metals B ions were unsuccessful for a long time. This prompted some authors to term the situation as a ferroelectric "d^0 mystery" [61–63], accompanied by a conclusion that nonzero spin states are detriment to ferroelectric polarization. However, more recently quite a number of ferroelectrics-multiferroics, mostly perovskites with configurations d^3–d^7, were obtained and studied" [3,4].

The origin of these special properties of perovskite ferroelectrics with d^n configurations does not follow directly from displacive theories and had no general explanation for a long time. The vibronic (PJTE) theory of ferroelectricity outlined above elucidates the role of spin states in the local polar instability, explains the origin of perovskite multiferroics with proper ferroelectricity, and formulates the necessary conditions that ABO$_3$ perovskites with a magnetic d^n configuration of the B ion may be ferroelectric [2–4]. Moreover, because of the spin implication, the multiferroics conditions that emerge from the PJTE for d^n ions with n = 3, 4, 5, 6, and 7 are directly influenced by the well-known transition metal high-spin/low-spin crossover, resulting in the coexistence of three phenomena: ferroelectricity (FE), magnetism (M), and spin crossover (SCO). This, in turn, leads to a quite novel phenomenon, magnetic-ferroelectric (multiferroics) crossover (MFCO), creating a rich variety of possible magnetoelectric and related effects [3,4]. The vibronic (PJTE) theory, exclusively, reveals the role of spin in the spontaneous polarization of crystals.

Referring to the typical MO energy scheme for the octahedral cluster BO_6^{8-} discussed above (Section 2) and shown in Figure 2 for the MO electron population of the d^0 configuration, e.g., when B ≡ Ti, we see that the HOMO in this case is t_{1u}, which is a three-fold degenerate odd-parity linear combination of mostly oxygen p_π orbitals, while the LUMO is t_{2g}, mostly atomic three d_π orbitals of the transition metal ion B, and the next excited MO is the double degenerate one e_g (the non-bonding oxygen b_{1g} MO is not shown as irrelevant). Using the arrows up and down to indicate the two spin states, we find for the d^0 case the HOMO configuration $(t_{1u})^6 = (t_{1u}\downarrow)^3(t_{1u}\uparrow)^3$, with the energy term $^1A_{1g}$. The excited state with opposite parity is formed by the one-electron, $(t_{1u}\uparrow) \to (t_{2g}\uparrow)$ or $(t_{1u}\uparrow) \to (e_g\uparrow)$, excitation, resulting in the lowest excited odd-parity term $^1T_{1u}$ at the energy gap 2Δ. In this case, the PJTE at the B center, under the condition of instability $\Delta < 8F^2/K_0$ (Equation (3)), produces a polar displacement of the B atom along [111]-type directions, which triggers the ferroelectric polarization (Section 3).

For other d^n configurations of the B ion instead of the d^0 one, the electronic structure of the ground and low-lying states changes drastically, and so does the possibility of formation of close in energy ground and excited states of opposite parity but equal multiplicity. In particular, for the $B(d^1)$ ions, the HOMO becomes $(t_{1u}\downarrow)^3(t_{1u}\uparrow)^3(t_{2g}\uparrow)^1$ with the term $^2T_{2g}$, while the LUMO (taking into account Hund's rule) is $(t_{1u}\downarrow)^2(t_{1u}\uparrow)^3(t_{2g}\uparrow)^2$ with the lowest excited odd-parity term $^4T_{1u}$. Hence, the two closest terms of different parity possess different spin multiplicity, and hence they do not mix by the vibronic coupling; the latter does not contain spin operators [9,10]. In principle, higher in energy electronic configurations of opposite parity with the same spin as the ground state one are quite possible, but they are at much larger energy gaps Δ and therefore less appropriate to satisfy the condition of instability (3) (numerical estimates show that the condition (3) may be very restricting). For the next possible d^2 configurations of the B ion, the two lowest terms of opposite parity are $^2T_{2g}$ and $^5T_{1u}$, which, again, do not satisfy the condition for the PJTE dipolar instability.

Moving to the case of $B(d^3)$ configuration, we find that the HOMO becomes $(t_{1u}\downarrow)^3(t_{1u}\uparrow)^3(t_{2g}\uparrow)^3$ with the ground state term $^4A_{1g}$, and in the low-spin (LS) conditions of the strong ligand fields (sufficiently large t_{2g}-e_g separation in Figure 2), the LUMO is $(t_{1u}\downarrow)^2(t_{1u}\uparrow)^3(t_{2g}\uparrow)^3(t_{2g}\downarrow)^1$ with the lowest odd-parity term $^4T_{1u}$. It follows that in perovskites with $B(d^3)$ ions in LS conditions of sufficiently strong ligand fields, the situation becomes again favorable for the PJTE and polar distortions. However, in this case, distinguished from the d^0 case, the system possesses also a magnetic moment created by three unpaired electrons. However, if the ligand field is weak and the separation t_{2g}-e_g is small, the high-spin (HS) arrangement of the excited electronic configuration takes place, and the excitation electron occupies the $e_g\uparrow$ orbital instead of $t_{2g}\downarrow$; the LUMO configuration under Hund's rule becomes $(t_{1u}\downarrow)^2(t_{1u}\uparrow)^3(t_{2g}\uparrow)^3(e_g\uparrow)^1$ with the lowest odd-parity state $^6T_{1u}$. Here, again, there is no PJTE on dipolar distortions and no ferroelectric instability. This is one of the examples which shows explicitly how the spin states interfere directly in the possible local polar displacement and ferroelectricity.

Repeating the above procedure for all the other d^n configurations with n = 0, 1, 2, ... , 10, it was shown that in perovskite ABO_3 crystals, only B ions with configurations d^0, d^3-LS, d^4-LS, d^5-LS, and HS, d^6-HS and intermediate-spin (IS), d^7-HS, d^8, and d^9 can, in principle, produce multiferroics, provided that the criterion of instability (3) is satisfied (see Table 2). If the contribution of higher excited states can be ignored, transition metal ions B with configurations d^1, d^2, d^3-HS d^4-HS, d^6-LS, d^7-LS, and d^{10} are not expected to produce multiferroics under this mechanism of proper ferroelectricity. Experimentally observed perovskite multiferroics with such B ions, for example, $Mn^{4+}(d^3)$, $Cr^{3+}(d^3)$,

$Mn^{3+}(d^4)$, $Fe^{3+}(d^5)$, $Fe^{2+}(d^6)$, $Co^{2+}(d^7)$, etc., fit well with this conclusion; there are no multiferroics with d^0, d^1, d^2, and d^{10} configurations.

Table 2. Necessary conditions that ABO_3 perovskites with the electronic d^n configuration of the B ion possess both ferroelectric and magnetic properties simultaneously; EC = electronic configuration, GS = ground state, LUES = lowest odd-parity excited state, FE = ferroelectric, MM = magnetic, MF = multiferroic, LS = low-spin, HS = high-spin; IS = intermediate spin; $(t_{1u})^6 = (t_{1u}\downarrow)^3(t_{1u}\uparrow)^3$; $(t_{1u})^5 = (t_{1u}\downarrow)^2(t_{1u}\uparrow)^3$; $(t_{2g})^6 = (t_{2g}\uparrow)^3(t_{2g}\uparrow)^3$; $(e_g)^4 = (e_g\uparrow)^2(e_g\downarrow)^2$ [3].

d^n	Example	HOMO EC and GS Term	LUMO EC and LUES Term	FE	MM	MF
d^0	Ti^{4+}	$(t_{1u})^6$, $^1A_{1g}$	$(t_{1u})^5(t_{2g}\uparrow)^1$, $^1T_{1u}$	yes	no	no
d^1	Ti^{3+}, V^{4+}	$(t_{1u})^6(t_{2g}\uparrow)^1$, $^2T_{2g}$	$(t_{1u})^5(t_{2g}\uparrow)^2$, $^4T_{1u}$	no	yes	no
d^2	V^{3+}, Cr^{4+}	$(t_{1u})^6(t_{2g}\uparrow)^2$, $^3T_{2g}$	$(t_{1u})^5(t_{2g}\uparrow)^3$, $^5T_{1u}$	no	yes	no
d^3, LS	Cr^{3+}, Mn^{4+}	$(t_{1u})^6(t_{2g}\uparrow)^3$, $^4A_{2g}$	$(t_{1u})^5(t_{2g}\uparrow)^3(t_{2g}\downarrow)^1$, $^4T_{1u}$	yes	yes	yes
d^3, HS		$(t_{1u})^6(t_{2g}\uparrow)^3$, $^4A_{2g}$	$(t_{1u})^5(t_{2g}\uparrow)^3(e_g\uparrow)^1$, $^6T_{1u}$	no	yes	no
d^4, LS	Mn^{3+}, Fe^{4+}	$(t_{1u})^6(t_{2g}\uparrow)^3(t_{2g}\downarrow)^1$, $^3T_{2g}$	$(t_{1u})^5(t_{2g}\uparrow)^3(t_{2g}\downarrow)^2$, $^3T_{1u}$	yes	yes	yes
d^4, HS		$(t_{1u})^6(t_{2g}\uparrow)^3(e_g\uparrow)^1$, $^5T_{2g}$	$(t_{1u})^5(t_{2g}\uparrow)^3(e_g\uparrow)^2$, $^7T_{1u}$	no	yes	no
d^5, LS	Mn^{2+}, Fe^{3+}	$(t_{1u})^6(t_{2g}\uparrow)^3(t_{2g}\downarrow)^2$, $^2T_{2g}$	$(t_{1u})^5(t_{2g})^6$, $^2T_{1u}$	yes	yes	yes
d^5, HS		$(t_{1u})^6(t_{2g}\uparrow)^3(e_g\uparrow)^2$, $^6A_{1g}$	$(t_{1u})^5(t_{2g})^4(e_g\uparrow)^2$, $^6T_{1u}$	yes	yes	yes
d^6, LS	Fe^{2+}, Co^{3+}	$(t_{1u})^6(t_{2g})^6$, $^1A_{1g}$	$(t_{1u})^5(t_{2g})^6(e_g\uparrow)^1$, $^3T_{1u}$	no	no	no
d^6, IS$_1$		$(t_{1u})^6(t_{2g})^5(e_g\uparrow)^1$, $^3T_{1g}$	$(t_{1u})^5(t_{2g})^6(e_g\uparrow)^1$, $^3T_{1u}$	yes	yes	yes
d^6, IS$_2$		$(t_{1u})^6(t_{2g})^5(e_g\uparrow)^1$, $^3T_{1g}$	$(t_{1u})^5(t_{2g})^5(e_g\uparrow)^2$, $^5T_{1u}$	no	yes	no
d^6, HS		$(t_{1u})^6(t_{2g})^4(e_g\uparrow)^2$, $^5T_{2g}$	$(t_{1u})^5(t_{2g})^5(e_g\uparrow)^2$, $^5T_{1u}$	yes	yes	yes
d^7, LS	Co^{2+}, Ni^{3+}	$(t_{1u})^6(t_{2g})^6(e_g\uparrow)^1$, ^2E_g	$(t_{1u})^5(t_{2g})^6(e_g\uparrow)^2$, $^4T_{1u}$	no	yes	no
d^7, HS		$(t_{1u})^6(t_{2g})^5(e_g\uparrow)^2$, $^4T_{2g}$	$(t_{1u})^5(t_{2g})^6(e_g\uparrow)^2$, $^4T_{1u}$	yes	yes	yes
d^8	Ni^{2+}, Cu^{3+}	$(t_{1u})^6(t_{2g})^6(e_g\uparrow)^2$, $^3A_{1g}$	$(t_{1u})^5(t_{2g})^6(e_g)^3$, $^3T_{1u}$	yes	yes	yes
d^9	Cu^{2+}	$(t_{1u})^6(t_{2g})^6(e_g)^3$, ^2E_g	$(t_{1u})^5(t_{2g})^6(e_g)^4$, $^2T_{1u}$	yes	yes	yes
d^{10}	Zn^{2+}	$(t_{1u})^6(t_{2g})^6(e_g)^4$, ^1A_g	$(t_{1u})^5(t_{2g})^6(e_g)^4(ns\uparrow)^1$, $^3T_{1u}$	no	no	no

Consider now that some d^n ions with n = 3, 4, 5, 6, and 7, dependent on the ligands of the octahedral environment, may produce two types of magnetic centers, high-spin (HS) and low-spin (LS), and in the d^6 case, there maybe also intermediate spin (IS) states (d^3 has two spin configurations in the one-electron excitation). According to the analysis [2–4], only d^5 systems follow the necessary condition of potential multiferroics in both spin states, but the PJTE conditions of instability and the magnetic moments are different in these two cases. For d^3, d^4, d^6, and d^7 ions, only one of the two spin states may serve as a candidate of potential multiferroics.

On the other hand, in many cases, the two spin states are close in energy, producing the well-known phenomenon of transition metal spin crossover (SCO), in which case the system can be relatively easily transferred from one spin state to another by external perturbations like heat, light, and magnetic fields [3,4]. As shown above, the change in the spin state changes also the possibility of ferroelectric polarization; hence, the SCO in some perovskite crystals is simultaneously a magnetic–ferroelectric (multiferroic) crossover (MFCO). This coexisting magnetic, ferroelectric, and spin-crossover phenomenon opens a variety of new possibilities to manipulate the properties of the system with novel functionalities to electronics and spintronics [3,4]: (1) For d^3 and d^4 (Cr^{3+}, Mn^{4+}, Mn^{3+}, Fe^{4+}, etc.) ferroelectrics in the LS state in conditions of MFCO, magnetic fields facilitate the LS→HS transition that destroys the ferroelectricity (and multiferroicity), while an electric field in the HS non-ferroelectric state may transfer the system to the ferroelectric (multiferroic) LS state, thus realizing electric demagnetization; (2) For d^5 ferroelectrics in conditions of MFCO, if the ferroelectricity is (most probably) different in the two spin states, an electric field may change the spin state (electric magnetization or demagnetization); (3) For d^6 and d^7 (Fe^{2+}, Co^{3+}, Co^{2+}, Ni^{3+}, etc.) in the non-ferroelectric LS state under conditions of MFCO, magnetic fields facilitate the LS→HS transition that induces ferroelectricity and hence multiferroicity in a strong magnetoelectric effect (the d^6 LS state is nonmagnetic); in the non-ferroelectric LS state in MFCO conditions, an electric field may transfer the system to the multiferroic state (electric magnetization); (4) The SCO phenomenon is well known to be influenced also by stress, heat, light, and cooperative effects in crystals [46], hence these perturbations can be used to manipulate the MFCO and all the consequent properties including those mentioned above. The dependence of the MFCO on pressure adds a ferroelastic order to the magnetic and ferroelectric ones; (5) There is already a long history of attempts to use transition metal SCO systems as units of magnetic bistability. The difficulty is in the fast relaxation (short lifetime) of the higher in energy spin state [64–66]. By choosing a system in the MFCO condition, one can increase the lifetime of the excited dipolar (multiferroic) state by applying an external electric field; (6) An important feature of the revealed MFCO is that it is of local origin and hence it does not necessarily require strong cooperative interactions, meaning that, in principle, it may take place as a magnetic-dipolar effect in separate molecular systems, clusters, thin films, etc., provided that the condition of instability (3) is obeyed.

In addition to the mentioned above examples that confirm the (outlined by the PJTE theory) necessary conditions of multiferroicity, several papers [67–69] reported observing also the predicted [3,4] magnetic–ferroelectric spin-crossover effect, while in [69], it is shown that in $BiCoO_3$, the ferroelectric polarization is greatly enhanced when the Co^{3+} ion is in the high-spin state, as compared to the nonmagnetic state with the Co^{3+} ion in the low-spin configuration. They demonstrated also the predicted electric magnetization [3] (see point 3 above), when, by means of induced polarization, the spin state changes from low-spin (S = 0) to high-spin (S = 2). The authors concluded that, contrary to the widespread belief, "*unpaired electron spins actually drive ferroelectricity, rather than inhibit it, which represents a shift in the understanding of how ferroelectricity and magnetism interact in perovskite oxides*" [69].

6. Origin of Polar Nanoregions and Relaxor Properties

The PJTE theory recently solved also another problem of ferroelectric crystals, the origin of polar nanoregions (PNR) producing relaxor properties [7], which remained unsolved (inconclusive) in spite of over 60 year of studies (see [70–74] and references therein). PNR in the non-polarized, paraelectric phases are observed arguably in all perovskite ferroelectrics. They are formed spontaneously in the paraelectric phases of ferroelectrics above the Curie temperature T_C (where the bulk crystal is cubic and no polarization is expected) in the form of small islands (nanoregions), containing a limited number n of unit cells, their size decreasing with increasing temperature T_n-T_C. Above the so-called Burns temperature T_B, the PNR disappear, and the crystal becomes regular, paraelectric. PNR also disappear under sufficiently strong electric fields. Significantly above T_C, the paraelectric phase remains ergodic. With cooling, PNR grow in size, and at a temperature T_f (closer to T_C, $T_C < T_f < T_B$), relaxor properties change again to a non-ergodic, glass-like state, which then undergoes the phase transition to the tetragonal polarized state at $T = T_C$. Distinguished from dipole glasses, this non-ergodic state of the crystal can be irreversibly transformed into a regular polarized state by strong external electric fields. These relaxor properties of ferroelectrics influence all their main properties, with direct applications in materials science.

Attempts to explain the formation of PNR as being due to basic structural disorder, or "random fields" produced by the differences in the active centers (especially in mixed perovskites), as well as by other crystal imperfections (see [70–74]), failed because they do not explain the origin of temperature-dependent size effects—in particular, the disappearance of PNR above T_B, or under polarizing electric fields. Moreover, PNR are present in ferroelectric perovskites, but they do not show up in very similar non-ferroelectric crystals with structural phase transitions.

The vibronic (PJTE) theory of ferroelectricity explains also the origin of PNR and relaxor properties of ferroelectric crystals [7] as due to the described above local dynamics of the dipolar distortions at the B centers, which are fully disordered (uncorrelated) in the paraelectric phase and partially ordered (correlated) in the ferroelectric phases. At temperatures above T_C, the Helmholtz free energy of the cubic phase Φ_{cub} is lower than its value in the tetragonal phase Φ_{tetr}. Accordingly, as $\Phi = U - TS$, where U is the internal energy and S is the entropy, in the temperature interval $T_C < T < T_B$, we have

$$T(S_{cub} - S_{tetr}) > U_{cub} - U_{tetr} \tag{13}$$

This means that at $T > T_C$, the energy gain $\Delta U = U_{cub} - U_{tetr}$ in lowering the potential energy from (average) cubic (in the paraelectric phase) to (average) tetragonal (in the PNR) is not enough to compensate the corresponding entropy loss, $T(S_{cub} - S_{tetr})$. Therefore, no displacive theory can explain their formation and specific properties.

Based on the basic features of the vibronic PJTE theory of ferroelectricity, outlined above, the local polarized isles (polar nanoregions, PNR, Figure 9) with a limited, relatively small number n of centers are formed in a non-equilibrium self-assembly process of alignment of the local dipolar displacements [7]. As shown below, at $T_n > T_C$, the inequality (13) is compensated by the transformation of a part of the tetragonal potential energy ΔU_{tetr} of the PNR into the work of the formation of its surface W_n and transfers heat Q to the environment. The gain of energy in the formation of the n-center PNR $\Delta U_n = U_n^{(cub)} - U_n^{(tetr)}$ consists of two contributions: $\Delta U_n^{(in)}$, the energy of the internal centers, i.e., the energy of ordering the n dipoles inside the PNR (all oriented, on average, along the polarization direction), and $\Delta U_n^{(surf)}$, the energy of the centers in the surface layer (the "domain wall"), which are influenced by the neighbor disordered centers of the cubic phase (Figure 9). Denoting by n and n' the total number of centers in the PNR and in its border surface, respectively, we get the following estimates for the ordering energies based on the mentioned above mean-field approximation (Section 3): $\Delta U_n^{(in)} = (n - n') \Delta U_0$ and $\Delta U_n^{(surf)} = n'(\Delta U_0 - g)$. Here, ΔU_0 is the per-unit energy gain by tetragonal ordering and g is the loss in this energy by the surface centers because of them being in a lower mean-field, $E' < E$, of the environment; the mean-field contribution from a part of their environment, by the disordered centers of the bulk cubic phase, is zero. A rough estimate yields $g \gtrsim (1/6) \Delta U_0$ [7].

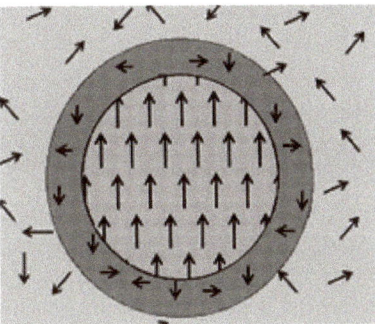

Figure 9. Polarized nanoregion inside the cubic perovskite with spherical form (center) and a border layer (dark) at temperatures T above the Curie one TC, but below the Burns temperature T_B, $T_C < T < T_B$. Arrows indicate the direction of the local averaged dipole moments, which are tetragonally ordered inside the PNR and fully disordered in the cubic phase. For the role of the PNR surface layer, see the text (reproduced with permission from [7]. Copyright 2018, American Physical Society).

For the system in a thermostat under consideration, the exchange of energy with the environment may take place without free energy conservation, $\Phi_i \neq \Phi_f$, the total energy balance being preserved by compensation of heat transfer, $Q = T(S_f - S_i) = T\Delta S$, with internal energy change ΔU and mechanical work of internal forces, W_{intern}. Similar to other processes with heat (entropy) transfer, the formation of PNR is a non-equilibrium thermodynamic process, described by Gibbs free energy change, $\Delta G = \Delta H - T\Delta S$, where ΔH is the change in enthalpy H. With growing size of PNR, their Gibbs free energy decreases and reaches a minimum value when n satisfies the condition of thermodynamic equilibrium with the environment. At this point, the shape-restoring force, $-dG/dr$, is close to zero, while according to the fist law of thermodynamics, $\Delta U = Q - W_{\text{intern}}$. This determines the size of the PNR with the number of units n at a given temperature T_n. Some relatively simple estimates based on the PJTE-induced APES in perovskite crystals briefly outlined above (Sections 2 and 3) yield the following temperature dependence of the size of PNR [7]:

$$\frac{T_n}{T_C} = 1 + \frac{4.8g}{\Delta U_0}n^{-1/3}, \quad T_n = T_C\left(1 + \frac{4.8g}{\Delta U_0}n^{-1/3}\right) \quad (14)$$

or by introducing the crystal constant $A = 36\pi\left(\frac{g}{\Delta U_0}\right)^3$, we obtain (Figure 10) the following:

$$n = A\left(\frac{T_n}{T_C} - 1\right)^{-3} \quad (15)$$

Figure 10 shows also the size of PNR by its diameter D, which is obtained under the assumption of its (most probably) approximately spherical form.

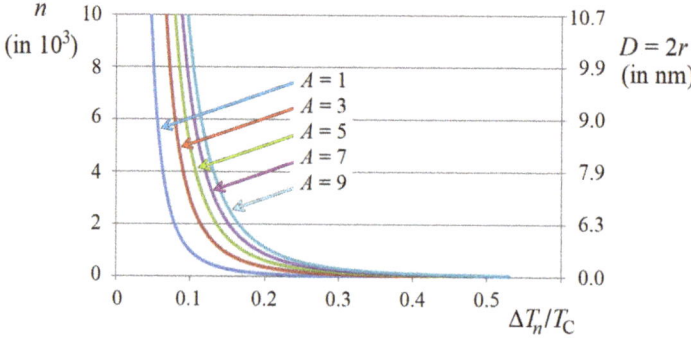

Figure 10. (Color online). Temperature dependence of the PNR size shown by the number of centers n (in 10^3, left scale), and the diameter of its spherical form D (in nm, right scale), as a function of the crystal parameter A (reproduced with permission from [7]. Copyright 2018, American Physical Society).

The most important cubic $(T_n - T_C)^{-3}$ dependence of the PNR size is confirmed experimentally. Subject to the power "-3", the increase in the size of the PNR with decrease in temperature is very fast, especially at smaller increments $\Delta T_n/T_C \approx 0.1$–$0.2$. This explains the formation of the nonergodic "glassy" state of relaxor ferroelectrics [7]. The estimates leading to Equation (14) show also [7] that, albeit small, an initial trigger (fluctuation) is needed to start the growth of a PNR. It clarifies the role of crystal imperfections or structural irregularities (particularly in mixed perovskites) in enhancing the formation of PNR. Another important point is that the formation of the PNR in time is strongly dependent on the speed of cooling [7].

With the PNR included, we can revisit the polarizability of ferroelectric perovskites. Under an external electric field **E**, the PNR turns as a whole, like a dipolar molecule, aligning with **E**, its total dipole moment $\langle p \rangle \approx np_0/\sqrt{6}$. Applying the Debye–Langevin equation to the ensemble of these "molecules," we find the actual PNR-induced polarizability,

$$\alpha = \frac{Nn^2 p_0^2}{18a^3 \varepsilon_0 kT} = \frac{NA^2 p_0^2}{18a^3 \varepsilon_0 kT}\left(\frac{T}{T_C} - 1\right)^{-6} \tag{16}$$

Here, N is the concentration of PNR, close to the number of precursors (imperfections) in the crystal lattice that trigger their formation. We neglect the contribution of the applied external field to N. Notably, as follows from Equation (16), the PNR contribution to the polarizability of perovskite relaxor ferroelectrics manifests non-Curie–Weiss behavior: due to the strong temperature dependence of the size of PNR, ferroelectric perovskites have a much stronger temperature dependence of α, proportional to $(T - T_C)^{-6}$, which may serve as an indication of the PNR contribution (experimental measured polarizabilities contain both PNR and bulk crystal contributions). Note, however, that the employed above Debye–Langevin equation is not applicable for temperatures near the phase transition where the crystal with PNR becomes non-ergodic and glassy-like, and the formation of the latter depends also on the speed of cooling [7]. For BaTiO$_3$, at temperatures above the glassy state, assuming (as above) that $p_0 \approx 7$ D, $a \approx 4$ Å and $T \approx 450$ K, with N of the order of 0.1% and $A \approx 5$ [7], we come to a rather realistic estimate, $\alpha(PNR) \approx 5.6 \times 10^4$ [71–74].

7. Conclusions

The main conclusion that emerges from the outlined above studies of perovskite crystals, notably BaTiO$_3$, is that its spontaneous polarization and all relevant properties are triggered by the local PJTE. This conclusion is in drastic contrast to the long dominated "displacive" theories, which are based on the assumption that the spontaneous polarization of the crystal occurs due to the compensation of the local repulsions (in the dipolar distortions) with the long-range dipole–dipole attractions. The predicted by the PJTE local dipolar distortions in BaTiO$_3$ are multiply confirmed experimentally; they take place in all its phases irrelevant to the phase transitions. The latter occur as a result of the temperature-dependent ordering of the local dipolar displacements. Remarkably, this basic finding in perovskite ferroelectrics affects all their main properties and predicts novel features in interaction with external perturbation, important in applications. The latter includes a qualitatively novel property, the orientational polarization of solids, which (similar to the orientational polarization of polar liquids) is orders of magnitude larger than their displacive polarization. The possible presence of ready-made dipoles in solids which, above "freezing temperature", behave similarly to those in polar liquids was first suggested by P. Debye more than a century ago but revealed only by the vibronic theory. Extension of the PJTE theory to other perovskites and other crystal structures seems to be an up-to-date problem.

Another conclusion from these works is related to the general approaches to the study of solid-state problems. The overwhelming way to explore crystal structures and their properties is to start with revealing the cooperative electronic and nuclear motions characterized by electron and phonon bands formed by equivalent atoms or atomic groups, which are uniformly distributed with translation symmetry. The vibronic (PJTE) theory of ferroelectricity shows (convincingly) that this approach to the problem may be inadequate, losing the main properties of the system. The traditional electron and phonon band approach to solid-state problems fails when the translation symmetry groups are equivalent in average, but not equivalent at each moment in time. This takes place when there are additional local nuclear dynamics induced by the JTE or PJTE, which are uncorrelated (before their cooperative ordering). There are no general theories or computer programs to reveal the band structure of such systems. Our works are the first to reveal this problem and to handle it in the particular case of local PJTE that triggers ferroelectricity and related properties in perovskite crystals.

A short note about our meetings with K. A. Muller and his visits to USSR is available on Supplementary Materials

Supplementary Materials: A short note about our meetings with K. A. Muller and his visits to USSR is available online at http://www.mdpi.com/2410-3896/5/4/68/s1.

Author Contributions: Conceptualization, I.B.B. and V.P.; Data curation, V.P.; Investigation, I.B.B. and V.P.; Methodology, I.B.B. and V.P.; Project administration, I.B.B. All authors have read and agreed to the published version of the manuscript.

Funding: This research received no funding.

Conflicts of Interest: The authors declare no conflict of interest.

References and Note

1. Bersuker, I.B. On the origin of ferroelectricity in perovskite type crystals. *Phys. Lett.* **1966**, *20*, 589–590. [CrossRef]
2. Bersuker, I.B. The vibronic (pseudo Jahn-Teller) theory of ferroelectricity. Novel aspects and applications. *Ferroelectrics* **2018**, *536*, 1–59. [CrossRef]
3. Bersuker, I.B. Pseudo Jahn-Teller Origin of Perovskite Multiferroics, Magnetic-Ferroelectric Crossover, and Magnetoelectric Effects. The d^0–d^{10} problem. *Phys. Rev. Lett.* **2012**, *108*, 137202. [CrossRef] [PubMed]
4. Bersuker, I.B. A Local Approach to Solid State Problems: Pseudo Jahn-Teller origin of Ferroelectricity and Multiferroicity. *J. Phys. Conf. Ser.* **2013**, *428*, 012028. [CrossRef]
5. Polinger, V.Z.; Garcia-Fernandez, P.; Bersuker, I.B. Pseudo Jahn-Teller Origin of Ferroelectric Instability in BaTiO3 Type Perovskites. The Green's Function Approach and Beyond. *Phys. B. Condens. Mater* **2015**, *457*, 296–309. [CrossRef]
6. Polinger, V.Z. Ferroelectric phase transitions in cubic perovskites. *J. Phys. Conf. Ser.* **2013**, *428*, 012026. [CrossRef]
7. Polinger, V.; Bersuker, I.B. Origin of Polar Nanoregions and Relaxor properties of ferroelectrics. *Phys. Rev. B* **2018**, *98*, 214102. [CrossRef]
8. Öpik, U.; Pryce, M.H.L. Studies of the Jahn-Teller effect I. A survey of the static problem. *Proc. R. Soc. London* **1957**, *238*, 425–447. [CrossRef]
9. Bersuker, I.B. The Pseudo Jahn-Teller Effect—A Two-State Paradigm in Formation, Deformation, and Transformation of Molecular Systems and Solids. *Chem. Rev.* **2013**, *113*, 1351–1390. [CrossRef] [PubMed]
10. Bersuker, I.B.; Polinger, V.Z. *Vibronic Interactions in Molecules and Crystals*; Springer: Berlin/Heidelberg, Germany, 1989; ISBN 3-540-19259-X.
11. Bersuker, I.B. *The Jahn-Teller Effect*; Cambridge University Press: Cambridge, UK, 2006; ISBN 978-0-521-82212-1.
12. Bersuker, I.B. Spontaneous Symmetry Breaking in Matter Induced by Degeneracy and Pseudodegeneracy. In *Advances in Chemical Physics*; Rice, S., Dinner, R., Eds.; Wiley: New York, NY, USA, 2016; Volume 160, pp. 159–208, ISBN 978-1-1191-6514-9.
13. Bersuker, I.B. The Jahn-Teller and Pseudo Jahn-Teller Effect in Materials Science. In *Recent Studies in Materials Science*; Lind, P.R., Ed.; Nova Science Publishers: Hauppauge, NY, USA, 2019; pp. 1–95, ISBN 978-1-53615-270-8.
14. Bersuker, I.B. The Jahn-Teller and Pseudo Jahn-Teller Effect in Materials Science. *J. Phys. Conf. Ser.* **2017**, *833*, 012001. [CrossRef]
15. Müller, K.A. Perovskite-Type Oxides—The New Approach to High-T_c Superconductivity. Part 1. In *Nobel Lectures in Physics 1981—1990*; Ekspong, G., Ed.; World Scientific: Singapore, 1993; pp. 424–444, ISBN 978-981-02-0728-1.
16. Simon, A. Superconductivity—A Chemical Phenomenon? *Angew. Chem. Int. Ed. Engl.* **1987**, *26*, 579–580. [CrossRef]
17. Seino, Y.; Kotani, A.; Bianconi, A. Effect of Rhombic Distortion on the Polarized X-Ray Absorption Spectra in High Tc Superconductors. *J. Phys. Soc. Jpn.* **1990**, *59*, 815–818. [CrossRef]
18. Müller, K.A. High-Tc ferroelectrics and superconductors. *Phase Transit.* **1990**, *22*, 5–7. [CrossRef]
19. Bersuker, G.I.; Gorinchoi, N.N.; Polinger, V.Z.; Solonenko, A.O. Jahn-Teller pairing mechanism in high-T_C superconductors. *Supercond. Phys. Chem. Technol.* **1992**, *5*, 1005–1014.
20. Bianconi, A. Linear Arrays of Non-homogeneous Cu Sites in the CuO$_2$ Plane, a New Scenario for Pairing Mechanisms in a Corrugated-Iron-Like Plane. In *Phase Separation in Cuprate Superconductors*; Müller, K.A., Benedek, G., Eds.; World Scientific: Singapore, 1993; pp. 125–138, ISBN 978-981-4553-51-3. [CrossRef]

21. Bianconi, A. On the Fermi liquid coupled with a generalized Wigner polaronic CDW giving high Tc superconductivity. *Solid State Commun.* **1994**, *91*, 1–5. [CrossRef]
22. Zhou, J.S.; Bersuker, G.I.; Goodenough, J.B. Non-adiabatic electron-lattice interactions in the copper-oxide superconductors. *J. Supercond.* **1995**, *8*, 541–544. [CrossRef]
23. Bersuker, G.I.; Goodenough, J.B. Large low-symmetry polarons of the high-T_C copper oxides: Formation, mobility and ordering. *Phys. C* **1997**, *274*, 267–285. [CrossRef]
24. Müller, K.A. Large, small, and especially Jahn-Teller polarons. *J. Supercond.* **1999**, *12*, 3–7. [CrossRef]
25. Lanzara, A.; Zhao, G.M.; Saini, N.L.; Bianconi, A.; Conder, K.; Keller, H.; Müller, K.A. Oxygen-isotope shift of the charge-stripe ordering temperature in $La_{2-x}Sr_xCuO_4$ from x-ray absorption spectroscopy. *J. Phys. Condens. Matter* **1999**, *11*, L541–L546. [CrossRef]
26. Keller, H.; Bussmann-Holder, A.; Müller, K.A. Jahn–Teller physics and high-Tc superconductivity. *Mater. Today* **2008**, *11*, 38–46. [CrossRef]
27. Müller, K.A.; Berlinger, W. Microscopic probing of order-disorder versus displacive behavior in $BaTiO_3$ by Fe^{3+} EPR. *Phys. Rev. B* **1986**, *34*, 6130–6136. [CrossRef]
28. Müller, K.A.; Berlinger, W.; Blazey, K.W.; Albers, J. Electron paramagnetic resonance of Mn^{4+} in $BaTiO_3$. *Solid State Commun.* **1987**, *61*, 21–25. [CrossRef]
29. Volkel, G.; Müller, K.A. Order-disorder phenomena in the low-temperature phase of $BaTiO_3$. *Phys. Rev. B* **2007**, *76*, 094105. [CrossRef]
30. Zalar, B.; Laguta, V.V.; Blinc, R. NMR Evidence for the Coexistence of Order-Disorder and Displacive Components in Barium Titanate. *Phys. Rev. Lett.* **2003**, *90*, 037601. [CrossRef]
31. Zalar, B.; Labar, A.; Seliger, J.; Blinc, R. NMR study of disorder in $BaTiO_3$ and $SrTiO_3$. *Phys. Rev. B* **2005**, *71*, 064107. [CrossRef]
32. Ravel, B.; Stern, E.A.; Vedrinskii, R.I.; Kraisman, V. Local structure and the phase transitions of $BaTiO_3$. *Ferroelectrics* **1998**, *206–207*, 407–430. [CrossRef]
33. Stern, E. Character of order-disorder and displacive components in barium titanate. *Phys. Rev. Lett.* **2004**, *93*, 037601. [CrossRef] [PubMed]
34. Comes, R.; Lambert, M.; Guinner, A. The chain structure of $BaTiO_3$ and $KNbO_3$. *Solid State Commun.* **1968**, *6*, 715–719. [CrossRef]
35. Burns, G.; Dacol, F. The index of refraction of $BaTiO_3$ above T_c. *Ferroelectrics* **1981**, *37*, 661–664. [CrossRef]
36. Gervais, F. Displacive—Order-disorder crossover IFI ferroelectrics. *Ferroelectrics* **1984**, *53*, 91–98. [CrossRef]
37. Dougherty, T.P.; Wiederrecht, G.P.; Nelson, K.A.; Garrett, M.H.; Jensen, H.P.; Warde, C. Femtosecond resolution of soft mode dynamics in structural phase transitions. *Science* **1992**, *258*, 770–774. [CrossRef] [PubMed]
38. Sicron, N.; Ravel, B.; Yacoby, Y.; Stern, E.A.; Dogan, F.; Rehr, J.J. Nature of the ferroelectric phase transition in $PbTiO_3$. *Phys. Rev. B* **1994**, *50*, 13168–13180. [CrossRef] [PubMed]
39. Jeong, I.-K.; Lee, S.; Jeong, S.-Y.; Won, C.J.; Hur, N.; Llobet, A. Structural evolution across the insulator-metal transition in oxygen-deficient $BaTiO_{3-\delta}$ studied using neutron total scattering and Rietveld analysis. *Phys. Rev. B.* **2011**, *84*, 064125. [CrossRef]
40. Comes, R. Letter to the author of Ref. [1]: "*University of Paris, Orsay, January 23, 1969. Dear Dr Bersuker, we just discovered your note "On the Origin of Ferroelectricity in Perovskite type Crystals" published in Physics Letters (1. April 1966). From a completely different approach we came to conclusions, which are similar to yours. You will find enclosed our publications on BaTiO3 and KNbO3 ... *".
41. Gomez, M.; Bowen, S.P.; Krumhansl, J.A. Physical properties of an off-center impurity in the tunneling approximation. I. Statics. *Phys. Rev.* **1967**, *153*, 1009–1024. [CrossRef]
42. Harada, J.; Axe, J.D.; Shirane, G. Neutron-Scattering Study of Soft Modes in Cubic $BaTiO_3$. *Phys. Rev. B* **1971**, *4*, 155–162. [CrossRef]
43. Luspin, Y.; Servoin, J.L.; Gervais, F.J. Soft mode spectroscopy in barium titanate. *J. Phys. C Solid State* **1980**, *13*, 3761–3774. [CrossRef]
44. Vogt, H.; Sanjurjo, J.A.; Rossbroich, G. Soft-mode spectroscopy in cubic $BaTiO_3$ by hyper-Raman scattering. *Phys. Rev. B* **1982**, *26*, 5904–5910. [CrossRef]
45. Bussmann-Holder, A.; Beige, H.; Volkel, G. Precursor effects, broken local symmetry, and coexistence of order-disorder and displacive dynamics in perovskite ferroelectrics. *Phys. Rev. B.* **2009**, *79*, 184111. [CrossRef]

46. Kristoffel, N.N.; Konsin, P.I. Pseudo-Jahn-Teller effect and second-order phase transitions in crystals. *Phys. Status Solidi* **1967**, *21*, K39–K43. [CrossRef]
47. Kristoffel, N.N. Vibronic interaction and ferroelectricity. *Czech. J. Phys.* **1984**, *34*, 1253–1263. [CrossRef]
48. Girshberg, Y.G.; Tamarchenko, V.I. The instability and phase transitions in systems with interband interaction. *Fiz. Tverd. Tela* **1976**, *18*, 1066–1072. (In Russian)
49. Mailyan, G.L.; Plakida, N.M. Fluctuations of order parameter and anharmonic interaction in the vibronic model of ferroelectrics. *Phys. Status Solidi b* **1977**, *80*, 543–547. [CrossRef]
50. Girshberg, Y.G.; Yacoby, Y. Off-centre displacements and ferroelectric phase transition in dilute KTa$_{1-x}$Nb$_x$O$_3$. *J. Phys. Condens. Matter* **2001**, *13*, 8817–8830. [CrossRef]
51. Bersuker, I.B. Pseudo Jahn-Teller effect in the origin of enhanced flexoelectricity. *Appl. Phys. Lett.* **2015**, *106*, 022903. [CrossRef]
52. Bersuker, I.B. Giant permittivity and electrostriction induced by dynamic Jahn-Teller and pseudo Jahn-Teller effects. *Appl. Phys. Lett.* **2015**, *107*, 202904. [CrossRef]
53. Polinger, V.Z.; Bersuker, I.B. Pseudo Jahn-Teller effect in permittivity of ferroelectric perovskites. *J. Phys. Conf. Ser.* **2017**, *833*, 012012. [CrossRef]
54. Debye, P. Eine Resultate einer Kinetischen Theorie der Isolatoren. *Phys. Zeitschr.* **1912**, *13*, 97–100.
55. Feynman, R.P.; Leighton, R.B.; Sands, M. *The Feynman Lectures on Physics*; Pearson: London, UK, 2012; Volume 2, Section 11-3, ISBN 978-9332580954.
56. Kogan, S.M. Piezoelectric effect under inhomogeneous strain and acoustic scattering of electric carriers in crystals. *Sov. Phys. Solid State* **1964**, *5*, 2069–2070.
57. Yudin, P.V.; Tagantsev, A.K. Fundamentals of flexoelectricity in solids. *Nanotechnology* **2013**, *24*, 432001. [CrossRef]
58. Hong, J.; Vanderbilt, D. First-principles theory and calculation of flexoelectricity. *Phys. Rev. B* **2013**, *88*, 174107. [CrossRef]
59. Eliseev, E.A.; Morozovska, A.N. Hidden symmetry of flexoelectric coupling. *Phys. Rev. B* **2018**, *98*, 094108. [CrossRef]
60. Eliseev, E.A.; Morozovska, A.N.; Khist, V.V.; Polinger, V. Effective flexoelectric and flexomagnetic response of ferroics. In *Recent Advances in Topological Ferroics and Their Dynamics*; Stamps, R.L., Schultheiss, H., Eds.; Elsevier: London, UK, 2019; pp. 238–289, ISBN 978-0-08-102920-6.
61. Hill, N.A. Why are there so few magnetic ferroelectrics? *J. Phys. Chem.* **2000**, *104*, 6694–6709. [CrossRef]
62. Spaldin, N.A.; Fiebig, M. The Renaissance of Magnetoelectric Multiferroics. *Science* **2005**, *309*, 391–392. [CrossRef] [PubMed]
63. Barone, P.; Picozzi, S. Mechanisms and origin of multiferroicity. *Comptes Rendus Phys.* **2015**, *16*, 143–152. [CrossRef]
64. Gütlich, P.; Goodwin, H.A. (Eds.) *Spin Crossover in Transition Metal Compounds*; Springer: Berlin/Heidelberg, Germany, 2004; Volume III, ISBN 978-3-540-40395-1.
65. Hauser, A.; Enachescu, M.L.; Daku, D.; Vargas, A.; Amstutz, N. Low-temperature lifetimes of metastable high-spin states in spin-crossover and in low-spin iron(II) compounds: The rule and exceptions to the rule. *Coord. Chem. Rev.* **2006**, *250*, 1642–1652. [CrossRef]
66. Renz, F.; Oshio, H.; Ksenofontov, V.; Waldeck, M.; Spiering, H.; Gütlich, P. Strong field iron (II) complex converted by light into a long-lived high-spin state. *Angew. Chem. Int. Ed.* **2000**, *39*, 3699–3700. [CrossRef]
67. Domracheva, N.E.; Pyataev, A.V.; Vorobeva, V.E.; Zueva, E.M. Detailed EPR study of spin-crossover dendrimeric iron(III) complex. *J. Phys. Chem. B* **2013**, *117*, 7833–7842. [CrossRef]
68. Raymond, O.; Ostos, C.; Font, R.; Curiel, M.; Bueno-Baques, D.; Machorro, R.; Mestres, L.; Portelles, J.; Siqueiros, J.M. Multiferroic properties and magnetoelectric coupling in highly textured Pb(Fe$_{0.5}$Nb$_{0.5}$)O$_3$ thin films obtained by RF sputtering. *Acta Mater.* **2014**, *66*, 184–191. [CrossRef]
69. Weston, L.; Cui, X.Y.; Ringer, S.P.; Stampfl, C. Multiferroic crossover in perovskite oxides. *Phys. Rev. B* **2016**, *93*, 165210. [CrossRef]
70. Bokov, A.A.; Ye, Z.-G. Recent progress in relaxor ferroelectrics with perovskite structure. *J. Mater. Sci.* **2006**, *41*, 31–52. [CrossRef]
71. Bokov, A.A.; Ye, Z.-G. Dielectric relaxation in relaxor ferroelectrics. *J. Adv. Dielectr.* **2012**, *02*, 1241010. [CrossRef]

72. Wang, D.; Bokov, A.A.; Ye, Z.-G.; Hlinka, J.; Bellaiche, L. Subterahertz dielectric relaxation in lead-free Ba(Zr,Ti)O3 relaxor ferroelectrics. *Nat. Commun.* **2016**, *7*, 11014. [CrossRef] [PubMed]
73. Helal, M.A.; Aftabuzzaman, M.; Tsukada, S.; Kojima, S. Role of polar nanoregions with weak random fields in Pb-based perovskite ferroelectrics. *Sci. Rep.* **2016**, *7*, 44448. [CrossRef] [PubMed]
74. Burns, G.; Dacol, F.H. Glassy polarization behavior in ferroelectric compounds Pb(Mg$_{13}$Nb$_{23}$)O$_3$ and Pb(Zn$_{13}$Nb$_{23}$)O$_3$. *Solid State Commun.* **1983**, *48*, 853–856. [CrossRef]

Publisher's Note: MDPI stays neutral with regard to jurisdictional claims in published maps and institutional affiliations.

© 2020 by the authors. Licensee MDPI, Basel, Switzerland. This article is an open access article distributed under the terms and conditions of the Creative Commons Attribution (CC BY) license (http://creativecommons.org/licenses/by/4.0/).

Article

Phase Separation and Pairing Fluctuations in Oxide Materials

Andreas Bill [1], Vladimir Hizhnyakov [2], Reinhard K. Kremer [3], Götz Seibold [4,*], Aleksander Shelkan [2] and Alexei Sherman [2]

[1] Department of Physics & Astronomy, California State University Long Beach, Long Beach, CA 90840, USA; abill@csulb.edu
[2] Institute of Physics, University of Tartu, 1 W. Ostwaldi Street, 50411 Tartu, Estonia; hizh@ut.ee (V.H.); aleksander.shelkan@ut.ee (A.S.); alekseis@ut.ee (A.S.)
[3] MPI for Solid State Research Heisenbergstraße 1, 70569 Stuttgart, Germany; rekre@fkf.mpg.de
[4] Institut für Physik, BTU Cottbus, P.O. Box 101344, 03013 Cottbus, Germany
* Correspondence: seibold@b-tu.de; Tel.: +49-355-693006

Received: 12 September 2020; Accepted: 13 October 2020; Published: 19 October 2020

Abstract: The microscopic mechanism of charge instabilities and the formation of inhomogeneous states in systems with strong electron correlations is investigated. We demonstrate that within a strong coupling expansion the single-band Hubbard model shows an instability towards phase separation and extend the approach also for an analysis of phase separation in the Hubbard-Kanamori hamiltonian as a prototypical multiband model. We study the pairing fluctuations on top of an inhomogeneous stripe state where superconducting correlations in the extended s-wave and d-wave channels correspond to (anti)bound states in the two-particle spectra. Whereas extended s-wave fluctuations are relevant on the scale of the local interaction parameter U, we find that d-wave fluctuations are pronounced in the energy range of the active subband which crosses the Fermi level. As a result, low energy spin and charge fluctuations can transfer the d-wave correlations from the bound states to the low energy quasiparticle bands. Our investigations therefore help to understand the coexistence of stripe correlations and d-wave superconductivity in cuprates.

Keywords: phase separation; cuprate superconductors; electronic correlations

1. Introduction

Already in their groundbreaking paper on 'Possible High T_c Superconductivity in the Ba-La-Cu-O System' [1] Bednorz and Müller discussed the possibility of 'superconductivity of percolative nature' to explain their observation. It may be that chemical inhomogeneity was in their immediate line of sight but they also discussed granularity and 2D fluctuations associated with the superconducting wave function [1]. The discovery that high-temperature superconductivity results from hole doping of a 2D antiferromagnet stimulated Sigmund and his group at the University of Stuttgart in close collaboration with Hizhnyakov from the University of Tartu to study the problem of how doped charge carriers behave in a 2D magnetic insulating lattice. According to their initial ideas, doped charge carriers are stabilized in the dilute limit as localized magnetic polarons in a 2D fluctuating antiferromagnetic environment. On increasing doping concentration, such polarons condense to form larger clusters ('droplets') and above a critical concentration a percolating phase is formed, which then becomes superconducting [2–4]. This scenario got early support (see Figure 1) from experiments on lanthanum cuprate phases which showed that an antiferromagnetic and a superconducting phase can exist simultaneously and their ratio can favorably be modified by thermal quenching experiments [5,6]. In particular, the comparison of field and zero field cooled magnetization curves of $La_2CuO_{4+\delta}$ and $La_{2-x}Sr_xCuO_4$ demonstrated that it is rather the electronic component (i.e., magnetic polarons) which is affected by the thermal treatment.

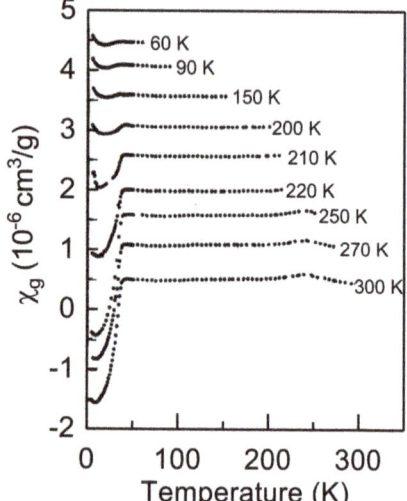

Figure 1. Gram-susceptibility of an 'as-prepared' $La_2CuO_{4+\delta}$ ($\delta \sim 0.01$) sample as a function of temperature. The sample was rapidly quenched from room temperature to the indicated temperatures and subsequently the magnetization ($B_{ext} \sim 9$ mT) was collected by slowly cooling the sample. Beginning from the lowest data set each curve was shifted upwards by a value of 5×10^{-7} cm^3/g compared to the preceding one. (Adapted from Figure 1a, Ref. [5,6] by permission from Springer/Nature/Palgrave).

X-ray scattering experiments on analogously quenched La_2CuO_{4+y} single crystals show that ordering of the oxygen interstitials in the layers of La_2CuO_{4+y} is characterized by a fractal distribution of dopants up to a maximum limiting size of 400 µm which appears with the dopants enhancing superconductivity to high temperatures [7]. Evidence for charge segregation on a local scale came first from NMR [8] and NQR [9–11] investigations (cf. also Baranov and Badalyan as well as Hammel et al. in Refs. [12,13]). Independently, Emery and Kivelson emphasized that 'clumping' of the holes is an important feature of cuprate superconductors' [14]. Since this early experimental and theoretical evidence numerous experimental and theoretical accounts have appeared, discussing the importance of electronic inhomogeneity ('electronic phase separation') for high-T_c superconductivity.

Instead of analyzing the formation of electronic inhomogeneities from the low doping side, an alternative theoretical approach is to investigate the phase separation instability of a correlated metal from the overdoped side, eventually supplemented with an electron-phonon interaction (see for example Refs. [15–20]). In this context, it was proposed [14,21,22] that the inclusion of long-range Coulomb interactions is a crucial ingredient since they suppress long-wavelength charge density fluctuations associated with phase separation favoring shorter-wavelength density fluctuations, giving rise either to dynamical slow density modes [14] or to incommensurate charge density waves [22].

Such incommensurate structures have been observed in $La_{1.48}Nd_{0.4}Sr_{0.12}CuO_4$ by Tranquada and collaborators who detected a splitting of both spin and charge order peaks by elastic neutron scattering experiments [23]. Their finding suggested that the doped holes arrange themselves in quasi-one-dimensional aggregates, 'stripes', which simultaneously constitute antiphase domain walls for the antiferromagnetic order. While the neutron scattering experiments only provide indirect evidence for charge ordering via the coupling to the lattice, bulk evidence for charge stripe order in the lanthanum cuprates has been found in $La_{1.875}Ba_{0.125}CuO_4$ and $La_{1.8-x}Eu_{0.2}Sr_xCuO_4$ by resonant X-ray scattering (RXS) experiments [24,25]. The rapid improvement and development of this technique has meanwhile led to the detection of charge order in a large variety of cuprate compounds, including YBCO, [26,27] Bi2212, [28] and Bi2011 [29]. Moreover, charge order was also measured in YBCO by high-energy X-ray diffraction [30] and quantum oscillations in both transport and thermodynamic experiments in magnetic fields [31–33] sufficient to suppress superconducting long-range order.

Whereas there appears to exist consensus on the formation of stripes in high-T_c materials its relation to the mechanism of superconductivity is controversial. In fact, long before the discovery of high-T_c Balseiro and Falicov [34] have shown that static charge-density waves (CDW) and superconductivity mutually suppress each other. Moreover, one-dimensional electronic correlations do not seem to be compatible with two-dimensional superconductivity in the high-T_c compounds. On the other hand STM and ARPES experiments on LaBaCuO [35] suggest the existence of a d-wave gap below the stripe ordering temperature which is most pronounced for $\delta = 1/8$, when T_c tends to zero. A subsequent study of the same compound presented evidence from the temperature dependence of the in-plane resistivity that this d-wave gap originates from superconducting fluctuations above a Kosterlitz-Thouless transition [36]. The authors conclude that the static stripe order is therefore fully compatible with two-dimensional superconducting fluctuations.

The essential role of electronic heterogeneities for superconductivity in hole-doped cuprates and the coexistence of multiple electronic components has been frequently pointed out by Alex Müller [37–39] in particular related to the formation and ordering of (bi)polarons [40,41]. For the particular case of stripes or CDW's there have been several attempts to link them to the pairing mechanism in high-T_c superconductors. In a series of papers the Bianconi group has investigated pairing in a superlattice of quantum stripes where they found an amplification of superconductivity when the chemical potential is tuned towards a so-called shape resonance [42–47] and the multiband electronic structure can also induce an anomalous isotope effect [48]. In fact, formation of a CDW with the concomitant multiband structure can significantly enhance the intraband pairing scattering while suppressing the interband pairing [49,50]. However, inclusion of local Coulomb correlations has a strong impact on the renormalization of the electron-phonon vertex so that the interplay with CDW scattering can lead to both an enhancement or suppression of the pairing interaction [50]. Also the choice of the cutoff in the pairing interaction ('original' electrons vs. quasiparticles) plays a role in this regard.

Emery and coworkers have proposed a pairing mechanism [51] where holes on a charge stripe acquire a spin gap via pair hopping into the adjacent Mott insulating environment. Long-range superconducting phase coherence is then generated by Josephson coupling between the stripes. An alternative scenario has been put forward by Castellani et al. [22] It relies on the existence of a quantum critical point (QCP) near optimal doping. The QCP separates a homogeneous Fermi-liquid (in the overdoped regime) from a symmetry-broken ground state on the underdoped side of the phase diagram. The low doping phase was associated with incommensurate charge-density waves (ICDW). However, more exotic phases have also been proposed in this context. The singular fluctuations in the particle-hole channel generated in the vicinity of the QCP are reflected as divergent pairing correlations in the particle-particle channel. As has been shown in Ref. [52] an ICDW-QCP is compatible with a d-wave superconducting order parameter. More recently it was proposed [53,54] that superconductivity in the striped state occurs at a non-zero wave vector ('pair density wave') which results in the suppression of the inter-layer Josephson coupling and thus a dimensional reduction in agreement with transport measurements on $La_{1.875}Ba_{0.125}CuO_4$. [36]

The aim of the present paper is twofold. First, we review in Section 2 the phase separation mechanism due to the formation and attraction of spin polarons. Section 3 is devoted to the problem how a phase separation instability in the Hubbard model can be realized without the additional involvement of phonons. Phonons (or other bosonic degrees of freedom) rather support the energy equilibration between the two phases which allows the phase separated state to be realized as a thermal state. Moreover, we show that the same mechanism can also be invoked to understand phase separation in multiband models including Hund exchange which is relevant for other oxide materials as for example manganites (cf. Ref. [55]). Furtheron, we show in Section 4 how isotropic superconducting correlations can be realized on top of an inhomogeneous electronic ground state. For this purpose we first review the pairing mechanism due to long-range optical phonon modes as proposed by Hizhnyakov and Sigmund [56–58]. We then exemplify the isotropy of superconducting correlations for a striped system where it turns out that for both d- and extended s-wave symmetry the corresponding vertex contribution has only a marginal orientational dependence.

2. Phase Separation in the Mean-Field Approximation

In the case of a homogeneous lattice, one of the sources of inhomogeneous charge distribution and lattice distortions (stripes) may be strong electron correlations. It was shown [59,60] that this phenomenon already takes place in the mean-field (Hartree-Fock) approximation of the three band Hubbard model.

In Refs. [59,60] we studied hole states in the antiferromagnetically (AF) ordered CuO_2 planes of cuprate perovskites with a self-consistent calculation of the Cu spin polarization. Both the Cu-O hybridization and the O-O transfer are taken into account. We used the following Hamiltonian for charge carriers (holes) in the CuO_2 plane, which follows from the original Hubbard Hamiltonian in the Hartree-Fock (HF) approximation:

$$H = \sum_\sigma H^\sigma_{MF} - U \sum_m \langle n^d_{m\uparrow}\rangle \langle n^d_{m\downarrow}\rangle, \qquad (1)$$

where

$$H^\sigma_{MF} = \sum_n \left[\epsilon_d + U\langle n^d_{n-\sigma}\rangle\right] n^d_{n\sigma} + \epsilon_p \sum_m n^p_{m\sigma}$$
$$+ T \sum_{nm} (d^+_{n\sigma} p_{m\sigma} + h.c.) + t \sum_{mm'} \left(p^+_{m\sigma} p_{m'\sigma} + h.c\right), \qquad (2)$$

d (d^+) and p (p^+) are electronic annihilation (creation) operators on Cu and O orbitals, $U \approx 8$ eV, $T \approx 1$ eV, $t \approx 0.3$ eV, $\epsilon = \epsilon_p - \epsilon_d \approx 3$ eV. In the AF ordered CuO_2 plane the elementary cell is doubled (the magnetic unit cell contains two CuO_2 units). The copper on-site energies are given by $\epsilon_{1\sigma} = \epsilon_d + U\langle n^d_{1-\sigma}\rangle$, $\epsilon_{2\sigma} = \epsilon_d + U\langle n^d_{2-\sigma}\rangle$.

In what follows we are interested in the behavior of a large-size wave packet of extra holes added to the AF ground state. The Hamiltonian (1) does not take into account the Coulomb repulsion of these holes, assuming that it is compensated by attraction with sufficiently mobile doping ions.

In the AF-ordered state Cu_2O_4–elementary cells form a simple square lattice with the lattice constant $a' = a\sqrt{2}$ and with main directions along $x' = (x+y)/\sqrt{2}$ and $y' = (x-y)/\sqrt{2}$. Therefore, it is convenient to use the site vectors $\vec{m}' = (m_{x'}, m_{y'})$ which count the elementary cells in the x' and y' directions $(m_{x'}, m_{y'} = 0, \pm 1, \pm 2, ...;$ $\vec{m}'$ corresponds to the cell with coordinates $x' = a'm_{x'}, y' = a'm_{y'})$. Within this choice the second hole band (empty in the undoped case) has 4 minima at the points $(\pm \pi/a', 0)$

and $(0, \pm\pi/a')$ in the Brillouin zone. The wave functions of the minima contain only negligibly small ($< 10^{-6}/\sqrt{N}$) amplitudes of the first states $|d_1\rangle_{\vec{m}'}$, corresponding to the Cu with the opposite spin; neglecting these contributions, the wave function can be presented in the form [for the minimum at $\vec{k}' = (\pi/a', 0)$]:

$$|\psi_{min}\rangle = \frac{1}{\sqrt{N}} \sum_{\vec{m}'} |\psi\rangle_{\vec{m}'}, \tag{3}$$

where N is the number of elementary cells,

$$|\psi\rangle_{\vec{m}'} = (-1)^{m_{x'}} (\sin\alpha |d_2\rangle_{\vec{m}'} + \cos\alpha |P_1\rangle_{\vec{m}'}), \tag{4}$$

$$|P_1\rangle_{\vec{m}'} = \frac{1}{2}(|p_1\rangle_{\vec{m}'} - |p_2\rangle_{\vec{m}'} + i|p_3\rangle_{\vec{m}'} - i|p_4\rangle_{\vec{m}'}), \tag{5}$$

and $\sin\alpha \approx 0.39$ (for $U = 8T$, $t = 0.3T$, $\epsilon = 3T$); $|p_n\rangle_{\vec{m}'}$ denote the states of the 4 oxygens surrounding the second Cu ion in the $\vec{m}'$-th elementary cell, counted counterclockwise starting from the right position.

We construct the wave packet from the states close to $(\pi/a', 0)$, the minimum of the hole band. This wave-packet can be presented in the form

$$|\psi_L\rangle = \sum_{\vec{m}'} c_{\vec{m}'} a^+_{\vec{m}'} |0\rangle,$$

where $|0\rangle$ is the state with a filled lower Hubbard band, $a^+_{\vec{m}'}$ is a creation operator of the hole state $|\psi\rangle_{\vec{m}'}$, $c_{\vec{m}'}$ is the corresponding probability amplitude. We choose $c_{\vec{m}'}$ in the exponential form:

$$c_{\vec{m}'} = A_L \exp[-2\left(|m_{x'}| + |m_{y'}|\right) a'/L + i\pi m_{x'}], \tag{6}$$

where $A_L = \tanh(2a'/L)$; the oscillating multiplier $\exp(i\pi m_{x'} a'/L)$ accounts for the wave vector $\vec{k}' = (\pi/a', 0)$, of the $(\pi/2a, \pi/2a)$ minimum of the hole-band. This shape of the wave-packet is close to that of the soliton-type ($\sim \text{sech}(x/L)$) packet of the minimal energy for the given size $L = (\int |\psi|^4 dx)^{-1/2}$.

The expectation values of polarization are obtained from the self-consistent equations

$$\langle n^d_\sigma \rangle = \sum_k |\phi_{ik}|^2, \tag{7}$$

where ϕ_{ik} is the eigenvector of the Hamiltonian matrix, corresponding to the eigenvalues E_k. The second band (empty in the undoped case) has 4 minima at the $(\pm\pi/2a, \pm\pi/2a)$ points in the Brillouin zone.

Our solutions of the self-consistent Equation (7) show that for a single hole the lowest energy solution corresponds to a spin-polaron of small size [61,62]. The free hole state is about 0.15 eV higher in energy. We also found that in order to obtain such spin-polaron state from the state of a free hole it is necessary to overcome an energy barrier of about 0.05 eV before the formation of the polaron can occur [62]. However, at finite doping the spin-polaron states become less favorable and at a critical concentration $c \sim 0.5$ they turn out to be metastable.

We also have observed that already at small hole concentrations their spatial distribution changes from homogeneous to a domain type. Such behavior of holes is expected from general considerations. Indeed, in a two-dimensional lattice the self-energy of a large size ($\sim L$) hole wave packet caused by the interaction with the surrounding Cu spins is $\propto |\psi|^4$. It depends on L as $(-L^{-2})$, i.e., in the same way as its kinetic energy, but with different sign [63]. Therefore, in case of a high effective hole mass the attractive self-action dominates, thus leading to the formation of domains. According to our calculations the local hole concentration in domains is ~ 0.5–0.6. This result is demonstrated in Figure 2, where, for a system containing hole-enriched stripes, the dependence of the total energy on the concentration of holes in the stripes is shown for the case of a total hole concentration $c = 0.05$. The free energy of the domain only weakly depends on its shape. Consequently, the formation of stripe domains already takes place in the HF approximation. The optimum hole concentration in domains obtained via this approximation is close to the observed value of $c = 0.5$.

Similar results on phase separation were obtained using slave-boson, slave-fermion and large-N expansions [16,64–67]. Within the tJ-model it has been shown [68] that phase separation supersedes superconducting instabilities for large enough exchange coupling. Mechanisms of phase separation in solids different from cuprates (e.g., manganites) were considered in Refs. [55,69–71].

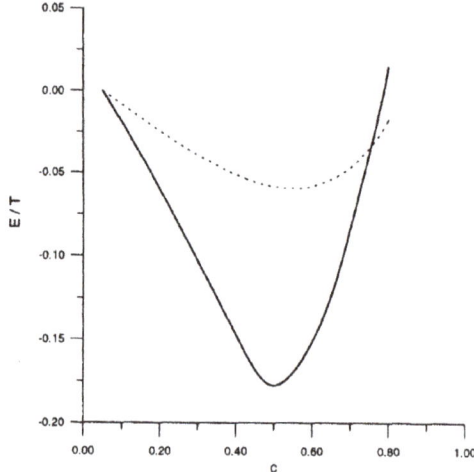

Figure 2. Full energy of the crystal E vs hole concentration c in the hole-rich (stripe) region. Initial (mean) hole concentration is 0.05. The dotted line corresponds to the rigid AF lattice.

3. Phase Separation and Fluctuations

In systems with strong electron correlations, phase separation takes place with the inclusion of charge and spin fluctuations. This result was recently demonstrated in the one-band repulsive Hubbard model on a two-dimensional square lattice [72]. It was shown that at low temperatures, regions of negative electron compressibility (NEC), $\kappa = x^{-2}(dx/d\mu) < 0$, arise near certain values of the chemical potential μ. Here x is the electron concentration. The source of this unusual behavior of κ is the crossing of the energy levels in the Hubbard atom at these μ. A power series expansion around the atomic limit is the natural investigation method in the case of strong correlations. This approach is called the strong coupling diagram technique (SCDT) [73–77]. The convergence of the series expansion is confirmed by the summation of infinite sequences of diagrams. Its validity follows from the successful comparison of its results with data from Monte Carlo simulations, exact diagonalization and experiments with ultracold atoms in optical lattices [76,77].

The Hamiltonian of the Hubbard atom reads

$$H_1 = \sum_\sigma [(U/2)n_{1\sigma}n_{1,-\sigma} - \mu n_{1\sigma}], \tag{8}$$

where U is the on-site repulsion, $n_{l\sigma} = a_{l\sigma}^\dagger a_{l\sigma}$ is the occupation-number operator on the lattice site l with the spin projection $\sigma = \pm 1$, $a_{l\sigma}^\dagger$ and $a_{l\sigma}$ are electron creation and annihilation operators. As seen from Equation (8), the Hamiltonian has four eigenvectors $|\lambda\rangle$: the empty state $|0\rangle$ with the eigenenergy $E_0 = 0$, two singly occupied degenerate states $|\sigma\rangle$ with the energy $E_1 = -\mu$, and the doubly occupied state $|2\rangle$ with the energy $E_2 = U - 2\mu$. As follows from the energy expressions, with the change of μ, these states become alternately the ground states of the atom: for $\mu < 0$ it is $|0\rangle$, for $0 < \mu < U$ the degenerate singly occupied states are the lowest ones, and for $U < \mu$ the ground state is $|2\rangle$.

The terms of the SCDT series expansion for Green's functions are products of hopping constants and on-site cumulants [78] of electron creation and annihilation operators. In particular, the first-order cumulant $C^{(1)}(\tau) = \langle \mathcal{T} a_{l\sigma}^\dagger a_{l\sigma}(\tau) \rangle_0$, i.e., the first term of the expansion for the one-particle Green's function, after Fourier transformation reads

$$C^{(1)}(j) = \frac{1}{Z} \sum_{\lambda\lambda'} \frac{e^{-\beta E_\lambda} + e^{-\beta E_{\lambda'}}}{i\omega_j + E_\lambda - E_{\lambda'}} \langle \lambda | a_{l\sigma} | \lambda' \rangle \langle \lambda' | a_{l\sigma}^\dagger | \lambda \rangle, \tag{9}$$

where $\mathcal{T}$ is the time ordering operator, the subscript 0 of the averaging brackets indicates that the averaging and time dependence are determined by Hamiltonian (8), the partition function $Z = \sum_\lambda \exp(-\beta E_\lambda)$ with $\beta = 1/T$ the inverse temperature, and j is the integer defining the Matsubara frequency $\omega_j = (2j-1)\pi T$. At low temperatures, due to the Boltzmann factors $e^{-\beta E_\lambda}$ in Equation (9), the cumulant changes drastically as μ goes from a region with one of the mentioned ground states to another one. Namely, for $\mu \ll -T$, $C^{(1)}(j) \approx 1/(i\omega_j + \mu)$, while for $T \ll \mu$ and $T \ll U - \mu$, $C^{(1)}(j) \approx (1/2)[1/(i\omega_j + \mu) + 1/(i\omega_j + \mu - U)]$. In the third region, $T \ll \mu - U$, $C^{(1)}(j) \approx 1/(i\omega_j + \mu - U)$. Similar sharp changes occur in other cumulants. Since they enter into irreducible diagrams composing the irreducible part $K(\mathbf{k}, j)$, which defines the one-particle Green's function, [76]

$$G(\mathbf{k}, j) = \left\{ [K(\mathbf{k}, j)]^{-1} - t_\mathbf{k} \right\}^{-1}, \tag{10}$$

sharp changes occur in spectral functions, densities of states, and band dispersions. Here $\mathbf{k}$ is the wave vector and $t_\mathbf{k}$ the Fourier transform of hopping constants. The drastic variation of electron bands near $\mu \approx 0$ and $\mu \approx U$ can be characterized as their pronounced non-rigidity—a strong dependence of the electron dispersion on the chemical potential/electron concentration. This non-rigidity is the origin of the NEC observed near these values of the chemical potential.

Figure 3 exhibits a cartoon image of one of the NEC regions. In the one-band Hubbard model, the topmost point of this dependence may be close to $x = 1$ at low temperatures [72]. Let us suppose that the crystal is divided into two parts with the electron concentration x_1 and chemical potential μ_1 in one of them, and $x_2 < x_1$ and $\mu_2 > \mu_1$ in another. Representative points of these two parts are shown in Figure 3. Both parts are considered to be macroscopic crystals. Hence the dependence $x(\mu)$ is described by the curve in this figure. Let us suppose that we transfer an electron from part 2, with a smaller concentration, to part 1, with a larger concentration. Therefore, the concentration difference between the two parts is further increased. Such a transfer is energetically favorable, since $\mu_2 > \mu_1$. Therefore, if there is a subsystem in contact with the electron subsystem, which can absorb the energy $\mu_2 - \mu_1$, such an electron separation will proceed spontaneously until the concentration and chemical potential in part 1 reach the topmost point in the curve in Figure 3, while part 2 attains the lowermost point. The character of the curve prohibits further separation.

In crystals, such an energy-absorbing subsystem is provided by phonons. Let us consider the simplest model for vibrations – local lattice distortions q_l linearly interacting with the electron density. This interaction is described by the term $H_i = -v \sum_{l\sigma} q_l n_{l\sigma}$. We use the adiabatic approximation. The two parts of the crystal are large enough, which allows us to suppose that at the bottom of the adiabatic potential all local distortions are equal, $q_1 = q$. In this case, vq in H_i becomes a correction to the electron chemical potential, and the adiabatic potential is easily calculated. Its minimization gives $q = -vx/u$ with u the elastic stiffness constant. Thus, the distortion in the electron-rich part is larger than in the electron-poor one. We arrive at a state in which both electronic components and distortions are inhomogeneous. If there is a predominant interaction of electrons with distortions of special symmetry, especially connected with a softening phonon mode, one can expect the formation of stripes.

Hence in our approach with a fixed chemical potential, the assistance of phonons is needed for phase separation. Ground states featuring phase separation were found in many works using different optimization procedures (see, e.g., Refs. [79–81]). These procedures do not fix the energy of a solution, but rather minimize it. Therefore, in such works, phase-separated ground states are obtained in purely electronic systems, without the involvement of phonons.

The charge instability in the form of the NEC is observed in other models of strongly correlated systems as well. In particular, we found such instability in the Hubbard-Kanamori (HK) model described by the Hamiltonian [82,83]

$$\begin{aligned} H = & -t \sum_{\langle ll' \rangle i\sigma} a^\dagger_{l'i\sigma} a_{li\sigma} + \sum_{li\sigma} \left[-\mu n_{li\sigma} \right. \\ & + \frac{U}{2} n_{li\sigma} n_{li,-\sigma} + \frac{U-2J}{2} n_{li\sigma} n_{l,-i,-\sigma} \\ & + \frac{U-3J}{2} n_{li\sigma} n_{l,-i,\sigma} \\ & + \frac{J}{2} \left(a^\dagger_{li\sigma} a^\dagger_{li,-\sigma} a_{l,-i,-\sigma} a_{l,-i,\sigma} \right. \\ & \left. \left. - a^\dagger_{li\sigma} a_{li,-\sigma} a^\dagger_{l,-i,-\sigma} a_{l,-i,\sigma} \right) \right], \end{aligned} \qquad (11)$$

where the subscript $i = \pm 1$ labels two degenerate site orbitals and J is the Hund coupling. Hamiltonians of this type are used for the description of transition metal oxides, iron pnictides, and chalcogenides. In contrast to the simpler one-band Hubbard model with only two NEC regions, in the HK model with two orbitals, there are four such regions. Two of them at $\mu = 0$ and at $\mu = U - 3J = 1.5t$ are seen in Figure 4. This dependence $x(\mu)$ was also obtained using the SCDT. The mechanism of the appearance of these NEC regions is the same, as in the one-band model, i.e., the level crossing in the on-site terms of the Hamiltonian (11) leads to sharp changes of electron bands (for more details see Ref. [84]). Indeed, phase separation is inherent in crystals, for the description of which the HK Hamiltonian is applied [85–87].

Hence the appearance of NEC regions corresponds to the common mechanism of phase separation in crystals with strong electron correlations.

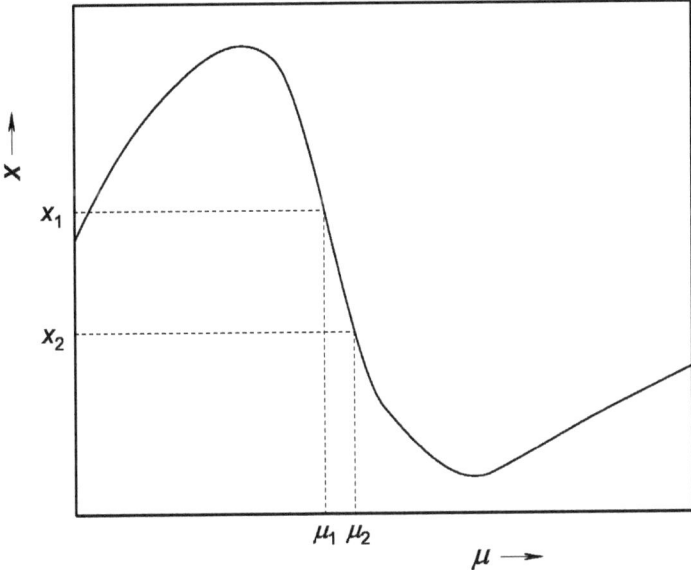

Figure 3. The dependence $x(\mu)$ near one of the NEC regions.

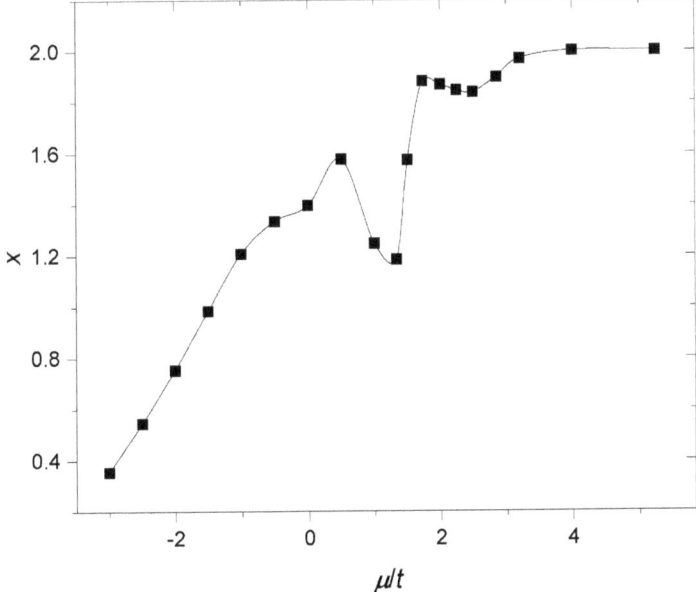

Figure 4. The dependence $x(\mu)$ in the Hubbard-Kanamori model with two orbitals. $U = 6t$, $J = 1.5t$, and $T = 0.13t$.

4. Pair Correlations In Cuprates

The presence of electronic inhomogeneities resulting from competing interactions at the microscopic level not only affects the normal state properties of strongly correlated materials. They also play a major role for the existence of the superconducting state. There have been many attempts at determining the pairing mechanism but no consensus has been reached in the community. As stated in the introduction, Sigmund and Hizhnyakov pointed out early on the importance of taking into account electronic inhomogeneities in the description of the superconducting state [2–4]. However, contrary to many they did not consider strong correlations or electronic inhomogeneities as the driving mechanism of superconductivity *per se*, but rather as modifying the conventional picture of superconducting pairing in ways that make the superconducting state of these materials unique.

At the time when Alex Müller visited Sigmund and Hizhnyakov in Stuttgart, he was very interested in the pairing mechanism proposed by the group and resulting from the transitive electron-phonon interaction in one-dimensional, percolative, stripe-like inhomogeneities [2–4]. With time, Alex moved his research in the direction of more local pair correlations, favoring the formation of bipolarons. He also viewed the symmetry of the order parameter as resulting from the coexistence of two condensates with s- and d-wave symmetry [88]. On the other hand, the authors developed a theory that involves electronic inhomogeneities and long range interactions in the formation of Cooper pairs [56–58]. As explained below, that model leads to a single condensate with anisotropic s-wave symmetry that changes into d-wave symmetry when accounting for the local Coulomb repulsion [56–58,89].

In the following we consider two aspects of superconducting pairing in a strongly correlated system. In the first, electronic inhomogeneities cause the electron-phonon coupling to have an essential long-range component that sustains a superconducting state while the balancing between the resulting effective attractive interaction with the local Coulomb repulsion determines the symmetry of the order parameter. In the second, the superconducting pair fluctuations are considered on top of a symmetry-broken ground state of stripes. The phase separated configuration and underlying interactions lead to pair correlations that appear very isotropic despite the anisotropic inhomogeneous electronic state. Common to both aspects of pair fluctuations is that they occur at low momentum transfer **q** but result from local electronic correlations which determine the symmetry of the superconducting order parameter.

4.1. Pairing from Long Range Electron-Phonon Interaction

Electronic inhomogeneities are the expression of static or dynamic phase separation into hole rich and hole poor regions. An essential aspect of these considerations is the time scale associated with different components of the phase separated state. It is less the absolute time scale that matters: the electronic inhomogeneities need not be static to result in the pairing mechanism discussed here. They may fluctuate but with a time scale much larger than the motion of quasiparticles within the metallic component of the inhomogeneous state. Furthermore, the inhomogeneities are on a microscopic scale and only in specific situations become more macroscopic as for example the stripe phase at $x = 1/8$ in LBCO.

The particular model used to describes the inhomogeneous electronic structure is not crucial for the considerations of this subsection. The important aspect is the reduction of screening effects. This reduction is reinforced by the layered nature and anisotropic transport properties of high temperature superconductors. As a result, there is poor screening along the c-axis, and the screening of the Coulomb interaction has an important dynamic contribution. Unlike conventional three-dimensional metals, low energy electronic collective modes appear in correlated layered materials [90–93]. These modes are acoustic ($\omega \sim q$), although they may display a small gap at $q \to 0$ that depends on the interlayer hopping parameter. The gap, however, does not affect the physics of low energy collective modes in an essential way [91–93]. Calculations demonstrate that low-energy plasmons contribute constructively to the pairing mechanism in a variety of novel superconductors [91–93]. Only very recently have low-energy plasmons been observed experimentally using resonant inelastic X-ray scattering (RIXS) in electron doped cuprates [94,95].

The unusual plasmon spectrum resulting from electronic inhomogeneities, the layered structure and anistropic transport, have another implication that is essential for understanding the pairing mechanism in high temperature superconductors. The attractive electron-ion interaction screening is reduced in the hole poor region as compared to the hole rich phase. Hence, one expects two components to the electron phonon interaction, $H_{eph} = H_{eph,S} + H_{eph,L}$. The usual short-range, screened interaction $H_{eph,S}$ is dominant in most metals. The long-range, weakly screened interaction $H_{eph,L}$ is reminiscent of what occurs in a polar crystal, though high-T_c materials are rather in an intermediate case between these two extremes. The long-range interaction $H_{eph,L}$ is unique to systems such as high-temperature superconductors since it results from the presence of strong-correlation induced dynamic inhomogeneities of the electronic ground state.

Breathing and buckling modes, which are the A_{1g} in-plane and the B_{1g}, B_{2g} out-of-plane motion of oxygen atoms in the CuO_2 planes, respectively, were shown to contribute most to the phonon-mediated pairing interaction [96,97]. It was argued that correlations suppress charge fluctuations, which leads to a reduction of the electron-phonon interaction [98,99]. This result was obtained for a local interaction. Calculations using the time-dependent Gutzwiller approximation have shown that correlations can enhance the transitive coupling to these optical phonon modes at small momenta q [100,101]. Nevertheless, the amplitude of the coupling to the A_{1g} mode is too small in absolute value to lead to high critical temperatures. These considerations were made without taking into account the fact that the screening along the c-axis is reduced and allows for a three-dimensional coupling of charge carriers to long wave length optical phonons.

Sigmund and Hizhnyakov considered the pairing interaction resulting from long wave length optical phonons, noticing three essential features. First, the long range interaction implies an averaging over the smaller scale of electronic inhomogeneities and the interlayer distance, rendering the superconducting state truly three-dimensional. Second, the anisotropy of the superconducting order parameter does not result from the anisotropy of the pairing interaction which mixes states with close momenta k, but rather from the band structure and in particular the anisotropic density of states at the Fermi surface [56–58]. Third, the pairing interaction determines the stability of the superconducting state and magnitude of the superconducting order parameter. The symmetry of the latter is, however, not solely determined by the pairing interaction. Account of the local, Hubbard like, Coulomb repulsion and the relative magnitude of these two-particle interactions determines the symmetry; a relatively modest Coulomb repulsion transforms an (anisotropic) s-wave symmetry into a d-wave symmetry [89]. These three features are in stark contrast with most alternative models for the description of the superconducting state in high temperature superconductors.

To be more specific, the BCS gap equation requires the knowledge of the electronic dispersion relation and the pairing potential. The in-plane parametrization of the dispersion for conduction holes, ε_k, is well-established [56–58,102]. Neglecting here the interlayer hopping, the density of states $\rho_F = v_F^{-1}$ displays strong maxima along $\phi = n\pi/2$ ($n = 0, 1, 2, 3$) in the $(k_x; k_y)$-plane [56–58]. The strong anisotropy bears resemblance to the superconducting order parameter anisotropy.

The pairing interaction resulting from the long range coupling to A_{1g}, B_{1g} and B_{2g} optical phonon modes was derived in Ref. [56–58] and take the form

$$V_{A_{1g}}(\mathbf{q}_\parallel) \simeq \frac{U_{A_{1g}}}{\kappa^2 + q_x^2 + q_y^2}, \tag{12}$$

$$V_{B_{1g}}(\mathbf{q}_\parallel) = \frac{U_{B_{1g}}}{\kappa^2 + q_x^2 + q_y^2}(\cos k_x - \cos k_y)^2. \tag{13}$$

Here $\mathbf{q} = \mathbf{k} - \mathbf{k}'$ and $\mathbf{k} = (k_x, k_y)$ is the wave-vector component *parallel* to the CuO$_2$ plane. The coupling to the B$_{2g}$ mode is obtained from $V_{B_{1g}}$ rotating the k-space basis by $\pi/4$. Because of this rotation the squared parenthesis in Equation (13) is small in the antinodal regions and vice versa. As a result, the contribution of this mode can be neglected. Hence, the maximum of the gap, Δ_{max} and T_c are determined by the A$_{1g}$ and B$_{1g}$ contributions to pairing. $U_{A_{1g},B_{1g}}$ and κ were estimated in Ref. [89] to be $U_{A_{1g},B_{1g}} \sim 100$ meV and $\kappa \sim 0.3$ and are the same parameters that allow describing the experimentally observed phonon renormalization at the superconducting transition [103–105].

To solve the BCS gap equation one needs also to add the pairing interaction resulting from the short-range part of the electron-phonon interaction $H_{e-ph,S}$ [89]. Their expression is the standard dominant pairing interaction found in conventional superconductors with its attractive and repulsive parts.

Using both the long-range and short-range contributions of the electron-phonon interaction and Coulomb repulsion we solved the gap Equation [56–58,89]. Two important conclusions were obtained. First, the *magnitude* Δ_{max} and *anisotropy* of the superconducting gap $\Delta_\mathbf{k}$ are determined by the coupling of charge carriers to the long range electron-phonon interactions at small **q**. The anisotropic density of states indeed determines the **k**-dependence of the gap. The results are in excellent agreement with the experimental determination of the order parameter. Second, the *symmetry* of the order parameter is not determined by the pairing interaction but by the relative weight of competing attractive and repulsive interactions. The long-range electron-phonon interactions lead to an anisotropic s-wave gap. Accounting for a relatively modest local Coulomb repulsion transforms the s-wave gap to a d-wave gap, without fundamentally affecting the magnitude or the overall **k**-dependence of the gap [89].

4.2. Pair Fluctuations in the Symmetry-Broken Ground State

Motivated by the coexistence of stripe order and a two-dimensional d-wave like gap in the stripe phase of LBCO we proceed by investigating the structure of pairing fluctuations for static, metallic stripes, i.e., deep in the symmetry broken ground state. In this regard our present study is in some sense complementary to the work of Ref. [52] where SC has been obtained from ICDW fluctuations on top of a *homogeneous* ground state. The problem is complex because the formation of stripes also alters the spectrum of low energy charge and spin fluctuations which contribute to the correlations in the pairing channel. Due to this complexity we will use the time-dependent Gutzwiller approach, [106] instead of the SCDT, which conveniently allows also considering symmetry-broken solutions.

Our investigations are based on the one-band Hubbard model with hopping restricted to nearest ($\sim t$) and next nearest ($\sim t'$) neighbors

$$H = -t \sum_{\langle ij \rangle, \sigma} c^\dagger_{i,\sigma} c_{j,\sigma} - t' \sum_{\langle\langle ij \rangle\rangle, \sigma} c^\dagger_{i,\sigma} c_{j,\sigma} + U \sum_i n_{i,\uparrow} n_{i,\downarrow}. \tag{14}$$

Here $c^{(\dagger)}_{i,\sigma}$ destroys (creates) an electron with spin σ at site i, and $n_{i,\sigma} = c^\dagger_{i,\sigma} c_{i,\sigma}$. U is the on-site Hubbard repulsion.

As a starting point we treat the model Equation (14) within an unrestricted Gutzwiller approximation (GA) as in Ref. [106]. Basically one constructs a Gutzwiller wave function $|\Psi\rangle$ by applying a projector to a Slater determinant $|SD\rangle$ which reduces the double occupancy. The Slater determinant is allowed to have an inhomogeneous charge and spin distribution describing generalized spin and charge density waves determined variationally [107]. The advantage of the GA in the present context is that our saddle point solutions reproduce several features of experiments [106,108] while the same would not be true if the starting point where HF [106] for which stripes are not even the ground state for realistic parameters.

The parameters were fixed by requiring that (a) the linear concentration of added holes is $1/(2a)$ according to experiment [23,109,110] and (b) a TDGA computation of the undoped AF insulator reproduces the experimental magnon dispersion [111] observed by inelastic neutron scattering. Condition (a) was shown to be very sensitive to t'/t [106] whereas condition (b) is sensitive to U/t and t, the former parameter determining the observed energy splitting between magnons at wave-vectors $(1/2,0)$ and $(1/4,1/4)$. Ref. [111] Indeed, the splitting vanishes within spin-wave theory applied to the Heisenberg model which corresponds to $U/t \to \infty$. We find that both conditions are met by $t'/t = -0.2$, $U/t = 8$ and $t = 354$ meV.

The results shown in this paper are for $d = 4$ bond-centered stripes oriented along the y-direction calculated on a 40×40 lattice; see Figure 5 for a visualization of the charge and spin structure. Figure 6 shows the corresponding band structure. Stripe formation induces two bands (B_1 and B_2, cf. Figure 6) in the Mott-Hubbard gap and the chemical potential is located in the (half-filled) band labeled B_1.

Dynamical pairing fluctuations are computed on top of the inhomogeneous solutions within the time-dependent GA [112,113] (TDGA). This scheme allows for the incorporation of particle-particle correlations in a similar manner as the traditional ladder approximation based on Hartree-Fock (HF) ground states. At the same time, it starts from a solution which incorporates correlations already at mean-field level. In the particle-hole channel the TDGA has previously been shown to provide an accurate description of magnetic fluctuations [114,115] and the optical conductivity [116] in cuprates.

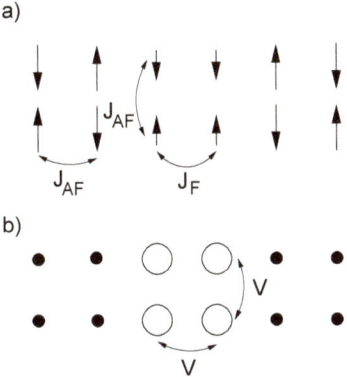

Figure 5. Spin (a) and charge (b) structure of $d = 4$ bond centered stripes at $\delta = 1/8$. Also shown are the interaction parameters used in the spin (J_{AF}, J_F) and charge (V) channel which determine the pairing fluctuations in Equation (16).

Here we are interested in the dynamical order parameter correlation function

$$P_q(\omega) = \frac{1}{i}\int_{-\infty}^{\infty} dt e^{i\omega t} \langle T \Delta_q(t) \Delta_q^\dagger(0) \rangle, \tag{15}$$

where $\Delta_q = 1/\sqrt{N} \sum_k \gamma_k c_{-k-q,\downarrow} c_{k,\uparrow}$ and γ_k specifies the symmetry of the order parameter fluctuations. We focus on $\gamma_k = (\cos(k_x) + \cos(k_y))/2$ (extended s-wave) and $\gamma_k = (\cos(k_x) - \cos(k_y))/2$ (d-wave) symmetries. For $\omega > 0$ ($\omega < 0$) the imaginary part of $P_q(\omega)$ yields the order parameter correlations for two-particle addition (removal). Please note that a pole in $P_q(\omega \to 0)$ signals the occurence of a long-range ordered SC state for a given symmetry. Of interest is also the vertex contribution $\Delta P_q(\omega) = P_q(\omega) - P_q^0(\omega)$ where P^0 corresponds to the non-interacting pair-correlation function calculated with the bare Green's function on the GA level. For a given symmetry $\Delta P_q(\omega)$ yields information on whether the order parameter fluctuations at a given momentum and frequency are attractive or repulsive for the GA quasiparticles.

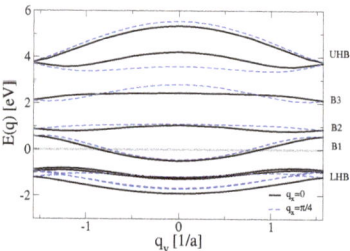

Figure 6. GA band structure for $d = 4$ bond-centered stripes oriented along the y-direction. Shown are the dispersions for $q = 0$ (black, solid) and $q = \pi/4$ (dashed). The energy is measured with respect to the chemical potential (horizontal grey line) and the lattice constant is set to $a \equiv 1$. LHB/UHB: Lower and upper Hubbard band.

Our ladder-type approach incorporates the pair correlations on a scale of U but unfortunately does not properly take into account the energetically low lying collective excitations in the charge and spin particle-hole channels. As shown for the case of ICDW scattering [22] these excitations couple back to the particle-particle channel and thus can strongly enhance the low energy SC fluctuations. We incorporate this effect by adding the operator

$$H_{Fluc} = -g \sum_{ij} J_{ij} \delta \langle c_{i,\uparrow}^\dagger c_{j,\downarrow}^\dagger \rangle \delta \langle c_{i,\downarrow} c_{j,\uparrow} \rangle$$
$$+ g \sum_{ij} V_{ij} \delta \langle c_{i,\uparrow}^\dagger c_{j,\downarrow}^\dagger \rangle \delta \langle c_{j,\downarrow} c_{i,\uparrow} \rangle . \qquad (16)$$

to the system which generates particle-particle scattering in the spin ((J_{ij})) and charge ((V_{ij})) channel but does not alter the ground state solution. The parameter g is introduced to model vertex corrections which for simplicity are assumed to be constant. The interaction parameters J_{ij} in the spin channel are obtained by calculating the magnetic excitations in the stripe phase within the TDGA [114,115]. The low energy Goldstone mode emerging from the incommensurate wave-vectors can be fitted by linear spin-wave theory applied to a Heisenberg model with site dependent interactions as shown in Figure 5a. We find $J_{AF} = 0.4t$ between antiferromagnetically ordered spins and $J_F = -0.2 J_{AF}$ between the ferromagnetically ordered bonds on the stripe legs. To obtain information about the interaction parameters in the charge channel we calculate the charge profile of bond centered stripes within the following HF approximated spinless fermion model

$$H = -t \sum_{\langle ij \rangle} f_i^\dagger f_j - t' \sum_{\langle \langle ij \rangle \rangle} f_i^\dagger f_j + \sum_{ij} V_{ij}(1 - n_i)(1 - n_j).$$

where the kinetic part is the same as in Equation (14) and V_{ij} is a nearest-neighbor attraction acting on holes on the stripe legs (Figure 5b). We find that a value of $V_{ij} = -0.25t$ reproduces the charge profile obtained within the full GA calculation.

Figure 7a shows the instantaneous pairing correlations in the d-wave channel obtained from the removal part $\langle \Delta_q^{\dagger,d} \Delta_q^d \rangle = \int_{-\infty}^0 d\omega P_q(\omega)$ for coupling parameter $g = 1$. The correlations calculated from the addition spectra ($\omega > 0$) are similar. The shape reflects the quasi one-dimensionality of the underlying ground state with the most pronounced correlations along $(q_x, q_y) \approx 0$ and the maximum at $(0,0)$. However, when we substract the contribution of the non-interacting GA quasiparticles the resulting vertex contribution (cf. Figure 7b) takes a different and much more isotropic shape. The correlations at $\mathbf{q} = (0,0)$

are still attractive but the maxima now occur at $(\pm\pi,0)$ and $(0,\pm\pi)$. The vertex contribution for extended s-wave symmetry (cf. Figure 7c) is also rather isotropic but displays the maximum attraction at $\mathbf{q}=(0,0)$. Also the (instantaneous) attraction is by a factor of 3–4 larger than in the d-wave channel. The strong reduction of the d-wave attractive fluctuations for small momenta can be traced back to the form factor $\gamma_k = \cos(k_x) - \cos(k_y)$ and occurs also in the homogeneous state in the absence of J_{AF}, J_F, and V (for finite interaction parameters the homogeneous state has a pole at $\mathbf{q}=0$ and $\omega=0$ reflecting an instability towards superconductivity).

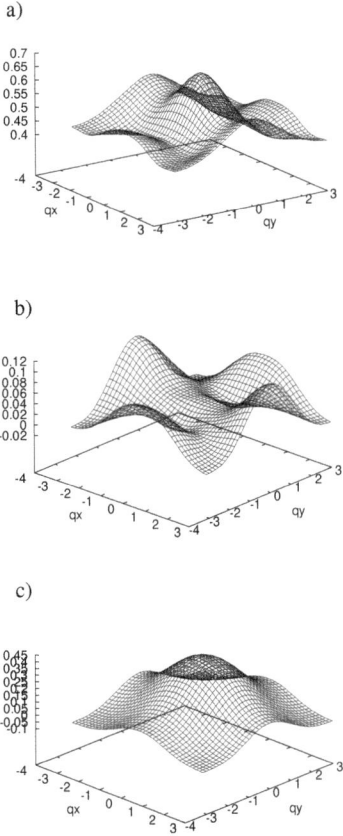

Figure 7. Instantaneous order parameter fluctuations for d-wave (**a**,**b**) and extended s-wave symmetry (**c**) obtained from the removal part ($\omega<0$). Panel (**a**) corresponds to the TDGA ladder result for $\langle \Delta_q^d \Delta_q^d \rangle = \int d\omega P_q(\omega)$ whereas panels (**b**) and (**c**) show the vertex contributions $\int d\omega \Delta P_q(\omega)$. Coupling parameter $g=1$.

It is important to know in which frequency range $P_q(\omega)$ contributes most to the instantantaneous correlations. One can expect that a system with low energy order parameter fluctuations is more susceptible towards the transition to a superconducting state than if these would occur at higher frequencies. In fact, we find pronounced differences between *d*-wave and extended *s*-wave symmetry as can be seen from Figure 8. In the main panels we show the **q** = 0 pairing fluctuations for two-particle removal ($\omega < 0$) and addition ($\omega > 0$) and a coupling parameter ($g = 1$). Also shown are the bare GA two-particle spectra which correspond to a convolution of the single-particle bands shown in Figure 6. In the *d*-wave sector (panel a) one observes two features in the uncorrelated removal part. The lower (higher) energy one is due to the removal of two particles from the band B_1 (B_2) whereas interband correlations (which have weight in between) are suppressed. For extended *s*-wave fluctuations (panel b) the GA removal part has only weight in the energy range which corresponds to the removal of two particles from B_2.

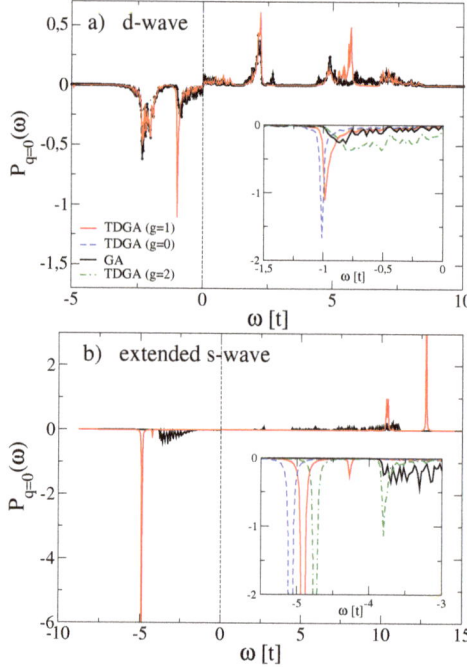

Figure 8. Order parameter fluctuations at **q** = 0 for *d*-wave (**a**) and extended *s*-wave symmetry (**b**). Red: TDGA; Black: GA. The inset covers the energy range around the bound states of the removal spectra. Also shown (blue dashed) is the TDGA result without spin and charge attractive interactions ($g = 0$) and the strong coupling case (green dashed dotted) for $g = 2$.

Including correlations within the TDGA ladder scheme leads to the appearance of bound ($\omega < 0$) and antibound ($\omega > 0$) states similar to the physics of local pairs in the Hubbard model as relevant for e.g., Auger spectroscopy [112,117,118]. In case of the *d*-wave removal fluctuations the bound state forms at the bottom of the convoluted band B_1. As can be seen from the inset to Figure 8a this bound state exists even without the inclusion of additional interactions in the spin and charge channel ($g = 0$). The effect of J_F, J_{AF} and V is to push the bound state toward lower energy, thus enhancing the mixing with the B_1 band states and strengthening the *d*-wave correlations inside the active band. In fact, for $g = 2$ the bound state is no more visible as a separate feature and instead one observes a convoluted B_1 band with increased *d*-wave correlations as compared to the $g = 0$ case.

By contrast, the bound state in the extended s-wave removal part occurs below the convoluted lower Hubbard band (LHB, cf. Figure 6). Also in this case inclusion of J_F, J_{AF} and V leads to a shift of the bound state to lower energies and induces also a smaller satellite. Upon increasing the coupling ($g = 2$) this satellite increases in intensity and approaches the energy of the LHB. However, the formation of the extended s-wave bound state is accompanied by a strong suppression of two-hole band states in contrast to the d-wave case so that there are no low energy band states with extended s-wave correlations.

5. Conclusions

In this work, we investigated the mechanism of phase separation in systems with strong electron correlations. The perturbation series expansion around the atomic limit is a reasonable approach for the description of such systems. Therefore, processes of atomic-level crossing occurring at certain values of the chemical potential play a key role in the evolution of their band structure with doping. An extreme non-rigidity of the bands near these peculiar chemical potentials leads to the appearance of regions of negative electron compressibility. Hence these regions are an inherent property of all strongly correlated systems. We have demonstrated their occurrence in the one-band Hubbard and Hubbard-Kanamori models widely used for the description of cuprate perovskites, transition metal oxides, and some other crystals. The existence of these regions gives rise to the charge instability—the segregation of the crystal into electron-rich and electron-poor domains. The energy released by their formation has to be absorbed by phonons. This leads to different lattice distortions in the above domains. In the case of a predominant interaction of electrons with distortions of special symmetry, for example, with a softening phonon mode, the separation into two domains acquires the shape of stripes.

For such textures we have investigated the frequency and momentum structure of pairing correlations in the d-wave and extended s-wave channels. It turns out that depsite the underlying quasi one-dimensional electronic structure these correlations are quite isotropic, similar to the isotropy of spin fluctuations which arise from stripe textures, [114,115] which therefore is compatible with the observation of an isotropic superconducting gap in stripe ordered compounds [35]. While the present approach was restricted to the evaluation of pairing fluctuations on top of the normal state it would be interesting in future work to investigate directly the structure of superconducting order in the symmetry broken stripe state.

Author Contributions: Conceptualization, R.K.K. and G.S.; formal analysis, A.B., V.H., G.S., A.S. (Aleksander Shelkan) and A.S. (Alexei Sherman); software, A.S. (Aleksander Shelkan); writing—original draft, G.S.; writing—review & editing, A.B., V.H., R.K.K., G.S. and A.S. (Alexei Sherman). All authors have read and agreed to the published version of the manuscript.

Funding: V.H.'s contribution was supported by the grant of the Estonian Scientific Council PRG347. G.S. acknowledges support from the Deutsche Forschungsgemeinschaft.

Conflicts of Interest: The authors declare no conflict of interest.

References

1. Bednorz, J.G.; Müller, K.A. Possible high Tc superconductivity in the Ba-La-Cu-O system. *Z. Physik B* **1986**, *64*, 189. [CrossRef]
2. Hizhnyakov, V.; Sigmund, E. High-Tc superconductivity induced by ferromagnetic clustering. *Phys. C* **1988**, *156*, 655. [CrossRef]
3. Hizhnyakov, V.; Kristoffel, N.E.; Sigmund, E. On the percolation induced conductivity in high-Tc superconducting materials. *Phys. C* **1989**, *160*, 119. [CrossRef]
4. Hizhnyakov, V.; Sigmund, E.; Schneider, M. Magnetic interactions and dynamics of holes in CuO_2 planes of high-Tc superconducting materials. *Phys. Rev. B* **1991**, *44*, 795. [CrossRef] [PubMed]

5. Kremer, R.; Sigmund, E.; Hizhnyakov, V.; Hentsch, F.; Simon, A.; Müller, K.A.; Mehring, M. Percolative phase separation in La2CuO$_{4+\delta}$ and La$_{2-x}$Sr$_x$CuO$_4$. *Z. Physik B* **1992**, *86*, 319. [CrossRef]
6. Kremer, R.; Hizhnyakov, V.; Sigmund, E.; Simon, A.; Müller, K.A. Electronic phase separation in La-cuprates. On the role of hole and oxygen diffusion. La$_2$CuO$_{4+y}$; La$_{2-x}$Sr$_x$CuO$_{4+y}$. *Z. Physik B* **1993**, *94*, 17.
7. Fratini, M.; Poccia, N.; Ricci, A.; Campi, G.; Burghammer, M.; Aeppli, G.; Bianconi, A. Scale-free structural organization of oxygen interstitials in La2CuO$_{4+y}$. *Nature* **2010**, *466*, 841–844. [CrossRef] [PubMed]
8. Haase, J.; Slichter, C.P.; Milling, C.T. Static Charge and Spin Inhomogeneity in La$_{2-x}$Sr$_x$CuO$_4$ by NMR. *J. Supercond.* **2002**, *15*, 339. [CrossRef]
9. Krämer, S.; Mehring, M. Low-Temperature Charge Ordering in the Superconducting State of YBa$_2$Cu$_3$O$_{7-\delta}$. *Phys. Rev. Lett.* **1999**, *83*, 396. [CrossRef]
10. Teitel'baum, G.B.; Büchner, B.; de Gronckel, H. Cu NQR Study of the Stripe Phase Local Structure in the Lanthanum Cuprates. *Phys. Rev. Lett.* **2000**, *84*, 2949. [CrossRef]
11. Singer, P.M.; Hunt, A.W.; Imai, T. ^{63}Cu NQR Evidence for Spatial Variation of Hole Concentration in La$_{2-x}$Sr$_x$CuO$_4$. *Phys. Rev. Lett.* **2002**, *88*, 047602. [CrossRef]
12. Müller, K.A.; Benedek, G. (Eds.) *Proceedings of the First Workshop on Phase Separation in Cuprate Superconductors*; World Scientific: Singapore, 1993.
13. Sigmund, E.; Müller, K.A. (Eds.) *Proceedings of the Second Workshop on Phase Separation in Cuprate Superconductors*; Springer: Berlin/Heidelberg, Germany, 1994.
14. Emery, V.J.; Kivelson, S.A.; Lin, H.Q. Phase separation in the t-J model. *Phys. Rev. Lett.* **1990**, *64*, 475. [CrossRef] [PubMed]
15. Visscher, P.B. Phase separation instability in the Hubbard model. *Phys. Rev. B* **1974**, *10*, 943. [CrossRef]
16. Grilli, M.; Raimondi, R.; Castellani, C.; Castro, C.D.; Kotliar, G. Superconductivity, phase separation, and charge-transfer instability in the $U = \infty$ limit of the three-band model of the CuO$_2$ planes. *Phys. Rev. Lett.* **1991**, *67*, 259. [CrossRef]
17. Moreo, A.; Scalapino, D.; Dagotto, E. Phase separation in the Hubbard model. *Phys. Rev. B* **1991**, *43*, 11442. [CrossRef]
18. Van Dongen, P.G.J. Phase Separation in the Extended Hubbard Model at Weak Coupling. *Phys. Rev. Lett.* **1995**, *74*, 182. [CrossRef] [PubMed]
19. Capone, M.; Sangiovanni, G.; Castellani, C.; Castro, C.D.; Grilli, M. Phase Separation Close to the Density-Driven Mott Transition in the Hubbard-Holstein Model. *Phys. Rev. Lett.* **2004**, *92*, 106401. [CrossRef] [PubMed]
20. Sherman, A.; Schreiber, M. Fluctuating charge-density waves in the Hubbard model. *Phys. Rev. B* **2008**, *77*, 155117. [CrossRef]
21. Löw, U.; Emery, V. J.; Fabricius, K.; Kiverlson, S.A. Study of an Ising model with competing long- and short-range interactions. *Phys. Rev. Lett.* **1994**, *72*, 1918. [CrossRef]
22. Castellani, C.; Castro, C.D.; Grilli, M. Singular Quasiparticle Scattering in the Proximity of Charge Instabilities. *Phys. Rev. Lett.* **1995**, *75*, 4650. [CrossRef]
23. Tranquada, J.M.; Sternlieb, B.J.; Axe, J.D.; Nakamura, Y. Evidence for stripe correlations of spins and holes in copper oxide superconductors. *Nature* **1995**, *375*, 56. [CrossRef]
24. Abbamonte, P.; Rusydi, A.; Smadici, S.; Gu, G.D.; Sawatzky, G.A.; Feng, D.L. Spatially modulated 'Mottness' in La$_{2-x}$Ba$_x$CuO$_4$. *Nat. Phys.* **2005**, *1*, 155. [CrossRef]
25. Fink, J.; Schierle, E.; Weschke, E.; Geck, J.; Hawthorn, D.; Soltwisch, V.; Wadati, H.; Wu, H.-H. Charge ordering in La$_{1.8-x}$Eu$_{0.2}$Sr$_x$CuO$_4$ studied by resonant soft x-ray diffraction. *Phys. Rev. B* **2009**, *79*, 100502. [CrossRef]
26. Ghiringhelli, G.; Le Tacon, M.; Minola, M.; Blanco-Canosa, S.; Mazzoli, C.; Brookes, N.B.; De Luca, G.M.; Frano, A.; Hawthorn, G.; He, F.; et al. Long-Range Incommensurate Charge Fluctuations in (Y,Nd)Ba$_2$Cu$_3$O$_{6+x}$. *Science* **2012**, *337*, 821. [CrossRef] [PubMed]

27. Chang, J.; Blackburn, E.; Holmes, A.T.; Christensen, N.B.; Larsen, J.; Mesot, J.; Liang, R.; Bonn, D.A.; Hardy, W.N.; Watenphul, A.; et al. Direct observation of competition between superconductivity and charge density wave order in YBa$_2$Cu$_3$O$_{6.67}$. *Nat. Phys.* **2012**, *8*, 871. [CrossRef]
28. Neto, E.H.d.; Aynajian, P.; Frano, A.; Comin, R.; Schierle, E.; Weschke, E.; Gyenis, A.; Wen, J.; Schneeloch, J.; Xu, Z.; et al. Ubiquitous Interplay Between Charge Ordering and High-Temperature Superconductivity in Cuprates. *Science* **2014**, *343*, 393. [CrossRef]
29. Comin, R.; Frano, A.; Yess, M.M.; Yoshida, Y.; Eisaki, H.; Schierle, E.; Weschke, E.; Sutarto, R.; He, F.; Soumyanarayanan, A.; et al. Charge Order Driven by Fermi-Arc Instability in Bi$_2$Sr$_{2-x}$La$_x$CuO$_{6+\delta}$. *Science* **2014**, *343*, 390. [CrossRef]
30. Hücker, M.; Christensen, N.B.; Holmes, A.T.; Blackburn, E.; Forgan, E.M.; Liang, R.; Bonn, D.A.; Hardy, W.N.; Gutowski, O.; Zimmermann, M.V.; et al. Competing charge, spin, and superconducting orders in underdoped YBa$_2$Cu$_3$O$_y$. *Phys. Rev. B* **2014**, *90*, 054514. [CrossRef]
31. Doiron-Leyraud, N.; Proust, C.; LeBoeuf, D.; Levallois, J.; Bonnemaison, J.-B.; Liang, R.; Bonn, D.A.; Hardy, W.N.; Taillefer, L. Quantum oscillations and the Fermi surface in an underdoped high-Tc superconductor. *Nature* **2007**, *447*, 565. [CrossRef]
32. Sebastian, S.E.; Harrison, N.; Lonzarich, G.G. Towards resolution of the Fermi surface in underdoped high-Tc superconductors. *Rep. Prog. Phys.* **2012**, *75*, 102501. [CrossRef]
33. Laliberté, F.; Chang, J.; Doiron-Leyraud, N.; Hassinger, E.; Daou, R.; Rondeau, M.; Ramshaw, B.J.; Liang, R.; Bonn, D.A.; Hardy, W.N.; et al. Fermi-surface reconstruction by stripe order in cuprate superconductors. *Nat. Commun.* **2011**, *2*, 432. [CrossRef] [PubMed]
34. Balseiro, C.A.; Falicov, L.M. Superconductivity and charge-density waves. *Phys. Rev. B* **1979**, *20*, 4457. [CrossRef]
35. Valla, T.; Federov, A.V.; Lee, J.; Davis, J.C.; Gu, G.D. The Ground State of the Pseudogap in Cuprate Superconductors. *Science* **2006**, *314*, 1914. [CrossRef]
36. Li, Q.; Hücker, M.; Gu, G.D.; Tsvelik, A.M.; Tranquada, J.M. Two-Dimensional Superconducting Fluctuations in Stripe-Ordered La$_{1.875}$Ba$_{0.125}$CuO$_4$. *Phys. Rev. Lett.* **2007**, *99*, 067001. [CrossRef] [PubMed]
37. Müller, K.A. *Proceedings of the ICTP Workshop: Intrinsic Multiscale Structure And Dynamics in Complex Electronic Oxides*; Bishop, A.R., Shenoy, S.R., Sridhar, S., Eds.; World Scientific Publishing: Singapore, 2003; pp. 1–5.
38. Müller, K.A. *Superconductivity in Complex Systems*; Müller, K.A., Bussmann-Holder, A., Eds.; Springer: Berlin/Heidelberg, Germany, 2005; pp. 1–11
39. Müller, K.A. *Handbook of High-Temperature Superconductivity*; Schrieffer, J.R., Brooks, J.S., Eds.; Springer: New York, NY, USA, 2007; pp. 1–18.
40. Bussmann-Holder, A.; Keller, H.; Bishop, A.R.; Simon, A.; Müller, K.A. Polaron coherence as origin of the pseudogap phase in high temperature superconducting cuprates. *J. Supercond. Nov. Magn.* **2008**, *21*, 353. [CrossRef]
41. Bussmann-Holder, A.; Keller, H.; Bishop, A.R.; Simon, A.; Micnas, R.; Müller, K.A. Unconventional isotope effects as evidence for polaron formation in cuprates. *EPL* **2005**, *72*, 423. [CrossRef]
42. Perali, A.; Bianconi, A.; Lanzara, A.; Saini, N.L. The gap amplification at a shape resonance in a superlattice of quantum stripes: A mechanism for high Tc. *Solid State Commun.* **1996**, *100*, 181. [CrossRef]
43. Bianconi, A.; Valletta, A.; Perali, A.; Saini, N.L. High Tc superconductivity in a superlattice of quantum stripes. *Solid State Commun.* **1997**, *102*, 369. [CrossRef]
44. Bianconi, A.; Innocenti, D.; Valletta, A.; Perali, A. Shape Resonances in superconducting gaps in a 2DEG at oxide-oxide interface. *J. Phys. Conf. Ser.* **2014**, *529*, 012007. [CrossRef]
45. Innocenti, D.; Poccia, N.; Ricci, A.; Valletta, A.; Caprara, S.; Perali, A.; Bianconi, A. Resonant and crossover phenomena in a multiband superconductor: Tuning the chemical potential near a band edge. *Phys. Rev. B* **2010**, *82*, 184528. [CrossRef]
46. Innocenti, D.; Valletta, A.; Bianconi, A. Shape resonance at a Lifshitz transition for high temperature superconductivity in multiband superconductors. *J. Supercond. Nov. Magn.* **2011**, *24*, 1137. [CrossRef]

47. Innocenti, D.; Caprara, S.; Poccia, N.; Ricci, A.; Valletta, A.; Bianconi, A. Shape resonance for the anisotropic superconducting gaps near a Lifshitz transition: The effect of electron hopping between layers. *Supercond. Sci. Technol.* **2010**, *24*, 015012. [CrossRef]
48. Perali, A.; Innocenti, D.; Valletta, A.; Bianconi, A. Anomalous isotope effect near a 2.5 Lifshitz transition in a multi-band multi-condensate superconductor made of a superlattice of stripes. *Supercond. Sci. Technol.* **2012**, *25*, 124002. [CrossRef]
49. Seibold, G.; Varlamov, S. Relationship between incommensurability and superconductivity in Peierls- distorted charge-density-wave systems. *Phys. Rev. B* **1999**, *60*, 13056. [CrossRef]
50. Mierzejewski, M.; Zielinski, J.; Entel, P. Phonon-Induced Contributions to Superconductivity in the Presence of Charge-Density Waves. *J. Supercond. Nov. Magn.* **2001**, *14*, 449. [CrossRef]
51. Emery, V.J.; Kivelson, S.A.; Zachar, O. Spin-gap proximity effect mechanism of high-temperature superconductivity. *Phys. Rev. B* **1997**, *56*, 6120. [CrossRef]
52. Perali, A.; Castellani, C.; Castro, C.D.; Grilli, M. d-wave superconductivity near charge instabilities. *Phys. Rev. B* **1996**, *54*, 16216. [CrossRef]
53. Berg, E.; Fradkin, E.; Kim, E.-A.; Kivelson, S.A.; Oganesyan, V.; Tranquada, J.M.; Zhang, S.C. Dynamical Layer Decoupling in a Stripe-Ordered High-Tc Superconductor. *Phys. Rev. Lett.* **2007**, *99*, 127003. [CrossRef]
54. Berg, E.; Fradkin, E.; Kivelson, S.A.; Tranquada, J.M. Striped superconductors: how spin, charge and superconducting orders intertwine in the cuprates. *New J. Phys.* **2009**, *11*, 115004. [CrossRef]
55. Dagotto, E. *Nanoscale Phase Separation and Colossal Magnetoresistance*; Springer Series in Solid-State Sciences; Springer: Berlin/Heidelberg, Germany, 2003.
56. Hizhnyakov, V.; Sigmund, E. Anisotropic pairing caused by unscreened long-range interactions. *Phys. Rev. B* **1996**, *53*, 5163. [CrossRef]
57. Hizhnyakov, V.; Sigmund, E. Anisotropic pairing caused by unscreened long-range interactions. *J. Supercond.* **1996**, *9*, 335. [CrossRef]
58. Sigmund, E.; Hizhnyakov, V.; Nevedrov, D.; Bill, A. Anisotropy of the Superconducting Gap in the Presence of Electron-Phonon and Coulomb Interactions. *J. Supercond.* **1997**, *10*, 441. [CrossRef]
59. Shelkan, A.; Hizhnyakov, V.; Sigmund, E. Self-Consistent Calculation of the Self-Trapping Barrier for a Hole in the CuO_2 Plane. *J. Supercond.* **1998**, *11*, 677. [CrossRef]
60. Shelkan, A.; Hizhnyakov, V.; Sigmund, E. Free and Spin-Polaron States in High TC Superconductors. In *Vibronic Interactions: Jahn-Teller Effect in Crystals and Molecules*; NATO Science Series (Series II: Mathematics, Physics and Chemistry); Kaplan, M.D., Zimmerman, G.O., Eds.; Springer: Dordrecht, The Netherlands, 2001.
61. Hizhnyakov, V.; Sigmund, E.; Zavt, G. Existence of a barrier between free and ferron-type (self-trapped) hole states in high- Tc cuprates. *Phys. Rev. B* **1991**, *44*, 12639. [CrossRef]
62. Shelkan, A.; Zavt, G.; Hizhnyakov, V.; Sigmund, E. Effect of O-O transfer upon a hole in CuO_2 planes of HTS. *Z. Phys. B* **1997**, *104*, 433. [CrossRef]
63. Shelkan, A.; Hizhnyakov, V.; Sigmund, E. Self-Trapping Barrier for Phonon Polaron in Anisotropic Crystal. *J. Supercond.* **2000**, *13*, 21. [CrossRef]
64. Ivanov, T.I. Phase separation in the slave-fermion mean-field theory of the t-J model. *Phys. Rev. B* **1991**, *44*, 12077. [CrossRef]
65. Gimm, T.; Salk, S.H.S. Phase separation based on a U(1) slave-boson functional integral approach to the t-J model. *Phys. Rev. B* **2000**, *62*, 13930. [CrossRef]
66. Koch, E.; Zeyher, R. Charge density waves as the origin of dip-hump structures in the differential tunneling conductance of cuprates: The case of d-wave superconductivity. *Phys. Rev. B* **2004**, *70*, 094519.
67. Irkhin, V.Y.; Igoshev, P.A. Electron States and Magnetic Phase Diagrams of Strongly Correlated Systems. *Phys. Met. Metallogr.* **2018**, *119*, 1267. [CrossRef]

68. Kagan, M.Y.; Rice, T.M. Superconductivity in the two-dimensional t-J model at low electron density. *J. Phys. Condens. Matter* **1994**, *6*, 3771. [CrossRef]
69. Kagan, M.Y.; Kugel, K.I. Inhomogeneous charge distributions and phase separation in manganites. *Phys. Uspekhi* **2001**, *44*, 553. [CrossRef]
70. Kugel, K.I.; Khomskii, D.I. The Jahn-Teller effect and magnetism: transition metal compounds. *Sov. Phys. Uspekhi* **1982**, *25*, 231. [CrossRef]
71. Sboychakov, A.O.; Kugel, K.I.; Rakhmanov, A.L. Phase separation in a two-band model for strongly correlated electrons. *Phys. Rev. B* **2007**, *76*, 195113. [CrossRef]
72. Sherman, A. Negative electron compressibility in the Hubbard model. *Phys. Scr.* **2020**, *95*, 015806. [CrossRef]
73. Vladimir, M.I.; Moskalenko, V.A. Diagram technique for the Hubbard model. *Theor. Math. Phys.* **1990**, *82*, 301. [CrossRef]
74. Metzner, W. Linked-cluster expansion around the atomic limit of the Hubbard model. *Phys. Rev. B* **1991**, *43*, 8549. [CrossRef]
75. Pairault, S.D.; Tremblay, A.-M.S. Strong-coupling perturbation theory of the Hubbard model. *Eur. Phys. J. B* **2000**, *16*, 85. [CrossRef]
76. Sherman, A. Influence of spin and charge fluctuations on spectra of the two-dimensional Hubbard model. *J. Phys. Condens. Matter* **2018**, *30*, 195601. [CrossRef] [PubMed]
77. Sherman, A. Magnetic properties and temperature variation of spectra in the Hubbard model. *Eur. Phys. J. B* **2019**, *92*, 55. [CrossRef]
78. Kubo, R. Generalized Cumulant Expansion Method. *J. Phys. Soc. Jpn.* **1962**, *17*, 1100. [CrossRef]
79. Chang, C.; Zhang, S. Spatially inhomogeneous phase in the two-dimensional repulsive Hubbard model. *Phys. Rev. B* **2008**, *78*, 165101. [CrossRef]
80. Heiselberg, H. Hubbard-model calculations of phase separation in optical lattices. *Phys. Rev. A* **2009**, *79*, 063611. [CrossRef]
81. White, S.R.; Scalapino, D.J. Doping asymmetry and striping in a three-orbital CuO_2 Hubbard model. *Phys. Rev. B* **2015**, *92*, 205112. [CrossRef]
82. Kanamori, J. Electron Correlation and Ferromagnetism of Transition Metals. *Prog. Theor. Phys.* **1963**, *30*, 275. [CrossRef]
83. Georges, A.; Medici, L.d.; Mravlje, J. Strong Correlations from Hund's Coupling. *Annu. Rev. Condens. Matter Phys.* **2013**, *4*, 137. [CrossRef]
84. Sherman, A. Hubbard-Kanamori model: Spectral functions, negative electron compressibility, and susceptibilities. *Phys. Scr.* **2020**, *95*, 095804. [CrossRef]
85. Shenoy, V.B.; Rao, C.N.R. Electronic phase separation and other novel phenomena and properties exhibited by mixed-valent rare-earth manganites and related materials. *Phil. Trans. R. Soc. A* **2008**, *366*, 63. [CrossRef]
86. Dai, P.; Hu, J.; Dagotto, E. Magnetism and its microscopic origin in iron-based high-temperature superconductors. *Nat. Phys.* **2012**, *8*, 709. [CrossRef]
87. Lang, G.; Grafe, H.-J.; Paar, D.; Hammerath, F.; Manthey, K.; Behr, G.; Werner, J.; Büchner, B. Nanoscale Electronic Order in Iron Pnictides. *Phys. Rev. Lett.* **2010**, *104*, 097001. [CrossRef]
88. Müller, K.A. Possible coexistence of s- and d-wave condensates in copper oxide superconductors. *Nature* **1995**, *377*, 133. [CrossRef]
89. Bill, A.; Hizhnyakov, V.; Nevedrov, D.; Seibold, G.; Sigmund, E.Z. Electronic inhomogeneities, electron-lattice and pairing interactions in high-T c superconductors. *Phys. B* **1997**, *104*, 753. [CrossRef]
90. Kresin, V.Z.; Morawitz, H. Layer plasmons and high-Tc superconductivity. *Phys. Rev. B* **1988**, *37*, 7854. [CrossRef] [PubMed]
91. Bill, A.; Morawitz, H.; Kresin, V.Z. Electronic collective modes and superconductivity in layered conductors. *Phys. Rev. B* **2003**, *68*, 144519. [CrossRef]
92. Bill, A.; Morawitz, H.; Kresin, V.Z. Acoustic Plasmons in Layered Systems and the Phonon-Plasmon Mechanism of Superconductivity. *J. Low Temp. Phys.* **1999**, *117*, 283. [CrossRef]

93. Bill, A.; Morawitz, H.; Kresin, V.Z. Dynamical screening and superconducting state in intercalated layered metallochloronitrides. *Phys. Rev. B* **2002**, *66*, 100501. [CrossRef]
94. Hepting, M.; Chaix, L.; Huang, E.W.; Fumagalli, R.; Peng, Y.Y.; Moritz, B.; Kummer, K.; Brookes, N.B.; Lee, W.C.; Hashimoto, M.; et al. Three-dimensional collective charge excitations in electron-doped copper oxide superconductors. *Nature* **2018**, *563*, 374. [CrossRef] [PubMed]
95. Lin, J.; Yuan, J.; Jin, K.; Yin, Z.; Li, G.; Zhou, K.J.; Lu, X.; Dantz, M.; Schmitt, T.; Ding, H.; et al. Doping evolution of the charge excitations and electron correlations in electron-doped superconducting $La_{2-x}Ce_xCuO_4$. *NPJ Quantum Mater.* **2020**, *5*, 4. [CrossRef]
96. Heid, R.; Bohnen, K.-P.; Zeyher, R.; Manske, D. Momentum Dependence of the Electron-Phonon Coupling and Self-Energy Effects in Superconducting $YBa_2Cu_3O_7$ within the Local Density Approximation. *Phys. Rev. Lett.* **2008**, *100*, 137001. [CrossRef]
97. Heid, R.; Zeyher, R.; Manske, D.; Bohnen, K.-P. Phonon-induced pairing interaction in $YBa_2Cu_3O_7$ within the local-density approximation. *Phys. Rev. B* **2009**, *80*, 024507. [CrossRef]
98. Gunnarsson, O.; Rösch, O. Interplay between electron–phonon and Coulomb interactions in cuprates. *J. Phys. Condens. Matter* **2008**, *20*, 043201. [CrossRef]
99. Capone, M.; Castellani, C.; Grilli, M. Electron-Phonon Interaction in Strongly Correlated Systems. *Adv. Condens. Matter Phys.* **2010**, *2010*, 920860. [CrossRef]
100. von Oelsen, E.; Ciolo, A.D.; Lorenzana, J.; Seibold, G.; Grilli, M. Phonon renormalization from local and transitive electron-lattice couplings in strongly correlated systems. *Phys. Rev. B* **2010**, *81*, 155116. [CrossRef]
101. Seibold, G.; Grilli, M.; Lorenzana, J. Influence of correlations on transitive electron-phonon couplings in cuprate superconductors. *Phys. Rev. B* **2011**, *83*, 174522. [CrossRef]
102. Norman, M.R. Relation of neutron incommensurability to electronic structure in high-temperature superconductors. *Phys. Rev. B* **2000**, *61*, 14751. [CrossRef]
103. Bill, A.; Hizhnyakov, V.; Sigmund, E. Phonon renormalization and symmetry of the superconducting order parameter. *Phys. Rev. B* **1995**, *52*, 7637. [CrossRef]
104. Bill, A.; Hizhnyakov, V.; Sigmund, E. q-Dependent Neutron Scattering: A Signature of the Gap Anisotropy in High-Tc Superconductors. *J. Supercond.* **1996**, *10*, 493. [CrossRef]
105. Maly, J.; Lin, D.Z.; Levin, K. Superconducting order parameter symmetry in multilayer cuprates. *Phys. Rev. B* **1996**, *53*, 6786. [CrossRef]
106. Seibold, G.; Lorenzana, J. Stability of metallic stripes in the one-band extended Hubbard model. *Phys. Rev. B* **2004**, *69*, 134513. [CrossRef]
107. Gebhard, F. Gutzwiller correlated wave functions in finite dimensions d: A systematic expansion in $1/d$. *Phys. Rev. B* **1990**, *41*, 9452. [CrossRef]
108. Lorenzana, J.; Seibold, G. Metallic Mean-Field Stripes, Incommensurability, and ChemicalPotential in Cuprates. *Phys. Rev. Lett.* **2002**, *89*, 136401. [CrossRef] [PubMed]
109. Yamada, K.; Lee, C.H.; Kurahashi, K.; Wada, J.; Wakimoto, S.; Ueki, S.; Kimura, H.; Endoh, Y.; Hosoya, S.; Shirane, G.; et al. Doping dependence of the spatially modulated dynamical spin correlations and the superconducting-transition temperature in $La_{2-x}Sr_xCuO_4$. *Phys. Rev. B* **1998**, *57*, 6165. [CrossRef]
110. Tranquada, J.M.; Woo, H.; Perring, T.G.; Goka, H.; Gu, G.D.; Xu, G.; Fujita, M.; Yamada, K. Quantum magnetic excitations from stripes in copper oxide superconductors. *Nature* **2004**, *429*, 534. [CrossRef] [PubMed]
111. Coldea, R.; Hayden, S.M.; Aeppli, G.; Perring, T.G.; Frost, C.D.; Mason, T.E.; Cheong, S.-W.; Fisk, Z. Spin Waves and Electronic Interactions in $LaCuO_4$. *Phys. Rev. Lett.* **2001**, *86*, 5377. [CrossRef] [PubMed]
112. Seibold, G.; Becca, F.; Lorenzana, J. Theory of Antibound States in Partially Filled Narrow Band Systems. *Phys. Rev. Lett.* **2008**, *100*, 016405. [CrossRef]
113. Seibold, G.; Lorenzana, J. Time-Dependent Gutzwiller Approximation for the Hubbard Model. *Phys. Rev. Lett.* **2001**, *86*, 2605. [CrossRef]
114. Seibold, G.; Lorenzana, J. Magnetic Fluctuations of Stripes in the High Temperature Cuprate Superconductors. *Phys. Rev. Lett.* **2005**, *94*, 107006. [CrossRef]

115. Seibold, G.; Lorenzana, J. Doping dependence of spin excitations in the stripe phase of high-Tc superconductors. *Phys. Rev. B* **2006**, *73*, 144515. [CrossRef]
116. Lorenzana, J.; Seibold, G. Dynamics of Metallic Stripes in Cuprates. *Phys. Rev. Lett.* **2003**, *90*, 066404. [CrossRef]
117. Sawatzky, G.A. Quasiatomic Auger Spectra in Narrow-Band Metals. *Phys. Rev. Lett.* **1977**, *39*, 504. [CrossRef]
118. Cini, M.; Verdozzi, C. Photoemission and Auger CVV spectra of partially filled bands: A cluster approach. *Solid State Commun.* **1986**, *57*, 657. [CrossRef]

Publisher's Note: MDPI stays neutral with regard to jurisdictional claims in published maps and institutional affiliations.

 © 2020 by the authors. Licensee MDPI, Basel, Switzerland. This article is an open access article distributed under the terms and conditions of the Creative Commons Attribution (CC BY) license (http://creativecommons.org/licenses/by/4.0/).

Creative

Color Centers and Jahn-Teller Effect in Ionic Crystals—My Scientific Encounters with Alex Müller

Hans Bill

Département de Chimie Physique, Université de Genève, 1211 Genève 4, Switzerland; Hans.Bill@unige.ch

Received: 14 September 2020; Accepted: 29 September 2020; Published: 15 October 2020

Abstract: This contribution presents a personal account of the influence Karl Alex Müller had on the early stages of my career and the scientific questions about which we exchanged our views over the years. While both our research branched into a variety of topics, the common experimental technique, Electron Paramagnetic Resonance, and the Jahn-Teller effect led to fruitful exchanges of ideas on these matters in semiconducting, metallic and ionic crystals.

Keywords: Electron Paramagnetic Resonance (EPR); ENDOR; Jahn-Teller; color centers; $3d$ impurities; perovskite

---

Preface: Unexpected, and often rather ordinary encounters determine the path of a young scientist. My encounter with Karl Alex Müller (Alex in the following), winner with G. Bednorz of the 1987 Nobel prize in physics for the discovery of high-temperature superconductivity, was such an encounter when I started my PhD at the University of Geneva. A short time before leaving the Geneva Battelle Institute in 1963, Alex Müller presented a seminar to the group I was a member of. The meeting guided the use of my first experimental tool in research (Electron Paramagnetic Resonance, EPR) and the first steps of my scientific career. Here is a brief recollection of my encounters with Alex over the course of our careers, and some of the scientific questions about which we shared our views.

1. A Group Seminar at the University of Geneva

Alex studied Physics at the Eidgenössischen Technischen Hochschule in Zürich (ETHZ), and completed his doctorate in 1957 in the Institute of Professor Busch with a thesis entitled Paramagnetische Resonanz von Fe^{3+} in $SrTiO_3$ - Einkristallen. It was published in 1958 [1]. More about it below.

There was one anecdote mentioned to me by Charles Enz that is revealing of Alex's approach to physics. In 1956 Charles Enz completed his doctorate in theoretical physics and became (the last) assistant of Wolfgang Pauli. One of Charles' duties as an assistant consisted in taking care of the library for theoretical physics. Pauli was a frequent user of the library and assumed that Charles was there for his scientific work. The main door to the library was quite noisy. Alex Müller had many questions to ask a theoretician. Having met Charles during his first year as a PhD student at the ETHZ, he looked for ways to obtain some time from him to answer his questions. Alex succeeded in meeting Charles in the library. During one of these meetings they heard the squeaky noisy door. Charles' reaction was immediate: he asked Alex to pack his items and disappear through a side door. Charles knew he had work to do for Wolfgang Pauli. This anecdote reveals that already as a PhD student Alex had the right approach to succeed as a physicist; he did not shy off speaking with colleagues, including theoreticians, to help his understanding of the topic he was working on. An approach we share.

Alex left Zürich in 1958 having been nominated as group leader at the Solid State Physics Division of the Swiss Battelle Centre in Carouge, Geneva. At that institution he created the Electron Paramagnetic Resonance (EPR) group, realized and supervised the construction of an 18 GHz EPR spectrometer to do research on ferro- and antiferroelectric systems. He stayed at Battelle until 1963.

I completed my diploma in experimental physics on the realisation and analysis of an Omegatron mass spectrometer at the University of Bern in 1962. In October of the same year I moved with my wife to Geneva to work as a PhD student in the group of Professor Lacroix at the Physics Institute of the University of Geneva. In Bern I had become much interested in the strongly upcoming area of solid state physics and Geneva offered this field as one of the main streams of its physics curriculum. The topic of my thesis was the study of a color center in natural fluorites with the aid of Paramagnetic Resonance. In 1958 a female student, Anne-Marie Germanier, working in the physics department had observed a peculiar EPR spectrum in rose natural fluorites. My first task was to reproduce this result and to analyze the spectrum. I rapidly realized that the investigation of that specific color center is a challenging undertaking. But I also felt that widening the study, which concentrated mainly on structural aspects of a specific center, could bring new insight on the physics of color centers. This approach would also provide insight into the more general topic of solid state physics for example, by studying the microscopic interactions of electronic and vibrational degrees of freedom in a solid. It would indeed be much more interesting to extend the investigation by selecting individual ions (or molecular entities) that can serve as model systems for studying specific aspects of solid state theory and quantum mechanics. In short, to find such model systems.

My PhD advisor was a theoretician who performed calculations of the level structure of $3d$ ions in crystals. His specialty led him to be a consultant for the physics group at Battelle. In 1963, a few months before Alex left Battelle Roger Lacroix asked him to give an informal seminar to our group, which consisted at that time in Roger and myself.

2. What I Learned at the Seminar

I fondly remember the seminar held in the laboratory space I was occupying. At that time Alex was studying members of the ABO_3 family (Figure 1) containing intentionally introduced $3d$ impurities, using his 18 GHz EPR spectrometer and crystallographic techniques.

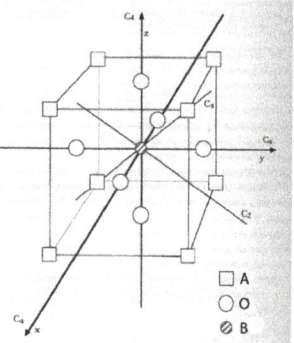

Figure 1. With permission from Reference [2]. Schematic structure of ABO_3.

He began his talk by presenting EPR results about $Cr^{3+/4+}$ $Mn^{2+/3+}$ and Fe ions introduced as local probes into the crystals [1,3,4]. The g-values, crystal field splittings, hyperfine structure constants (see e.g., Reference [5]) and effects of temperature and external fields on the non-Kramers $3d^2$, $3d^4$, $3d^6$ ground states are signatures of the electronic structure and bonding in the ferroelectric host. Hence, their studies provided a way to monitor structural and electronic properties of the ferroelectric host. Next, Alex analyzed the observed relations between g-values and crystal field effects in these matrices using the theory of Tanabe and Sugano [6]. It occurred to me during Alex's presentation that, indeed, the benefits these "impurities" provided for the investigation of the ferroelectric hosts could, *mutatis mutandi*, also benefit the field of color centers I was interested in. Specific color centers might be used as "spy ions" or "spy molecules" for the local electronic and vibronic excitations of the host crystal.

There was another insight that would spark my interest for many years to come. In that meeting Alex showed that the Jahn-Teller effect could be responsible for some of his results since degenerate electronic ground states had potentially to be considered as Jahn-Teller active. A concept new to me at the time. When asked about it, he presented the effect and recent developments of the theory of the Jahn-Teller effect. In particular results published by Longuet-Higgins et al. [7] on the static effect and by Frank Ham and Isaac B. Bersuker on the dynamic effect [8–12]. See also Reference [13]. Alex Müller's talk was a revelation to me as he provided new insight into solid state physics and, at the same time, confirmed and reinforced my point of view adopted at the beginning of my thesis work as mentioned above with respect to the possibility that color centers could provide fundamental insight on the quantum physics of ionic solids.

3. Interaction with Alex Müller during His Activity at IBM Rüschlikon

Visit at IBM in Rüschlikon. In 1975 I had the opportunity to meet Alex in his laboratory. I saw with much interest the magnetic resonance spectrometers he and his group had built. To me it has always been very inspiring to see the solutions other researchers had realized in their experimental setups. In particular, we both had a setup that allowed the application of unaxial stress and external electric fields. However, the experimental implementations were very different because Alex's group was studying essentially strongly covalent, semiconducting and metallic samples, while we focused on ionic crystals.

Among the results obtained with his equipment, Alex showed me data from $SrTiO_3$ doped with Fe^{3+} obtained under uniaxial stress at different temperatures [14]. The electron cloud of the free Fe^{3+} is approximately 12% smaller than the one of Ti^{3+} and has a $^6S_{5/2}$ ground state (no orbital degeneracy). It replaces a Ti^{3+} host ion. The EPR results and extracted spin Hamiltonian parameters correspond to a covalent structure with a rather small band gap ($\sim$3 eV). Their systems had more explicit many-body effects compared to what we observed in our experiments with the more ionic insulators Na_2S, Li_2S. In the latter, Gd^{3+} ($^8S_{7/2}$ ground state) introduced in pro mille concentrations served as the spy for ionic diffusion studies. Our experiments were realized beginning 1969 and also included the application of uniaxial stress [15–17].

Most interesting was to learn about his use of the angular overlap model of Newman [18–20]. This model describes the ligand field acting on appropriately chosen paramagnetic spies (e.g., 3d-ions, Rare Earths with electron half-integer spin and orbital-singlet ground state) in a crystal as a linear superposition of the individual spy-ligand interactions. This work uses spies which substitute for a host cation and ideally necessitates precise crystallographical information about the host crystal geometry. The procedure Alex had chosen, as he explained to me, was to use $SrTiO_3$ and $BaTiO_3$ crystals. As both componds show phase transitions and correspondingly different local symmetries of the cation coordination polyeders several local geometries are available and can be examined. The spies he used were Cr^{3+} (goes on both cation sites) or Fe^{3+} (on the Ti site). The EPR experiments they performed provided ligand field terms in the spin Hamiltonian as a function of stress and temperature. By assuming the power-law distance dependency proposed by Newman, experimental parameters for the energy as a function of the individual ligand-spy distance could be obtained. The crystal symmetry properties of the hosts were thereby included. As there are only oxygen anions and the crystals are rather covalent (the local ligand field is predominant) the experiments obtained such dependence of the energy on the cation-oxygen distance for both the A and B cations of the ABO_3 structure. This is, by the way, a beautiful application of EPR. As Alex explained, the application of this method to selected members is an essential step for the understanding of the origin of the phase transitions and electronic level structure of the ABO_3 group (see also Reference [21]).

We applied the Newman model to EPR on the layer perovskites MeFX (Me = Sr, Ba; X = Cl, Br) doped with Gd^{3+} or Eu^{2+}. These systems are quite different from those studied by Alex. First, they are much more ionic (with the exception of BaFI). Second, two different anionic ligands (e.g., F, Br) instead of one are involved. Hence, two constants had to be determined in the Newman function. Third, the ligand field is made of a superposition of several strong long-range components of partly opposite signs complicating the choice of the power law. Fourth, these hosts are tetragonal (no phase transitions at $p \sim 1$ bar); fortunately, much precise crystallographic information was available. The use of this analysis was motivated by the finding illustrated in Figure 2 (top) that SrFCl:Eu^{2+} showed a large crystal field splitting of almost 6 kG, whereas the Eu^{2+} EPR spectra had practically no splitting in the BaFCl host (bottom of Figure 2), indicating subtle geometric compensations between the individual ligand-spy contributions [22].

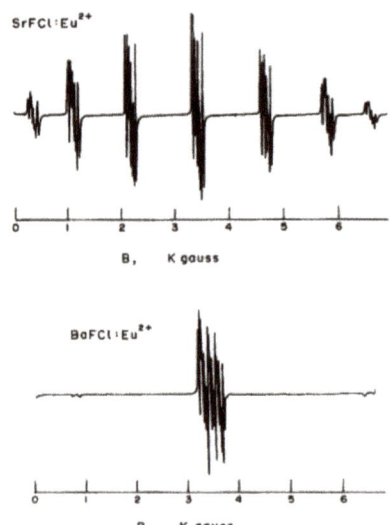

Figure 2. With permission, from Reference [22]. Electron Paramagnetic Resonance (EPR) spectra of Eu^{2+} in SrFCl (**top**) and BaFCl (**bottom**). Measurements were made at 4.2 K with magnetic field parallel to the C$_4$ axis.

External expert for a PhD on doped CaF$_2$. During the early research activity on color centers in CaF$_2$ I had noticed a publication by O'Connor and Chen which presented results related to my research (see References [23,24]). Discussions with Alex Müller at the Swiss Physical Society Meeting in Zürich (1970) and at IBM Rüschlikon strengthened my opinion that this publication deserved further work.

The in-house developed crystal growth setup (see below) allowed to grow a variety of doped CaF$_2$ crystals and to study the corresponding samples by Paramagnetic Resonance methods at temperatures as low as 1.6 K. A systematic study began in 1983 with one of my physics PhD students, Georges Magne. Alex Müller was one of the external experts for his thesis. The extended investigation was carried out with EPR and Electron Nuclear Double Resonance (ENDOR), both including uniaxial stress or monochromatic light applied to the sample. The dominant center studied consisted of an Y^{2+} substituting for a lattice Ca^{2+} ion. One of the eight F$^-$ ions of the surrounding fluorine cube is shifted from its equilibrium position along the [111] direction into a cation free neighboring F$^-$ cube, as observed by ENDOR. This results in a trigonal ligand field. Since the ground multiplet of Y^{2+} transforms as Γ_3 in C$_{3v}$ symmetry the trigonal field does not lift this degeneracy of the orbital electronic ground state, resulting in a trigonal E $\otimes$ e Jahn-Teller effect. The experimental facts that below 20 K the EPR spectrum has symmetry C$_s$ and reacts to uniaxial stress applied along specific directions indicate that a trigonal deformed "Mexican Hat" type potential with three equivalent minima is present. The potential barrier between two adjacent minima is approximately 0.35 eV. The EPR spectrum, "static" below 20 K, gradually converts into a "dynamical" structure and becomes trigonal around 40 K. A rich superhyperfine structure consisting of 34 lines is observed in the EPR spectrum of the pseudo-static center. The wavefunction of the ith Y $4d$ electron has admixed contributions of nine F^-, nearest and next-nearest neighbors as explored by ENDOR [25,26].

Several exchanges with Alex Müller regarding the model developed for this system and the experimental results gave way to new insights and orientations to the study. For example, the Frenkel versus Wannier exciton models was debated; the Frenkel exciton was found to be a more adequate description of the system. A still open question about the $CaF_2:Y^{2+}$ system is related to the optical absorption spectrum of the Y^{2+} center. There are absorption bands at 630, 490, 330 nm, with oscillator strengths of the order of 0.2–0.4. The short wavelength peaks were identified as arising from CaF_2 V_K, V_F centers but the other band had not clearly been assigned [27]. Tentatively we assigned it to a d-f transition of the $4d$ electron.

Graduate lectures by K.A. Müller in Geneva. Thanks to the *Convention Inter-Cantonale Romande pour l'Enseigement Du 3ème Cycle en Chimie* I could invite Alex for a series of lectures in Physical Chemistry on "Examples of the Application of Paramagnetic Resonance In Solids" (original title in french). The lectures took place between November 1985 and March 1986 in Geneva, and one at the University of Berne, hosted by Prof. Güdel. The invitation for this series was motivated by the many publications Alex wrote on paramagnetic resonance applied to ferroelectrics, and his focus on Jahn-Teller systems. These very well attended lectures presented the rich work of his group on local and collective (especially ferro-distortive) systems monitored through EPR and where the Jahn-Teller effect plays an essential role. Several invited talks by Alex followed during the next decade, including one to receive the honoris causa doctorate of the Univerity of Geneva.

The Jahn-Teller effect movement. Theoretical and experimental work on the Jahn-Teller effect in systems with a degenerate electronic ground state started in its modern form around 1954. British groups studied model cases for the static lifting of the electronic state degeneracy [7]. Another important progress was made by F.S. Ham and I.B. Bersuker who independently published seminal papers on the dynamic Jahn-Teller effect [8–12]. The vibrant activities around the static and dynamic Jahn-Teller effects motivated the creation of an interdisciplinary symposium, starting in Bad Honneff, Germany, in 1976. The "Jahn-Teller Symposium" gathers theorists and experimentalists, physicists and chemists who present and discuss new developments related to the Jahn-Teller effect. Alex was one of the founding members of the movement. The symposium met on a regular basis in Bad Honneff (Germany), Leoni (Italy), Trento (Italy), Chantilly (France), Oxford (UK), Nijmegen (Netherlands), Liblice (Czech Rep.), Marburg (Germany), Nottingham (UK), Kishinev (Moldova), Ovronnaz (Switzerland), Tartu (Estonia), Berlin (Germany), Erice (Sicily), Boston (USA), Leuven (Belgium), Beijing (China), Trieste (Italy), Heidelberg (Germany), Fribourg (Switzerland), Tsukuba (Japan), Graz (Austria) and Santander (Spain).

The symposium in Ovronnaz (Valais, Switzerland) took place at the end of August 1992 and was organized by the author. Alex agreed to participate. His presentation dealt with the Jahn-Teller polaron idea to explain the electronic level structure of high-temperature superconductors. Figure A1 is the so far unpublished official picture of participants at the Ovronnaz Jahn-Teller symposium, though not all of them are on the picture. Alex is in the first row, sixth from the left, with his wife on his left, and the author and his wife are next on his right. Other persons mentioned in this paper are R. Lacroix at the right end of the first row, F.S. Ham at the right end of the second row and I.B. Bersuker third from the left in the first row. The Appendix A lists the names of participants to the 1992 Jahn-Teller symposium.

4. A Few Common Scientific Interests

Over the years Alex Müller and I worked on quite different aspects of condensed matter physics. Nevertheless, our paths merged time and again on some topics of shared interest. I mention here three that generated fruitful exchanges.

Oxygen centers in the cubic MeF$_2$. Around 1965–1966 together with Eric Walker, one of R. Lacroix's diploma student that I was advising, we realized a crystal growth furnace of the Bridgman type which allowed to grow crystals in a controlled atmosphere containing an extremely low fraction of oxygen (in quantities less than a part per million). The resulting monocrystals of 2 to 8 cm^3 volume allowed to study oxygen centers, among others, which could be produced under controlled conditions (e.g., by using rare isotopes). Alex showed interest in the setup we had built. Initial work with this furnace and EPR/ENDOR experiments identified the origin of the color of natural rose Fluorites: an ($Y^{3+}O_2^{3-}$) molecule [28]. Another system of interest to Alex was the paramagnetic oxygen ion O^- ion substituting for an F^- ion in the CaF$_2$ lattice. It undergoes a $T_{2g} \otimes e$ Jahn-Teller effect below ~40 K as measured with EPR, and reacts strongly to uniaxial stress applied to the sample crystal [29]. The strongest effects were observed on the O^- ion in SrF$_2$ [30]. In 1969 we successfully realized exploratory experiments to use this oxygen ion as a cooling engine starting at 1.6 K.

CaF$_2$:La—charge compensation and exciton effects. In 1985, during Alex's lectures at the University of Geneva we had an interesting discussion on "when is the identification of a Jahn-Teller center in a given host established". The La ion in CaF$_2$ is doubtlessly a system where this question applies. As-grown crystals from our furnace never showed any Lanthanum EPR signal, indicating the high purity of the samples. But the La^{2+} ion (5d^1) appears as a Jahn-Teller system (cubic, $E \otimes e$), for example, when these crystals are subsequently weakly hydrolized and then X-irradiated at 255 K [31]. EPR shows an isotropic spectrum above 16 K and an axial one below about 6K. EPR realized with uniaxial stress applied to the sample demonstrated that the system reorients between the three tetragonal minima of a warped "Mexican Hat" potential. In addition, the EPR spectrum showed the O^- ion mentioned above with all the dynamical features present. Clearly this signaled that the O^- is indeed the Jahn-Teller ion described above. The two centers are mutually independent and the equation of the electron transfer is $O^{2-} + La^{3+} \xrightarrow{X-rays} O^- + La^{2+}$. Thus, there is sufficient distance between the two ions that no mutual interaction shows up in the spectra.

Non-hydrolized CaF$_2$:La. Crystals grown from highest quality powders and excluding any Oxygen, present the most detailed information regarding the Jahn-Teller effect that the La^{2+} ion undergoes. At 4.2 K the EPR spectrum shows 10 lines of 28 Gauss peak-to-peak linewidth (first derivative spectra) due to the Lanthanum ion. Depending on the sample and its treatment a few narrow and weak lines are observed on the low field part of the spectrum when the magnetic field is parallel to a C$_4$ axis. EPR under uniaxial stress applied to the CaF$_2$:La^{2+} crystal at a temperature of 1.7 K allowed to clarify this situation. At an applied pressure of 5.2×10^8 Pa along a [110] direction and with the applied DC magnetic field **B** parallel to the [001] direction, perpendicular to [110], a strong group of 64 narrow lines appeared due to the hyperfine interaction of the electron spin with the La-nucleus and super-hyperfine interaction with a number of F^- nuclei. The g-tensor is axial with $g_{zz} > g_{xx} = g_{yy}$, resulting in only partial overlap of the spectral components. Indeed, the parallel spectrum is on the low magnetic field side (the intense group of 64 lines) and the one perpendicular to **B** on the high field side of the spectrum. Both groups overlap, thereby covering 4 of the 8 lines of the perpendicular spectrum. The chosen geometry of the fields applied to the crystal corresponds to the situation where the minimum of the deformed "Mexican hat" potential associated with the parallel component of the EPR spectrum is lowered in energy while the other two minima are pushed up, thereby remaining mutually degenerate. These latter ones are responsible for the perpendicular (8-line) spectral component. This indicates that the parallel spectrum behaves as a Frenkel exciton [32].

Fluoro-perovskites doped with Cu^{2+}. At the XXII Ampère Congress in Zürich my group showed a poster of preliminary results on KZnF$_3$ doped with Copper. The as-grown crystals presented EPR spectra of Cu^{2+} the symmetry of which depends on growth conditions. The ones shown in Zürich were isotropic at 20 K and "static" at 1.6 K sample temperature. Uniaxial stress was applied to the crystal. Alex passed by and was interested in the Copper data as well as in the details of the crystal preparation. At the time we only had first EPR results. But encouraged by his interest I enlarged the scope of the research to several members of this family (KZnF$_3$, CsCdF$_3$ and RbCdF$_3$), using Cu^{2+} and

Ag^{2+} as spy ions. The crystals of the $KZnF_3:Cu^{2+}$ system remains cubic, the dynamical effects the Cu ion present in the ground state are of the Jahn-Teller type $E \otimes e$ [2,33,34].

5. Concluding Remarks

The recollections in this article span several decades, but only a portion of the scientific questions both Alex Müller and myself addressed in our respective laboratories. They underline the importance of personal scientific exchanges. Our interactions were determinant at the beginning of my career and always constructive thereafter. These fruitful conversations, spread over the years, dealt with details of our common experimental techniques as we applied them to solids with very different physical properties, but also with the physics they revealed and how theoretical works could be applied to describe our experimental findings. I am grateful to Alex for the enriching exchanges on our shared interests.

Funding: This research received no external funding.

Acknowledgments: I thank Andreas Bill ("der junge Bill" as Alex Müller calls him) for feedback on the manuscript.

Conflicts of Interest: The author declares no conflict of interest.

Appendix A. The 1992 Jahn-Teller Symposium in Ovronnaz, Switzerland

Figure A1 is a picture of most participants of the 1992 Jahn-Teller symposium in Ovronnaz, Switzerland, which the author of the paper organized. Following is the list of participants in alphabetic order (not including spouses). The numbers in parenthesis (X, Y) following a name indicate the row and column on the picture for those my recollections allowed to identify. For example, Alex Müller is the sixth person (from the left) in the first row (from the bottom) and thus has the index $(1, 6)$. His wife Inge Müller is the person on his left and has the index $(1, 5)$. The author and his wife are on Alex's right and have indices $(1, 7)$ and $(1, 8)$, respectively.

Figure A1. Participants of the 1992 Jahn-Teller symposium organized by the author in Ovronnaz, Switzerland. Alex Müller is in the first row, sixth from the left. His wife Inge is on his left. The author and his wife are next on his right. The complete list of participants is in the appendix.

M. Atanasov, C.A. Bates (2,7), I.B. Bersuker (1,3), A. Bill (3,13), H. Bill (1,7), O. Brandt, C.C. Chancey (4,4), F.Th. Chaudron, T.S. Dod, M. Dominoni, J.L. Dunn (2,10), H. Eiermann (3,9), M. Eremin (3,3), A. Furlan, H. Hagemann (3,1), F. S. Ham (2,15), S. Jamila (2,9), Z. Jirak, K.H. Johnson, M. Kaplan (3,4), P.J. Kirk (2,8), R. Lacroix (1,14), W.J.A. Maaskant (4,6), E. Mayerniekova (3,14), R. Meiswinkel, K.A. Müller (1,6), M.C.M. O'Brien (3,5), I. Ohnari (4,5), E.P. Pearl (2,8), H. Reik (3,8), D. Reinen (4,2), U. Schotte, K.D. Schotte (2,3), H.J. Schulz (2,12), E. Sigmund (3,11), Ch. Simmons (1,9), D.R. Taylor (2,5), H. Thomas (4,3), I. Vallin, H. Van d. Waals (4,9), J. Van Tol (4,10), A.M. Vasson (1,2), A. Vasson (1,1), M. Wagner (3,6), R. Willett (4,10), O.A. Yakovleva (1,11).

References

1. Müller, K.A. Paramagnetische Resonanz von Fe^{3+} in $SrTiO_3$ - Einkristallen. *Helv. Phys. Acta* **1958**, *31*, 173.
2. Minner, E. Etude Spectroscopique des Ions Jahn-Teller Cuivre et Argent Bivalents dans des Monocristaux de Fluoro-Perovskites de Composition Chimique AMF_3. Ph.D. Thesis, University of Geneva, Geneva, Switzerland, 1993; Thesis N° 2635.
3. Müller, K.A. Electron Paramagnetic Resonance of Manganese IV in $SrTiO_3$. *Phys. Rev. Lett.* **1959**, *2*, 341; See also his list of references published on the internet. [CrossRef]
4. Wysling, P.; Müller, K.A.; Hochli, U. Paramagnetic Resonance and Relaxation of Ag^{2+} and Pd_3P in MgO and CaO. *Helv. Phys. Acta* **1965**, *38*, 358.
5. Müller, K.A.; Simanek, E.S. *Covalency and Hyperfine Structure Constant A of Iron Group Impurities in Crystals*; RZ 296 (# 11513); IBM Thomas J. Watson Research Center: New York, NY, USA, 1969.
6. Tanabe, Y.; Sugano, S. On the Absorption Spectra of Complex Ions. I. *J. Phys. Soc. Jpn.* **1954**, *9*, 753–766. [CrossRef]
7. Longuet-Higgins, H.C.; Öpik, U.; Pryce, M.H.L.; Sack, R.A. Studies of the Jahn-Teller effect. II. The dynamical problem. *Proc. Roy. Soc. Lond.* **1958**, *A244*, 1.
8. Ludwig, G.W.; Ham, F.S. Effects of Electric Fields in Electron Spin Resonance. *Bull. Am. Phys. Soc.* **1962**, *7*, 29.
9. Ham, F.S. Dynamical Jahn-Teller effect in paramagnetic resonance spectra: orbital reduction factors and partial quenching of spin-orbit interaction. *Phys. Rev. A* **1965**, *138*, 1727. [CrossRef]
10. Bersuker, I.B. Inversion Splitting of Levels in Free Complexes of Transition Metals. *Zh. Eksperim. i Teor. Fiz.* **1962**, *43*, 1315; [English transl.: *Soviet Phys. JETP* **1963**, *16*, 933].
11. Bersuker, I.B. Spin-inversion Levels in a Magnetic Field and the EPR Spectrum of Octahedral Cu^{2+} Ion Complexes. **1963**, *44*, 1239; [English transl.: *Soviet Phys. JETP* **1963**, *17*, 836].
12. Bersuker, I.B.; Vekhter, G.B. *Fiz. Tverd. Tela* **1963**, *5*, 2432; [English transl.: *Soviet Phys. Solid State* **1964**, *5*, 1772].
13. Müller, K.A. *Magnetic Resonance and Relaxation*; Blinc, R., Ed.; North-Holland: Amsterdam, The Netherlands, 1967; Volume 11, p. 192–208.
14. Von Waldkirch, T.; Müller, K.A. EPR of Fe^{3+} in $SrTiO_3$ Under Uniaxial (100) Stress. *Helv. Phys. Acta* **1973**, *46*, 331–339.
15. Moret, J.M.; Bill, H. Experimental determination of the spin-stress coefficients of Gd^{3+} in the new host crystal Na. *Phys. Stat. Sol. A* **1977**, *41*, 163–171. [CrossRef]
16. Montaner, A.; Galtier, M.; Benoit, C.; Bill, H. Optical constants of sodium sulphide. *Phys. Stat. Sol. A* **1979**, *52*, 597. [CrossRef]
17. Carron, P.-L. Etude Expérimentale des Effets de Désordre à Hautes Températures dans des Cristaux de Sulfure Alcalins de Structure Anti-Fluorine. Ph.D. Thesis, University of Geneva, Geneva, Switzerland, 1990; Thesis N° 2444.
18. Newman, D.J. Theory of lanthanide crystal fields. *Adv. Phys.* **1971** *20*, 197. [CrossRef]
19. Newman, D.J.; Urban, W. Interpretation of S-state ion EPR spectra. *Adv. Phys.* **1975**, *24*, 793. [CrossRef]
20. Newman, D.J.; Siegel, E. Superposition model analysis of Fe^{3+} and Mn^{2+} spin-Hamiltonian parameters. *J. Phys. C* **1976**, *9*, 4285. [CrossRef]
21. Ritz, D; Boatner, LA; Chatelain, A; Höchli, U.T.; Müller, K.A. Octahedral Potential in $KTA_{1-X}NB_XO_3$ for an ION 6S5/2-Fe^{3+}. *Helv. Phys. Acta* **1978**, *51*, 430–433.
22. Nicollin, D.; Bill, H. Gd^{3+}, Eu^{2+} in SrFCl and BaFCl single crystals: EPR results. *Sol. State Comm.* **1976**, *20*, 135. [CrossRef]

23. O'Connor, J.R.; Chen, J.H. Color Centers in Alkaline Earth Fluorides. *Phys. Rev.* **1963**, *130*, 1790. [CrossRef]
24. Alig, R.G. Theory of Photochromic Centers in CaF$_2$. *Phys. Rev. B* **1971**, *3*, 536. [CrossRef]
25. Bill, H.; Magne, G.; Balestra, C.; Lovy, D. A trigonally charge-compensated Y^{2+} ion in CaF$_2$: A Jahn-Teller system. *J. Phys. C* **1986**, *19*, L19. [CrossRef]
26. Magne, G. Etude Par résonance Magnétique de Centres Photochromiques dans les Fluorures Alcalino-Terreux. Ph.D. Thesis, University of Geneva, Geneva, Switzerland, 1986; Thesis N° 2189.
27. Hayes, W.; Twidell, J.W. The self-trapped hole in CaF$_2$. *Proc. Phys. Soc.* **1962**, *79*, 1295. [CrossRef]
28. Bill, H. Investigation of Color Centers in Alkaline Earth Fluorides. Ph.D. Thesis, University of Geneva, Geneva, Switzerland, 1968; Thesis N° 1475.
29. Bill, H.; Silsbee, R.H. Dynamical Jahn-Teller and reorientation effects in the EPR spectrum of CaF$_2$:O$^-$. *Phys. Rev. B* **1974**, *10*, 2697. [CrossRef]
30. Bill, H. Chapter 13. In *The Dynamical Jahn-Teller Effect In Localized Systems*; Perlin, Y.E., Wagner, M., Eds.; Elsevier: Amsterdam, The Netherlands, 1984; pp. 709–817.
31. Hayes, W.; Twidell, J.W. Paramagnetic resonance of divalent lanthanum in irradiated CaF$_2$. *Proc. Phys. Soc.* **1963**, *82*, 330. [CrossRef]
32. Bill, H. University of Geneva, Geneva, Switzerland. Unpublished, 2020.
33. Minner, E.; Bill, H. The Jahn-Teller systems Ag^{2+} and Cu^{2+} in RbCdF$_3$, an EPR investigation. In Proceedings of the XII International Conference on Defects in Insulating Materials, Schloss Nordkirchen, Germany, 16–22 August 1992; Kanert, O., Spaeth, J.-M., Eds.; Volume 1, p. 571.
34. Minner, E.; Lovy, D.; Bill, H. Electron paramagnetic resonance and relaxation study of copper (II) and silver (II) in CsCdF$_3$ single crystals. *J. Chem. Phys.* **1993**, *99*, 6378. [CrossRef]

Publisher's Note: MDPI stays neutral with regard to jurisdictional claims in published maps and institutional affiliations.

© 2020 by the authors. Licensee MDPI, Basel, Switzerland. This article is an open access article distributed under the terms and conditions of the Creative Commons Attribution (CC BY) license (http://creativecommons.org/licenses/by/4.0/).

Article
A Lattice Litany for Transition Metal Oxides

Alan R. Bishop

Los Alamos National Laboratory, Los Alamos, NM 87545, USA; arb@lanl.gov

Received: 23 June 2020; Accepted: 3 July 2020; Published: 13 July 2020

Abstract: In this tribute to K Alex Müller, I describe how his early insights have influenced future decades of research on perovskite ferroelectrics and more broadly transition metal oxides (TMOs) and related quantum materials. I use his influence on my own research journey to discuss impacts in three areas: structural phase transitions, precursor structure, and quantum paraelectricity. I emphasize materials functionality in ground, metastable, and excited states arising from competitions among lattice, charge, and spin degrees of freedom, which results in highly tunable landscapes and complex networks of multiscale configurations controlling macroscopic functions. I discuss competitions between short- and long-range forces as particularly important in TMOs (and related materials classes) because of their localized and directional metal orbitals and the polarizable oxygen ions. I emphasize crucial consequences of elasticity and metal–oxygen charge transfer.

Keywords: transition metal oxides; lattice–spin–charge landscapes; elasticity

1. Introduction

This article is a personal perspective on aspects of perovskites, particularly transition metal oxides (TMOs). It is written through the (limited) lens of my own research journey, as a tribute to K. Alex Müller and the lessons I have accumulated from his prescient insights into this remarkably tunable class of materials. I first met Alex during a several-month visit to IBM Ruschlikon in 1977 to collaborate on nonlinear excitations and structural phase transitions, on which I had begun research in 1974 with Jim Krumhansl at Cornell. I have been privileged by numerous interactions with Alex since then. Science advances relentlessly, but some pioneers are able to perceive the truth beyond the limits of current techniques. Alex is such an individual. I similarly recall many conversations with Heinrich Rohrer during the 1977 visit concerning prospects for Scanning Tunneling Micsoscopy (STM). In particular, discommensurations were only indirectly suggested by k-space scattering techniques at that time and were a topic of strong disagreements regarding data interpretation. STM directly imaged these structures, resolved many of the disagreements, and contributed to important future research on commensurate–incommensurate phase transitions.

In this spirit of important insights, I will highlight just three (among many in his illustrious career) from Alex in the context of his decades of research on TMOs, including ferroelectrics (SrTiO$_3$, BaTiO$_3$, etc.) and of course cuprate superconductors. These insights have resonated though research history as experimental, theoretical, and simulation capabilities have improved.

(i) *Structural phase transitions*

Neutron and X-ray scattering in the 1970s were beginning to have sufficient resolution to suggest two low-frequency scattering components as the structural phase transition was approached: "central peaks" and "soft modes." Alex recognized the need for judicious experiments to probe different *timescales* [1] and was thereby able to separate phonon oscillations from slow cluster dynamics. This was powerful input to theory and simulation attempting to distinguish mean-field self-consistent phonon approximations from true critical behavior in double-well Landau–Ginzburg phase-transition theories. Advanced time-resolved experiments

(ii) *Precursor Structure*

Local structure measurements and inferences of the 1970/80s had relatively poor resolution (spectroscopy, NMR, diffuse scattering, etc.). However, Alex used them [1] to suggest that local distortions appear as precursor structure as Tc is approached in ferroelectric TMOs. Importantly, he showed that these precursors onset at temperatures significantly beyond critical regimes above and below Tc, and are tunable with strain, electric fields, etc. Fifty years later, I can suggest that many of these properties in TMOs and related materials are *elastic* microstructures. Indeed, as I discuss below, tuning phases and functions through elasticity is now emerging as an important focus in quantum materials [2].

(iii) *Quantum Paraelectricity*

Unlike, e.g., $BaTiO_3$, $SrTiO_3$ does not undergo a ferroelectric phase transition, but one can be induced with appropriate doping or pressure (strain). Alex understood the importance of this and advocated the concept of quantum tunneling between orientations (e.g., octahedral orientations), which had to be frozen out to stabilize a permanent ferroelectric state. He and others created the term "quantum paraelectric" to describe this situation and designed elegant experiments to probe the dynamics [2,3]. This concept now has many diverse analogs, including the internal dynamics of small polarons (below) and excitons, Kondo spin singlets, dynamic magnetism in Pu, and concepts for computational qubits such as Josephson junctions.

In the next sections, I will describe connections of these lessons from Alex to some of my own research on TMOs and related materials.

2. TMOs and Their Lattice

It has long been appreciated (see, e.g., [4]) that TMOs exhibit a striking variety of broken-symmetry ground states, including magnetic, Peierls, Mott, spin-Peierls (SP), charge density wave, (CDW) bond order wave (BOW), superconductivity, etc. We can ascribe this variety to sensitive coupling among degrees of freedom—spin, charge, orbital, and lattice. Much research over recent decades has emphasized the metal d-orbitals, arguing that since these are substantially localized, competitions occur between localized states or flat electronic k-space bands, and, through hybridization, wider (e.g., Op) bands. Many fascinating electronic/magnetic many-body states emerge from modeling these competitions, some of which surely occur in actual current, future, or engineered materials. However, here I will deliberately play the devil's advocate and emphasize the explicit importance of the *lattice* degrees of freedom, the oxygen ion polarizability, and functional multiscales beyond asymptotic scaling limits.

It is a fascinating feature of science sociology that "solid-state and correlated electron physics" and "materials" research separated so much in the last several decades. In part, this was the result of experimental, theoretical, and simulation limitations. For example, quantum mechanics could only be implemented for periodic (and small supercell) structures. This led to the creation of a comprehensive conceptual framework in which lattice variations were a linear perturbation used to describe extended phonons, and similarly magnons, etc. In contrast, materials science recognized the functional role of microstructure (dislocations, grains, twinning, etc.) but often omitted important entropic contributions and built interpretative frameworks around the observed microstructure. This situation has begun to change because of decisive advances in experimental probes and their resolution capabilities (real-space, k-space, time, frequency) and concomitant advances in computing power (for all-atom system sizes, ab initio electronic methods, non-adiabatic effects, etc.), as well as data analytics and visualization at scale. Equally importantly, the technologies now based on nanoscales and on active surfaces, internal surfaces, and multilayer interfaces require the unification of the disciplines and the explicit roles of all the degrees of freedom and their coupling.

Under-representing the functional role of the lattice is an oversight for several reasons:

1. Although the d-orbital (and even more so, the f-orbital) is indeed electronically localized, resulting in localization–delocalization electronic competition, it is also highly directional. This results in symmetry constrained unit cell structural distortions and a "network" competition for ground and metastable structural patterns (multiscale "landscapes"). This is not the case for extended and symmetrical (e.g., s) orbital materials, where dynamic screening dominates. The constraints are the origin of measured strongly anisotropic elastic constants in these materials, with intrinsically coupled configuration scales from unit cell to long-range, optic to acoustic—and hence high tunability by both local and global perturbations (doping, pressure, external fields).

2. Neglecting or "integrating out" the oxygen degrees of freedom in TMOs and related materials is a significant over-simplification for many properties. The O polarizability and metal–O charge transfer (and associated bond length/buckling/rotation changes) must be treated explicitly [5]. Pioneers such as Heinz Bilz (Alex's professional peer and colleague) appreciated this by augmenting "shell models" of the TMO electronic structure to capture effects of M–O charge transfer and polarizability [6]. We return to such "nonlinear shell models" below, including a successful prediction of the observed quantum paraelectric to ferroelectric transition in O_{18}-doped $SrTiO_3$.

3. The electron–lattice coupling strength is typically *not* weak in TMOs. It may appear so if measured by conventional spatial averaging techniques. However, because of the delicate energy balances affecting electronic/magnetic orders, even rather weak average electron–lattice coupling can dominate globally and locally. We illustrate this with examples below. In addition, exotic (e.g., topological) singularities are usually energetically costly, and nature avoids or smooths them by engaging additional weaker degrees of freedom—as in dislocations, vortices, superconducting flux line cores, etc. Even when the globally averaged el-lattice strength is weak, it can be locally strong around dopants and defects. The formation of small (coupled spin–charge–lattice) polarons is an important example that we return to below. A finite densities of such polarons can order into secondary mesoscopic patterns (clumps, stripes, filaments, checkerboard phases, etc.) because of the long-range, directional elasticity noted above. Small polaron center-of-mass dynamics is very slow (because of Peierls–Nabarro lattice pinning), but unless dissipation is strong, their *internal* dynamics is a fundamental quantum tunneling property, and the internal charge oscillation is necessarily accompanied by non-adiabatic lattice (e.g., bond–length), and sometimes spin, oscillation. This coupled charge–lattice dynamics is familiar elsewhere, including quantum chemistry (e.g., [7]), "macroscopic quantum tunneling" [8], and perhaps also in the context of quantum paraelectric tunneling [9].

4. Lattice *anharmonicity* is typically important in TMOs and assuming linear lattice dynamics is incomplete. Anharmonicity is the result of slaving among lattice, electronic, and magnetic degrees of freedom, proximity to a structural phase transition, impurities, interfaces, surfaces, etc. The M–O charge transfer is a particularly important source of nonlinear lattice dynamics. Among many interesting emergence and complexity consequences are multi-phonon bound states ("intrinsic local modes," ILMs). Modern neutron scattering has indeed resolved modes outside linear phonon bands and attributed them to ILMs (e.g., [10]). Below, we will introduce ILMs embedded self-consistently in a sea of extended modes as a description of, e.g., relaxor ferroelectrics.

3. Small Polarons, Filamentary Landscapes, and Local Modes

As noted, TMOs as a materials class exhibit a remarkable variety of often competing broken symmetry ground states, which have been the subject of intensive research using a huge range of experimental, numerical, and analytical techniques [11,12]. For example, ab initio (e.g., density functional theory) and even fully non-adiabatic quantum electronic calculations have advanced significantly in recent years because of great strides in algorithms and computational power, e.g., [13,14]. However, they remain limited for many properties by technique-imposed assumptions such as system or supercell size, periodicity, treatment of polarizability, adiabaticity, etc. This means that accurate TMO description remains a challenge for the inclusion of O polarizability and multiscale (including elastic) lattice patterns. We are typically driven to separate the mechanisms creating mesoscopic structures (e.g., stripes) from modeling the signatures and functionality of those structures. I will give examples of both of these steps below.

Some useful insights can be gained from *real-space* Hartree–Fock (HF) numerical solutions of 3-band (M–d and O–p) M–O charge and spin models, *including* lattice degrees of freedom treated adiabatically (e.g., in a Su–Schrieffer–Heeger inter-site form for el–lattice coupling). For example, [15] uses such a model Hamiltonian with parameters appropriate to a Cu–O plane but allowing parameter strengths (electronic hopping, electron correlation, el–lattice coupling) to vary. A variety of broken-symmetry ground states are found numerically—including CDW, BOW, SP, AF, co-existing SP and anti-ferromagnet (AF)—as magnetism, covalency, and lattice distortion compete. In particular, as observed in Section 2, relatively weak el–lattice coupling is found to induce a zero-temperature ground state transition.

The Hamiltionian studied in [15] is

$$\mathcal{H} = \sum_{i \neq j, \sigma} t_{ij}(\{u_k\}) c_{i\sigma}^\dagger c_{j\sigma} + \sum_{i,\sigma} e_i(\{u_k\}) c_{i\sigma}^\dagger c_{i\sigma} + \sum_i U_i c_{i\uparrow}^\dagger c_{i\downarrow}^\dagger c_{i\downarrow} c_{i\uparrow}$$
$$+ \sum_{\langle i \neq j \rangle, \sigma, \sigma'} U_{ij} c_{i\sigma}^\dagger c_{j\sigma'}^\dagger c_{j\sigma'} c_{i\sigma} + \sum_l \frac{1}{2M_l} p_l^2 + \sum_{k,l} \frac{1}{2} K_{kl} u_k u_l. \tag{1}$$

Here, $c_{i\sigma}^\dagger$ creates a *hole* with a spin σ at the site i in the Cu $d_{x^2-y^2}$ or the O $p_{x,y}$ orbital. In the lattice part, for simplicity only, the motion of O ions along the Cu–O bonds are includeed, and it is assumed that only diagonal components of the spring constant matrix are finite, $K_{kl} = \delta_{k,l} K$. For electron–lattice coupling, the nearest neighbor Cu–O hopping is modified by the O-ion displacement u_k as $t_{ij} = t_{pd} \pm \alpha u_k$, where the +(−) applies if the bond shrinks (stretches) with positive u_k. The Cu-site energy is assumed to be modulated by the O-ion displacements u_k linearly, $e_i = \epsilon_d + \beta \sum_k (\pm u_k)$, where the sum extends over the four surrounding O ions; here, the sign takes +(−) if the bond becomes longer (shorter) with positive u_k. The other electronic matrix elements are O–O hopping $(-t_{pp})$ for t_{ij}, O-site energy (ϵ_p) for e_i, with $\Delta = \epsilon_p - \epsilon_d$, Cu-site (U_d) and O-site (U_p) repulsions for U_i, and the nearest-neighbor Cu–O repulsion (U_{pd}) for U_{ij}. Parameter values are used in regimes relevant to the copper oxides from local density approximation (LDA) calculations: $t_{pd} = 1$, $t_{pp} = 0.5$, $\Delta = 3$, $U_d = 8$, $U_p = 3$, and $U_{pd} = 1$. These parameters and $\lambda_\alpha = \lambda_\beta = 0$ are used as a reference parameter set: $\lambda_\alpha = \alpha^2/(Kt_{pd})$, $\lambda_\beta = \beta^2/(Kt_{pd})$. λ_α, λ_β, U_{pd}, and U_d are varied and Δ changed with U_{pd} and U_d so as to maintain a constant renormalized energy difference between Cu and O levels in the undoped case. A comparison of results for local lattice distortion and reduced Cu magnetic moments accompanied by added holes with generalized, inhomogeneous LDA calculations is consistent, e.g., with values of $\lambda_\alpha = 0.28$, $\lambda_\beta = 0$, and $K = 32 t_{pd}/\text{Å}^2$.

It is natural to examine whether electron or hole polarons result upon doping into the various broken-symmetry ground state: a generalization of the much studied Holstein polarons (see [16]) to coupled spin–charge–lattice local quenching of the broken symmetry. As mentioned above, among the essential properties of polarons (and excitons) are the center-of-mass translation of the composite local deformation (resulting in polaronic electronic bands) and the internal (charge, lattice, spin) dynamics *within* the composite wave function. "Large" polarons (deforming many lattice sites coherently) have wide electronic k-space bands and incoherent internal dynamics, and they can translate (tunnel or, with temperature, diffuse) relatively easily, scattering off phonons, impurities, etc. In contrast, "small" polarons (deforming few lattice sites) have narrow electronic bands and transport very slowly from site to site by quantum tunneling or thermal hopping: they are easily pinned by the lattice's Peierls–Nabarro barrier, weak impurities, surfaces, or fields. However, the small polaron internal dynamics exhibit coherent quantum tunneling oscillations. Again, accurate quantum simulations, even for model Hamiltonians, are limited to very small systems, despite modern computing power and efficient algorithms (e.g., [17]). Nevertheless, we can gain important insights by studying the limit of a single charge added to, e.g., a single O–M–O unit (an example of a Boson–Fermion composite and relevant to small polarons on cuprates, bismuthates, and nickelates, for example). This limit is, numerically, exactly solvable, including fully non-adiabatic (i.e., multi-quanta) treatment of lattice, charge, and spin. This limit is also relevant, since the extremely slow center-of-mass translation can be essentially decoupled. As described in detail in [18], the total energy of such a unit is a double well with two degenerate lowest energy configurations, corresponding to the added charge occupying the left or right O, and the charge oscillates (quantum tunnels) periodically between them. Figure 1 shows the numerically exact ("quantum-entangled") ground-state wave function as the strength of el–lattice coupling is varied. As a result of el–lattice (M–O bond length) and spin coupling, the lattice and spin also oscillate. The symmetric ground state captures the double-well probability, and the gap to the anti-symmetric first excited state quantifies the tunneling frequency. We can visualize this situation as an M–O bond length oscillating between short and long M–O bond lengths as the charge tunnels (a "charged lattice vibration" in some earlier literature). This picture is useful but an oversimplification, since non-adiabaticity means that the total energy double well is effectively changing shape as the charge tunnels and thus the frequencies of charge, lattice, and spin are, although related, not the same: indeed, many interesting resonances are possible [18,19]. In fact, this is a precise description of lattice/spin-assisted C–T (i.e., dynamic polarization) in this small unit. There are many relevant issues such as polaronic excited states, decoherence, spectroscopy, etc. Ref. [18] calculates the tunneling frequencies for parameters relevant to cuprates. These frequencies are indeed similar to "anomalous" ones measured in cuprates (optical, polarizability, magnetic) and the Cu–O bond length differences from average Cu–O bond lengths are also similar to those measured by, e.g., XAFS and neutron pair-distribution functions [18,20]. These results illustrate the differences between measurement techniques as a function of their time resolution—*recall Alex Müller!* An important additional consideration is the dissipation of energy during tunneling through coupling to the medium in which the tunneling unit is embedded. If the tunneling is slow, the dissipation freezes the polaron into a permanent polar distortion. This is similarly important to Caldeira–Leggett [8] or Kondo singlet freeze-out, to quantum paraelectrics [9], and probably to other TMO functionalities (below).

Although the unit above is directly relevant to small polarons in 1D chains and quasi-1D structures embedded in higher dimensions, fully quantum, non-adiabatic calculations remain rather limited, despite continuous progress; for example, Ref [21] reports excellent advances in the first-principle characterization of a single polaron in WOx. In particular, as the doping level increases in a full M–O system, we are faced with the question of how multiple polarons interact. This is another problem beyond current analytical or simulation capability except in rather small systems [17,22]. Qualitatively, we can expect that at low density, small polarons localize independently; if they are close enough, they may bind into bi/multi-polaron clumps (depending on lattice deformation and Coulomb energies), and at densities where the average polaron spacing is near the polaron size, they transform their broken-symmetry host (probably through a sequence of commensurate–incommensurate transitions) to a metallic unbroken symmetry with a new Fermi energy ("quantum melting").

We can again numerically explore aspects of this scenario using the same 3-band Peierls–Hubbard model (1), treated in a real-space HF approximation but now including electron or hole doping. This cannot directly describe either center-of-mass or internal tunneling dynamics. However, it is a useful guide to the (positive) feedbacks leading to polaron formation and multi-polaron patterning, and the coupled roles of lattice, charge, and spin. Reference [23] shows that single electron and hole small polarons occur in the various broken-symmetry ground states, as illustrated in Figure 2. At finite doping, bipolarons can be found [22] but also ordering into filamentary "stripes." As discussed in [24], these filamentary patterns share oxygens and thus minimize potential energy. At sufficient doping, a hysteretic insulator to metal transition is found. Recent progress beyond HF with quantum MC simulations also suggest the possibility of multi-polaron liquid-like states [22].

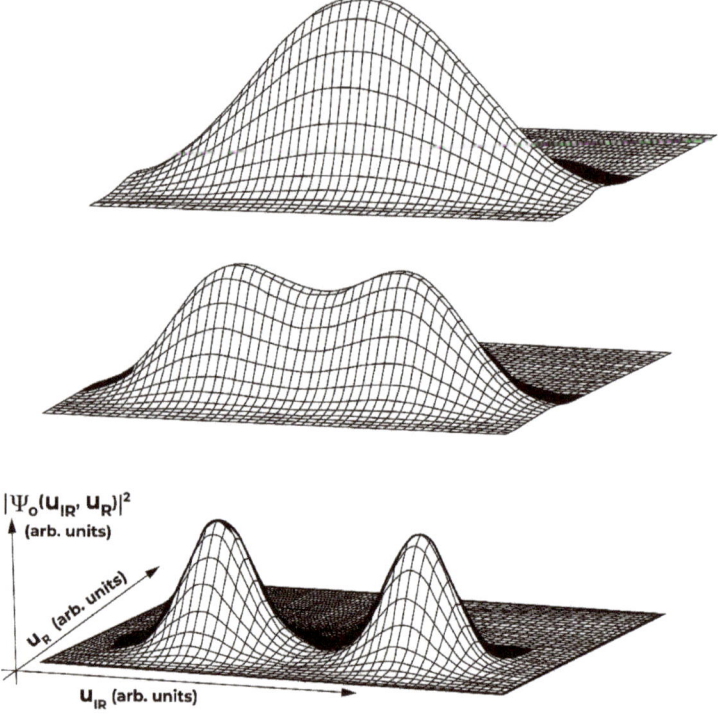

Figure 1. Squared many-body ground-state wave function of an O–Cu–O cluster as a function of the Raman (U_R) and infrared (U_{ir}) coordinates for three IR coupling strengths (weak, intermediate, strong). When the phonon coordinate is at the left bump, the extra hole is predominately on the leftmost oxygen, and vice versa. The scale of the coordinates is fixed to be 3 times the root-mean-square fluctuation of the quantum wave function, and therefore it is different for the three panels (see [18]).

Numerically, a landscape of low-energy polaron configurations can be identified, metastable but pinned by the lattice, which can be accessed by external fields, photoexcitation, impurity pinning centers, internal microstructure and interfaces, surfaces, etc. Note that the stripe interfaces are sharp on the unit-cell lattice scale: as for single small polarons, gradient energy costs are small compared to those from deviations of the undeformed broken symmetry energy. The broken symmetry host (e.g., AF moment) is locally quenched at the stripe.

We can study the lattice, spin, charge *fluctuations* around these polarons, and multi-polaron patterns within a real-space random field approximation (RPA) [25,26]. Since the polarons are small and the stripe interfaces sharp, the stripes are topologically "protected" and the fluctuations include ones that are spatially localized around the polarons or stripes and thus substantially separated in frequency from the band of k-space modes in the broken-symmetry host: i.e., symmetry-determined transverse and longitudinal "edge mode" vibrations of spin, charge, and lattice [25]. As illustrated in Figure 3, they are all dominated by O ion motion, resulting in Cu–O bond length fluctuations coupled to charge and spin fluctuations. Of course, any slow diffusion of polarons or stripes introduces a cut-off or broadening for the local mode frequencies, and any stripe curvature introduces a wave-vector cut-off.

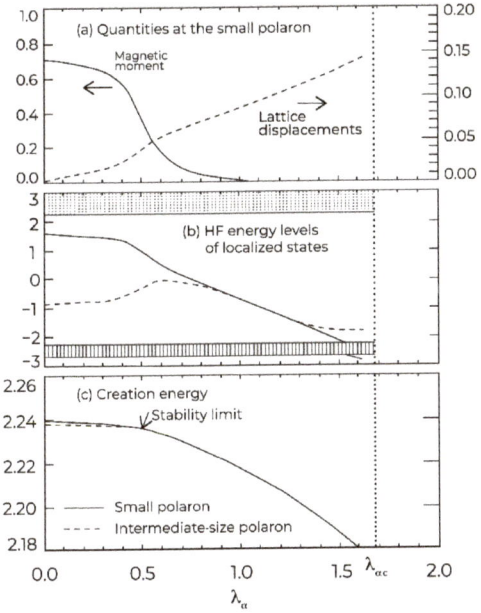

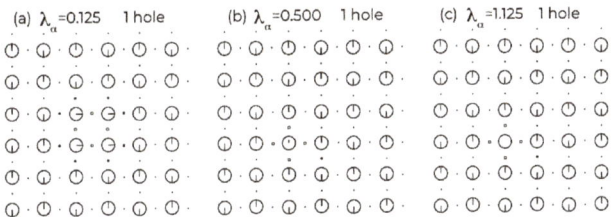

Figure 2. Polarons calculated in a 2D 3-band Peierls–Hubbard model (see Ref.[21]). **Upper Graph:** (a) Magnetic moment of the central Cu site of a calculated small polaron and ratio of lattice displacement of the surrounding O to Cu–O distance (1.89 Å), (b) gap energy levels, and (c) creation energy $\epsilon_1 = E^{N+1} - E^N - \bar{\mu}$, for the small polaron state as a function of λ_α. Other energy levels in the gap close to the bands (shaded areas) are not shown. E^{N+i} is the total energy with i added holes and $\bar{\mu}$ is the midgap energy. In addition, (c) also shows ϵ_1 for the intermediate-size ferromagnetic polaron. All energies are in units of t_{pd}. Parameters are $t_{pd} = 1$, $t_{pp} = 0.5$, $\Delta = 4$, $U_d = 10$, $U_p = 3$, $U_{pd} = 1$, and $K = 32 t_{pd}/\text{Å}^2$. The dotted line at $\lambda_{ac} \simeq 1.68$ marks the stability limit of the anti-ferromagnet (AF) ground state at stoichiometry. **Lower Graph:** Charge (radii of the circles) and spin (arrows) densities in the one-hole doped systems, (a) with the intermediate-size polaron; and (b,c) with the small polaron. The arrows are normalized so as to touch the circle if completely polarized. Big (small) circles are Cu (O).

The frequencies of these local modes are again similar to experimentally observed "anomalous" signatures in doped cuprates and also some other doped TMOs. For example, in doped nickelates, the intensity of anomalous modes observed with neutron scattering tracks the doping density [27], and interesting pressure effects have been proposed [28]. Checkerboard ("liquid crystal") patterns have been indicated in doped cuprates [29] and are predicted to have their own signatures of specific local modes [30].

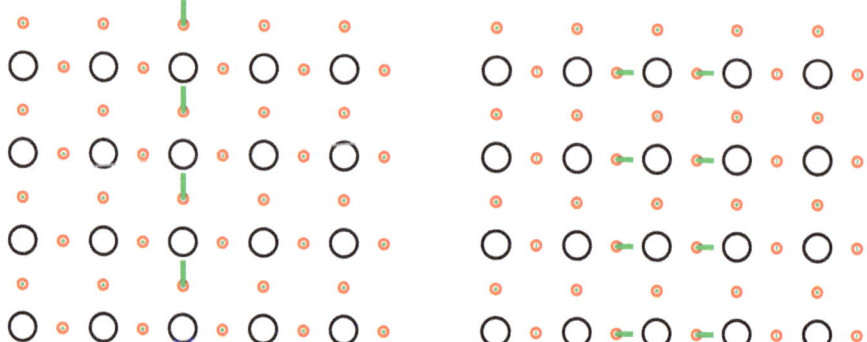

Figure 3. Lattice vibrational eigenmodes for the case of a stripe in a 20%-doped CuO2 plane computed in a 3-band Peierls–Hubbard model. The stripe is centered along the middle of the vertical Cu row. There are two branches localized at the stripe: one corresponds to the oxygen vibration parallel to the stripe (low frequency, $E = 14.8$ meV, **left**), and the other corresponds to the oxygen displacements perpendicular to the stripe (high frequency, $E = 68.5$ meV, **right**) (see Ref [25]).

Note that the Cu–O stripes share O ions coherently in the above results (see Figure 3). If the longitudinal O vibrations are extended to strongly nonlinear amplitudes, so that there is full CT, they will correspond to a coherent filament of O–Cu–O units (the small polaron described above), periodically tunneling charge between neighboring Os, with concomitant oscillation between short and long Cu–O bond lengths. This is equivalent to a dynamical charge density wave (CDW) along the stripe. It is clear conceptually that an infinitesimal external electric field can then coherently transfer charge along the stripe—a "sliding CDW" avoiding impurity or lattice pinning (cf. [31]). In this scenario, there is no polaron center-of-mass motion, only resonant transfer of charge. In addition, the transverse O vibrations at the stripe edge provide a natural local-to-extended mode coupling with the broken-symmetry host, e.g., anti-ferromagnet (AF) [32]. These observations are intriguing in terms of conductivity mechanisms. A similar scenario applies to $Bi_xK_{1-x}TiO_3$, where a window of superconductivity appears around a specific x (doping) value with a transition from a broken-symmetry CDW (i.e., charge-disproportionated) insulator to metal—in this case, polarons, stripes, and precursor local structure will be with respect to the CDW, instead of AF, host. We note that if the dissipation from the coupling to the host medium for the stripes is too strong, then the CT/bond-length vibrations will freeze into a static CDW/BOW. This sets a lower frequency limit for the CT frequency, which suggests consequences for models of superconductivity, dynamic versus static magnetic moments (e.g., in Pu [33]), etc. For instance, higher superconducting Tc would be aided by a lower CT tunneling frequency. Partially charged stripes can also be found with the above formalism by tuning the band-filling. Similarly, topological (including fractionally charged) edge states resulting from boundary conditions in finite systems, as has been extensively studied in 2D Dirac metals, can be captured.

4. Elasticity and Short–Long-Range Competitions

As suggested earlier, elasticity in TMOs (and related materials such as f-electron and many organic compounds) is very important because of their localized, directional d-orbitals. This consideration complements electronic/spin focuses and cannot be ignored. In fact, the elasticity in these materials is strong (because it is weakly screened and therefore must be integrated over long ranges) and directional, resulting in constraints and competitions for ground and metastable states. Indeed, the origin of elasticity is a coupling of optic and acoustic modes [34]: unit cell optic lattice distortions constrain next-neighbor unit cell distortions and sequentially to acoustic long-range "elastic" fields. This provides a self-consistent complex *network* situation, with multiscale microstructure (twinning, tweed, etc.) as intrinsic sub-grain textures, which is the result of the self-consistent short–long-range field competition. Importantly, the network adjusts globally to either local or global perturbations. Furthermore, solid–solid phase transitions are accompanied by extended regimes of lattice microstructure around the transition, as well as sensitivity to strain and external fields (electric, magnetic, etc., depending on the specific material). *Recall Alex Müller!*

For illustration, consider the class of materials known as ferroelastic martensites. Ferroelasticity is the existence of two or more stable orientation states of a crystal that correspond to different arrangements of the atoms but are structurally identical (enantiomorphous). These orientation states are degenerate in energy in the absence of mechanical stress. The term martensitic usually refers to a diffusionless first-order phase transition that can be described in terms of one or successive shear deformations from a parent to a product phase. Schematic illustrations of symmetry-allowed 2D transitions are shown in Figure 4. The morphology and kinetics of the transition are dominated by the strain energy, and the transition results in characteristic lamellar (twinned) microstructures. Features observed in proper ferroelastic crystals include mechanical hysteresis and mechanically switchable domain patterns. Ferroelasticity usually occurs as a result of a phase transition from a nonferroelastic high-symmetry parent phase and is associated with the softening of an elastic modulus with decreasing temperature or increasing pressure in the parent phase. The ferroelastic transition can be described by Landau theory with spontaneous strain or deviation of a given ferroelastic orientation state from the parent phase as the order parameter (OP). The strain can be coupled to other fields, such as electric polarization and magnetic moment; thus, the crystal can have more than one transition.

A comprehensive Ginzburg–Landau (GL) theory of this lattice elasticity can be built using so-called St. Venant lattice compatibility constraints to capture the symmetry-constrained short–long-range framework described above. This theory is detailed in [35,36] for many 2D and 3D cases. The theory successfuly describes key observed features of solid–solid phase transitions, including twins and twinning periodicity system (e.g., grain) size-dependence; twinning hierarchies at high–low symmetry interfaces and boundaries; extensive precursor (nucleation/spinodal) microstructure regimes around the solid–solid phase transition; global, multiscale structural effects of local dopants; ("local stress"); and global, multiscale, including local, structural effects of external stress and other applied fields. Figures 5 and 6 show a few examples of the results, illustrating the intrinsic and sensitive landscapes of microstructure. There is some optimism that this sensitivity might be represented in statistical configuration ensembles, including the use of Machine Learning/Neural Network techniques for relevant feature capture (see [37]). The above GL theory (extended to time-dependent GL (TDGL) [35,36]) also self-consistently predicts multi-timescale dynamics and relaxation of the multi-lengthscale microstructure, including glassy phenomena.

There are several important lessons from this elasticity description. The precursor microstructure extends over *large* parameter regimes of temperature, pressure, etc., around solid–solid phase transitions—recall Alex Müller. The functionality of this microstructure is often as important as the transition itself. (For example the tetragonal-orthorhomic transition in cuprate superconductors or hierarchical ordering in shape-memory alloys.) Throughout the precursor regime, local doping sites, because of their long-range elastic effects, have significant influence on the transition and nucleate structured precursor domains of the incipient phase. We emphasize that the sensitive microstructural landscapes are intrinsic free-energy states. Attempts to eliminate the microstructure (e.g., de-twinning), especially in the extended vicinity of solid–solid transitions, are unhelpful, since they will simply reform unless pinned into non-equilibrium configurations. With this in mind, it is natural to ask whether the intrinsic microstructure provides the template on which electronic, magnetic, etc., degrees of freedom can act, i.e., can the microstructure be the driver of electronic or magnetic properties?

This question has been explored through several examples. For example, [38] examines the electronic signature of twin boundaries (TB) and anti-phase boundaries in 2D. Here, the boundary centers constitute a local (e.g., square) lattice symmetry smoothly joining domains (twins) of a different (e.g., rectangular) lattice symmetry (see Figure 4). This results in locally metallic filaments in a semiconducting host. In fact, the TB can be considered a 2D Dirac metal with flat electronic bands of edge states [39]. WOx provides an excellent example of quasi-periodic patterns of shear planes [40,41], and intriguingly, in a certain x regime, it is also a high-temperature superconductor, with the superconductivity probably strongest on the shear planes—recall the comments on possible filamentary superconductivity above, since dopant (polaronic) charges will preferentially decorate shear planes (and TBs).

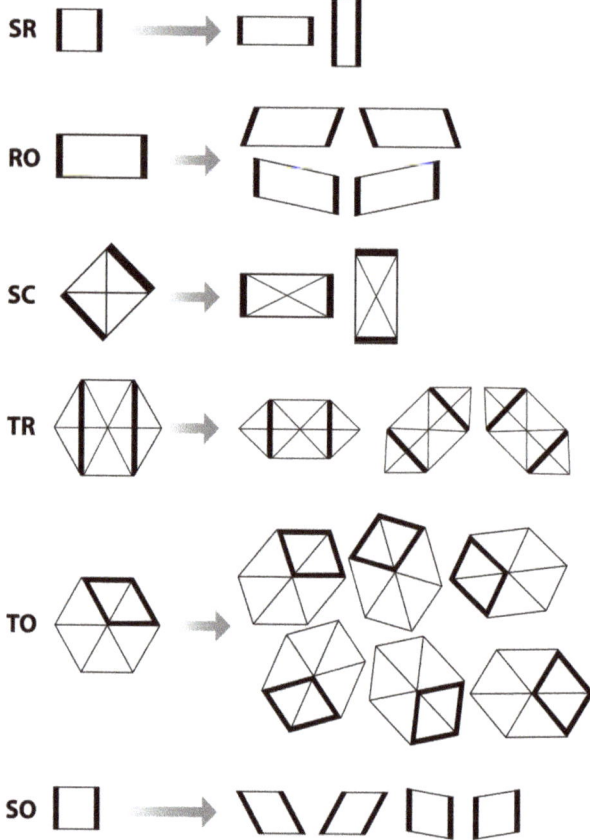

Figure 4. Symmetry-allowed transitions in 2D for four crystal systems. (See Ref. [35]). The dark lines are visual guides to indicate deformations. There is a one-component strain order parameter (OP) for the square-to-rectangle (SR) case, driven by deviatoric strain, ϵ_2: the rectangle-to-oblique (RO) case, driven by ϵ_2; and the square-to-centered rectangle (SC) case, driven by shear strain, ϵ_3. A two-component OP, or two one-component OPs, leads to the triangular-to-centered rectangle (TR) case, driven by ϵ_2, ϵ_3; the triangle-to-oblique (TO) case, driven by ϵ_2, ϵ_3; and the square-to-oblique (SO) case, driven by ϵ_2 and ϵ_3 independently. Copyright 2020 American Physical Society.

In [42], the effects of long-range anisotropic elastic deformations on electronic structure in conventional BCS superconductors are analyzed within the framework of Bogoliubov–de Gennes equations. Cases of TBs and isolated defects are considered there as illustrations. The calculated local density of states suggests that the electronic structure is strongly modulated in response to lattice deformations and propagates to longer distances because of the elasticity. In particular, this allows the trapping of low-lying quasiparticle states around defects. Some of these predictions could be directly tested by scanning tunneling microscopy.

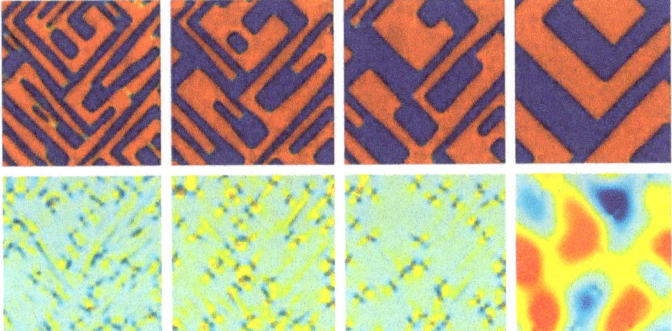

Figure 5. Multiscale texture evolution computed in a 2D TDGL elasicity theory (see Ref. [35]). Square-to-rectangle (Figure 4) case, showing simulated interface propagation. The rows show (left to right) temporal sequences at 40, 80, 160, and 1000 ps. **Top Row**: The order parameter (OP) deviatoric strain, showing domain walls propagating under the repulsive long-range lattice compatibility potential. **Bottom Row**: Non-OP shear strain, propagating outward with interfaces, concentrated at the corners. Copyright 2020 American Physical Society.

Figure 6. Same square-to-rectangle case as Figure 5, demonstrating global effects of a local stress: simulated strain evolution, with an added fixed, time-independent, Lorentzian-profile local stress (see Ref [33]). The sequence (left to right) is shown for time$t$ = 40, 60, 76, and 106 ps. **Top Row**: Dynamic texturing of deviatoric strain. The system reduces the energy from the imposed single-sign strain by elastic photocopying, or adaptive screening, of the long-range elastic interaction, generating higher multipoles. **Bottom Row**: The non-order parameter (OP) shear strain follows the OP propagation. Copyright 2020 American Physical Society.

Reference [43] uses the St Venant lattice compatibility constraint theory to predict the role of elasticity in colossal magnetoresistance (CMR) perovskite manganites. The intrinsic coexistence of distinct metallic and insulating electronic phases in perovskite manganites, such as $La_{1-x-y} Pr_y Ca_x MnO_3$, is predicted, which presents opportunities for sensitively tuning the electronic properties. In particular, the CMR in these materials is closely related to the observed texture owing to coexisting nanometer- and micrometer-scale inhomogeneities. Extensive data from various high-resolution probes show the existence of such inhomogeneities. Experimental results also support the presence of metastable states in manganites. For example, magnetic fields or X-rays have been used to convert insulating regions into ferromagnetic metallic ones, which are stable even when the fields are removed. Explanations based on electronic mechanisms and chemical disorder have not been sufficient to describe the multiscale, multiphase coexistence within a unified picture. However, lattice distortions and strain are known to be important in the manganites [44], and indeed the resistivity transition can be tuned with pressure as well as magnetic field. In [43], it is shown how the texturing can be due to the intrinsic complexity of a system with coupling between the electronic and elastic degrees of freedom. This leads to landscapes of energetically favorable configurations and provides a self-organizing mechanism for the observed inhomogeneities. Since the domain formation is self-sustained, external stimulii such as optical lasers, X-rays, or ultrasonic standing waves can be used to sensitively manipulate patterns of metallic and insulating regions, thus making the control of nanoengineered functional structures feasible and technologically important.

The same mechanisms should be applicable to describing intrinsic inhomogeneities in other materials with strong bonding constraints, such as relaxor ferroelectrics and high transition temperature superconducting oxides, f-electron materials (including heavy-fermions), organics (including superconductors), multiferroics, and 2D Dirac materials (such as graphene, dichalcogenides), etc., where the functionalities may also be mediated through self-organized lattice distortions. We return to the probable role of polarons below. More generally, there are many materials where the above approach to coupling lattice, spin, charge and orbital degrees of freedom, including elastically driven transitions, can be applied. For instance, epitaxial oxide layers and multilayers provide extensive functional tunability [45], and magnetocalorics [46] are studied for their potential application as efficient refrigeration and waste-recovery materials. Reference [47] uses the above elasticity framework to argue that certain metal–insulator transitions are accompanied by precursor elasticity microstructure. Indeed, early neutron scattering around the metal–insulator transition in VO_2 found short magnetic correlation lengths even very close to the transition, suggesting that the lattice microstructure limits the electronic/magnetic properties. Modern time-resolved crystallography can now resolve substantial lattice contributions to several M–I transitions, including VO_2, as anticipated in early theory [48]. I suggest that similar roles for elastic fluctuations can be expected around many broken-symmetry transitions, including many quantum critical points [49]. Note that the elasticity discussion here was for a single grain. Multigrain interactions are beyond the scope of the present discussion and require coarser-scale (e.g., phase-field, finite-element) modeling for homogenization and constitutive equations.

Reference [50] uses an extended Holstein polaron model to suggest how elastic interactions can control the organization of small polaron patterns. This modeling leads to a landscape of filamentary (stripe) polaron patterns, similarly to those outlined in Section 3. The small polarons have very localized electronic cores but act as local impurities in the multiscale, directional elastic field, creating anisotropic elastic fields and driving the filamentary ordering. Figure 7 shows examples calculated for a 2D square lattice hosting a finite density of small polarons.

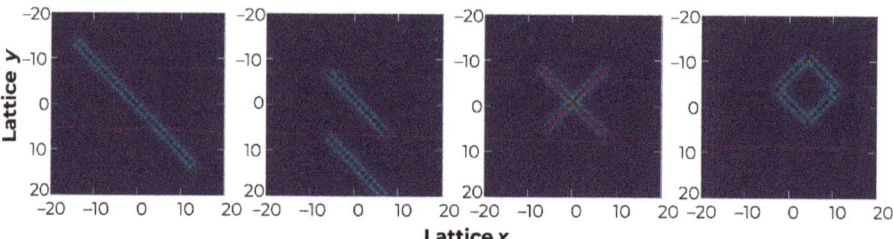

Figure 7. Computed configurations of small polarons embedded in an elastic 2D medium, illustrating the angular dependence of the strain field favoring diagonal strings. Four examples of metastable minima are shown for different numbers of polarons (see Reference [50] for details). Copyright 2020 American Physical Society.

5. Hybridized Bands and more examples of Short–Long-Range Competitions

The coexistence of narrow (d,f) and wider (s,p) electronic bands has motivated a great variety hybridization studies. Many of these are theoretically rich, and some are surely exhibited in various TMOs and related electronic and magnetic materials. Below, we summarize a few examples but emphasize the additional features and competing phases resulting from the inclusion of lattice degrees of freedom.

Multi-band superconductivity is a natural candidate. Substantial Tc enhancement and tunability is possible, which is controlled by the interband coupling strengths and locations (in k-space) (e.g., [51]). Resonant coupling (e.g., Fermi resonance) is especially intriguing and is well-covered in this volume and elsewhere [52]. Clearly, the mesoscopic phase separation of lattice, charge, spin broken- and unbroken-symmetry regions described above provide environments for these scenarios, with stripes and TBs sources of flat electronic bands.

Intrinsic local modes (ILMs) are spatially localized lattice modes (multi-phonon bound states) resulting from sufficiently anharmonic lattice vibrations (e.g., [10]), where the anharmonicity is often the consequence of feedback from electron or spin coupling to the lattice. In fact, as discussed above, polarons are the result of such a (positive) feedback from *linear* coupling between charge/spin and lattice. When this coupling is strong enough to create small polarons, the lattice distortion is *locally* anharmonic and can create an ILM, resulting in a composite polaron–ILM state [53]. In addition, doping sites can locally distort the lattice into an anharmonic regime and create a ILM. An appealing scenario is that ILMs can be induced in this way by doping and then embedded self-consistently in an undoped background. Reference [54] uses this approach to model ferroelectrics, including anomalous phonons and glassy/relaxor phases. This approach has also been used [55] within nonlinear shell models, which, as as noted above, are an approximate but effective description of O polarizability and M–O charge transfer. For example, the quantum paraelectric phase of $SrTiO_3$ and the onset of ferroelectricity with O_{18} isotope substitution are well explained [56], as are the differences between ferroelectrics $BaTiO_3$ and $EuTiO_3$ [57]. More generally, this approach provides an excellent basis for understanding the T-doping-strain phase-diagram of these materials—the doping-induced superconductivity phase-diagram [58] is reminiscent of cuprates. I note again that nonlinear shell models owe much to Heinz Bilz and his ability to combine insights from solid-state physics, quantum chemistry, and nonlinear statistical mechanics long before sufficient ab initio quantum methods were available.

In [59], dimerized AF (homogeneous SP) and inhomogeneous-lattice AF (inhomogeneous SP) ground states are predicted in both 1D and 2D periodic Anderson models when el–lattice coupling is included, as shown schematically in Figure 8. Coexistence and mutual enhancement of the Peierls distortion and the AF long-range order are found. The stoichiometric phase diagrams are strongly dependent on the relative hybridization and el–lattice coupling strengths. For non-stoichiometric fillings, coupled spin–charge–lattice polarons are found containing precursor textures of neighboring phases. Relations to Ce-based heavy-fermion systems, volume collapse, and inorganic SP materials are discussed in [59].

Returning to the importance of short–long-range field competitions (Section 4), RKKY is a familiar magnetic long-range (and oscillatory) interaction. In [60], the magnetic properties of a system of coexisting localized spins and conduction electrons are investigated within an extended version of the 1D Kondo lattice model in which effects of el–lattice and on-site Coulomb interactions are explicitly included. It is found that intrinsic inhomogeneities with the statistical scaling properties of a Griffiths phase appear and determine the spin structure of the localized impurities. The appearance of the inhomogeneities is enhanced by appropriate phonons and acts destructively on the spin ordering. The inhomogeneities appear on well-defined length scales and can be compared to the formation of intrinsic mesoscopic metastable patterns found in two-fluid phenomenologies. A mapping to an effective random field transverse field Ising model is found to be instructive. The RKKY system can indeed be viewed as intrinsically frustrated [61].

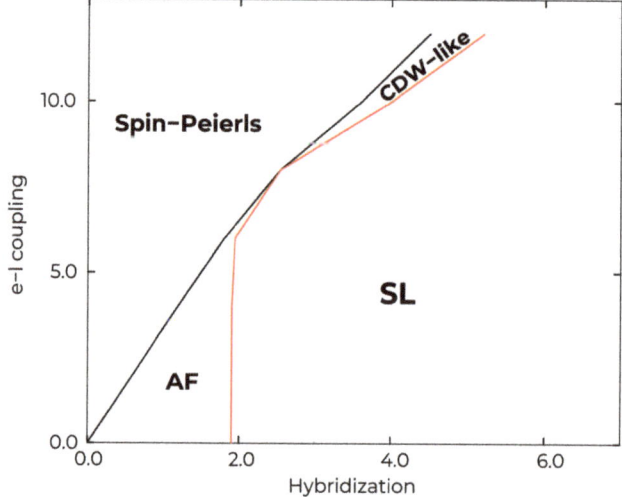

Figure 8. Computed phase diagram in a 2D periodic Anderson model with el–lattice coupling, as a function of hybridization and el-lattice coupling strengths. (See [59]). Ground states found are: long-period spin–Peierls; antiferromagnet (AF); charge density wave (CDW); spin liquid (SL).

I conclude this section with a more general perspective on multiscale landscapes resulting from coexisting short–long-range interactions. In particular, the appearance of glassy filamentary phases in windows of intermediate doping concentration ("intermediate phases") [62,63] is demonstrated in [64,65] as a model of doped 2D anti-ferromagnets. I believe these intermediate phases are very important functionally and should be realized in many materials, including doped TMOs. For example, using efficient numerical methods to handle long-range interactions, references [65,66] study a quasi-classical model for the charge ordering of holes in TMOs, in which the particles have a Coulomb repulsion and a dipolar attraction. As a function of hole density (doping), an extended soft phase comprising partially ordered filaments is found (see Figure 9). Ordered clumps form for low densities and ordered stripes (Wigner crystal-like phases) at high densities (see Figure 10). The soft filamentary structures persist to high temperatures. Within the soft phase region, there is an onset at low T of motion along the filaments: i.e., the filaments act as a template for the correlated percolation of particle motion. When the particle positions are averaged over long times, the filaments form a checkerboard pattern (see Figure 11). All of this rich multiscale patterning and dynamics arises from a deceptively simple 2D model in which the effective interaction between two holes, 1 and 2, a distance r apart is given by

$$V(r) = \frac{q^2}{r} - Ae^{-\frac{r}{a}} - B\cos(2\theta - \phi_1 - \phi_2)e^{-\frac{r}{\xi}}. \tag{2}$$

Here, $q = 1$ is the hole charge, θ is the angle between **r** and a fixed axis, and $\phi_{1,2}$ are the angles of the magnetic dipoles relative to the same fixed axis, which is allowed to take an arbitrary value. A is the strength of a short-range anisotropic interaction, and B is that of a magnetic dipolar interaction [$B \approx A/(2\pi\,\xi/2)$], which were assumed to be independent variables. Here, for simplicity, we take $A = 0$ and assume a scaling for the magnetic correlation length $\xi \sim 1/\sqrt{n}$, with n the hole density. Reference [67] describes characteristic noise and hysteresis associated with the stripe, clump, and checkerboard phases. Reference [68] demonstrates how both commensurate and incommensurate checkerboard configurations are possible.

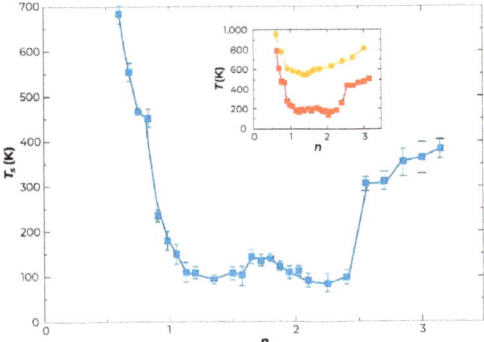

Figure 9. Onset temperature T_s of a modulated square liquid state as a function of hole density n in model Equation (2). **INSET**: Melting temperature of pattern T_m (yellow circles) and T_s (red squares). Copyright 2020 American Physical Society.

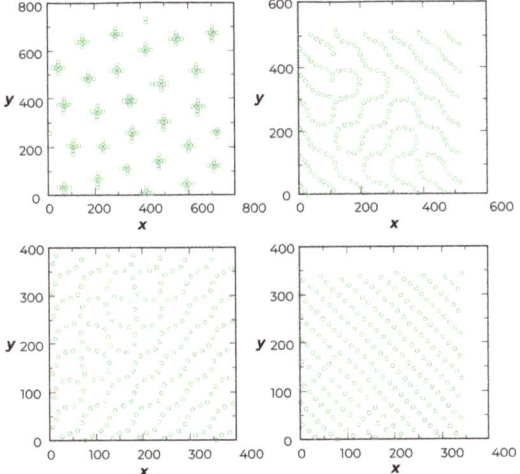

Figure 10. Simulated static positions of holes for different densities, n, in model Equation (2): (**a**) clump phase, $n = 0.6$; (**b**) soft phase, density $n = 1.2$; (**c**) soft phase, $n = 2.1$; and (**d**) anisotropic Wigner crystal phase, $n = 2.7$. Copyright 2020 American Physical Society.

TMOs, and related materials, exhibit complex interplays of spin, charge, orbital, and lattice so that a simplified model such as the above should not be overinterpreted. For instance, as noted earlier, there are essential roles of polarizable oxygen or equivalent ions [displacements, rotations, charge transfer, etc. (e.g., [5,12])]. Nevertheless, some implications are suggestive. Namely, charge-ordered states may persist up to very high temperatures, but signatures of disordered filamentary states occur at much lower temperatures with a transition to a checkerboard state at intermediate temperatures. The coexisting short- and long-range interactions will appear only upon the (polaronic) localization of holes, which happens below a characteristic temperature. Above this temperature, a more metallic-like electronic state is expected. The soft phase also shows similarities with the inhomogeneous states observed in manganite oxides (above) between the true critical temperature and a higher temperature at which short-range order first appears. In addition, as we note below, some stripe-based theories for superconductivity require fluctuating stripes. An important feature of the soft phase in Figure 9 is that the fluctuations are predominantly *on* (correlated) percolating filaments rather than meandering of the filaments themselves. Therefore, the fluctuating filamentary and checkerboard states should provide a reasonable starting point for introducing detailed quantum-mechanical and oxygen effects.

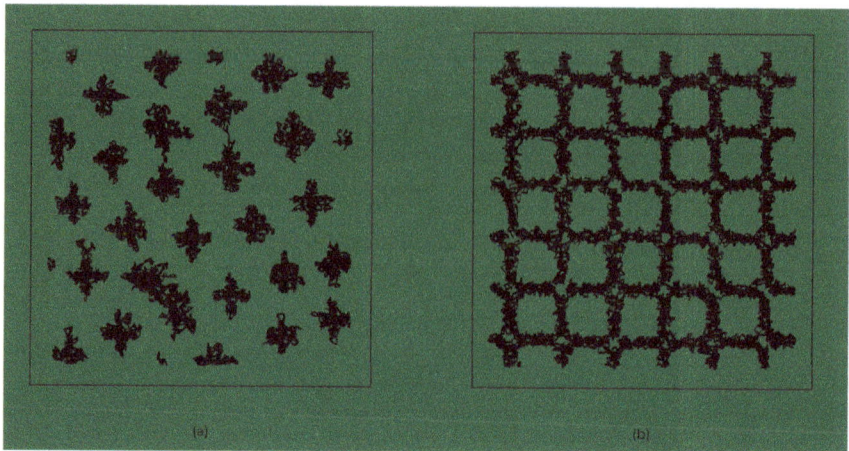

Figure 11. Image of system above the filamentary melting transition, with lines indicating the motion of the holes between consecutive simulation frames. (**a**) $n=0.6$ in the clump phase, where the particles remain predominantly localized. (**b**) $n=1.2$, showing the square modulated liquid phase in the filamentary regime (see [65]).

Although, a full simulation has not yet been performed, we should expect that similar soft intermediate phase behavior will be found for the multi-polaron patterns introduced in Section 4. Indeed, in that case, the long- and short-range fields are provided self-consistently by the same elasticity. In this context, it is worth recalling that elasticity has its origins in the total energy of a solid including the Coulomb contributions. Hence, if atomic configurations are allowed to relax self-consistently in a Coulomb field, the elasticity features above must be recovered. Finally, we note that a quite general framework for intermediate phases can be expressed in a theory describing network adaptability and rigidity transitions based on the degree of lattice connectivity and the number of bonding constraints [62,63,69]. The next material frontiers must include understanding *reactive networks*. Viz, the network structure adapts to the functioning (lattice, spin, charge) of (or collections of) nodes because this changes local- and multiscale network connections, and the new structure then initiates new functioning. This feedback cycle underpins the complementary aspects of learning and aging in complex materials, similarly to emerging views of hierarchical landscapes and substates in quantitative biology and other fields adopting network scenarios [70].

6. Conclusions and Discussion

The competitions among spin, charge, orbital, and lattice degrees of freedom in TMOs (and related materials with directional bonding and coupled localized/narrow band and delocalized/wide band electronic orbitals) leads to a richness of ground (and excited) states because of delicate free-energy balances. This is a wonderful situation for tunable functionality and applications, but it challenges conceptual frameworks created for simpler classes of materials. Asymptotic length and time-scaling descriptions, although certainly useful, are incomplete, and the community has yet to create a full palette of experimental, theoretical, and computational tools to understand the resulting multiscale, multiphysics landscapes of states. However, it is clear that understanding and controlling the functionality of this "dark matter"(!) will require combining techniques from many disciplines (solid state, quantum chemistry, correlated electrons, field theory, mathematical physics, nonlinear science, high-performance computing, data analytics such as ML, etc.). This is surely the interdisciplinary path to addressing the challenges and opportunities of "quantum materials" [11].

It is always true that new experimental and modeling techniques motivate new conceptual frameworks, and vice versa. This also motivates new interpretations of existing data. For example, I mentioned in Section 1 the birth of STM and its impact. It is striking to trace STM's history since then through mesoscopics, Bose–Einstein condensates, to TMOs [71], DNA manipulation, etc. No less impressive are similar journeys for neutron and X-ray resolution, angle-resoved photoemission, nuclear magnetic resonance, resonant ultrasound, pump-probe spectroscopy, time-resolved crystallography, and more. These are all now critical tools for understanding the multiple spatial and temporal scales that Alex Müller so presciently anticipated in TMOs. They are essential capabilities to combine with equally impressive revolutions in high-performance computing, visualization, and data analytics on the path to the holy grail of understanding, controlling, and using materials' synthesis–structure–property–performance relationships.

I finish with a few open questions.

1. I have deliberately emphasized lattice, including elastic, effects in TMOs (and related d- and f- orbital materials) because these have often been under-represented in research on "strongly correlated" quantum matter. The recent workshop [2] is an encouraging community step. However, clearly, it remains to be understood when lattice, spin, charge, and orbital degrees of freedom individually dominate and then the others are slaved. There are likely to be classes of materials that should be so categorized in this fascinating collusion among degrees of freedom. This is a quantum mechanical adiabaticity question that must be addressed on appropriate spatial and temporal scales and, as with the importance of el–lattice coupling, not simply in terms of average properties. Functionality through charge, spin, and lattice at active interfaces, including between TMOs, is an important direction for applications (e.g., perovskite-based solar cells and detectors) [45,72]. Similarly, domain boundary engineering, e.g., in multiferroic materials [73], is a very attractive direction.

2. The recent emphasis on quantum entanglement and the increasing recognition of geometry and topology in electronic and magnetic materials research raises important questions. For instance, (1) above in the context of when does the lattice simply renormalize parameters for quantum phases and when do topological *lattice* configurations act as the driving template for quantum mechanics? For example, this is an important issue in materials design for qubits and quantum information research.

3. I have not focused here on high-temperature (HTC) mechanisms. The wonderful discovery of HTC in cuprates by Bednorz and Müller certainly propelled remarkable advances in synthesis, experimental, theoretical and simulation capabilities for complex electronic materials and whole new classes of materials have benefitted: multilayer TMOs [45], Dirac–Weyl materials, organic and heavy-fermions SCs, pnictides, multiferroics, over-doped HTC cuprates [74], etc. After these many years, a generally accepted theory of HTC remains elusive. Which of the multitude of measured perovskite features are directly relevant to the superconductivity mechanism has yet to be understood. However, much research falls into the framework of multiscale, coexisting charge-rich and charge-poor regions [29,75]. Inhomogeneous superconductivity is familiar, e.g., in granular superconductors. However, the possibility of inhomogeneity being an intrinsic template for the SC mechanism. For example, Reference [76] proposes micro (i.e., bond-length) strain (cf, Section 4) as a primary control parameter in cuprate SCs. The polaron patterns discussed above are also examples: precursors to CDWs and equilibria with multi-polaron (including bi-polaron) bound states. Such CDW-like phases embedded in various undoped broken-symmetry hosts are attractive scenarios, but the AF broken-symmetry host state is not unique. Rather, the main issue becomes: How do the charge-rich regions communicate? Magnetic, charge, and elastic fluctuations are all feasible [77]. There is accumulating evidence [29] for a checkerboard period-4 CDW-like configuration in cuprates. Various small perturbations can

isolate such a specific periodicity for the ordering illustrated in Section 5 (e.g., [68]). However, qualitatively different origins are also interesting (e.g., [78]). The charge-transfer (polarizability) features of perovskites we have discussed lend themselves to W.A. Little's scenario of enhanced SC from off-chain/plane polarizable material [79,80]. Finally, an intriguing consideration is that (correlated) percolating charged filaments organize *fractally* as doping increases, before finally over-packing and quantum melting. This would certainly be the optimal space filling to maximize percolating filamentary properties if those properties are desirable—e.g., for high Tc, as suggested for many years by J.C. Phillips [81].

Materials play fundamental roles in the health and prosperity of society. It is not accidental that many new technologies and their impacts on society are rapidly accelerating in this dynamic era of complexity science: they are results of remarkable new experimental, simulation, and visualization tools and the resulting data explosion. As throughout history, the new technologies are also producing new societal challenges—in this century for health, energy, natural resources, climate, national security, space, cyber, social media, etc., sectors—but now at a very accelerated pace. It is fortunate that this century is generating the tools for the science, technology, engineering, and mathematics workforce to play its part in addressing these societal opportunities and challenges, making this an exciting time for the historic cycle of science–society evolution. Happily, the many disciples of Alex Müller are well supplied with opportunities to conceive new ideas and create tools to test them as the future history of quantum materials is written.

Funding: This research received no external funding.

Acknowledgments: Although the views expressed in this article are personal, I have very many colleagues to thank for partnering with me through the research I have referenced, including Miklos Gulacsi and Zlatko Tesanovic who are sadly no longer with us. Since there are so many to recognize individually, I hope that the references will suffice and I apologize to all those colleagues whose important research I have not directly referenced. *However, this is a tribute to Alex Müller.* I wish to acknowledge him in particular for his inspiration and to mention two of his contemporaries who deeply influenced my scientific life—Heinz Bilz and Jim Krumhansl. Alex, Heinz, and Jim are wonderful icons of a great science generation.

Conflicts of Interest: The author declares no conflict of interest.

References

1. Müller, K.A. Microscopic Probing of BaTiO$_3$ Ferroelectric Phase Transitions by EPR. In *Nonlinearity in Condensed Matter, Proceedings of the Sixth Annual Conference, Center for Nonlinear Studies, Los Alamos, NM, USA, 5–9 May 1986*; Bishop, A.R., Campbell, D.K., Kumar, P., Trullinger, S.E., Eds.; Springer: Berlin/Heidelberg, Germany, 1987; pp. 234–245.
2. Schmalian, J.; Sinova, J.; Valenti, R. Elastic Tuning and Response of Electronic Order. In Proceedings of the Spin Phenomena Interdisciplinary Center (SPICE) Workshop, Mainz, Germany, 9–11 December 2019.
3. Müller, K.A.; Berlinger, W.; Tosatti, E. Indication for a novel phase in the quantum paraelectric regime of SrTiO$_3$. *Z. Phys. B Condens. Matter* **1991**, *84*, 277–283. [CrossRef]
4. Tokura, Y. Correlated-electron physics in transition-metal oxides. *Phys. Today* **2003**, *56*, 50–55. [CrossRef]
5. Krumhansl, J.A. Fine Scale Mesostructures in Superconducting and Other Materials. In Proceedings of the Lattice Effects in High-Temperature Superconductors, Santa Fe, NM, USA, 13–15 January 1992.
6. Migoni, R.; Bilz, H.; Bauerle, D. Origin of raman scattering and ferroelectricity in oxidic perovskites. *Phys. Rev. Lett.* **1976**, *37*, 1155. [CrossRef]
7. Dernadis, K.D. The localized-to-delocalized transition in mixed-valence chemistry. *J. Chem. Rev.* **2001**, *101*, 2655–2685.
8. Caldeira, A.O.; Leggett, A.J. Quantum Tunneling in a Dissipative System. *Ann. Phys.* **1983**, *149*, 374–456. [CrossRef]
9. Martonák, R.; Tosatti, E. Path-integral Monte Carlo study of a model two-dimensional quantum paraelectric. *Phys. Rev. B* **1994**, *49*, 596–613. [CrossRef]
10. Manley, M.E.; Lynn, J.W.; Abernathy, D.L.; Specht, E.D.; Delaire, O.; Bishop, A.R.; Sahul, R.; Budai, J.D. Phonon localization drives polar nanoregions in a relaxor ferroelectric. *Nat. Commun.* **2014**, *5*, 3683. [CrossRef]

11. Awschalom, D.; Christen, H. Opportunities for basic research for next generation quantum system. In Proceedings of the Basic Energy Sciences Roundtable on Opportunities for Basic Research for Next-Generation Quantum Systems, Gaithersburg, MD, USA, 30–31 October 2017; Oak Ridge National Laboratory: Oak Ridge, TN, USA, 2017.
12. Porer, M.; Fechner, M.; Kubli, M.; Neugebauer, M.J.; Parchenko, S.; Esposito, V.; Narayan, A.; Spaldin, N.A.; Huber, R.; Radovic, M.; et al. Ultrafast transient increase of oxygen octahedral rotations in a perovskite. *Phys. Rev. Res.* **2019**, *1*, 012005. [CrossRef]
13. Nelson, T.R.; White, A.J.; Bjorgaard, J.A.; Sifain, A.E.; Zhang, Y.; Nebgen, B.; Fernandez-Alberti, S.; Mozyrsky, D.; Roitberg, A.E.; Tretiak, S. Non-adiabatic Excited-State Molecular Dynamics: Theory and Applications for Modeling Photophysics in Extended Molecular Materials. *Chem. Rev.* **2020**, *120*, 2215–2287. [CrossRef]
14. Kreutzer, M.; Ernst, D.; Bishop, A.R.; Fehske, H.; Hager, G.; Nakajima, K.; Wellein, G. Chebyshev Filter Diagonalization on Modern Manycore Processors and GPGPUs. In *High Performance Computing, Proceedings of the 33rd International Conference, ISC High Performance 2018, Frankfurt, Germany, 24–28 June 2018*; Yokota, R., Weiland, M., Keyes, D., Trinitis, C., Eds.; Springer International Publishing AG: Cham, Switzerland, 2018; pp. 329–349.
15. Yonemitsu, K.; Bishop, A.R.; Lorenzana, J. Sensitivity of Doping States in the Copper Oxides to Electron-Lattice Coupling. *Phys. Rev. Lett.* **1992**, *69*, 965–968. [CrossRef]
16. Emin, D. *Polarons*; Cambridge University Press: New York, NY, USA, 2013.
17. Fehske, H.; Wellein, G.; Bishop, A.R. Spatiotemporal evolution of polaronic states in finite quantum systems. *Phys. Rev. B* **2011**, *83*, 075104. [CrossRef]
18. Salkola, M.I.; Bishop, A.R.; Trugman, S.A.; de Leon, J.M. Correlation-function analysis of nonlinear and nonadiabatic systems: Polaron tunneling. *Phys. Rev. B.* **1995**, *51*, 8878. [CrossRef] [PubMed]
19. Raghavan, S.; Bishop, A.R.; Kenkre, V.M. Quantum versus semiclassical description of self-trapping: Anharmonic effects. *Phys. Rev. B* **1999**, *59*, 9929–9932. [CrossRef]
20. Bishop, A.R.; Mihailovic, D.; Mustre de León, J. Signatures of mesoscopic Jahn–Teller polaron inhomogeneities in high-temperature superconductors. *J. Phys. Condens. Matter* **2003**, *15*, L169–L175. [CrossRef]
21. Bousquet, E.; Hamdi, H.; Aguado-Puente, P.; Salje, E.K.; Artacho, E.; Ghosez, P. First-principle characterization of single-electron polaron in WO3. *Phys. Rev. Res.* **2020**, *2*, 012052. [CrossRef]
22. Li, S.; Johnston, S. Quantum Monte Carlo study of lattice polarons in the two-dimensional multi-orbital Su-Schrieffer-Heeger model. *arXiv* **2019**, arXiv:1901.07612.
23. Yonemitsu, K.; Bishop, A.R.; Lorenzana, J. Magnetism and covalency in the two-dimensional three-band Peierls-Hubbard model. *Phys. Rev. B* **1993**, *47*, 8065–8075. [CrossRef]
24. Yu, Z.G.; Zang, J.; Gammel, J.T.; Bishop, A.R. Charge localization and stripes in a two-dimensional three-band Peierls-Hubbard model. *Phys. Rev. B* **1998**, *57*, R3241–R3244. [CrossRef]
25. Martin, I.; Kaneshita, E.; Bishop, A.R.; McQueeney, R.J.; Yu, Z.G. Vibrational edge modes in intrinsically heterogeneous doped transition metal oxides. *Phys. Rev. B* **2004**, *70*, 224514. [CrossRef]
26. Yonemitsu, K.; Batistić, I.; Bishop, A.R. Random-phase-approximation approach to collective modes around inhomogeneous Hartree-Fock states: One-dimensional doped Hubbard model. *Phys. Rev. B* **1991**, *44*, 2652–2663. [CrossRef]
27. McQueeney, R.J.; Bishop, A.R.; Yi, Y.-S.; Yu, Z.G. Charge localization and phonon spectra in hole-doped La_2NiO_4. *J. Phys. Condens. Matter* **2000**, *12*, L317–L322. [CrossRef]
28. Kaneshita, E.; Tohyama, T.; Bishop, A.R. Modeling of pressure effects in striped nickelates. *Physica C* **2010**, *470*, S247–S248. [CrossRef]
29. Agterberg, D.F.; Séamus Davis, J.C.; Edkins, S.D.; Fradkin, E.; Van Harlingen, D.J.; Kivelson, S.A.; Lee, P.A.; Radzihovsky, L.; Tranquada, J.M.; Wang, Y. The Physics of Pair-Density Waves: Cuprate Superconductors and Beyond. *Annu. Rev. Condens. Matter Phys.* **2020**, *11*, 231–270. [CrossRef]
30. Kaneshita, E.; Martin, I.; Bishop, A.R. Local Edge Modes in Doped Cuprates with Checkerboard Polaronic Heterogeneity. *J. Phys. Soc. Jpn.* **2004**, *73*, 3223–3226. [CrossRef]
31. Kaneshita, E.; Martin, I.; Bishop, A.R. Pressure-induced phase transition and bipolaronic sliding in a hole-doped Cu_2O_3-ladder system. *Phys. Rev. B* **2006**, *73*, 094514. [CrossRef]
32. Chernyshev, A.L.; Castro Neto, A.H.; Bishop, A.R. Metallic Stripe in Two Dimensions: Stability and Spin-Charge Separation. *Phys. Rev. Lett.* **2000**, *84*, 4922–4925. [CrossRef]

33. Janoschek, M.; Das, P.; Chakrabarti, B.; Abernathy, D.L.; Lumsden, M.D.; Lawrence, J.M.; Thompson, J.D.; Lander, G.H.; Mitchell, J.N.; Richmond, S.; et al. The valence-fluctuating ground state of plutonium. *Sci. Adv.* **2015**, *1*, e1500188. [CrossRef] [PubMed]
34. Zener, C. *Elasticity and Anelasticity of Metals*; University of Chicago Press: Chicago, IL, USA, 1948.
35. Lookman, T.; Shenoy, S.R.; Rasmussen, K.Ø.; Saxena, A.; Bishop, A.R. Ferroelastic dynamics and strain compatibility. *Phys. Rev. B* **2003**, *67*, 024114. [CrossRef]
36. Planes, A.; Lloveras, P.; Castán, T.; Saxena, A.; Porta, M. Ginzburg–Landau modelling of precursor nanoscale textures in ferroelastic materials. *Contin. Mech. Thermodyn.* **2012**, *24*, 619–627. [CrossRef]
37. Holm, E.A.; Cohn, R.; Gao, N.; Kitahara, A.R.; Matson, T.P.; Lei, B.; Yarasi, S.R. Overview: Computer vision and machine learning for microstructural characterization and analysis. *arXiv* **2020**, arXiv:2005.14260.
38. Ahn, K.H.; Lookman, T.; Saxena, A.; Bishop, A.R. Electronic properties of structural twin and antiphase boundaries in materials with strong electronic-lattice coupling. *Phys. Rev. B* **2005**, *71*, 212102. [CrossRef]
39. Zhu, L.; Prodan, E.; Ahn, K.H. Flat energy bands within antiphase and twin boundaries and at open edges in topographical materials. *Phys. Rev. B* **2019**, *99*, 041117. [CrossRef]
40. Aird, A.; Salje, E.K.H. Sheet conductivity in twin walls; experimental evidence of WO3-x. *J. Phys. Cond. Matt.* **1998**, *10*, L377–L380. [CrossRef]
41. Shengelaya, A.; Conder, K.; Müller, K.A. Signatures of Filamentary Superconductivity up to 94 K in Tungsten Oxide $WO_{2.90}$. *J. Supercond. Nov. Magn.* **2020**, *33*, 301–306. [CrossRef]
42. Zhu, J.-X.; Ahn, K.H.; Nussinov, Z.; Lookman, T.; Balatsky, A.V.; Bishop, A.R. Elasticity-Driven Nanoscale Electronic Structure in Superconductors. *Phys. Rev. Lett.* **2003**, *91*, 057004. [CrossRef]
43. Ahn, K.H.; Seman, T.F.; Lookman, T.; Bishop, A.R. Role of complex energy landscapes and strains in multiscale inhomogeneities in perovskite manganites. *Phys. Rev. B* **2013**, *88*, 144415. [CrossRef]
44. Millis, A.J. Lattice effects in magnetoresistive manganese perovskites. *Nature* **1998**, *392*, 147–150. [CrossRef]
45. Chen, A.; Su, Q.; Han, H.; Enriquez, E.; Jia, Q. Metal Oxide Nanocomposites: A Perspective from Strain, Defect, and Interface. *Adv. Mater.* **2019**, *31*, 1803241. [CrossRef]
46. Gottschall, T.; Benke, D.; Fries, M.; Taubel, A.; Radulov, I.A.; Skokov, K.P.; Gutfleisch, O. A Matter of Size and Stress: Understanding the First-Order Transition in Materials for Solid-State Refrigeration. *Adv. Funct. Mater.* **2017**, *27*, 1606735. [CrossRef]
47. Guzmán-Verri, G.G.; Brierley, R.T.; Littlewood, P.B. Cooperative elastic fluctuations provide tuning of the metal-insulator transition. *Nature* **2019**, *576*, 429–432. [CrossRef]
48. Shi, J.; Bruinsma, R.; Bishop, A.R. Theory of Vanadium Dioxide. *Synth. Met.* **1991**, *43*, 3527–3530. [CrossRef]
49. She, J.-H.; Zaanen, J.; Bishop, A.R.; Balatsky, A.V. Stability of Quantum Critical Points in the Presence of Competing Orders. *Phys. Rev. B* **2010**, *82*, 165128. [CrossRef]
50. Maniadis, P.; Lookman, T.; Bishop, A.R. Elastically driven polaron patterns: Stripes and glass phases. *Phys. Rev. B* **2011**, *84*, 024304. [CrossRef]
51. Bussmann-Holder, A.; Micnas, R.; Bishop, A.R. Enhancements of the superconducting transition temperature within the two-band model. *Eur. Phys. J. B* **2004**, *37*, 345–348. [CrossRef]
52. Bussmann-Holder, A.; Keller, H.; Simon, A.; Bianconi, A. Multi-Band Superconductivity and the Steep Band/Flat Band Scenario. *Condens. Matter* **2019**, *4*, 4040091. [CrossRef]
53. Cuevas, J.; Kevrekidis, P.G.; Frantzeskakis, D.J.; Bishop, A.R. Existence of bound states of a polaron with a breather in soft potentials. *Phys. Rev. B* **2006**, *74*, 064304. [CrossRef]
54. Bussmann-Holder, A.; Bishop, A.R.; Egami, T. Relaxor ferroelectrics and intrinsic inhomogeneity. *Europhys. Lett.* **2005**, *71*, 249–255. [CrossRef]
55. Bishop, A.R.; Bussmann-Holder, A.; Kamba, S.; Maglione, M. Common characteristics of displacive and relaxor ferroelectrics. *Phys. Rev. B* **2010**, *81*, 064106. [CrossRef]
56. Bussmann-Holder, A.; Büttner, H.; Bishop, A.R. Polar-Soft-Mode-Driven Structural Phase Transition in $SrTiO_3$. *Phys. Rev. Lett.* **2007**, *99*, 167603. [CrossRef]
57. Bettis, J.L.; Whangbo, M.-H.; Köhler, J.; Bussmann-Holder, A.; Bishop, A.R. Lattice dynamical analogies and differences between $SrTiO_3$ and $EuTiO_3$ revealed by phonon-dispersion relations and double-well potentials. *Phys. Rev. B* **2011**, *84*, 184114. [CrossRef]
58. Edge, J.M.; Kedem, Y.; Aschauer, U.; Spaldin, N.A.; Balatsky, A.V. Quantum Critical Origin of the Superconducting Dome in $SrTiO_3$. *Phys. Rev. Lett.* **2015**, *115*, 247002. [CrossRef] [PubMed]

59. Yi, Y.-S.; Bishop, A.R.; Röder, H. Spin-Peierls ground states in an electron-lattice periodic Anderson model. *J. Phys. Condens. Matter* **1999**, *11*, 3547–3554. [CrossRef]
60. Gulacsi, M.; Bussman-Holder, A.; Bishop, A.R. Spin and lattice effects in the Kondo lattice model. *Phys. Rev. B* **2005**, *71*, 214415. [CrossRef]
61. She, J.-H.; Bishop, A.R. RKKY Interaction and Intrinsic Frustration in Non-Fermi-Liquid Metals. *Phys. Rev. Lett.* **2013**, *111*, 017001. [CrossRef]
62. Phillips, J.C.; Saxena, A.; Bishop, A.R. Pseudogaphs, dopants, and strong disorder in cuprate high-temperature superconductors. *Rep. Prog. Phys.* **2003**, *66*, 2111–2182. [CrossRef]
63. Chakravarty, S.; Almutairi, B.S.; Chbeir, R.; Chakraborty, S.; Bauchy, M.; Micoulaut, M.; Boolchand, P. Progress, Challenges, and Rewards in Probing Melt Dynamics, Configurational Entropy Change, and Topological Phases of Group V- and Group IV-Based Multicomponent Sulfide Glasses. *Phys. Status Solidi B* **2020**, 2000116. [CrossRef]
64. Stojković, B.P.; Yu, Z.G.; Chernyshev, A.L.; Bishop, A.R.; Castro Neto, A.H.; Grønbech-Jensen, N. Charge ordering and long-range interactions in layered transition metal oxides: A quasiclassical continuum study. *Phys. Rev. B* **2000**, *62*, 4353–4369. [CrossRef]
65. Olson Reichhardt, C.J.; Reichhardt, C.; Bishop, A.R. Fibrillar Templates and Soft Phases in Systems with Short-Range Dipolar and Long-Range Interactions. *Phys. Rev. Lett.* **2004**, *92*, 016801. [CrossRef]
66. Olson Reichhardt, C.J.; Reichhardt, C.; Bishop, A.R. Structural transitions, melting, and intermediate phases for stripe- and clump-forming systems. *Phys. Rev. E* **2010**, *82*, 041502. [CrossRef]
67. Reichhardt, C.; Olson Reichhardt, C.J.; Bishop, A.R. Noise and hysteresis in charged stripe, checkerboard, and clump forming systems. In *SPIE Proceedings Vol. 6600, Proceedings of the SPIE 6600, Noise and Fluctuations in Circuits, Devices, and Materials, Florence, Italy, 20–24 May 2007*; Macuci, M., Vandamme, L.K.J., Ciofi, C., Weissman, M.B., Eds.; SPIE: Bellingham, WA, USA, 2007; p. 66001B.
68. Reichhardt, C.; Olson Reichhardt, C.J.; Bishop, A.R. Commensurate and incommensurate checkerboard charge ordered states. *Physica C* **2007**, *460–462*, 1178–1179. [CrossRef]
69. Barré, J.; Bishop, A.R.; Lookman, T.; Saxena, A. Random bond models of the intermediate phase in network forming glasses. In *Rigidity and Boolchand Intermediate Phases in Nanomaterials*; Micolaut, M., Popescu, M., Eds.; INOE Publishing House: Bucharest, Romania, 2009; Volume 3, pp. 105–128.
70. Frauenfelder, H.; Bishop, A.R.; Garcia, A.; Perelson, A.; Schuster, P.; Sherrington, D.; Swart, P.J. (Eds.) *Landscape Paradigms in Physics and Biology*; North-Holland: Amsterdam, The Netherlands, 1997.
71. She, J.-H.; Fransson, J.; Bishop, A.R.; Balatsky, A.V. Inelastic Electron Tunneling Spectroscopy for Topological Insulators. *Phys. Rev. Lett.* **2013**, *110*, 026802. [CrossRef] [PubMed]
72. Haraldsen, J.T.; Wölfle, P.; Balatsky, A.V. Understanding the electric-field enhancement of the superconducting transition temperature for complex oxide interfaces. *Phys. Rev. B* **2012**, *85*, 134501. [CrossRef]
73. Salje, E.K.H. Multiferroic domain boundaries as active memory devices: Trajectories towards domain boundary engineering. *Chem. Phys. Chem.* **2010**, *11*, 940–950. [CrossRef] [PubMed]
74. Conradson, S.D.; Geballe, T.H.; Jin, C.; Cao, L.; Baldinozzi, G.; Jiang, J.M.; Latimer, M.J.; Mueller, O. Local structure of $Sr_2CuO_{3.3}$, a 95 K cuprate superconductor without CuO_2 planes. *Proc. Natl. Acad. Sci. USA* **2020**, *117*, 4565–4570. [CrossRef] [PubMed]
75. Barzykin, V.; Gor'kov, L.P. Inhomogeneous stripe phase revisited for surface suoerconductivity. *Phys. Rev. Lett.* **2002**, *89*, 227002. [CrossRef]
76. Agrestini, S.; Saini, N.L.; Bianconi, G.; Bianconi, A. The strain of CuO2 lattice: The second variable for the phase diagram of cuprate perovskites. *J. Phys. A Math. Gen.* **2003**, *36*, 9133. [CrossRef]
77. Eroles, J.; Ortiz, G.; Balatsky, A.V.; Bishop, A.R. Inhomogeneity-induced superconductivity? *Europhys. Lett.* **2000**, *50*, 540–546. [CrossRef]
78. Mazumdar, S. Valence transition model of the pseudogap, charge order, and superconductivity in electron-doped and hole-doped copper oxides. *Phys. Rev. B* **2018**, *98*, 205153. [CrossRef]
79. Bishop, A.R.; Martin, R.L.; Müller, K.A.; Tesanovic, Z. Superconductivity in oxides: Toward a unified picture. *Z. Phys. B* **1989**, *76*, 17. [CrossRef]

80. Shenoy, S.R.; Subrahmanyam, V.; Bishop, A.R. Quantum paraelectric model for layered superconductor. *Phys. Rev. Lett.* **1997**, *79*, 4657. [CrossRef]
81. Phillips, J.C. Ted Geballe and HTSC. *J. Supercond. Nov. Mag.* **2020**, *33*, 11–13. [CrossRef]

© 2020 by the author. Licensee MDPI, Basel, Switzerland. This article is an open access article distributed under the terms and conditions of the Creative Commons Attribution (CC BY) license (http://creativecommons.org/licenses/by/4.0/).

Creative

A Retrospective of Materials Synthesis at the Paul Scherrer Institut (PSI)

Kazimierz Conder [1], Albert Furrer [2] and Ekaterina Pomjakushina [1,*]

1   Laboratory for Multiscale Materials Experiments, Paul Scherrer Institut, CH-5232 Villigen PSI, Switzerland; kazimierz.conder@psi.ch
2   Laboratory for Neutron Scattering, Paul Scherrer Institut, CH-5232 Villigen PSI, Switzerland; albert.furrer@psi.ch
*   Correspondence: ekaterina.pomjakushina@psi.ch; Tel.: +41-56-310-3207

Received: 14 September 2020; Accepted: 17 September 2020; Published: 23 September 2020

1. Establishing Material Synthesis at PSI—A. Furrer

The availability of high-quality and well characterized materials is a key factor for condensed-matter research. The Laboratory for Neutron Scattering at PSI has profited for a long time from extended collaborations with international groups to receive samples for neutron scattering experiments in different fields of science. The situation changed in the year 1986, with the discovery of high-temperature superconductivity in copper-oxide perovskites by Müller and Bednorz [1]. In high-T_C research, the systematic variation of the materials properties is an essential feature, e.g., the variation of the oxygen content. Consequently, the experiments had to be carried out for a large number of systematically varied samples, so that the cooperation with partner groups became rather difficult. It was therefore decided to start in-house efforts for the production of high-T_C materials.

The driving force behind these efforts was Peter Allenspach, who set up a materials synthesis laboratory based on a procedure developed at the Jülich Research Center [2]. In order to obtain sufficient material for neutron scattering experiments, the necessary devices were enlarged up to capacities of 300 g and 30 g for the sintering and oxygen loading furnaces, respectively [3]. Subsequently, the instrumentation of the materials synthesis laboratory was extended to include further devices (e.g., a physical properties measurement system), supported by funds from the Swiss National Science Foundation, the PSI, the ETH Zurich, and the University of Zurich.

Initially, the materials synthesis laboratory operated in a self-service mode. However, this operational mode turned out to be ineffective due to the increasing complexity of the instruments. In the year 1998, Kazimierz Conder took over the responsibility for the materials synthesis laboratory. He further developed the instrumentation, e.g., by crystal-growth devices. Early requests to promote the status of the materials synthesis laboratory to the level of a PSI-group failed, but thanks to the invaluable support by Alex Müller, the group "Materials Synthesis" was installed at PSI in the year 2003 with Kazimierz Conder as group leader. At the same time, Ekaterina Pomjakushina joined the group.

Around the year 2000, the main activities of the materials synthesis laboratory focused on high-T_C superconductors. The concept which led Alex Müller to the discovery of superconductivity in the cuprates was the vibronic property of the Jahn–Teller effect. Therefore, looking for an isotope effect for both oxygen and copper ions was an obvious task. Neutron crystal-field studies provide relevant information on the pseudogap temperature T* through the linewidth of the crystal-field transitions. This method was applied to the compound $HoBa_2Cu_4O_8$ for both oxygen isotope substitution with $T_c(^{16}O)$ = 79.0(1) K and $T_c(^{18}O)$ = 78.5(1) K [4] and copper isotope substitution with $T_c(^{63}Cu)$ = 79.0 K and $T_c(^{65}Cu)$ = 78.6(1) K [5]. The corresponding samples were synthesized in the materials synthesis laboratory.

Figure 1 shows the temperature dependence of the intrinsic linewidth (HWHM) corresponding to the lowest-lying crystal-field transition at energy 0.6 meV. There is evidence for large isotope shifts

$\Delta T^*(O) \approx 50$ K and $\Delta T^*(Cu) \approx 25$ K. The corresponding isotope coefficients α^* defined by the relation $T^* \propto 1/M^{\alpha^*}$ (M is the mass of the O or Cu ion) turn out to be $\alpha^*(O) = -2.2$ and $\alpha^*(Cu) = -4.9$. Giorgio Benedek highlighted the latter coefficient in his lecture presented at an international symposium in Zurich in the year 2006 and suggested to call it the Alfa Romeo number, since Alex Müller is an enthusiastic driver of Alfa Romeo cars, such as the model Alfa Romeo Montreal 4.9.

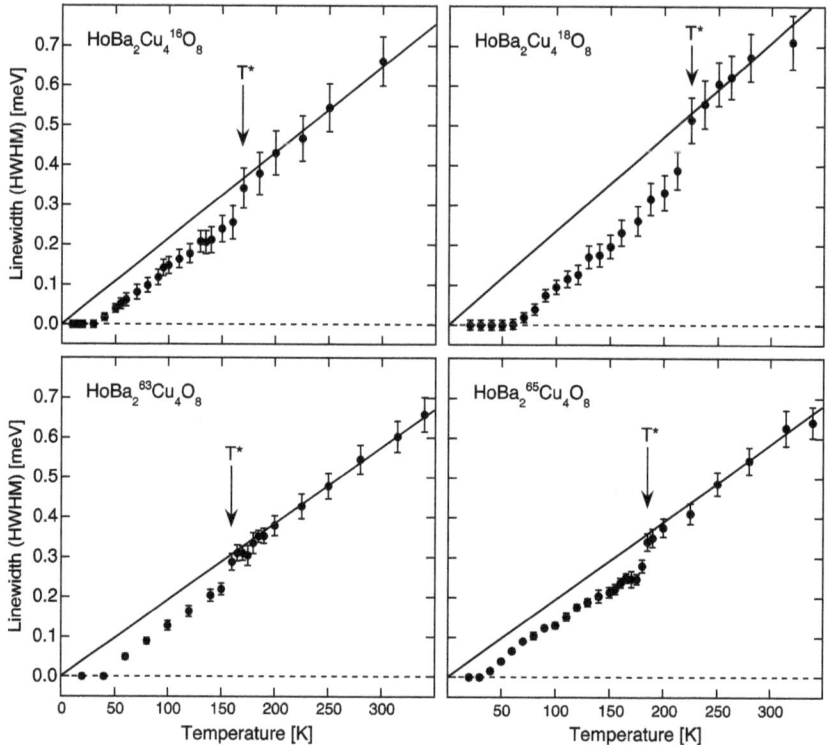

Figure 1. Temperature dependence of the intrinsic linewidth W (HWHM) corresponding to the lowest-lying crystal-field transition in HoBa$_2$Cu$_4$O$_8$ for different oxygen and copper isotopes [4,5]. The lines denote the linewidth in the normal state calculated from the Korringa law.

2. Materials Synthesis at PSI from the Past to Nowadays—K. Conder and E. Pomjakushina

2.1. HTc and Copper Age at ETH

In 1986, K.A. Müller and G. Bednorz announced discovery of the high-temperature (HTc) superconductivity. Very quickly, many research groups around the world started scientific work on this subject. In the same year one of us (K.C) started an annual scholarship in the group of Professor E. Kaldis in the Laboratory of Solid State Physics (Laboratorium für Festkörperphysik) of the Federal University of Technology (ETH) in Zurich. In addition, also in the group of prof. Kaldis, superconductors quickly displaced other subjects. Among others, a pioneering work of Dr. Karpinski concerning synthesis of superconductors under high oxygen pressure was at the forefront in the field of HTc superconductors. Additionally, in late 1980s, Prof. K.A. Müller proposed a research project concerning studies of oxygen isotope effects in the HTc superconductors. The project was implemented jointly by the University of Zurich (physical property measurements—group of Prof. H. Keller) and ETH (sample preparation—group of Prof. E. Kaldis). In this project, sample preparation topic was entrusted to one of us (K.C), creating a unique opportunity to work with the Nobel prize laureate. This collaboration lasted for many years, resulting in over 40 publications.

2.2. Oxygen Isotope Effect in $YBa_2Cu_3O_{7-x}$

The discovery of the isotope effect in superconductors in the 1950s created an experimental foundation for the BCS (Bardeen, Cooper, and Schrieffer) theory, which is a microscopic theory of superconductivity announced in 1957 (Nobel Prize in 1972). In a classic experiment, a linear relationship between the temperature of the transition to the superconducting state T_C with the reciprocal square root of the mass of the mercury isotopes was achieved. This pointed out that the appearance of superconductivity is correlated with the energy of the crystal lattice vibrations, which in turn is determined by the masses of atoms creating crystal lattice. This led to the conclusion that the charge carriers in the superconducting state are formed by pairs of electrons (Cooper pairs). Thus, the electrons interact each other through the crystal lattice (electron-phonon interaction). For the HTc superconductors, Prof. K.A. Müller suggested studies of the critical temperature of the "most famous" superconducting material, i.e., $YBa_2Cu_3O_{7-x}$ (YBCO) synthesized with oxygen isotopes ^{16}O and ^{18}O. At this time, several papers on the isotope effect in HTc superconductors were already published, but the results presented were inconclusive; either very small or no effect were reported. In addition, Prof. K.A. Müller was interested in the selective isotopic effect, i.e., the effect caused by the substitution of the oxygen isotope at different sites in the unit cell (in the structure of YBCO there are three symmetrically independent oxygen sites). From the beginning, it was clear that the task is very difficult. First of all, the isotope effect was expected to be very small (a fraction of Kelvin). Additionally, it was known that T_C of YBCO decisively depends on the oxygen stoichiometry in the compound. The key issue was, therefore, to achieve an extremely high reproducibility of the sample preparation procedure. The samples could not differ in any property, besides the oxygen isotope used. Thus, it was necessary to develop a method of extremely accurate determination the oxygen content in YBCO. Apart from that, it was unavoidably to check whether the site-selective isotopic exchange is at all possible. This required comprehensive studies of the kinetics of the oxygen isotope exchange in YBCO. Based on such studies, a series of YBCO samples completely and selectively substituted with the oxygen isotope ^{18}O were made. Magnetization measurements conducted at the University of Zurich (D. Zech in the group of Prof. H. Keller) showed that the overall isotope effect although very small (0.2 K) comes entirely from the CuO_2 planes in the structure. The results were published in Nature in 1994 [6]. Following this work, oxygen isotope effect in other high-temperature superconductors and also related materials, e.g., magnetoresistive or exhibiting a metal-insulator transition were performed. This scientifically fruitful period ended in the late nineties with a retirement of Prof. E. Kaldis and a dissolution of his group. It was again initiative and a great commitment of Prof. K.A. Müller to transfer the "superconductivity-isotope" research from ETH to Paul Scherrer Institute (PSI). Fortunately, thanks to the interest and great efforts of Prof. A. Furrer, the head of the Laboratory for Neutron Scattering (LNS) of the time, it became possible. Over the years, we were able to endow our laboratory with equipment for syntheses of ceramic materials, crystal growth (zone melting method using a mirror furnace), as well as characterization instruments (thermal analysis, X-ray diffractometers, micro X-ray fluorescence). The staff of the group also increased over time.

2.3. HTc Projects at PSI

One of the first tasks at PSI was inspired by an announcement by Reich and Tsabba [7] about a possible nucleation of superconducting regions with T_C = 90 K on the surface of Na-doped WO_3 crystals. It would the only superconducting material containing no copper, with T_C higher than boiling point of liquid nitrogen at normal pressure. Our attempts to examine these materials continued (with many interruptions!) for several years. Unfortunately, due to the possibly filamentary character of the superconductivity in these materials, we were unable to draw decisive conclusions. A recent work [8] summarizes the results obtained.

The other subjects of the research, carried out at PSI were initially mostly associated with the isotope effect. For example, it was found that oxygen isotope substitution very strongly influences the temperature of the so-called "pseudogap" opening [4]. The isotope effect on pseudogap was also detected by the substitution of natural copper with 63 and 65 cooper isotopes [5]. In collaboration with other laboratories (LMS PSI, Physik Institut Universität Zürich), we found that in the case of the HTc-superconductors the isotopic substitution of oxygen also influences other properties, such as magnetic penetration depth, the Néel temperature and the linewidth of the paramagnetic resonance EPR signal.

2.4. Establishing Traveling Solvent Floating Zone Crystal Growth

Later, more and more, the work was concentrated on synthesis, crystal growth and investigating of oxide materials with interesting electrical and magnetic properties. Such materials are, among others, layered cobaltites $RBaCo_2O_{5+x}$ (R = rare earth metal) interesting family of complex oxides with strongly correlated electrons. Layered cobaltites exhibit a number of interesting properties associated with a possibility to adopt different oxidation states by the cobalt cations (II, III, IV) and also different spin states: low, intermediate and high. Thus, these materials exhibit metal-insulator transition at temperatures near room temperature and interesting magnetic properties including giant magnetoresistivity (GMR). All these properties decisively dependent on the oxygen content ($0 < x < 1$), being therefore graceful working subject for chemists. The most interesting were studies of magnetic properties of these compounds. On the basis of the measurements of the spin polarization and the muon spin rotation method (µSR), different magnetic phases have been identified characterized by the arrangement of the cobalt in the various spin states. A phase diagram was developed showing temperature ranges of existence of different magnetic phases, as well as ranges of their coexistence [9]. We also reported on oxygen isotope effect on metal-insulator transition temperature in $RBaCo_2O_{5.5}$ (R = Pr, Dy, Ho and Y) [10].

The TSFZ (Traveling Solvent Floating Zone) method of the crystal growth quickly became a specialty of our laboratory. Starting from 2003, we were able to synthesize many materials in a single crystalline form. These were, among others, superconductor $La_{2-x}Sr_xCuO_4$, frustrated magnet $YMnO_3$, $LaCoO_3$—showing high, medium and low-spin states of cobalt, multiferroic $LuFe_2O_4$, two-dimensional antiferromagnetic $SrCu_2(BO_3)_2$ and $YFeO_3$ showing magneto-optic Kerr effect. All these crystals and many other materials obtained in the crystalline and polycrystalline forms were often produced on the "order" of physicists from PSI or other research institutions in Switzerland. The materials were used for scientific studies, usually using available at PSI research facilities as large-scale neutron diffraction and synchrotron sources.

It is believed that in cuprates, the two-dimensional square lattice CuO_2 layers in the structure are responsible for the superconductivity. Therefore, cuprates having in the structure Cu-O layers of different geometry and stoichiometry are of particular interest. Some years ago, great interest was awaked with the cuprate family of the formula $Sr_{14-x}Ca_xCu_{24}O_{41}$ containing in the structure Cu-O layers of a geometry resembling a ladder (known as "spin ladders") and stoichiometry Cu_2O_3. These are the only cuprates, besides those containing CuO_2 layers, which exhibit superconductivity although only at high pressures and large calcium contents ($x > 10$). In these compounds, due to the certain geometry of the Cu-O layers, anisotropy of the transport properties in normal and the superconducting state are particularly important. Together with a partner from India, we performed during 2009–2014 scientific project focused on studies of properties of this compound under uniaxial pressure (i.e., applied along different crystallographic directions). The success of this project was dependent on an availability of single crystals of these compounds. We managed to grow a series of $Sr_{14-x}Ca_xCu_{24}O_{41}$ crystals by TSFZ method under elevated oxygen pressure [11]. The studies of electrical and magnetic properties, made by our partner, indicate that the mechanical pressure applied along the c-direction changes the nature of the conductivity from one- to the two-dimensional. This increases the critical temperature of superconductivity from previously reported 12 K (using hydrostatic pressure) to 25 K [12]. We also elaborated $BiCu_2O_2PO_4$, another compound having in the structure Cu-O layers with a special geometry [13].

2.5. Iron Era

The discovery of superconductivity in 2008 in iron compounds (chalcogenides, pnictides and phosphides) was absolutely unexpected, because of the presence in the structure of magnetic iron ions. Iron selenide FeSe has, among these compounds, the simplest structure but also the lowest critical temperature (approx. 8 K). We performed many syntheses experiments [14] and studies of structural, superconducting [15] and magnetic [16] properties. We also synthesized a series of samples with iron isotopes. Here, as in the case of cuprates, reproducibility of preparation was essential for the reliability of the results obtained. It turned out that the isotope effect is very similar to that expected for the conventional superconductors [17].

For iron selenides, substituting half selenium with tellurium increases the critical temperature from 8 to 14 K. In addition, such mixed compounds melt congruently and can be crystallized using the Bridgman method. We have grown a series of crystals of $Fe_{1.03}Se_xTe_{1-x}$ with $0 < x < 0.5$. Based on measurements of magnetization and muon spin rotation, a coexistence of magnetism and superconductivity in these compounds was shown [18].

In 2010, superconductivity with a critical temperature of $T_C \approx 31$ K in iron chalcogenide intercalated with potassium $K_xFe_2Se_2$ was reported. In a short time, we managed to synthesize the next member of this "family": $Cs_{0.8}(FeSe_{0.98})_2$ [19]. For this family of superconducting compounds, we found that superconductivity coexist with magnetism [20]. Detailed structural studies [21,22], muon spin rotation investigations [23] and microscopic examinations [24] showed a coexistence (at temperatures below 420 K) of two phases with compositions similar to AFe_2Se_2 and $A_2Fe_4Se_5$ (A = K, Rb or Cs). The first one, comprising only approx. 10% of the volume of the sample, becomes superconducting below $\approx$30 K, the latter is the antiferromagnetic showing ordered iron vacancy pattern. Our work on alkali metal intercalated iron chalcogenides is presented in a review paper [25].

2.6. Stabilization of Transition Metals in High Oxidation States under High Oxygen Pressure

In 2013, our group became a partner in the SNF Sinergia project "Mott Physics Beyond the Heisenberg Model (MPBH)", which was lasted for seven years and connected researchers across Switzerland in a collective effort to explore new aspects of Mott-physics, such as how a spin-orbit interaction in concert with strong electron correlations can lead to new and exotic quantum materials. The network consisted of researchers hosted by the Ecole Polytechnique Fédérale Lausanne (EPFL), the Paul Scherrer Institute (PSI) and University of Zurich (UniZH). The solid state chemistry group was working in materials discovery part of the project with a focus on iridate compounds. Iridates containing iridium at +VI oxidation state are of great interest in order to reveal the relative contribution of Spin Orbit Coupling effect in 5d compounds. The stabilization of Ir in the high oxidation state often needs high pressure. For the planned studies, a unique oxygen high pressure system relocated from Solid State Physics Laboratory ETH in Zürich [26] was used. The system allows synthesis, crystal growth and thermodynamic measurements at oxygen pressures up to 2000 bars and temperature up to 1200 °C. In contrast to cubic anvil devices, the accessible pressure range is an order of magnitude smaller; however, the possible maximum sample volume is much larger (10 cm^3 in comparison with mm^3 as applicable using the anvil system). This is especially important for a preparation of samples useful for neutron scattering experiments. Additionally, oxygen gas pressure apparatus in the existing form allows precise control of a pressure and temperature during synthesis and can also be used for studies of thermodynamics: e.g., measurements of a chemical stability and oxygen stoichiometry under high oxygen pressure.

One of the results was a successful synthesis of the double perovskite Sr_2BIrO_6 (B = Ni, Cu and Zn) in the high oxygen pressure furnace and determination of a novel magnetic structure with an incommensurate propagation vector along the body diagonal, which has not been seen in double perovskites yet [27]. The incommensurate structure can be explained by the intrinsic frustration between the ferromagnetic coupling of nickel-iridium bonds and the antiferromagnetic iridium-iridium coupling.

With the above-mentioned high-pressure setup, we also have found a new chemical route to synthesize $Tm_2Mn_2O_7$ pyrochlore, which is thermodynamically unstable at ambient pressure. Differently from the reported in the past high-pressure synthesis of the same compound applying oxides as starting materials, we obtained a pure $Tm_2Mn_2O_7$ phase by a converting $TmMnO_3$ at 1100 °C and an oxygen pressure of 1300 bar [28].

The same year (2015), we started a new project: "High pressure synthesis of iron complex oxides in high oxidation state (Fe^{4+}, Fe^{5+}): mapping between localized and itinerant behavior". A series of oxygen stoichiometric polycrystalline materials of $Re_{1/3}Sr_{2/3}FeO_3$ (Re = La, Pr, Nd, Sm, Gd, Dy, Y) were synthesized and studied. In this family of compounds iron adopts an average oxidation level higher than Fe^{3+}. The preparation of such materials requires high oxidation potential during synthesis, which we obtained using our high-pressure equipment. The aims of this project were to understand crystal and magnetic structure of $Re_{1/3}Sr_{2/3}FeO_3$ (R = La, Pr, and Nd), and to provide more information on the metal-insulator transition accompanied by the magnetic transition observed in this system. To disentangle their crystal and magnetic structures we used a combination of neutron powder and single crystal diffraction, neutron spherical polarimetry, and symmetry analysis [29]. One of the main problems when studying these compounds was the fact that two different magnetic structures with possible presence/or absence of iron charge ordering fit equally well neutron diffraction data. Further investigations of doped materials, suppressing possible iron charge ordering are now on the way.

2.7. New Techniques for Novel Materials with Topological States

Over time, the limitations of the TSFZ crystal growth system (CSC, Japan), installed in 2003 with a support of the R'EQUIP, were becoming more and more relevant to the further development of our science. First of all, the maximum temperature of growth is limited to 2100 °C. Additionally, many materials, especially oxides, decompose or become oxygen nonstoichiometric at high temperatures. Therefore, to stabilize them, high oxygen pressure during crystal growth is indispensable. In a new R'EQUIP project (2015) we proposed an acquisition of a high temperature crystal growth optical furnace offered by the Scientific Instruments Dresden GmbH (SciDre). This instrument comprises demands of performing crystallization processes at very broad growth temperatures from several hundred up to 2900 °C with simultaneously applied gas pressure up to 150 bars. The project was supported and after the delivery to PSI the HKZ-furnace was assembled and tested on site in 2017 and finally brought into operation in 2018.

Recently discovered new quantum states of matter, such as topological insulators and Weyl semi-metals, attract huge attention from condensed matter physics community, driving a demand for single crystals of these materials. The chemistry of these materials called for new synthesis approaches. Thus, we have implemented two new crystal growth methods into our group: growth from high temperature metal fluxes and chemical transport. Over the past few years, the collaboration established by the group and the large-scale research facilities of PSI has resulted in a good set of scientific publications in very high ranked journals, based on science performed on our single crystals [30]. In a recent work on crystal growth of magnetic Weyl semi-metals, we proved that TSFZ technique is crucial to obtain perfectly stoichiometric RAlGe crystals (R:rare earth metals) [31]. The availability of large crystals grown with a new TSFZ furnace (HKZ, SciDre) allows the possibility to reveal the interplay between magnetism and the properties of the Weyl state and will result in the discovery of new materials with interesting and potentially applicable electronic and magnetic properties.

Soon after we found the way of crystal growth of RAlGe, we reported the discovery of topological magnetism in the candidate magnetic Weyl semimetal CeAlGe. Using neutron scattering, we find this system to host several incommensurate, square-coordinated multi $-\vec{k}$ magnetic phases below T_N. The topological properties of a phase stable at intermediate magnetic fields parallel to the c-axis are suggested by observation of a topological Hall effect. Our findings highlight CeAlGe as an exceptional system for exploiting the interplay between the nontrivial topologies of magnetization in real space and Weyl nodes in momentum space [32]. This work is continued with a support of SNF project entitled "Tailoring materials with novel quantum states—exploring magnetic Weyl semimetals". In this project, we aim to test recent theoretical predictions and search experimentally for materials with novel quantum states. Such ambitious work requires a synergy between different aspects of solid state physics, from theory to experiment linked by material science, requiring elaborate work in the field of synthetic chemistry and crystal growth.

During all these years, the solid state chemistry group at PSI (see Figure A1) was working on the cutting-edge of fundamental science, making our own research in the field of novel materials and providing materials of interest for many groups within and outside PSI. Dozens of PhD students form PSI, ETH, University of Zürich, have made their thesis on materials synthesized in the group. The group was always supported by heads of PSI laboratories, especially LNS, LDM, LMX, and University of Zürich, especially Prof. Hugo Keller. Together with gaining experience in material synthesis, the number of instruments and techniques was incredibly increased, which brings the group to a very high competitive international level.

Author Contributions: A.F., K.C. and E.P.—Writing—Review & Editing. All authors have read and agreed to the published version of the manuscript.

Funding: This research received no external funding.

Conflicts of Interest: The authors declare no conflict of interest.

Appendix A

Below aresome statistics. Number of group members from 2000 including present (4) and former (13):

- 5 senior researchers—Kazik, Katja, Marisa, Hans, Darek
- 7 PostDocs—Guochu, Anka, Romain, Kathi, Junye, Pascal, Tian
- 5 PhD students—Marian, Ruggero, Shuang, Mickael, Fei
- ~300 papers published

Figure A1. Solid State Chemistry Group in April 2016. From left to right: Dariusz Gawryluk, Mickaël Morin, Fei Li, Romain Sibille, Ekaterina Pomjakushina, Marisa Medarde, Kazimierz Conder, Katharina Rolfs.

References

1. Bednorz, J.G.; Müller, K.A. Possible highT_C superconductivity in the Ba–La–Cu–O system. *Z. Phys. B Condens. Matter* **1986**, *64*, 189–193. [CrossRef]
2. Meuffels, P.; Rupp, B.; Pörschke, E. The solid combustion synthesis of small REBa$_2$Cu$_3$O$_x$ samples (RE = Y, Er). *Physica C* **1988**, *156*, 441. [CrossRef]
3. Allenspach, P. *Progress Report LNS-150*; ETH Zurich & PSI: Villigen, Switzerland, 1990; pp. 116–117.
4. Rubio Temprano, D.; Mesot, J.; Janssen, S.; Conder, K.; Furrer, A.; Mutka, H.; Müller, K.A. Large isotope effect on pseudogap in the high-temperature superconductor HoBa$_2$Cu$_4$O$_8$. *Phys. Rev. Lett.* **2000**, *84*, 1990. [CrossRef] [PubMed]
5. Rubio Temprano, D.; Mesot, J.; Janssen, S.; Conder, K.; Furrer, A.; Sokolov, A.; Trounov, V.; Kazakov, S.M.; Karpinski, J.; Müller, K.A. Large copper isotope effect on the pseudogap in the high-temperature superconductor HoBa$_2$Cu$_4$O$_8$. *Eur. Phys. J. B* **2001**, *19*, 5–8. [CrossRef]
6. Zech, D.; Keller, H.; Conder, K.; Kaldis, E.; Liarokapis, E.; Poulakis, N.; Müller, K.A. Site-selective oxygen isotope effect in optimally doped YBa$_2$Cu$_3$O$_x$. *Nature* **1994**, *371*, 681–683. [CrossRef]
7. Reich, S.; Tsabba, Y. Possible nucleation of a 2D superconducting phase on WO$_3$ single crystals surface doped with Na$^+$. *Eur. Phys. J. B* **1999**, *9*, 1–4. [CrossRef]
8. Shengelaya, A.; Conder, K.; Müller, K.A. Signatures of Filamentary Superconductivity up to 94 K in Tungsten Oxide WO$_{2.90}$. *J. Supercond. Nov. Magn.* **2020**, *33*, 301–306. [CrossRef]
9. Luetkens, H.; Stingaciu, M.M.; Pashkevich, Y.G.; Conder, K.; Pomjakushina, E.; Gusev, A.A.; Lamonova, K.V.; Lemmens, P.; Klauss, H.H. Microscopic Evidence of Spin State Order and Spin State Phase Separation in Layered Cobaltites RBaCo$_2$O$_{5.5}$ with R = Y, Tb, Dy, and Ho. *Phys. Rev. Lett.* **2008**, *101*, 017601. [CrossRef]
10. Conder, K.; Pomjakushina, E.; Pomjakushin, V.; Stingaciu, M.; Streule, S.; Podlesnyak, A. Oxygen isotope effect on metal-insulator transition in layered cobaltites RBaCo$_2$O$_{5.5}$ (R = Pr, Dy, Ho and Y). *J. Phys. Condens. Mater* **2005**, *17*, 5813. [CrossRef]
11. Deng, G.; Radheep, D.M.; Thiyagarajan, R.; Pomjakushina, E.; Wang, S.; Nikseresht, N.; Arumugam, S.; Conder, K. High oxygen pressure single crystal growth of highly Ca-doped spin ladder compound Sr$_{14-x}$Ca$_x$Cu$_{24}$O$_{41}$ (x > 12). *J. Cryst. Growth* **2011**, *327*, 182–188. [CrossRef]
12. Radheep, D.M.; Thiyagarjan, R.; Esakkimuthu, S.; Deng, G.; Pomjakushina, E.; Prajapat, C.L.; Ravikumar, G.; Conder, K.; Baskaran, G.; Arumugam, S. A spin ladder compound doubles its superconducting T_C under a gentle uniaxial pressure. *arXiv* **2013**, arXiv:1303.0921.
13. Wang, S.; Pomjakushina, E.; Shiroka, T.; Deng, G.; Nikseresht, N.; Rüegg, C.; Rønnow, H.M.; Conder, K. Crystal growth and characterization of thedilutable frustrated spin-ladder compound Bi(Cu$_{1-x}$Zn$_x$)$_2$PO$_6$. *J. Cryst. Growth* **2010**, *313*, 51. [CrossRef]
14. Pomjakushina, E.; Conder, K.; Pomjakushin, V.; Bendele, M.; Khasanov, R. Synthesis, crystal structure, and chemical stability of the superconductor FeSe$_{1-x}$. *Phys. Rev. B* **2009**, *80*, 024517. [CrossRef]
15. Khasanov, R.; Bendele, M.; Amato, A.; Conder, K.; Keller, H.; Klauss, H.H.; Luetkens, H.; Pomjakushina, E. Evolution of Two-Gap Behaviour of the Superconductor FeSe$_{1-x}$. *Phys. Rev. Lett.* **2010**, *104*, 087004. [CrossRef] [PubMed]
16. Bendele, M.; Amato, A.; Conder, K.; Elender, M.; Keller, H.; Klauss, H.H.; Luetkens, H.; Pomjakushina, E.; Raselli, A.; Khasanov, R. Pressure Induced Static Magnetic Order in Superconducting FeSe$_{1-x}$. *Phys. Rev. Lett.* **2010**, *104*, 087003. [CrossRef]
17. Khasanov, R.; Bendele, M.; Conder, K.; Keller, H.; Pomjakushina, E.; Pomjakushin, V. Iron isotope effect on the superconducting transition temperature and the crystal structure of FeSe$_{1-x}$. *New J. Phys.* **2010**, *12*, 073024. [CrossRef]
18. Khasanov, R.; Bendele, M.; Amato, A.; Babkievich, P.; Boothroyd, A.T.; Cervellino, A.; Conder, K.; Gvasaliya, S.N.; Keller, H.; Klauss, H.H.; et al. Coexistence of incommensurate magnetism and superconductivity in Fe$_{(1+y)}$Se$_x$Te$_{(1-x)}$. *Phys. Rev. B* **2009**, *80*, 140511. [CrossRef]
19. Krzton-Maziopa, A.; Shermadini, Z.; Pomjakushina, E.; Pomjakushin, V.; Bendele, M.; Amato, A.; Khasanov, R.; Luetkens, H.; Conder, K. Synthesis and crystal growth of Cs$_{0.8}$(FeSe$_{0.98}$)$_2$: A new iron-based superconductor with T_C = 27 K. *J. Phys. Condens. Matter* **2011**, *23*, 052203. [CrossRef]

20. Shermadini, Z.; Krzton-Maziopa, A.; Bendele, M.; Khasanov, R.; Luetkens, H.; Conder, K.; Pomjakushina, E.; Weyeneth, S.; Pomjakushin, V.; Bossen, O.; et al. Coexistence of Magnetism and Superconductivity in the Iron-based Compound $Cs_{0.8}(FeSe_{0.98})_2$. *Phys. Rev. Lett.* **2011**, *106*, 117602. [CrossRef]
21. Pomjakushin, V.Y.; Sheptyakov, D.V.; Pomjakushina, E.V.; Krzton-Maziopa, A.; Conder, K.; Chernyshov, D.; Svitlyk, V.; Shermadini, Z. Iron-vacancy superstructure and possible room-temperature antiferromagnetic order in superconducting $Cs_yFe_{2-x}Se_2$. *Phys. Rev. B* **2011**, *83*, 144410. [CrossRef]
22. Pomjakushin, V.Y.; Pomjakushina, E.V.; Krzton-Maziopa, A.; Conder, K.; Shermadini, Z. Room temperature antiferromagnetic order in superconducting $X_yFe_{2-x}Se_2$ (X = Rb, K): A neutron powder diffraction study. *J. Phys. Condens. Matter* **2011**, *23*, 156003. [CrossRef] [PubMed]
23. Shermadini, Z.; Luetkens, H.; Khasanov, R.; Krzton-Maziopa, A.; Conder, K.; Pomjakushina, E.; Klauss, H.-H.; Amato, A. Superconducting properties of single-crystalline $A_xFe_{2-y}Se_2$ (A = Rb, K) studied using muon spin spectroscopy. *Phys. Rev. B* **2012**, *85*, 100501(R). [CrossRef]
24. Speller, S.C.; Britton, S.C.; Hughes, G.M.; Krzton-Maziopa, A.; Pomjakushina, E.; Conder, K.; Boothroyd, A.T.; Grovenor, C.R.M. Microstructural analysis of phase separation in iron chalcogenide superconductors. *Supercond. Sci. Technol.* **2012**, *25*, 084023. [CrossRef]
25. Krzton-Maziopa, A.; Svitlyk, V.; Pomjakushina, E.; Puzniak, R.; Conder, K. Superconductivity in alkali metal intercalated iron selenides. *J. Phys. Condens. Matter* **2016**, *28*, 293002. [CrossRef] [PubMed]
26. Karpinski, J. High pressure in the synthesis and crystal growth of superconductors and III–N semiconductors. *Philos. Mag.* **2012**, *92*, 2662. [CrossRef]
27. Rolfs, K.; Tóth, S.; Pomjakushina, E.; Adroja, D.T.; Khalyavin, D.; Conder, K. Incommensurate magnetic order in a quasicubic structure of the double perovskite compound Sr_2NiIrO_6. *Phys. Rev. B* **2017**, *95*, 140403(R). [CrossRef]
28. Pomjakushina, E.; Pomjakushin, V.; Rolfs, K.; Karpinski, J.; Conder, K. New Synthesis Route and Magnetic Structure of $Tm_2Mn_2O_7$ Pyrochlore. *Inorg. Chem.* **2015**, *54*, 9092–9097. [CrossRef]
29. Li, F.; Pomjakushin, V.; Mazet, T.; Sibille, R.; Malaman, B.; Yadav, R.; Keller, L.; Medarde, M.; Conder, K.; Pomjakushina, E. Revisiting the magnetic structure and charge ordering in $La_{1/3}Sr_{2/3}FeO_3$ by neutron powder diffraction and Mössbauer spectroscopy. *Phys. Rev. B* **2018**, *97*, 174417. [CrossRef]
30. Xu, N.; Biswas, P.K.; Dil, J.H.; Dhaka, R.S.; Landolt, G.; Muff, S.; Matt, C.E.; Shi, X.; Plumb, N.C.; Radović, M.; et al. Direct observation of the spin texture in SmB_6 as evidence of the topological Kondo insulator. *Nat. Commun.* **2014**, *5*, 5566. [CrossRef]
31. Puphal, P.; Mielke, C.; Kumar, N.; Soh, Y.; Shang, T.; Medarde, M.; White, J.S.; Pomjakushina, E. Bulk single-crystal growth of the theoretically predicted magnetic Weyl semimetals RAlGe (R = Pr, Ce). *Phys. Rev. Mater.* **2019**, *3*, 024204. [CrossRef]
32. Puphal, P.; Pomjakushin, V.; Kanazawa, N.; Ukleev, V.; Gawryluk, D.J.; Ma, J.; Naamneh, M.; Plumb, N.C.; Keller, L.; Cubitt, R.; et al. Topological Magnetic Phase in the Candidate Weyl Semimetal CeAlGe. *Phys. Rev. Lett.* **2020**, *124*, 017202. [CrossRef] [PubMed]

© 2020 by the authors. Licensee MDPI, Basel, Switzerland. This article is an open access article distributed under the terms and conditions of the Creative Commons Attribution (CC BY) license (http://creativecommons.org/licenses/by/4.0/).

Perspective

The Role of the Short Coherence Length in Unconventional Superconductors

Guy Deutscher

School of Physics and Astronomy, Tel Aviv University, Ramat Aviv, Tel Aviv 69978, Israel; guyde@tauex.tau.ac.il

Received: 13 October 2020; Accepted: 27 November 2020; Published: 1 December 2020

Abstract: A short coherence length is a distinctive feature of many cases of unconventional superconductivity. While in conventional superconductors, it is many orders of magnitude larger than the basic inter-particle distance, a short coherence length is common to superconductors as diverse as the cuprates, the picnites and granular superconductors. We dwell particularly on the last, because their simple chemical structure makes them a favorable material for exploring fundamental phenomena such as the Bardeen-Cooper Schrieffer (BCS)-to-Bose–Einstein condensation cross-over and the effect of the vicinity of a Mott metal-to-insulator transition.

Keywords: coherence length; granular superconductivity; Mott transition; BCS–BEC cross-over

1. The Coherence Length in Conventional versus Unconventional Superconductivity

The importance of the role played by the short coherence length in the physical properties of the high Tc cuprates has been at the heart of my collaboration with Alex Muller [1].

In the framework of the Bardeen–Cooper–Schrieffer theory of superconductivity, Cooper pairing involves a very large number of pairs interacting within the radius of one pair. This radius is the coherence length of the superconducting state, over which the superconducting order can only vary slowly. The number of pairs within a coherence volume is of the order of the square of the Fermi energy divided by the energy gap. In a typical metal, this number is of the order of 1.10^6 to 1.10^8. This very large number is the reason why the superconducting state is so robust in the face of local perturbations of the crystal lattice, such as impurities or grain boundaries. It characterizes conventional superconductivity.

Can superconductivity persist if the number of pairs within a coherence volume is of order unity, and if yes, what are the properties of such an unconventional superconducting state? This question is important from both a theoretical and practical point of view. It is in fact at the heart of the possible applications of the high temperature superconductors discovered by Bednorz and Muller.

The cuprates are, of course, the most famous example of short-coherence-length superconductors. It is about 1.2 nanometers or three lattice spacings in the CuO planes. This immediately explains why superconducting properties such as the critical current density are so sensitive to defects such as grain boundaries that have little impact in conventional, long-coherence-length conventional superconductors [1]. The coherence length is even shorter along the direction perpendicular to these planes. At optimum doping, the number of pairs in a coherence volume is only of order 10, and the condensation energy per coherence volume is only a few times larger than the critical temperature. This puts severe limits on the possibility of designing effective pinning centers.

The short coherence length is also an indication regarding the type of interaction that is at the origin of the superconductivity in the cuprates. Regrettably, we do not yet have a full theoretical understanding of this mechanism. Is it basically of the same kind as that in conventional superconductors, namely, an interaction between charge and lattice displacement, or is it of a different nature, primarily involving the spin's degrees of freedom? Alternatively, is it a mixture of both? The short coherence length favors this possibility, as do local contractions of the Cu–O–Cu bond lengths [2,3].

In any case, the classification of conventional versus unconventional superconductivity according to the number of pairs in a coherence volume has the advantage of being very general. It does not presume anything regarding the pairing mechanism or the symmetry of the superconducting order.

A well-known transition from conventional to unconventional superconductivity is the cross-over regime between BCS and Bose–Einstein condensations. It indeed occurs when the number of pairs per coherence volume is of order unity. It is continuous, as shown by Nozieres and Schmitt-Rink [4], following the earlier work of Leggett [5]. In these theoretical papers, the control parameter is the strength of an attractive electron–electron interaction. Experimentally, we do not know how to control this parameter. However, we can use geometrical constraints to directly reduce the number of pairs per coherence volume and, in this way, experimentally study the cross-over regime. This has been achieved by studying granular superconductivity.

2. Granular Superconductivity

What happens to superconductivity when the volume of a grain becomes smaller than the coherence volume? At first, not much, as long as the number of pairs in the grain remains much larger than unity. However, if that number becomes of order unity, or smaller, the superconductivity is quenched. This happens when the distance between discrete electronic levels becomes smaller than the energy gap. For aluminum, for instance, this occurs below a size of about 5 nanometers.

A more interesting question is, what happens if we take such small, non-superconducting grains and start to couple them together through weak tunnel barriers? It turns out that this is, in fact, a very fundamental problem, which I have called "transition to zero dimensionality" [6].

As long inter-grain coupling is very weak, the granular system is non-superconducting and insulating. When the coupling is strong, namely, when the level broadening is much larger than the distance between the discrete levels in the individual grain, we expect to recover BCS superconductivity. However, what will happen in between? At what coupling strength will this happen? What will be the properties of the emerging superconductor?

This is not a gedanken experiment, but one that has actually been realized. When aluminum is evaporated in a vacuum system in the presence of a finite vapor pressure of oxygen, aluminum oxide forms and is segregated at Al grain boundaries, because the metal and the oxide are immiscible. Additionally, there exists only one aluminum oxide, Al_2O_3, which simplifies the situation. As the oxygen vapor pressure is increased, more oxide is formed, the tunnel barriers become less transparent and the electronic coupling between neighboring Al grains becomes weaker [6]. In this way, one can go continuously from weak to strong coupling.

The continuous transition predicted by Legget, Noziéres and Schmitt-Rink, and studied recently in more detail by Strinati and coworkers [7], does occur. It takes place when the coherence length is of the order of a few grain sizes. The ratio of the gap to the superconducting critical is higher in weakly coupled grains than in strongly coupled ones.

However, the emerging superconductor has some unexpected properties:

(i) The critical temperature of the granular superconductor goes through a maximum, which is higher than that of the bulk material [6];
(ii) The emerging superconductor allies a very small superfluid density and a relatively high critical temperature [8];
(iii) Additionally, free spins appear [9].

3. Granular versus Atomically Disordered Superconductors

These features are related to the nature of the metal-to-insulator transition, which is of the Mott type [9], while in atomically disordered superconductors, it is of the Anderson type.

Surprisingly, the effect of disorder is stronger in the case of "homogeneous" disorder than in the granular case. In the case of homogeneous disorder such as in NbN_x, where it is created by introducing N vacancies, Anderson localization drives the metal-to-insulator transition [10]. It occurs when the product $k_F l$, where k_F is the Fermi wave vector and l is the mean free path, is of order unity. Enhanced electron–electron interactions do occur, but disorder is the leading factor.

In the granular case, Coulomb interactions are the leading factor driving the metal-to-insulator transition. They result from the grain's electrostatic energy. As shown by Antoine Georges and co-workers, conduction through a weakly coupled grain is governed by the combined effect of this charging energy and of the splitting between the electronic levels of the isolated grain [11]. A Kondo resonance occurs when these two energy scales are of the same order. Surprising effects can then occur, for instance, when the charging energy is somewhat larger than the energy level splitting. At temperatures lower than the charging energy, a multi-level Kondo effect allowed by level broadening can restore metallic conduction. When the temperature becomes lower than the level broadening, the conduction becomes insulation-like. However, at even lower temperatures, lower than the energy level splitting, the Kondo resonance re-instates a metal-like conductivity. This predicted non-monotonous conductivity behavior has indeed been observed in nano-scale granular aluminum [12]. It can be considered as a precursor of a Mott transition. This is supported by the presence of free spins inferred from magneto-resistance measurements and, more directly, by muon spin rotation experiments.

4. Superconductivity near a Mott Transition

At a Mott transition, which occurs when the Coulomb energy is of the same order as the band width, the density of states at the Fermi level remains finite [13]. At the same time, the effective band-width reduces and tends towards zero. If the system is superconducting, the finite density of states allows the critical temperature to remain finite up to the transition and the gap to remain well defined, while the superfluid density tends towards zero.

These two features—the absence of sub-gap states and, at the same time, a very small superfluid density—are indeed observed in granular aluminum near the metal-to-insulator transition. They are *not* observed in atomically disordered superconductors near their transition, where the critical temperature collapses and many sub-gap states appear, because the transition is of the Anderson type. The vicinity of a Mott transition is what allows the successful use of thin film granular superconductors to produce circuit elements that have, at the same time, a large kinetic inductance and low losses [14].

It is interesting to note that a transition to zero dimensionality, first identified many years ago in granular aluminum in our original paper [6], has recently been invoked to explain the behavior of very thin NbN films [15]. These two-dimensional films, originally considered to be homogeneous, may in fact be granular. The same conclusion was reached in an extensive review of other nominally homogeneous two-dimensional films near the superconductor-to-insulator transition [16]. While the 2D homogeneous case was thought to be more accessible to theoretical analysis than the 3D granular "dirty" case, it may turn out in the end that the issue of granularity cannot be avoided near the superconductor-to-insulator transition.

5. Conclusions

The continuing in-depth study of granular superconductivity offers the possibility of studying a system that is chemically much simpler than the cuprates and, at the same time, shares with them some fundamental properties. Both are in a BCS-to-BEC cross-over regime, as attested to by their short coherence length (a few lattice spacings in the cuprates and a few grain sizes in granular Al) and the large value of the strong coupling factor. In both cases, one is close to a Mott insulating state—spins are present as expected [14]. The experimental results suggest that there is a close relationship between a Mott transition and the BCS-to-BEC cross-over. However, while the theories of the BCS–BEC cross-over [7] and of the Mott transition [13] are separately well developed, a comprehensive approach involving both of them is still lacking and needed for a comparison with experiments. While the cross-over regime is not favorable for some applications such as for high temperatures, high currents and high fields, it is favorable for some others such as high-kinetic-inductance elements for qubit circuits.

Funding: This research received no external funding.

Acknowledgments: It is a pleasure to thank Alex Muller for many years of fruitful interactions covering many aspects of unconventional superconductivity, from granular to high Tc.

Conflicts of Interest: The author declares no conflict of interedt.

References

1. Deutscher, G.; Müller, K.A. Origin of superconductive glassy state and extrinsic critical currents in high-T c oxides. *Phys. Rev. Lett.* **1987**, *59*, 1745. [CrossRef] [PubMed]
2. Deutscher, G.; de Gennes, P.G. A spatial interpretation of emerging superconductivity in lightly doped cuprates. *Comptes Rendus Phys.* **2007**, *8*, 937–941. [CrossRef]
3. Llordés, A.; Palau, A.; Gázquez, J.; Coll, M.; Vlad, R.; Pomar, A.; Arbiol, J.; Guzmán, R.; Ye, S.; Rouco, V.; et al. Nanoscale strain-induced pair suppression as a vortex-pinning mechanism in high-temperature superconductors. *Nat. Mater.* **2012**, *11*, 329–336. [CrossRef] [PubMed]
4. Nozières, P.; Schmitt-Rink, S. Bose condensation in an attractive fermion gas: From weak to strong coupling superconductivity. *J. Low Temp. Phys.* **1985**, *59*, 195. [CrossRef]
5. Leggett, A.J. Cooper pairing in spin-polarized Fermi systems. *J. Phys.* **1980**, *41*, C7–C19. [CrossRef]
6. Deutscher, G.; Fenichel, H.; Gershenson, M.; Grünbaum, E.; Ovadyahu, Z. Transition to zero dimensionality in granular aluminum superconducting films. *J. Low Temp. Phys.* **1973**, *10*, 231. [CrossRef]
7. Pisani, L.; Pieri, P.; Strinati, G.C. Gap equation with pairing correlations beyond the mean-field approximation and its equivalence to a Hugenholtz-Pines condition for fermion pairs. *Phys. Rev. B* **2018**, *98*, 104507. [CrossRef]
8. Moshe, A.G.; Farber, E.; Deutscher, G. Optical conductivity of granular aluminum films near the Mott metal-to-insulator transition. *Phys. Rev. B* **2019**, *99*, 224503. [CrossRef]
9. Bachar, N.; Lerer, S.; Levy, A.; Hacohen-Gourgy, S.; Almog, B.; Saadaoui, H.; Salman, Z.; Morenzoni, E.; Deutscher, G. Mott transitipn in granular aluminum. *Phys. Rev B* **2005**, *91*, 041123. [CrossRef]
10. Cheng, B.; Wu, L.; Laurita, N.J.; Singh, H.; Chand, M.; Raychaudhuri, P.; Armitage, N.P. Anomalous gap-edge dissipation in disordered superconductors on the brink of localization. *Phys. Rev. B* **2016**, *93*, 180511. [CrossRef]
11. Florens, S.; San José, P.; Guinea, P.; Georges, A. Coherence and Coulomb blockade in single-electron devices: A unified treatment of interaction effects. *Phys. Rev. B* **2003**, *68*, 2453111. [CrossRef]
12. Moshe, A.; Bachar, N.; Lerer, S.; Lereah, Y.; Deutscher, G. Multi-Level Kondo effect and enhanced critical temperature in nanoscale granular Al. *J. Supercond. Nov. Magn.* **2018**, *31*, 733–736. [CrossRef]
13. Georges, A.; Kotliar, G.; Krauth, W.; Marcelo, J.R. Dynamical mean-field theory of strongly correlated fermion systems and the limit of infinite dimensions. *Rev. Mod. Phys.* **1996**, *68*, 13. [CrossRef]
14. Glezer, M.A.; Farber, E.; Deutscher, G. Granular superconductors for high kinetic inductance and low loss quantum devices. *Appl. Phys. Lett.* **2020**, *117*, 062601. [CrossRef]
15. Carbillet, C.; Caprara, S.; Grilli, M.; Brun, C.; Cren, T.; Debontridder, F.; Vignolle, B.; Tabis, W.; Demaille, D.; Largeau, L.; et al. Confinement of superconducting fluctuations due to emergent electron inhomogeneities. *Phys. Rev. B* **2016**, *93*, 144509. [CrossRef]
16. Kapitulnik, A.; Kivelson, S.; Spivak, B. Colloquium: Anomalous metals: Failed superconductors. *Rev. Mod. Phys.* **2019**, *91*, 011002. [CrossRef]

Publisher's Note: MDPI stays neutral with regard to jurisdictional claims in published maps and institutional affiliations.

© 2020 by the author. Licensee MDPI, Basel, Switzerland. This article is an open access article distributed under the terms and conditions of the Creative Commons Attribution (CC BY) license (http://creativecommons.org/licenses/by/4.0/).

Review

Revival of Charge Density Waves and Charge Density Fluctuations in Cuprate High-Temperature Superconductors

Carlo Di Castro

Dipartimento di Fisica, Università "Sapienza". P.le A. Moro 2, I-00185 Rome, Italy; carlo.dicastro@roma1.infn.it

Received: 28 September 2020; Accepted: 28 October 2020; Published: 2 November 2020

Abstract: I present here a short memory of my scientific contacts with K.A. Müller starting from the Interlaken Conference (1988), Erice (1992 and 1993), and Cottbus (1994) on the initial studies on phase separation (PS) and charge inhomogeneity in cuprates carried out against the view of the majority of the scientific community at that time. Going over the years and passing through the charge density wave (CDW) instability of the correlated Fermi liquid (FL) and to the consequences of charge density fluctuations (CDFs), I end with a presentation of my current research activity on CDWs and the related two-dimensional charge density fluctuations (2D-CDFs). A scenario follows of the physics of cuprates, which includes the solution of the decades-long problem of the strange metal (SM) state.

Keywords: high-temperature superconductivity; cuprates; correlated Femi liquid; charge density wave; fluctuation; strange metal

1. Introduction

My scientific familiarity with Alex Müller goes back to the 1970s when I was involved in the research on critical phenomena. His contributions to this field could not be overlooked by any researcher active in the field.

His research was fundamental in bringing the static and dynamical behavior of structural phase transition and ferroelectrics within the mainstream of critical phenomena [1].

Again, in the 1970s, he together with Harry Thomas presented a theory of structural phase transitions in which the lowering of symmetry comes from the Jahn–Teller effect of the ionic constituents [2]. That paper, in addition to showing the versatile nature and the usual deep understanding of the physics by Alex, is also the paper to which he was often referring when discussing his approach in searching high-temperature superconductivity.

In 1986, his discovery with J. G. Bednorz [3] of high critical temperature superconductors (HTSCs) in barium-doped lanthanum copper oxides with the critical temperature T_C in the thirty Kelvin range realized the chimera of experimental physicists for decades. Their paper was quickly followed by the discovery of various families of cuprate superconductors, with critical temperature values well above the liquefaction temperature of nitrogen.

At the time, I was studying with C. Castellani another important problem in condensed matter—the generalization to correlated electrons of the metal–insulator transition resulting from the Anderson localization due to disorder. The problematic of interacting disordered electronic systems was still a very hot topic, but it was overshadowed by the new problematic of HTSCs. The importance of electron correlation for the study of these materials became immediately apparent, in particular under the input offered by P.W. Anderson [4]. Those who, like us, had been working on strongly correlated electron systems, could not escape the temptation to work on this new type of superconductor. Then, my adventure on HTSCs started and my friendly relations with Alex became direct and deeper.

It will be apparent how much our scientific community is indebted to Alex and how much our research activity in this field, mine in particular, was inspired and supported by him. I therefore consider it my duty and pleasure to dedicate this work to him. My aim is to retrace the line we followed over the years in Rome to produce a scenario of the physics of cuprates, for a recent review see [5]. I proceed through the years to end up with two recent contributions [6,7] produced with S. Caprara, M. Grilli and G. Seibold in a collaboration with the experimental group of the Politecnico of Milano on resonant X-ray scattering (RXS) (G. Ghiringhelli, L. Braicovich, R. Arpaia, among others) in which charge density fluctuations (CDFs) are thoroughly investigated leading to a deeper understanding of the physics of these systems. I have no claim of completeness; I only follow my line of reasoning for which no one else is responsible even though I am deeply indebted to Alex Müller for his continuous support in the first fifteen years of my research activity in this field and to all my collaborators throughout the years.

2. Results

2.1. General Properties

Let me recall the general properties of the cuprates by referring to their generic phase diagram (Figure 1a). All these materials are made up of copper–oxygen planes (CuO_2), a quasi-bidimensional structure intercalated with layers of rare earth (lanthanum, yttrium, barium, etc., depending on the various families of cuprates, Figure 1b). In their stoichiometric composition (e.g., La_2CuO_4), these materials are antiferromagnetic (AF) Mott insulators, despite the fact that they have one hole per CuO_2 unit cell in the copper–oxygen plane and should have been metals. Hence, there is the need for a strong correlation between charge carriers, holes in this case.

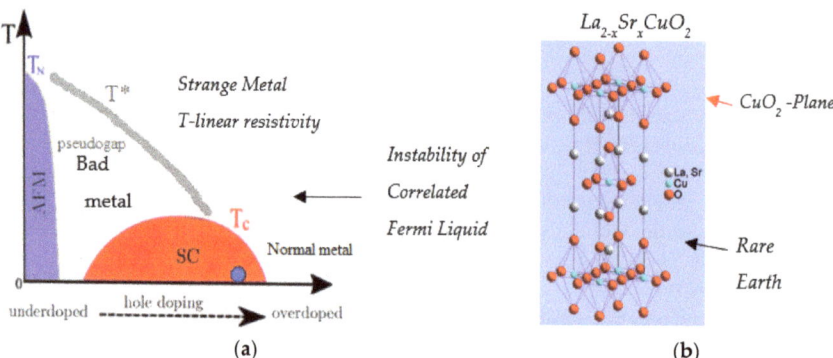

Figure 1. (a) Generic phase diagram of cuprates of temperature T vs. doping p. Blue region: antiferromagnetic (AF) phase. Red region: superconductive phase. T^* smeared gray line: pseudogap line. (b) Structure of cuprates, e.g., $La_{2-x}Sr_xCuO_2$ (LSCO).

Upon substitution of the rare earth with heterovalent dopants (e.g., strontium replacing lanthanum in LSCO), the number of holes increases in the CuO_2 plane and the system becomes a metal, albeit a bad metal. Nevertheless, the system becomes a superconductor, with d-wave symmetry, when the temperature is lowered below a dome-shaped doping-dependent critical temperature $T_C(p)$ (red region in Figure 1a). T_C reaches a maximum at the so-called optimal doping and decreases by lowering p in the underdoped region or increasing p in the so-called overdoped region, where the system is well described by the normal Fermi liquid (FL) theory. On the contrary, the metallic phase in the underdoped region and above optimal doping is not in line with ordinary metals and the predictions of the normal Fermi liquid theory. In the metal phase, a strong anisotropy is measured in the electrical conductivity that, in the CuO_2 plane, exceeds the transversal conductivity by orders of magnitude. Moreover, the resistance displays a linear in T dependence (strange metal) over a wide range of temperatures above the doping-dependent pseudogap line $T^*(p)$ [8]. $T^*(p)$ decreases with p and merges with T_C around optimal doping. Below $T^*(p)$, a pseudogap is formed with an even stronger violation of Fermi liquid (FL) behavior (Figure 1). The pseudogap formation is apparently connected [9] with the suppression of quasiparticle (QP) states at the Fermi surface (FS) along the so-called antinodal (0,0)–(0.5,0) direction, as shown in Figure 9, and the consequent formation of Fermi arcs in the nodal region.

In classic superconductors, the necessary attraction for the formation of electron pairs is mediated by phonons, as indicated by the so-called isotope effect, whereby the critical temperature for the onset of superconductivity depends on the isotopic mass of the lattice ion. As a further anomaly of these systems, the isotope effect is either absent or anomalous, for example, having a strong effect on T^* [10].

An international debate thus ensued among theoretical physicists about whether a strongly correlated hole (or electron) system, with an occupation of approximately one hole per lattice site of a quasi-two-dimensional system, can become a superconductor, what the new paired state would be, and by what mechanisms it might be formed. At the same time, another, perhaps even more stimulating theoretical problem, presented itself—understanding the anomalous metallic phase to which the superconductivity was linked. I embarked upon this twofold adventure together with the condensed matter theory group of Rome.

2.2. Phase Separation—Spin and Charge Order

The question we asked ourselves was how could a system, whose interaction between charges is strongly repulsive, generate the pairing between charges in general without referring to special symmetry conditions to reduce the repulsion? We then referred to the phase separation (PS), as it occurs in simple fluids. In cuprates, the separation occurs between charge-rich metallic regions and charge-poor zones. The strong local repulsion between charge carriers reduces their mobility and favors the possibility of phase separation. Theoretically, all of the models with strong local repulsion that were introduced to represent the copper–oxygen planes showed PS in charge-rich regions and in charge-poor regions (see, e.g., [11]). Besides the PS in various other models, we found that whenever PS is present, pairing then occurs in a nearby region of phase space, thus enforcing a possible connection between the two phenomena [12–14]. Charge inhomogeneity appeared to be a bridge that connects the low doping region (in the presence of strong repulsion necessary for an insulating antiferromagnet), the intermediate doping region (with a special attraction for superconductivity), up to a correlated FL at higher doping. Long-range Coulomb interaction, however, forbids the macroscopic phase separation. The system chooses a compromise and segregates charge on a shorter scale, while keeping charge neutrality at a large distance giving rise to the so-called frustrated phase separation [14–16].

Emery and Kivelson [15,16] achieved the frustrated PS concept starting from the low doping region near the antiferromagnetic insulating phase, in which the few present charges are expelled from the antiferromagnetic substrate and align in stripes. Spin fluctuations are dominating in this case.

Coming instead from the high doping region, *correlated FL in the presence of long-range Coulomb interaction undergoes to a Pomeranchuk instability towards* a charge density wave (CDW) state *with finite modulation vector* q_c [17]. This instability occurs along a line of temperature as a function of doping $T_{CDW}(p)$, ending at $T = 0$ in a quantum critical point (QCP) nearby optimal doping. In this case, charge fluctuations are prominent.

Both charge and spin modes may act as mediators of pairing [18,19].

Experimentally, spin and charge order have been observed in two different forms:

- Stripes with strong interplay between charge and spin degrees of freedom (Figure 2a) observed in the LSCO family by neutron scattering since 1995 [20]. Charge and spin incommensurate modulation vectors $q_c = 2\pi/\lambda_c$ and $q_s = 2\pi/\lambda_s$ are strictly related ($q_s = q_c/2$) one to another and align along the Cu-O bond.

However, both spin and charge modulation were observed simultaneously in co-doped LSCO only (e.g., $La_{1.48}Nd_{0.4}Sr_{0.12}CuO_4$); otherwise, there was no detection of charge modulation (only spin).

- Charge Density Wave (Figure 2b). Charge and spin degrees of freedom evolve independently one from another [21,22]. CDWs, foreseen since 1995 [17], were until 2011 elusive, contrasted, and confirmed only indirectly. Due to the improvement of resonant X-ray scattering (RXS), CDWs are now ubiquitously observed (see, e.g., [21–28]) in cuprate families, and most investigated among the orders competing with SC in countless papers following the pioneering paper on YBCO by Ghiringhelli et al. [23].

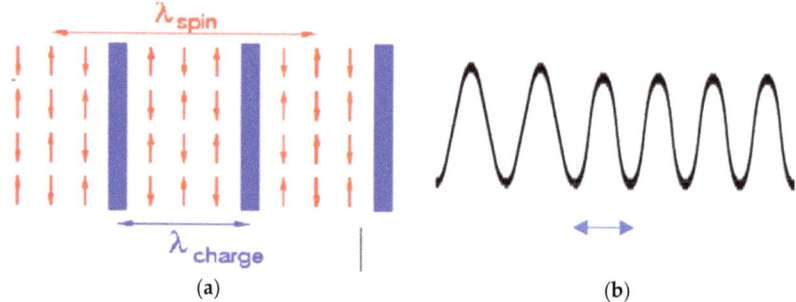

Figure 2. (a) Stripes: charge and spin modulation vectors q_c and q_s are connected. (b) The different modulation of charge density waves (CDWs) with respect to stripes is shown. In this case, CDWs and SDWs evolve independently one from another (see, e.g., [21,22]).

As shown in Figure 3, (i) energy-integrated CDW peaks are along the Cu-O bond, (0.31,0) and (0,031) in CuO_2 planes; (ii) the width decreases by lowering T and the correlation length increases but remains finite (less than 16a lattice constant), so that the fluctuating CDWs tend to become critical but never succeed, and for this reason they were called *Quasi-Critical dynamical fluctuating* CDWs (QC-CDWs); and (iii) the QC-CDW peak is increasing by lowering T down to Tc, while it is decreasing for $T < T_C$, i.e., the *incipient CDW phase transition is preempted by the SC transition*.

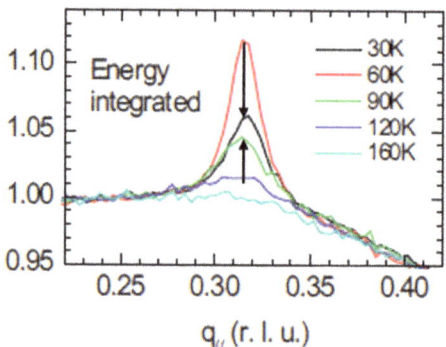

Figure 3. Adapted from [23]. Energy-integrated resonant X-ray scattering (RXS) for underdoped $YBa_2Cu_3O_{6.6}$ (T_c = 61 K) as a function of the in-plane momentum vector at various temperatures. The arrows indicate the increasing or decreasing of the peak by lowering the temperature above or below T_c, respectively.

The QC-CDW onset line $T_{QC\text{-}CDW}(p)$ starts around optimal doping and initially follows the pseudogap line $T^*(p)$ (see Figure 4).

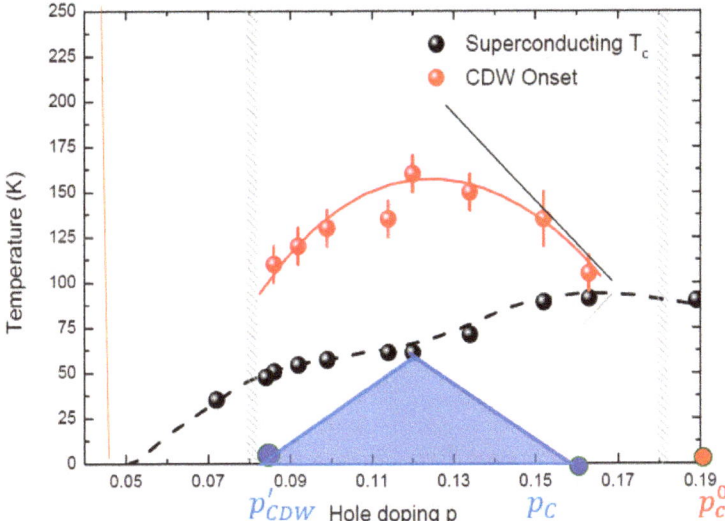

Figure 4. Adapted from [21,22]. Phase diagram T vs. doping for YBCO. Red line: Quasi-Critical-CDW T_{QCCDW}. Black line: pseudogap T^*. Black dashed line: superconductive critical temperature. Blue region below the superconductive dome: sketched 3D long-range CDW in the presence of high magnetic field. The two blue dots: 3D extrapolated quantum critical points (QCPs). Red dot: the extrapolated QCP of the QC-CDW.

In the presence of a sufficiently high magnetic field to suppress superconductivity, a *static 3D transition* is unveiled (blue region in Figure 4), mainly in YBCO, by NMR [29,30], quantum-oscillation, negative Hall, and FS reconstruction [31–38], and finally by *thermodynamic 3D transition* by pulsed ultrasound [39]. Zero field QC-CDW and 3D long-range charge order with two QCPs ($p_c' < p_c \neq p_c^0 \approx p^*$) have the same in-plane incommensurate vector q_c, suggesting a common origin.

3. A Digression for Alex Müller and My Memories

Actually, the community was and has been for decades strongly against the idea of phase separation and of non-uniform charge in general. Most researchers were referring to spin fluctuations due to the vicinity of the AF-QCP in the low-doped Mott insulator [4,19,40–43]. Charge order has been at most considered as a byproduct of spin order. For instance, at the first big international conference on HTSCs of 1988 held in Interlaken, following my remark on the possible interconnection between different charge density values that could coexist in these materials [44], P.W. Anderson argued that "simple calculations" show it to be impossible. Alex Müller, instead, trusted us from the beginning, and gave a strong support to the community that was inquiring about charge inhomogeneity in cuprates. He organized several conferences on PS (see, e.g., Erice in 1992 (Figures 5 and 6) and Cottbus in 1993).

In the proceedings [45,46], our work carried out in Rome, the one by Emery and Kivelson, and the one by Sigmund, Hiznyakov, and Seibold on PS of that early period were summarized. A comprehensive presentation of the experimental situation was also given. The task of keeping the members of this community linked was carried out also by the three editors of this volume, in particular by Antonio Bianconi with the Superstripes organization. Alex often recalled the scientific contributions of Annette Bussman-Holder, Hugo Keller, and Antonio Bianconi (see, e.g., the distorted octahedrons of Antonio that, according to Alex, can be assigned to Jahn–Teller polarons, whereas the undistorted ones are located within a metallic cluster or stripe [47]).

The idea of charge inhomogeneity corresponds, at least in part, to the idea of strong polarization that guided Müller in his discovery of high-temperature superconductivity.

I have been corresponding with Alex extensively on the subject, for many years. I presented the "elogio" on the occasion of his Laurea Honoris Causa in Rome in November 1990 (Figure 7). Furthermore, I was later called (1997) to Cottbus for the equivalent ceremony.

In Cottbus, we experienced an episode that is unusual for meetings among physicists. After the ceremony, the neo-laureate with his wife as well as my wife and myself were taken to a party in a villa–castle located in the outskirts of Cottbus. We were proceeding in a Mercedes along a highway, when the car stopped; we then got into a seventeenth century-like chariot with the coachmen in "livrea" (livery). Cars were speeding around us, and we finally arrived in the villa, where we were received by a string quartet playing classical-style music, organized for Alex by the noble family who had recently repossessed the villa.

Figure 5. Erice, 6–12 May 1992, Workshop on Phase Separation in cuprate Superconductors.

Figure 6. Erice, July 1995. E. Sigmund, C. Di Castro, and G. Benedek with K.A. Müller in the center, looking at a gift for him, a book on rare cars, one of his hobbies.

Figure 7. K.A. Müller. Laurea Honoris Causa, Rome 1990.

4. Correlated Fermi Liquid and CDW Fluctuations

4.1. Correlated Fermi Liquid Instability Producing a Charge Density Wave State

Let us see how the scenario presented in Section 2 arises from our old and new stories.

The proposal of a CDW instability in cuprates is quite old [17] and it is based on the mechanism of frustrated phase separation discussed above.

At high doping, as it has been already mentioned, strong correlations weaken the metallic character of the system giving rise to a correlated FL with a substantial enhancement of the QP mass m^* and a sizable repulsive residual interaction. The presence of weak or moderate additional attraction due to phonons reduces this residual interaction and may provide a mechanism that induces an electronic PS with charge segregation on a large scale, i.e., a diverging compressibility κ, the static density-density response function at $q = 0$:

$$\kappa = \lim_{q \to 0} \chi_{\rho\rho}(q, \omega = 0) = \lim_{q \to 0} < \rho(q, \omega = 0) \rho(-q, \omega = 0) > \to \infty \tag{1}$$

The FL expression of the compressibility κ in terms of the static Γ_q and dynamic Γ_ω scattering amplitudes is as follows:

$$\kappa = 2N(0)/(1 + 2N(0)\Gamma_\omega) = 2N(0)(1 - 2N(0)\Gamma_q) \to \infty \text{ as } q \to 0 \tag{2}$$

From equation (2), the Pomeranchuk condition for the instability of the FL reads as follows:

$$2N(0)\Gamma_\omega = -1, \; \Gamma_q \to -\infty \tag{3}$$

The static scattering amplitude diverges at $q = 0$.

With a long-range Coulomb interaction, it would cost too much in energy to fully segregate the charge and the system phase separates on a shorter scale. The Fermi liquid instability coming from the overdoped region occurs at *finite modulation vector* $q = q_c$ leading to an incommensurate charge density wave *state with wavelength* $\lambda_c = 2\pi/q_c$. q_c is not related to nesting but is determined by the balance between PS tendency and Coulomb energy cost [17].

A strong scattering increase has been indeed measured in YBCO at optimal doping [48], as it is required by Equation (3) but this time at a finite q.

4.2. Theoretical Realization

The general ideas of the previous section are implemented within a correlated FL theory described by a Hubbard–Holstein model with large local repulsion and a long-range Coulomb interaction $V_c(q)$. The values of the hopping parameters are derived from the quasiparticle band. Slave Boson technique was used to implement the condition of no double occupancy [17,49,50] with the result that strong correlation severely reduces the kinetic homogenizing term. This technical procedure was indeed shown [51] to produce a correlated FL with a quasiparticle dispersion having a reduced hopping and a residual density–density interaction $V(q)$ between the quasiparticles:

$$V(q) = V_c(q) + U(q) - \lambda \quad (4)$$

where $U(q)$ is the short-range residual repulsion stemming from the large repulsion of the one-band Hubbard model and λ is the coupling of the Holstein e-ph interaction leading to the thermodynamic PS in the absence of V_c. In the FL, $V(q)$ is screened in the random phase approximation via the polarization due to the repeated particle–hole virtual processes of the Lindhard polarization bubble $\Pi(q, \omega)$:

$$V_{scr} = \frac{V(q)}{1 + \Pi(q, \omega)V(q)}. \quad (5)$$

The effects of strong correlations are contained in the form of $V(q)$ and in the QP effective mass entering $\Pi(q, \omega)$. The dressed dynamical density–density correlation function $\chi_{\rho\rho} = \frac{\Pi}{1+\Pi(q,\omega)V(q)}$ has the same denominator of V_{scr}.

Within the present approximation, the FL instability condition (Equation (2)) towards the CDW is given by the vanishing of the denominator in Equation (5) at a finite modulation vector $q = q_c$, eventually marking a mean-field critical line as a function of doping p, $T^0_{CDW}(p)$ as it is sketched in Figure 8:

$$1 + \Pi(q, \omega)V(q) = 0 \quad (6)$$

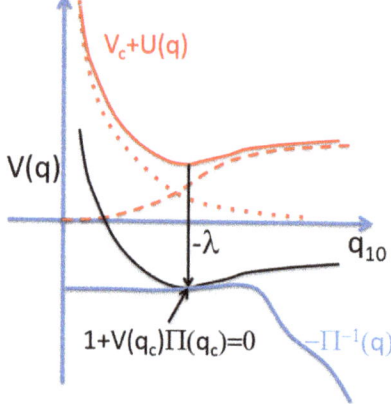

Figure 8. Courtesy of M. Grilli. Sketch of the instability condition of FL to form CDW. Dotted line: Coulomb interaction. Dashed line: residual repulsive interaction. Red line: sum of the two. Black line: red line shifted by the attraction λ. Blue line: inverse Lindhard polarization bubble. The FL instability condition is also shown.

$U(q)$ is less repulsive and the hopping largest along the Cu-O direction ((0.5,0) and (0,0.5)). Mean-field CDW instability of FL and *modulation vector* q_c are therefore in this direction and determine the hot regions of the Fermi surface (shown in Figure 9) where CDW fluctuations (CDFs) mediate strong scattering between Fermi quasiparticles and suppress quasiparticle states, forming the so-called Fermi arcs. We consider this as the beginning of the formation of a pseudogap (see also [9]); we then let the mean-field CDW critical line start at $T = 0$ at the same doping p^* of the extrapolated pseudogap line.

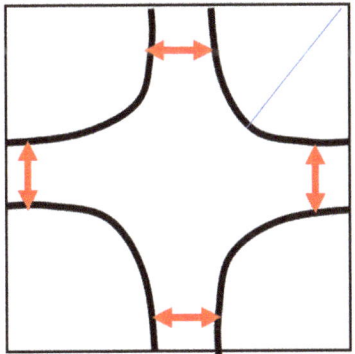

Figure 9. Red arrows: modulation vectors q_c of CDFs connecting two branches of the quasiparticle Fermi surface (FS) in the antinodal regions of the FS. Blue line: nodal direction.

The formation of gaped electron states near the antinodal regions suggests a pseudogap with d-wave symmetry as the superconductive gap [9]. The analysis of the superconductive transition in the presence of a doping-dependent normal state pseudogap with d-wave symmetry [52] accounts for a doping-dependent superconductive critical temperature $T_c(p)$ and a $T_c - T^*$ bifurcation near optimum doping, within a d-Density-Wave scheme a year before a more known proposal appeared [53].

Expanding in $q-q_c$ and ω around the mean-field critical line $T^0_{CDW}(p)$:

$$\chi^{-1}_{\rho\rho} \propto D^{-1} \equiv m_0 + v(q-q_c)^2 - i\omega\gamma,\ m_0 = v(\xi^0_{CDW})^{-2} \propto T - T^0_{CDW} \qquad (7)$$

where D in field theory is named the "propagator" of the fluctuation and m_0 is its "bare" mass.

Essentially, Equation (7) represents a time-dependent Landau–Ginzburg (TDLG) equation in the Gaussian approximation:

$$\frac{\partial \varphi(x,t)}{\partial t} = \gamma^{-1} \frac{\delta H\{\varphi\}}{\delta \varphi},\ H\{\varphi\} = \frac{1}{2}\int dx [m_0|\varphi|^2 + v|\nabla\varphi|^2], \qquad (7')$$

where φ is the CDW field (or fluctuation mode) and γ is the coefficient of the Landau damping term $i\omega$.

From an inspection of Equation (7), it is natural to define a characteristic energy for each q:

$$\omega_{ch}(q) = \frac{m_0}{\gamma} + \frac{v(q-q_c)^2}{\gamma} \sim \tau_q^{-1} \qquad (8)$$

which is minimum at $q = q_c$ and its inverse defines the relaxation time τ_q of the fluctuation mode φ_q to its equilibrium zero value.

4.3. Correction to Mean Field Due to Fluctuations

In systems with quasi-2D layered structures such as cuprates fluctuations reduce strongly the mean-field critical temperature. The first correction to m_0 due to fluctuations was calculated [54] in the mode–mode coupling of the CDFs, as shown diagrammatically in Figure 10. The loop of the fluctuation mode in the second diagram is a constant in momentum and gives a correction δm to the mean field m_0.

Figure 10. First correction to the CDF propagator at first order in the coupling u (blue dot). The blue dashed loop in the second diagram is the Hartree self-energy term, constant in q, and gives the first correction to m_0.

In the same paper [54], stimulated by discussion with Alex Müller, the anomalous isotope effect in cuprates [10] was also considered. Even though phonons do not directly pair the fermionic quasiparticles, they nevertheless act indirectly on various quantities as T^* and have a doping-dependent effect on T_c.

The expressions of the full mass m in terms of the corrected T_{CDW} and of the corrected correlation length ξ_{CDW} derived by the loop integral, shown in Figure 10, are as follows:

$$m(T) = m_0 + u\delta m \propto T - T_{CDW}(p), \ m = v\xi_{CDW}^{-2} \tag{9}$$

$m(T)$ must therefore vanish at criticality. m should replace m_0 in the Equation (8) of ω_{ch}.
There are three cases according to the value of the loop integration.

- In two dimensions (2D), the density of states $N(\omega)$ is constant and δm is ln-divergent. It is then impossible to satisfy the criticality condition of vanishing $m(T)$ and there is no transition at finite T, as in general for a continuous symmetry. A quantum critical point and charge order are possible only at $T = 0$. In 2D and finite T, dynamical charge fluctuations (CDFs) only can be present.
- The drastic role of fluctuations can be reduced by cutting their long-range effect in the loop integral with an infrared cut-off ω_{min} (possibly of the order of $\omega_{ch}(q_c) = \frac{m}{\gamma} = \frac{v}{\gamma}\xi_{CDW}^{-2}$) to simulate the quasi-2D structure of cuprates. Lowering T at fixed doping gives rise to a quasi-critical dynamical CDW (QC-CDW), which can be observed with sensitive experiments like RXS.
- In three dimension (3D), $N(\omega)$ vanishes as ω goes to zero and makes δm finite also at finite T. By lowering T further, the QC-CDW may cross over to a long-range 3D-CDW, which is however hidden by superconductivity.

4.4. Experimental Observation of the Scenario Presented Above

RXS experiments may test directly the dissipative part of the density–density dynamical response and can therefore be analyzed within the theoretical scenario presented above. In this case, the RXS intensity I_{RXS} is proportional to the $Im\chi_{\varrho\varrho} \propto ImD$, times the occupancy CDFs given by the Bose function $b(\omega/T)$ as follows:

$$I_{RXS} = AImD(q,\omega)b(\frac{\omega}{T}) \sim Ab(\frac{\omega}{T})\frac{\omega/\gamma}{(\omega_{ch})^2 + (\omega)^2}, \tag{10}$$

where A is a constant and the characteristic frequency ω_{ch} is specified by Equation (8) with m instead of m_0.

If $\omega_{ch}(q_c)$ never vanishes, its finite value determines the dynamical nature of the fluctuations that do not become critical. This is visualized in the experiment by a shift of the peak of high-resolution experiments. In Equation (10), $b(\frac{\omega}{T})$ can be classically approximated as T/ω and a T-factor appears in I_{RXS}, if $k_B T > \omega_{ch}(q_c)$.

A thorough RXS analysis in $(Y, Nd)Ba_2Cu_3O_7$ (YBCO, NBCO) at various doping values (0.11–0.19) and temperatures $20 < T < 270$ K was performed in [6].

After subtracting the background, the experimental results for the density–density part of the response can be summarized as follows:

At high T, the energy integrated RXS quasi-elastic intensity can be fitted by assuming *a single profile*. At lower T instead, *two peaks* are necessary. One peak (NP) has a narrow width at half maximum (FWHM ~ ξ^{-1}). The other, instead, is very broad with very short correlation length (BP).

In the underdoped YBCO sample at $p = 0.14$, the two peaks are centered at slightly different values of q. Both values, however, are around 0.3 r.l.u., a value proper of the modulation vector of CDW in the YBCO family at this doping. Moreover, the two vectors become equal ($q_c^{NP} = q_c^{BP} = 0.32$ r.l.u.) at optimal doping p=017. Both peaks are therefore related to the same phenomenon of CDW.

The NP has the following properties: (i) a dome-shaped onset temperature as a function of doping $T_{QC-CDW}(p)$ as shown in Figure 4; (ii) the peak *narrows and the correlation length increases as T decreases* down to T_c; and (iii) this ordering competes with superconductivity as T_c is reached (see Figure 11).

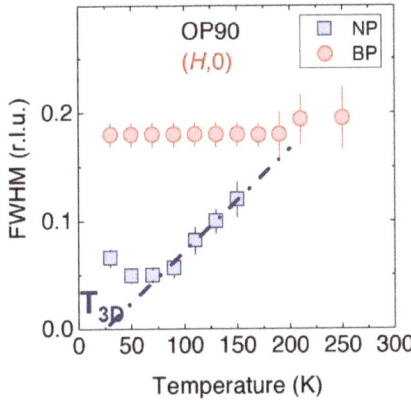

Figure 11. From [6]. Width vs. T in (H,0) scan of energy integrated quasi-elastic peaks for the optimally doped YBCO sample: NP (blue squares) and BP (red dots). The extrapolation of the NP width, marked by a dash-dotted line, tends to zero at the hypothetical T_{3D} and towards the BP width at higher T.

The NP presents all the characteristics of the quasi-critical dynamical CDW discussed above for the quasi-2D layered structure.

By extrapolating to zero the linear in T dependence of NP-FWHM before the competition with SC is reached, the long-range 3D onset temperature $T_{3D}(p)$ was determined at which NP-FWHM would vanish and ξ_{CDW} would diverge in the absence of superconductivity (Figure 11). The values determined in this way are in good agreement with the values detected in the presence of a strong magnetic field.

The new feature of the experiments is the very broad BP (FWHM~ξ^{-1}~0.16–0.18 r.l.u.), by now observed in many other families [55–60]. It has many peculiar properties (Figure 11), as follows: (i) it is present up to the highest measured T well above $T^*(p)$; (ii) the width (FWHM) and the correlation length are almost constant in T; the corresponding charge fluctuations are then *non-critical and do not show any tendency to become critical at any finite T*; and (iii) the characteristic energy ω_{ch} has been measured by fitting the dynamical high-resolution spectra on the OP90 (and UD60) samples at $T = 90$ K, 150 K, 250 K and $q = q_c$. At $T = 150$ K and 250 K, only the BP is present (no NP), $\omega_{ch} \approx 15$ meV of the order of T^* showing the dynamical nature of the fluctuations of the BP, which can be considered as the *2D dynamical charge fluctuations (CDFs) non-critical at any finite T*.

At the lower temperature, $T = 90$, the NP is also present and ω_{ch} has a finite but smaller value, reinforcing the idea of the quasi-critical nature of the CDW of the NP.

The *extrapolation* of NP-FWHM from $T > T_C$ to higher T (Figure 11) suggests that *the NP emerges from the BP*. It is not clear at the moment whether the two kinds of excitation coexist or are spatially separated, giving rise to an inhomogeneous landscape, as found in [61].

As a first conclusion, we can say that the three cases envisaged by the theory in Section 4.3. were found in the experiments with a continuous evolution *from pure 2D dynamical CDFs at high T and all doping, to a coexistence with quasi-critical CDWs below $T_{QC\text{-}CDW}$, to the static 3D CDW hindered by superconductivity and measured in a high magnetic field.*

4.5. Strange Metallic Behavior Above T*

The very broad spectrum (BP) of CDFs allows us [7] to approach *the old standing problem of strange metal (SM)* of cuprates, in particular the anomalous linear in T resistivity $\rho(T) \sim T$ from $T > T^*$ up to the highest attained temperatures. One of the prerequisites identified in the marginal FL phenomenological theory [62] to reproduce the strange metal is the presence of a strong *isotropic scattering* channel for the Fermi quasiparticles with all states on the Fermi surface nearly equally affected. This was accomplished by assuming *2D-CDFs of the BP as mediators of scattering among the Fermi QPs*. The width of the CDFs is such that a Fermi QP of one branch of the FS can always find a CDF that scatters it onto another region of FS. *The scattering among the Fermi quasiparticles is then almost uniform along the entire Fermi surface.*

I refer to the original article for the explicit calculation of the following quantities for an optimally doped (Tc = 93 K) NBCO sample and an overdoped (Tc = 83 K) YBCO sample: (i) the fermion self-energy, i.e., how the Fermi QPs are dressed by CDFs, *using the same set of parameters that was obtained by fitting the BP response*; (ii) the zero frequency quasiparticle scattering rate along the Fermi surface from the imaginary part of the self-energy; and (iii) finally, the resistivity within the standard Boltzmann equation approach using the calculated scattering. In both cases, the strange metal requirements are satisfied and linearity in T of the resistivity for $T > T^*$ is quantitatively matched.

I only consider here the following heuristic simple argument. Since the scattering is almost uniform over the entire Fermi surface, the scattering integral in the Boltzmann equation can be well approximated with an effective scattering time of a *single* scattering event. Its inverse then determines the effective characteristic frequency of the process, which depends on the characteristic frequency of the scatterers given by Equation (8) at $q = q_c$. In analogy with a disordered electron system, *the full inverse scattering time must be proportional to the one of the single event multiplied by the population of the scatterers*, i.e., the Bose function. As long as $kT > \hbar\omega_{ch}$, the semiclassical approximation of $b(\frac{\omega}{T})$ is valid and a linear T dependence is introduced:

$$\frac{\hbar}{\tau} = a(v, \xi, \ldots)kT \qquad (11)$$

where a is a dimensionless doping-dependent function of the parameters of the theory. In the present argument a can be considered as a fitting parameter, independent from T since the BP correlation length does not depend on T in any relevant way (Figure 11).

5. Conclusions

The following Figure 12 gives a graphic representation of the scenario described above.

The phase diagram visualizes the continuous evolution from a pure 2D dynamical CDF at high T and all doping, to a coexistence with quasi-critical CDWs below $T_{QC\text{-}CDW}$, to the static 3D-CDW hindered by superconductivity and measured in a high magnetic field.

Finally, the dynamical CDFs (BP), characterized by the same parameters used to fit RXS experiments, were chosen as mediators of isotropic scattering among Fermi quasiparticles and provide an understanding of the long-standing problem of strange metallic state, matching the resistivity data, in the whole range from room temperature down to T^* [7].

Some open problems deserve further investigation:

In Section 4.4., it was already mentioned that the present knowledge of the relation between the quasi-critical CDW (NP) and the dynamical CDF (BP) does not allow to distinguish between a spatial separation or a coexistence of the two.

At temperatures $T < T^*$, the role on scattering of the pseudogap state and of other intertwined incipient orders (CDWs, Cooper pairing, etc.) must be clarified.

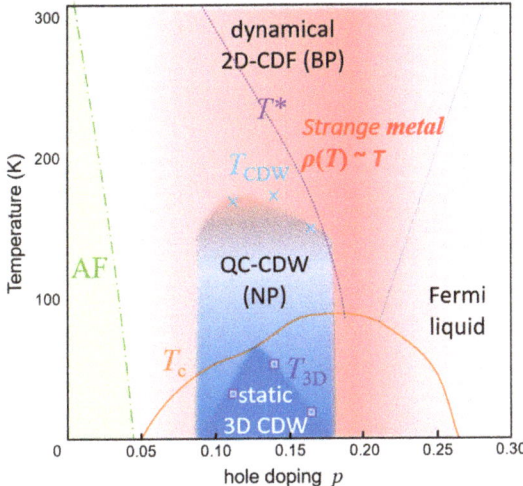

Figure 12. Adapted from [6]. CDW in the phase diagram T vs. doping p of YBCO. The green zone is the antiferromagnetic phase. The reddish zone represents the 2D charge density dynamical fluctuations associated with the BP (actually the experiments in [6] were performed for $0.11 < p < 0.19$). In the cone above optimal doping (between T^* and the Fermi liquid), CDFs produce the strange metal and the linear in T resistivity. Light blue zone: the quasi-critical CDWs associated with the NP. Dark blue zone: the static 3D long-range order hidden in the absence of a magnetic field by the superconductive dome.

Investigation is also required to see whether CDFs with a reduced value of ω_{ch} can also account for the so-called Planckian or better for the linear behavior observed at low temperatures and special doping in the presence of high magnetic fields to suppress superconductivity (see, e.g., [63]).

Funding: This research received no external funding.

Acknowledgments: I wish to warmly thank Claudio Castellani, Marco Grilli, Sergio Caprara, Lara Benfatto, Roberto Raimondi, undergraduates, doctoral students, and post-doctoral researchers who worked with us in Rome from time to time, particularly José Lorenzana (now permanently in Rome), Walter Metzner, Goetz Seibold, and more recently Giacomo Ghiringhelli, Lucio Braicovich Riccardo Arpaia (Chalmers University), and the whole Milano Politecnico Group for the present collaboration.

Conflicts of Interest: The author declares no conflict of interest.

References and Note

1. The importance of A.K. Müller's contribution in this field is stated e.g., by the quotations he received in the fundamental paper by Hohenberg, P.C.; Halperin, B. Theory of dynamic critical phenomena. *Rev. Mod. Phys.* **1977**, *49*, 435. [CrossRef]
2. Thomas, H.; Müller, K.A. Theory of a Structural Phase Transition Induced by the Jahn-Teller Effect. *Phys. Rev. Lett.* **1972**, *28*, 800. [CrossRef]
3. Bednorz, J.G.; Müller, K.A. Possible high-Tc superconductivity in the Ba-La-Cu-O system. *Z. Phys. B* **1986**, *64*, 189–193. [CrossRef]
4. Anderson, P.W. The Resonating Valence Bond State in La2CuO4 and Superconductivity. *Science* **1987**, *235*, 119. [CrossRef] [PubMed]
5. Caprara, S. The Ancient Romans' Route to Charge Density Waves in Cuprates. *Condens. Matter.* **2019**, *4*, 60. [CrossRef]

6. Arpaia, R.; Caprara, S.; Fumagalli, R.; De Vecch, G.; Peng, Y.Y.; Andersson, E.; Betto, D.; De Luca, G.M.; Brookes, N.P.; Lombardi, F.; et al. Dynamical charge density fluctuations pervading the phase diagram of a copper-based high T_c superconductor. *Science* **2019**, *365*, 906–9010. [CrossRef]
7. Seibold, G.; Arpaia, R.; Peng, Y.Y.; Fumagalli, R.; Braicovich, L.; Di Castro, C.; Grilli, M.; Ghiringhelli, G.; Caprara, S. Marginal Fermi Liquid behaviour from charge density fluctuations in Cuprates. *arXiv* **2019**, arXiv:1905.10232. (to be published in *Commun. Phys.*)
8. Ando, Y.; Komiya, S.; Segawa, K.; Ono, S.; Kurita, Y. Electronic Phase Diagram of High-Tc Cuprate Superconductors from a Mapping of the In-Plane Resistivity Curvature. *Phys. Rev. Lett.* **2004**, *93*, 26700. [CrossRef]
9. Timusk, T.; Statt, B. The pseudogap in high temperature superconductors: An experimental survey. *Rep. Prog. Phys.* **1999**, *62*, 61–122. [CrossRef]
10. Rubio Temprano, D.; Mesot, J.; Janssen, S.; Conder, K.; Furrer, A.; Mutka, H.; Müller, K.A. Large Isotope Effect on the Pseudogap in the High-Temperature Superconductor $HoBa_2Cu_4O_8$. *Phys. Rev. Lett.* **2000**, *84*, 1990–1993. [CrossRef] [PubMed]
11. Emery, V.J.; Kivelson, S.A.; Lin, H.Q. Phase separation in the t-J model. *Phys. Rev. Lett.* **1990**, *64*, 475. [CrossRef] [PubMed]
12. Cancrini, N.; Caprara, S.; Castellani, C.; Di Castro, C.; Grilli, M.; Raimondi, R. Phase separation and superconductivity in Kondo-like spin-hole coupled model. *Europhys. Lett.* **1991**, *14*, 597. [CrossRef]
13. Grilli, M.; Raimondi, R.; Castellani, C.; Di Castro, C.; Kotliar, G. Phase separation and superconductivity in the U=infinite limit of the extended multiband Hubbard model. *Int. J. Mod. Phys. B* **1991**, *5*, 309. [CrossRef]
14. Raimondi, R.; Castellani, C.; Grilli, M.; Bang, Y.; Kotliar, G. Charge collective modes and dynamic pairing in the three-band Hubbard model. II. Strong-coupling limit. *Phys. Rev. B* **1993**, *47*, 3331. [CrossRef] [PubMed]
15. Löw, U.; Emery, V.J.; Fabricius, K.; Kivelson, S.A. Study of an Ising model with competing long- and short-range interactions. *Phys. Rev. Lett.* **1994**, *72*, 1918. [CrossRef]
16. Emery, V.J.; Kivelson, S.A. Frustrated electronic phase separation and high-temperature superconductors. *Phys. C Supercond.* **1993**, *209*, 597–621. [CrossRef]
17. Castellani, C.; Di Castro, C.; Grilli, M. Singular Quasiparticle Scattering in the Proximity of Charge Instabilities. *Phys. Rev. Lett.* **1995**, *75*, 4650. [CrossRef] [PubMed]
18. Perali, A.; Castellani, C.; Di Castro, C.; Grilli, M. D-wave superconductivity near charge instabilities. *Phys. Rev. B* **1996**, *54*, 16216. [CrossRef]
19. Monthoux, P.; Balatsky, A.V.; Pines, D. Weak-coupling theory of high-temperature superconductivity in the antiferromagnetically correlated copper oxides. *Phys. Rev. B* **1992**, *46*, 14803. [CrossRef]
20. Tranquada, J.M.; Sternlieb, B.J.; Axe, J.D.; Nakzmura, Y.; Uchida, S. Evidence for stripe correlations of spins and holes in copper oxide superconductors. *Nature* **1995**, *375*, 561. [CrossRef]
21. Blaco-Caosa, S.; Frano, A.; Schierle, E.; Porras, J.; Loew, T.; Minola, M.; Bluschke, M.; Weschke, E.; Keimer, B.; Le Tacon, M. Resonant X-ray scattering study of charge-density- wave correlations in $YBa_2Cu_3O_{6+x}$. *Phys. Rev. B* **2014**, *90*, 54513. [CrossRef]
22. Hücker, M.; Christensen, N.B.; Holmes, A.T.; Blackburn, E.; Forgan, E.M.; Liang, R.; Bonn, D.A.; Hardy, W.N.; Gutowski, O.; Zimmermann, M.V.; et al. Competing charge, spin, and superconducting orders in underdoped YBa2Cu3Oy. *Phys. Rev. B* **2014**, *90*, 54514. [CrossRef]
23. Ghiringhelli, G.; Le Tacon, M.; Minola, M.; Blanco-Canosa, S.; Mazzoli, C.; Brookes, N.B.; De Luca, G.M.; Frano, A.; Hawthorn, D.G.; He, F.; et al. Long-range incommensurate charge fluctuations in(Y,Nd)Ba2Cu3O6+x. *Science* **2012**, *337*, 821. [CrossRef] [PubMed]
24. Comin, R.; Damascelli, A. Resonant X-Ray Scattering Studies of Charge Order in Cuprates. *Annu. Rev. Condens. Matter Phys.* **2016**, *7*, 369–405. [CrossRef]
25. Chang, J.; Blackburn, E.; Holmes, A.T.; Christensen, N.B.; Larsen, J.; Mesot, J.; Liang, R.; Bonn, D.A.; Hardy, W.N.; Watenphul, A.; et al. Direct observation of competition between superconductivity and charge density wave order in YBa2Cu3O6.67. *Nat. Phys.* **2012**, *8*, 871. [CrossRef]
26. Achkar, A.J.; Hawthorn, D.; Sutarto, R.; Mao, X.; He, F.; Frano, A.; Blanco-Canosa, S.; le Tacon, M.; Ghiringhelli, G.; Braicovich, L.; et al. Distinct charge orders in the planes and chains of ortho-III-ordered YBa2Cu3O6+_ superconductors identified by resonant elastic x-ray scattering. *Phys. Rev. Lett.* **2012**, *109*, 167001. [CrossRef]

27. Tabis, W.; Li, Y.; Le Tacon, M.; Braicovich, L.; Kreyssig, A.; Minola, M.; Dellea, G.; Weschke, E.; Veit, M.J.; Ramazanoglu, M.; et al. Charge order and its connection with Fermi-liquid charge transport in a pristine high-Tc cuprate. *Nat. Commun.* **2014**, *5*, 5875. [CrossRef] [PubMed]
28. Comin, R.; Frano, A.; Yee, M.M.; Yoshida, Y.; Eisaki, H.; Schierle, E.; Weschke, E.; Sutarto, R.; He, F.; Soumyanarayanan, A.; et al. Charge order driven by Fermi-arc instability in Bi2Sr2−xLaxCuO6+δ. *Science* **2014**, *343*, 390. [CrossRef]
29. Wu, T.; Mayaffre, H.; Krämer, S.; Horvatić, M.; Berthier, C.; Hardy, W.N.; Liang, R.; Bonn, D.A.; Julien, M.-H. Magnetic-field-induced charge-stripe order in the high-temperature superconductor YBa2Cu3Oy. *Nature* **2011**, *477*, 191. [CrossRef]
30. Wu, T.; Mayaffre, H.; Krämer, S.; Horvati´c, M.; Berthier, C.; Kuhn, P.L.; Reyes, A.P.; Liang, R.; Hardy, W.N.; Bonn, D.A.; et al. Emergence of charge order from the vortex state of a high-temperature superconductor. *Nat. Commun.* **2013**, *4*, 2113. [CrossRef]
31. Laliberté, F.; Frachet, F.; Benhabib, S.; Borgnic, B.; Loew, T.; Porras, J.; Le Tacon, M.; Keimer, B.; Wiedmann, S.; Proust, C.; et al. High field charge order across the phase diagram of YBa2Cu3Oy. *NPJ Quantum Mater.* **2018**, *3*, 11. [CrossRef]
32. Doiron-Leyraud, N.; Proust, C.; LeBoeuf, D.; Levallois, J.; Bonnemaison, J.; Liang, R.; Bonn, D.A.; Hardy, W.N. Taillefer Louis Quantum oscillations and the Fermi surface in an underdoped high-Tc superconductor. *Nature* **2007**, *447*, 565–568. [CrossRef]
33. Grissonnanche, G.; Cyr-Choinière, O.; Laliberté, T.; René de Cotret, S.; Juneau-Fecteau, A.; Dufour-Beauséjour, S.; Delage, M.È.; LeBoeuf, D.; Chang, J.; Ramshaw, B.J.; et al. Direct measurement of the upper critical field in cuprate superconductors. *Nat. Commun.* **2014**, *5*, 3280. [CrossRef] [PubMed]
34. LeBoeuf, D.; Doiron-Leyraud, N.; Vignolle, B.; Sutherland, M.; Ramshaw, B.J.; Levallois, J.; Daou, R.; Laliberté, F.; Cyr-Choinière, Q.; Chang, J.; et al. Lifshitz critical point in the cuprate superconductor YBa2Cu3Oy from high-field Hall effect measurements. *Phys. Rev. B* **2011**, *83*, 54506. [CrossRef]
35. Doiron-Leyraud, N.; Lepault, S.; Cyr-Choinière, O.; Vignolle, B.; Grissonnanche, G.; Laliberté, F.; Chang, J.; Barisic, N.; Chan, M.K.; Ji, L.; et al. Hall, Seebeck and Nernst coefficients of underdoped HgBa2CuO4+δ: Fermi-surface reconstruction in an archetypal cuprate superconductor. *Phys. Rev. X* **2013**, *3*, 21019.
36. Chang, J.; Daou, R.; Proust, C.; Leboeuf, D.; Doiron-Leyraud, N.; Laliberté, F.; Pingault, B.; Ramshaw, B.J.; Liang, R.; Bonn, D.A.; et al. Nernst and Seebeck coefficients of the cuprate superconductor YBa2Cu3O6.67: A study of fermi surface reconstruction. *Phys. Rev. Lett.* **2010**, *104*, 57005. [CrossRef]
37. Barišić, N.; Badoux, S.; Chan, M.K.; Dorow, C.; Tabis, W.; Vignolle, B.; Yu, G.; Béard, J.; Zhao, X.; Proust, C.; et al. Universal quantum oscillations in the underdoped cuprate superconductors. *Nat. Phys.* **2013**, *9*, 761. [CrossRef]
38. Grissonnanche, G.; Legros, A.; Badoux, S.; Lefrançois, E.; Zatko, V.; Lizaire, M.; Laliberté, F.; Gourgout, A.; Zhou, J.-S.; Pyon, S.; et al. Giant thermal Hall conductivity in the pseudogap phase of cuprate superconductors. *Nature* **2019**, *571*. [CrossRef]
39. Leboeuf, D.; Kramer, S.; Hardy, W.N.; Liang, R.; Bonn, D.A.; Proust, C. Thermodynamic phase diagram of static charge order in underdoped YBa2Cu3Oy. *Nat. Phys.* **2012**, *9*, 79–83. [CrossRef]
40. Sachdev, S.; Ye, J. Universal quantum-critical dynamics of two-dimensional antiferromagnets. *Phys. Rev. Lett.* **1992**, *69*, 2411. [CrossRef]
41. Sokol, A.; Pines, D. Toward a unified magnetic phase diagram of the cuprate superconductors. *Phys. Rev. Lett.* **1993**, *71*, 2813. [CrossRef] [PubMed]
42. Monthoux, P.; Pines, D. Nearly antiferromagnetic Fermi-liquid description of magnetic scaling and spin-gap behavior. *Phys. Rev. B* **1994**, *50*, 16015. [CrossRef] [PubMed]
43. Abanov, A.; Chubukov, A.; Schmalian, J. Quantum-critical theory of the spin-fermion model and its application to cuprates: Normal state analysis. *Adv. Phys.* **2003**, *52*, 119. [CrossRef]
44. Castellani, C.; Di Castro, C.; Grilli, M. Possible occurrence of band interplay in high Tc superconductors. In Proceedings of the International Conference on High-Temperature Superconductors and Materials and Mechanisms of Superconductivity Part II, Interlaken, Switzerland, 28 February–4 March 1988; *Physica C* 153–155. pp. 1659–1660.
45. Müller, K.A.; Benedeck, G. Phase Separation in Cuprate Superconductors. In Proceedings of the 3rd Workshop "The Science and Culture-Physics", Erice, Italy, 6–12 May 1992; World Scientific: Singapore, 1993.

46. Müller, K.A.; Sigmund, E. Phase Separation in Cuprate Superconductors. In Proceedings of the Second International Workshop on "Phase Separation in Cuprate Superconductors", Cottbus, Germany, 4–10 September 1993; Springer: Verlag, Germany, 1994.
47. Bianconi, A.; Saini, N.L.; Lanzara, A.; Missori, M.; Rossetti, T.; Oyanag, H.; Yamaguchi, H.; Oka, K.; Ito, T. Determination of the Local Lattice Distortions in the CuO_2 plane of La1.85Sr0.15CuO4. *Phys. Rev. Lett.* **1996**, *76*, 3412. [CrossRef]
48. Ramshaw, B.J.; Sebastian, S.E.; McDonald, R.D.; Day, J.; Tan, B.S.; Zhu, Z.; Betts, J.B.; Liang, R.; Bonn, D.A.; Hardy, W.N.; et al. Quasiparticle mass enhancement approaching optimal doping in a high-Tc superconductor. *Science* **2015**, *348*, 317–320. [CrossRef]
49. Becca, F.; Tarquini, M.; Grilli, M.; Di Castro, C. Charge-density waves and superconductivity as an alternative to phase separation in the infinite-U Hubbard-Holstein model. *Phys. Rev. B* **1996**, *54*, 12443–12457. [CrossRef]
50. Castellani, C.; Di Castro, C.; Grilli, M. Non-Fermi Liquid behavior and d-wave superconductivity near the charge density wave quantum critical point. *Zeit. Phys.* **1997**, *103*, 137–144. [CrossRef]
51. Seibold, G.; Becca, F.; Bucci, F.; Castellani, C.; Di Castro, C.; Grilli, M. Spectral properties of incommensurate charge-density wave systems. *Eur. Phys. J. B* **2000**, *13*, 87. [CrossRef]
52. Benfatto, L.; Caprara, S.; Di Castro, C. Gap and pseudogap evolution within the charge-ordering scenario for superconducting cuprates. *Eur. Phys. J. B* **2000**, *17*, 95. [CrossRef]
53. Chakravarty, S.; Laughlin, R.B.; Morr Dirk, K.; Nayak, C. Hidden order in the cuprates. *Phys. Rev. B* **2001**, *63*, 94503. [CrossRef]
54. Andergassen, S.; Caprara, S.; Di Castro, C.; Grilli, M. Anomalous Isotopic Effect Near the Charge-Ordering Quantum Criticality. *Phys. Rev. Lett.* **2001**, *87*, 5641. [CrossRef] [PubMed]
55. Yu, B.; Tabis, W.; Bialo, I.; Yakhou, F.; Brookes, N.; Anderson, Z.; Tan, G.Y.; Yu, G.; Greven, M. Unusual dynamic charge-density-wave correlations in HgBa2CuO4+. *arXiv* **2019**, arXiv:1907.10047.
56. Miao, H.; Fabbris, G.; Nelson, C.S.; Acevedo-Esteves, R.; Li, Y.; Gu, G.D.; Yilmaz, T.; Kaznatcheev, K.; Vescovo, E.; Oda, M.; et al. Discovery of Charge Density Waves in cuprate Superconductors up to the Critical Doping and Beyond. *arXiv* **2020**, arXiv:2001.10294.
57. Lin, J.Q.; Miao, H.; Mazzone, D.G.; Gu, G.D.; Nag, A.; Walters, A.C.; Garcia-Fernandez, M.; Barbour, A.; Pelliciari, J.; Jarrige, I.; et al. Nature of the charge-density wave excitations in cuprates. *arXiv* **2020**, arXiv:2001.10312.
58. Miao, H.; Lorenzana, J.; Seibold, G.; Peng, Y.Y.; Amorese, A.; Yakhou-Harris, F.; Kummer, K.; Brookes, N.B.; Konik, R.M.; Thampy, V.; et al. High-temperature charge density wave correlations in La1.875Ba0.125CuO4 without spin-charge locking. *Proc. Natl. Acad. Sci USA* **2017**, *114*, 12430. [CrossRef] [PubMed]
59. Miao, H.; Fumagalli, R.; Rossi, M.; Lorenzana, J.; Seibold, G.; Yakhou-Harris, F.; Kummer, K.; Brookes, N.B.; Gu, G.D.; Braicovich, L.; et al. Formation of incommensurate Charge Density Waves in Cuprates. *Phys. Rev.* **2019**, *9*, 031042. [CrossRef]
60. Wang, Q.S.; Horio, M.; Von Arx, K.; Shen, Y.; Mukkattukavil, D.J.; Sassa, Y.; Ivashko, O.; Matt, C.E.; Pyon, S.; Takayama, T.; et al. High-Temperature Charge-Stripe Correlations in $La_{1.675}Eu_{0.2}Sr_{0.125}CuO_4$. *Phys. Rev. Lett.* **2020**, *124*, 187002. [CrossRef]
61. Campi, G.; Bianconi, A.; Poccia, N.; Bianconi, G.; Barba, L.; Arrighetti, G.; Innocenti, D.; Karpinski, J.; Zhigadlo, N.D.; Kazakov, S.M.; et al. Inhomogeneity of charge-density-wave order and quenched disorder in a high-Tc superconductor. *Nature* **2015**, *525*, 359. [CrossRef]
62. Varma, C.M.; Littlewood, P.B.; Schmitt-Rink, S.; Abrahams, E.; Ruckenstein, A.E. Phenomenology of the normal state of Cu-O high-temperature superconductors. *Phys. Rev. Lett.* **1989**, *63*, 1996. [CrossRef]
63. Legros, A.; Benhabib, S.; Tabis, W.; Laliberté, F.; Dion, M.; Lizaire, M.; Vignolle, B.; Vignolles, D.; Raffy, H.; Li, Z.Z.; et al. Universal T-linear resistivity and Planckian dissipation in overdoped cuprates. *Nat. Phys.* **2019**, *15*, 142–147. [CrossRef]

Publisher's Note: MDPI stays neutral with regard to jurisdictional claims in published maps and institutional affiliations.

© 2020 by the author. Licensee MDPI, Basel, Switzerland. This article is an open access article distributed under the terms and conditions of the Creative Commons Attribution (CC BY) license (http://creativecommons.org/licenses/by/4.0/).

Review

Unconventional Magnetism in Layered Transition Metal Dichalcogenides

Zurab Guguchia

Laboratory for Muon Spin Spectroscopy, Paul Scherrer Institute, CH-5232 Villigen PSI, Switzerland; zurab.guguchia@psi.ch

Received: 18 May 2020; Accepted: 17 June 2020; Published: 20 June 2020

Abstract: In this contribution to the MDPI Condensed Matter issue in Honor of Nobel Laureate Professor K.A. Müller I review recent experimental progress on magnetism of semiconducting transition metal dichalcogenides (TMDs) from the local-magnetic probe point of view such as muon-spin rotation and discuss prospects for the creation of unique new device concepts with these materials. TMDs are the prominent class of layered materials, that exhibit a vast range of interesting properties including unconventional semiconducting, optical, and transport behavior originating from valley splitting. Until recently, this family has been missing one crucial member: magnetic semiconductor. The situation has changed over the past few years with the discovery of layered semiconducting magnetic crystals, for example CrI_3 and VI_2. We have also very recently discovered unconventional magnetism in semiconducting Mo-based TMD systems 2H-$MoTe_2$ and 2H-$MoSe_2$ [Guguchia et. al., *Science Advances* 2018, 4(12)]. Moreover, we also show the evidence for the involvement of magnetism in semiconducting tungsten diselenide 2H-WSe_2. These results open a path to studying the interplay of 2D physics, semiconducting properties and magnetism in TMDs. It also opens up a host of new opportunities to obtain tunable magnetic semiconductors, forming the basis for spintronics.

Keywords: transition metal dichalcogenides; magnetic semiconductor spintronics

PACS: 76.75.+i; 74.55.+v; 75.50.Pp

---

1. Introduction

Spintronics, or spin-based electronics, is one of the promising next generation information technology [1,2]. It makes use of the quantum property of electrons, such as spin, as information carriers and possesses potential advantages of speeding up the data processing, high circuit integration density, and low energy consumption. Magnetic semiconductors, combining the properties and advantages of both magnets and semiconductors, form the basis for spintronics. Not only ferromagnetic semiconductors, but also antiferromagnetic semiconductors were proposed to be natural candidates for integrating spintronics and traditional microelectronic functionalities in a single material [2]. All semiconductor spintronic devices act according to the following simple scheme: information is stored (written) into spins as a particular spin orientation; the spins, being attached to mobile electrons, carry the information along a wire; and the information is read and processed at a terminal. However, spintronics applications require novel magnetic semiconducting materials with high-temperature ferromagnetic or antiferromagnetic ordering of spins, which would simultaneously enable the conventional tunability of electronic properties and spintronic functionalities. Moreover, the materials should be produced as very stable thin layers to be incorporated in the devices.

Along these lines, transition metal dichalcogenides (TMDs), a family of two dimensional (2D) layered materials such as graphene, have appeared as the most promising platform due to their exciting mechanical, electronic and optoelectronic properties [3–9]. TMDs share the same formula, MX_2, where M is a transition metal and X is a chalcogen ion. They have a layered structure and crystallize in several polytypes, including 2H-, 1T-, 1T$'$- and T_d-type lattices. Much interest is focused on the cases where the transition metal M is either Molybdenum (Mo) or Tungsten (W). Hence, the 2H forms of these compounds are semiconducting and can be mechanically exfoliated to a monolayer. The unique properties of the TMDs, especially in the monolayer form, have triggered a wealth of device applications such as: magnetoresistance and spintronics, high on/off ratio transistors, optoelectronics, valley-optoelectronics, superconductors and hydrogen storage. Many of these interesting properties arise due to the strong spin-orbit interaction present in these materials arising from the presence of the heavy metal ion. While there are many studies focused on the spin-orbit coupling and the interesting consequences for electrical and optical properties in these materials, there are very limited, and mostly theoretical, studies on the intrinsic magnetism [10–17]. Theoretical work shows that, in the absence of crystalline imperfections, the Mo-based TMDs are nonmagnetic.

Combining a wealth of different technique, in particular the muon-spin rotation/relaxation (μSR) technique, we discovered novel magnetism in these very stable semiconducting materials: molybdenum ditelluride (MoTe$_2$) and molybdenum diselenide (MoSe$_2$) [9]. The results are published in a journal of the American Association for the Advancement of Science [9]. Here, I provide a short review of the previous results, show the new data for tungsten diselenide (2H-WSe$_2$) and discuss the importance of the presence of magnetism in semiconducting TMDs.

2. μSR Technique: Very Sensitive Microscopic Magnetic Probe

The acronym μSR stands for muon spin rotation, or relaxation, or resonance, depending respectively on whether the muon spin motion is predominantly a rotation (more precisely a precession around a magnetic field), or a relaxation towards an equilibrium direction, or a more complex dynamics dictated by the addition of short radio frequency pulses.

It is noteworthy that Prof. K. Alex Müller realised the importance and the strength of the μSR technique in the studies of high temperature superconductors (HTSs) at the very early stage of the era of high-T_c cuprates, as demonstrated in the pioneering papers [18–20]. Indeed, this technique allows us to study fundamental problems related to superconductivity [8,21–29].

Besides superconductivity, positive muons implanted into a sample serve as an extremely sensitive local probe to detect small internal magnetic fields and ordered magnetic volume fractions in the bulk of magnetic materials. μSR can distinguish the volume fraction effect from the ordered moment size and thus, it is a particularly powerful tool to study the thermal or quantum evolution of both magnetic moment and magnetically ordered volume fraction in solid materials [30–33]. μSR is also valuable for studying materials in which magnetic order is random or of short range. This makes μSR a perfectly complementary technique to scattering techniques such as neutron diffraction, which is used to determine crystallographic and magnetic structures. Moreover, the μSR technique has a unique time window (10^{-4} s to 10^{-11} s) for the study of magnetic fluctuations in materials, which is complementary to other experimental techniques such as neutron scattering, NMR, or magnetic susceptibility. With its unique capabilities, μSR should be considered to play a leading role in determining magnetic phase diagrams and elucidating the quantum evolution from the paramagnetic to magnetic state in the semiconducting TMDs, which are interesting due to both fundamental and practical aspects. A brief introduction to the μSR technique [34–36] is given below.

The μSR method is based on the observation of the time evolution of the spin polarization $\vec{P}(t)$ of the muon ensemble. A schematic layout of a μSR experiment is shown in Figure 1a–c. In a μSR experiments an intense beam (p_μ = 29 MeV/c) of 100 % spin-polarized muons is stopped in the sample (see Figure 1a). Currently available instruments allow essentially a background free μSR measurement at ambient pressure [37]. The positively charged muons thermalize in the sample at interstitial lattice sites, where they act as magnetic microprobes. In a magnetic material the muons spin precess in the local field B_μ at the muon site with the Larmor frequency $\nu_\mu = \gamma_\mu/(2\pi) B_\mu$ (muon gyromagnetic ratio $\gamma_\mu/(2\pi) = 135.5$ MHz T^{-1}). The muons μ^+ implanted into the sample will decay after a mean life time of $\tau_\mu = 2.2$ μs, emitting a fast positron e^+ preferentially along their spin direction. Various detectors placed around the sample track the incoming μ^+ and the outgoing e^+ (see Figure 1a). When the muon detector records the arrival of a μ in the specimen, the electronic clock starts. The clock is stopped when the decay positron e^+ is registered in one of the e^+ detectors, and the measured time interval is stored in a histogramming memory. In this way a positron-count versus time histogram is formed (Figure 1b). A muon decay event requires that within a certain time interval after a μ^+ has stopped in the sample an e^+ is detected. This time interval extends usually over several muon lifetimes (e.g., 10 μs). After a number of muons has stopped in the sample, one obtains a histogram for the forward (N_{e^+F}) and the backward (N_{e^+B}) detectors as shown in Figure 1b, which in the ideal case has the following form:

$$N_\alpha^{e^+}(t) = N_0 e^{-\frac{t}{\tau_\mu}}(1 + A_0 \vec{P}(t)\hat{n}_\alpha) + N_{bgr}. \quad \alpha = F, B \tag{1}$$

Here, the exponential factor accounts for the radioactive muon decay. $\vec{P}(t)$ is the muon-spin polarization function with the unit vector $\hat{n}_\alpha$ (α = F,B) with respect to the incoming muon spin polarization. N_0 is number of positrons at the initial time $t = 0$. N_{bgr} is a background contribution due to uncorrelated starts and stops. A_0 is the initial asymmetry, depending on different experimental factors, such as the detector solid angle, efficiency, absorption, and scattering of positrons in the material. Typical values of A_0 are between 0.2 and 0.3.

Since the positrons are emitted predominantly in the direction of the muon spin which precesses with ω_μ, the forward and backward detectors will detect a signal oscillating with the same frequency. In order to remove the exponential decay due to the finite life time of the muon, the so-called asymmetry signal $A(t)$ is calculated (see Figure 1c):

$$A(t) = \frac{N_{e^+F}(t) - N_{e^+B}(t)}{N_{e^+F}(t) + N_{e^+B}(t)} = A_0 P(t), \tag{2}$$

where $N_{e^+F}(t)$ and $N_{e^+B}(t)$ are the number of positrons detected in the forward and backward detectors, respectively. The quantities $A(t)$ and $P(t)$ depend sensitively on the spatial distribution and dynamical fluctuations of the magnetic environment of the muons. Hence, these functions allow us to study interesting physics of the investigated system.

In μSR experiments two different magnetic field configurations are used: (i) Transverse field (TF) μSR involves the application of an external field perpendicular to the initial direction of the muon spin polarization. The muon spin precesses around the transverse field, with a frequency that is proportional to the size of the field at the muon site in the material. (ii) In the longitudinal field (LF) configuration the magnetic field is applied parallel to the initial direction of the muon spin polarization. The time evolution of the muon spin polarization along its initial direction is measured in this configuration. Measurements are often carried out in the absence of external magnetic field, a configuration called zero-field (ZF) μSR. In this configuration the frequency of an obtained μSR signal is proportional to the internal magnetic field, from which the size of the ordered moment and thus the magnetic order parameter is calculated. The capability of studying materials in zero external field is a big advantage over other magnetic resonance techniques.

If the local magnetic field $\vec{B}(\vec{r})$ at the muon site is pointing under an angle θ with respect to the initial muon spin polarization, the decay positron asymmetry is given by [33]:

$$A(t) = A_0[\cos^2(\theta) + \sin^2(\theta)\cos(\gamma_\mu B t)], \tag{3}$$

where A_0 is the maximal value of the asymmetry. Further assuming that the random fields are isotropic and each component can be represented by a Gaussian distribution of width Δ/γ, then a statistical average of this distribution yields:

$$A(t) = A_0\left[\frac{1}{3} + \frac{2}{3}e^{-\Delta^2 t^2/2}(1 - \Delta^2 t^2)\right], \qquad (4)$$

This function was first obtained in a general stochastic treatment of Kubo and Toyabe [38]. The form of the distribution of internal magnetic fields influences the form of the μSR signal [30,31,33]. Thus, by analysing the observed muon-spin time evolution, the magnetic field distribution inside the sample can be obtained. For clarity, Figure 1d–f shows the expected time evolution of the muon spin polarisation for three different cases of magnetically ordered polycrystaline sample: fully magnetic and magnetically homogeneous (d), full volume magnetic and inhomogencous (e) and phase separation between magnetic and paramagnetic regions (f). The muons stopping in the homogeneous sample will sense the same magnetic field and their spin will precess around the internal field and the μSR signal is characterised by maximum amplitude and zero depolarisation (Figure 1d). If there is an inhomogeneous static internal field in the sample, different muons will precess at slightly different frequencies. This leads to a progressive dephasing of the μSR signal, and the oscillations in the μSR time spectra will be damped (see Figure 1e). In some cases the signal is strongly damped, so that the oscillation will not be observed, and the resulting muon spin polarization will be averaged out to zero. Then, at a magnetic phase transition, if no wiggles are observed in the μSR signal, one expects a drop in the effective initial asymmetry from A_0 in the paramagnetic state to $A_0 = 1/3$ in the ordered state [30]. However, this effect could also be due to fluctuations of the internal field. μSR is capable of distinguishing between these two possibilities by performing an LF-μSR experiment. In a longitudinal field inhomogeneous line broadening and fluctuations lead to different μSR time spectra. Since muons stop uniformly throughout a sample, and the amplitudes of the μSR signals arising from the different regions of the sample are proportional to the volume of the sample occupied by a particular phase, the presence of paramagnetic regions will result in the reduction of the signal amplitude as shown in Figure 1f. This schematics is a simple illustration of the fact that μSR is capable to provide quantitative information on coexisting and competing phases in a material.

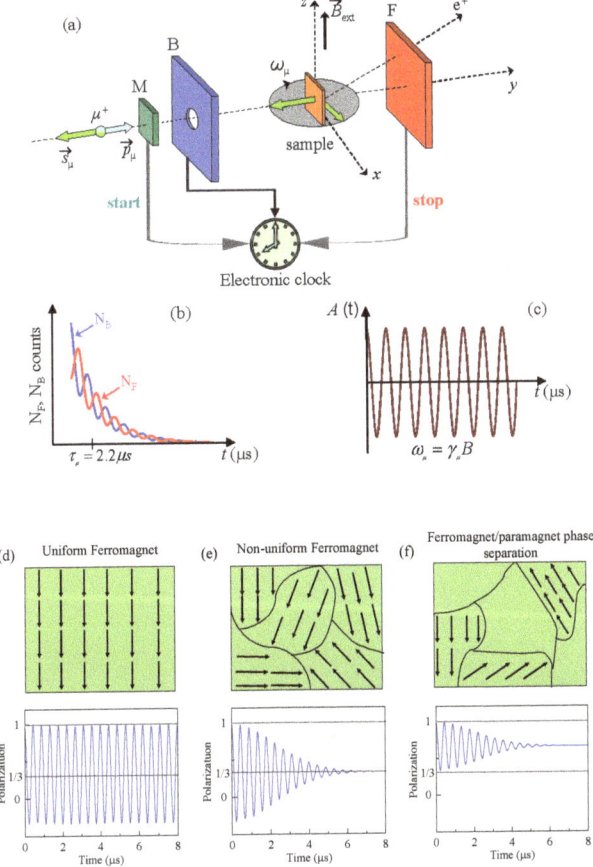

Figure 1. Principle of a μSR experiment. (**a**) Overview of the experimental setup. Spin polarized muons with spin S_μ antiparallel to the momentum p_μ are implanted in the sample placed between the forward (F) and the backward (B) positron detectors. A clock is started at the time the muon goes through the muon detector (M) and is stopped as soon as the decay positron is detected in the detectors F or B. (**b**) The number of detected positrons N_F and N_B as a function of time for the forward and backward detector, respectively. (**c**) The so-called asymmetry (or μSR) signal is obtained by essentially building the difference between N_F and N_B. (**d–f**) Schematic illustration of the magnetically homogeneous (i.e., full volume magnetic) (**d**), inhomogeneous (full volume magnetic, but with domains) and phase separated (i.e., part of the volume magnetic and part paramagnetic) polycrystalline samples and the corresponding μSR spectra. The 1/3 non-oscillating μSR signal fraction originate from the spatial averaging in powder samples where 1/3 of the magnetic field components are parallel to the muon spin and do not cause muon spin precession. Figure 1a–c is from [39,40].

3. Results

As we showed in our previous work, a spontaneous muon spin precession with a well-defined frequency was observed for both 2H-MoTe$_2$ and 2H-MoSe$_2$ at low temperatures (structural representation of the hexagonal 2H structure is shown in Figure 2). Moreover, muon spin precession was also observed for W-based TMD semiconductor 2H-WSe$_2$. The observation of a spontaneous muon spin precession is a clear signature of the occurrence of a static magnetic state in these semiconducting systems at low temperatures. Figure 3 shows the comparison of zero-field μSR time spectra, recorded at $T = 5$ K, between semimetallic orthorhombic T_d-MoTe$_2$ and semiconducting hexagonal 2H-MoTe$_2$ samples. It is interesting that while 2H-MoTe$_2$ showed coherent oscillations and thus the static magnetism, T_d-MoTe$_2$ gave a typical paramagnetic response from the whole sample volume. This indicates that static magnetism was a property for semiconducting 2H phase (Figure 2). The precession frequency was proportional to the local internal field $\mu_0 H_{int}$ at the muon site, which is shown in Figure 4a as a function of temperature for both 2H-MoTe$_2$ and 2H-MoSe$_2$ samples. A few points for 2H-WSe$_2$ were also shown. There was a smooth increase of $\mu_0 H_{int}$ below $T_M \simeq 40$ K and 100 K for 2H-MoTe$_2$ and 2H-MoSe$_2$, respectively. The internal field reached the saturated value of $\mu_0 H_{int} \simeq 200$ mT and 300 mT at low temperatures, for 2H-MoTe$_2$ and 2H-MoSe$_2$, respectively. For 2H-WSe$_2$, an even higher internal field value of $\mu_0 H_{int} \simeq 350$ mT was observed. A higher value of T_M and $\mu_0 H_{int}$ for 2H-MoSe$_2$ and 2H-WSe$_2$ as compared to values for 2H-MoTe$_2$ might be related to the distinct magnetic structures in these systems.

Although μSR measurements revealed the homogeneous internal magnetic fields in 2H-MoTe$_2$ and 2H-MoSe$_2$ below $T_M \simeq 40$ K and 100 K, respectively, we observed that short-range (inhomogeneous) magnetism occurred at much higher temperatures. Figure 4b displays the paramagnetic fraction V_{PM} as a function of temperature for both samples. In MoTe$_2$, at 450 K, V_{PM} exhibited nearly a maximum value and decreases with decreasing temperature and tends to saturate below 300 K. This ~30 % reduction of V_{PM} can arise for few different reasons and it was most likely due to the presence of a muonium fraction. In muonium, a bound state formed by a positive muon and an electron, may form in semiconductors [41]. In the bound state, the muon is much more sensitive to magnetic fields than as a free probe, because its magnetic moment couples to the much larger electron magnetic moment, thus amplifying the depolarization effects. Therefore, even small variations of the internal magnetic field may cause the loss of asymmetry such as that observed in 2H-MoTe$_2$ at high temperatures. However, there was an additional decrease of V_{PM} starting below $T_M^{onset} \sim 180$ K, which was due to the appearance of inhomogeneous short-range magnetism and V_{PM} continued to reduce until it reached the minimum value $V_{osc} \simeq 0.1$ at the long-range magnetic ordering temperature $T_M \simeq 40$ K, below which zero-field μSR showed a well defined uniform internal magnetic field. This implies that the long range magnetic order was achieved only below $T_M \simeq 40$ K, while the short range magnetism appeared at higher temperature $T_M^{onset} \sim 180$ K. The same was also observed in MoSe$_2$, i.e., the short range magnetism appeared below $T_M^{onset} \sim 250$ K, while the long-range order was achieved below $T_M \simeq 100$ K. The onset temperature of short range magnetism in 2H-WSe$_2$ was close to the one for 2H-MoSe$_2$ (see Figure 4b). This suggests the presence of magnetic correlations in 2H-MoSe$_2$ and 2H-WSe$_2$ almost at room temperature, which may render TMDs useful for novel optoelectronics–spintronics applications. Moreover, we determined that hydrostatic pressure[42,43] had a significant effect on the magnetic properties of these materials. Namely, we saw a large suppression of both the magnetically ordered fraction and the magnetic order temperature T_M of 2H-MoTe$_2$ as a function of pressure (Figure 4c). The magnetism was nearly fully suppressed within 2.5 GPa. On the other hand, 2H structure remained stable up to much higher pressure of $p \sim 11$ GPa (the results will be reported elsewhere), which implies that the suppression of magnetism in 2H-MoTe$_2$ did not originate from the structural modifications. This strong pressure dependence of magnetism is very encouraging, because it indicates that one can have quantum control, in addition to thermal control, over the magnetic properties.

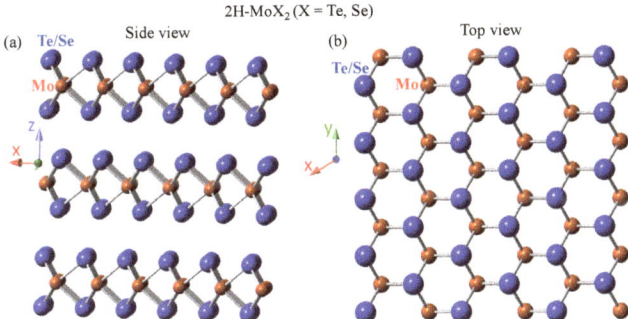

Figure 2. Structural representation of the hexagonal 2H structure for MoTe$_2$: (**a**) Side view. (**b**) Top view.

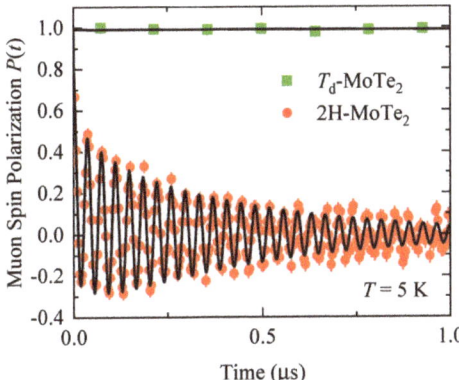

Figure 3. Zero-field (ZF) muon-spin rotation (μSR) time spectra for the single crystal samples of T_d-MoTe$_2$ and 2H-MoTe$_2$ recorded at $T = 5$ K. The data for 2H-MoTe$_2$ is from Guguchia et. al., Science Advances **4**: eaat3672 (2018).

The presence of magnetism in MoTe$_2$ and MoSe$_2$ was substantiated by temperature dependent macroscopic magnetisation measurements (see Figure 5a,b). A large difference between zero-field cooled (ZFC) and field-cooled (FC) magnetic moments was seen for both samples. The samples showed a combination of a temperature-independent diamagnetism, small van Vleck-type paramagnetism (which is determined by the energy separation of bonding and anti-bonding states E_a-E_b. E_a-E_b is proportional to the band gap E_g) and a magnetic contribution that onset near 230 K and 180 K for MoSe$_2$ and MoTe$_2$, respectively. The difference between the FC and ZFC response suggests that different magnetic domains tended to cancel out (anti-align) after ZFC. If we do field-cooling, then the domains align. The onset temperatures of hysteresis 230 K and 180 K for MoSe$_2$ and MoTe$_2$, respectively, were close to the temperature T_M^{onset} below which μSR experiments showed the appearance of inhomogeneous magnetism (see Figure 4b). This means that below T_M^{onset} some small magnetic domains (droplets of magnetic order) formed, which produced inhomogeneous magnetic fields and resulted in the absence of coherent oscillations in the μSR signal. Instead, the various magnetic fields produced in the sample gave rise to a strong damping of the muon asymmetry, which we clearly d at $T_M^{onset} = 180$ K in MoTe$_2$ and $T_M^{onset} = 250$ K in MoSe$_2$. μSR observed homogeneous magnetism below $T_M \simeq 40$ K and 100 K for MoTe$_2$ and MoSe$_2$, respectively, and the anomalies (such as an additional increase of the moment and of the difference) at around these temperatures could also be seen in magnetization data. We note that for the sample MoSe$_2$ the magnetic contribution dominated over the diamagnetism, gave rise to total positive moment. In contrast, for MoTe$_2$ diamagnetism dominated over the magnetic contribution. Assuming that the core diamagnetism is nearly the same in these materials, this difference might be related to a distinct magnetic structures of MoTe$_2$ and MoSe$_2$. Separation between ZFC and FC curves, observed at low temperatures, does not necessarily indicate the ferromagnetic (FM) order. It could be consistent with the canted antiferromagnetic (AFM) structure, i.e., AFM structure with some weak net FM moment. Since magnetic contribution dominates over the diamagnetism in MoSe$_2$, its magnetic structure most likely exhibits larger net FM moment than the one for MoTe$_2$.

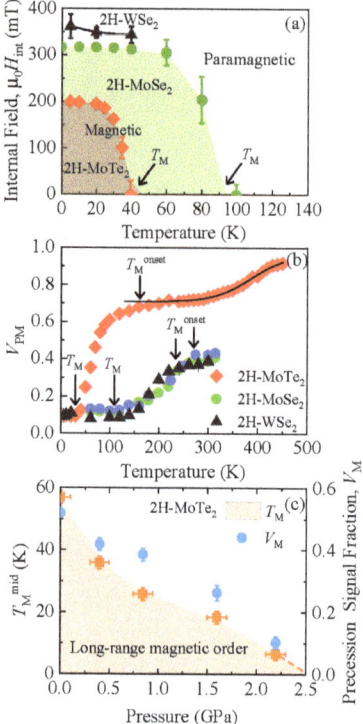

Figure 4. (a) Temperature dependence of the internal field $\mu_0 H_{int}$ of 2H-MoTe$_2$, 2H-MoSe$_2$ and 2H-WSe$_2$ as a function of temperature. (b) The temperature dependence of the paramagnetic fraction for 2H-MoTe$_2$, 2H-MoSe$_2$ and 2H-WSe$_2$. Arrows mark the onset and long-range ordered temperatures for magnetism. (c) Magnetic transition temperature T_M and magnetic volume fraction V_M for 2H-MoTe$_2$ as a function of pressure. Figure is adapted from Guguchia et. al., Science Advances **4**: eaat3672 (2018).

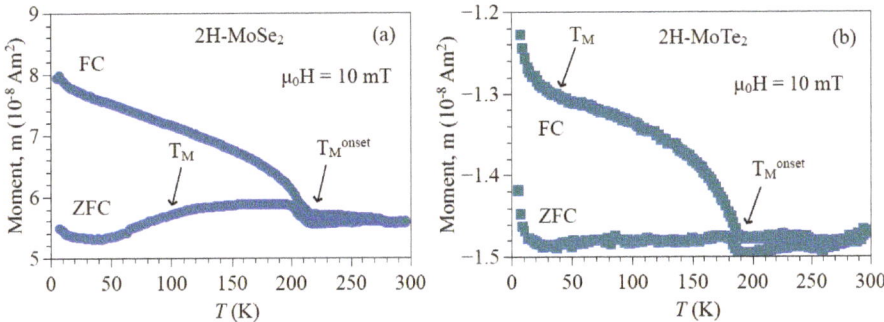

Figure 5. The temperature dependence of zero-field cooled (sample was cooled down to the base-T in zero magnetic field and the measurements were done upon warming) and field-cooled (the sample was cooled down to the base-T in an applied magnetic field and the measurements were done upon warming) magnetic moments for the polycrystalline samples of MoSe$_2$ (a) and MoTe$_2$ (b), recorded in an applied field of $\mu_0 H = 10$ mT. The arrows mark the onset of the difference between ZFC and FC moment as well as the anomalies seen at low temperatures. The Figure is from Guguchia et. al., Science Advances 4: eaat3672 (2018).

4. Summary and Discussion

The μSR measurements establish 2H-MoTe$_2$, 2H-MoSe$_2$ and 2H-WSe$_2$ as intrinsic magnetic, moderate bandgap semiconductors [9]. The μSR results demonstrate magnetic order below $T_M \simeq 40$ and 100 K for MoTe$_2$ and MoSe$_2$, respectively. These results came as a surprise since the previous theoretical work [11] and simple chemical bonding considerations indicate that the Mo atoms in these samples are in a nonmagnetic $4d^2$ configuration. To understand the origin of magnetism, we carried out [9] the detailed investigation of the presence of defects and examined their magnetic properties by combination of the high-resolution STM [44] and Hubbard corrected DFT+U calculations [45] for Mo-based systems. STM measurements demonstrate the presence of intrinsic dilute self-organized defects. Note that two major defects, i.e., metal-vacancies (Figure 6a) and chalcogen-antisites (Figure 6b) (where a molybdenum atom substitutes the Tellurium/Selenium atom) were found in these materials. The defect concentration is small (∼ 0.5–1%), but defects are found to have a large electronic impact. Moreover, DFT indicates that the chalcogen-antisite defects Mo$_{sub}$ are magnetic, while metal-vacancy defects Mo$_{vac}$ do not introduce a significant local moment. Although, DFT shows the magnetic defects in these systems, it is difficult to understand how the low-density of the chalcogen-antisite defects can give rise to homogeneous internal magnetic fields, observed in 2H-MoTe$_2$ and 2H-MoSe$_2$, indicative of long-range magnetic order. This may be possible if these defects have electronic coupling to the semiconductor valence electrons. The presence of such spin-polarized itinerant electrons would imply that these materials are dilute magnetic semiconductors. This idea may be partly supported by the recent report on the observation of hidden spin-polarized states in the bulk MoTe$_2$ [46]. Although the exact link between μSR and STM/DFT results [9] in 2H-MoTe$_2$ and 2H-MoSe$_2$ is not yet clear, both results together constitute a first strong evidence concerning the relevance of magnetic order in the TMDs physics. Defect induced, layer-modulated magnetism was recently reported for ultrathin metallic system PtSe$_2$ [47]. Ferromagnetism is also observed in VSe$_2$ monolayers. However, this system is characterized by a high density of states at the Fermi level [48], which is different from the systems 2H-MoTe$_2$, 2H-MoSe$_2$ and 2H-WSe$_2$, showing good semiconducting behavior. Thus, they open up a host of new opportunities to obtain tunable magnetic semiconductors, forming the basis for spintronics. Recently, there have been several reports on magnetism in W-based TMDs from bulk magnetisation measurements. Namely, the formation of ferromagnetism was reported for Vanadium doped WS$_2$ [49] and WSe$_2$ [50,51] monolayers with a small amount (∼0.5–4%) of V-content and the materials was classified as a dilute-magnetic semiconductor. We note that the mechanism behind the magnetic order in 2H-MoTe$_2$, 2H-MoSe$_2$ and 2H-WSe$_2$ with intrinsic magnetic defects and in WS$_2$ with incorporation of small amount of Vanadium could be similar. Besides Mo- and W-based TMDs, a very interesting magnetic semiconducting TMD system is CrI$_3$ [52]. Although the experimental investigations of bulk CrI$_3$ date back to the 1960s, the temperature dependent magnetic and structural properties have only recently been reported [53]. Standard magnetization measurements show that CrI$_3$ is an anisotropic ferromagnet below the T_C = 61 K, with its easy axis pointing perpendicular to the layers. Evidence for a second unexpected magnetic phase transition at T = 50 K is found both in the bulk and in atomically thin crystals of CrI$_3$ [52]. It is important to emphasize that neither the nature nor the details of the ensuing magnetic state in CrI$_3$ are currently understood. To date, the origin of the novel magnetic order in Mo/W-based TMDs and the nature of complex magnetism in CrI$_3$ is not understood. To fully exploit the magnetic properties of these TMD semiconductors, future works needs to address these important issues. It is essential to make use of pressure, strain, organic cation intercalation, light illumination [54], and particle irradiation as tuning parameters for bulk material to explore the precise role of itinerant carriers and the specific type of defects in the formation of magnetism in Mo- and W-based TMDs.

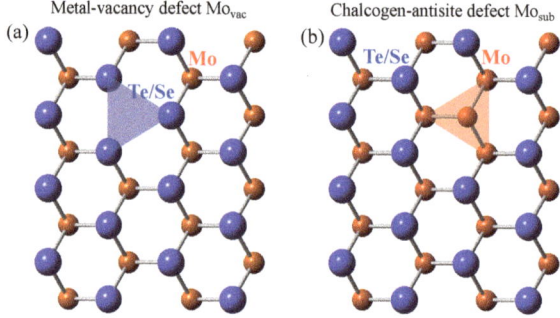

Figure 6. Schematic illustration of the metal-vacancy defect Mo_{vac} (**a**) and the chalcogen-antisite Mo_{sub} defect (**b**). Figure is adapted from Z. Guguchia et. al., Science Advances 4: eaat3672 (2018).

Previously, magnetic semiconductors have been synthesized in a range of thin film and crystal materials [55–57]. Much interest has been focused on the III-V (GaAs) [55,56] and I-II-V (LiZnAs) [57,58] semiconductor class, where a small concentration of some magnetic ions, in particular Mn^{2+}, can be incorporated by substituting for the group II (Zn^{2+},) and III (Ga^{3+}) cations of the host semiconductor. Numerous technical challenges in making uniform magnetic semiconducting materials have been overcome in recent years, but formidable challenges still remain in producing stable, high-quality materials with high T_M. For instance, GaMnAs has poor chemical solubility and can not be grown as a bulk material. Only MBE films are available and only as a p-type system. LiZnMnAs has high chemical solubility and can be grown as a bulk material. However, no films of LiZnMnAs are available. Our present systems offer an alternative route to synthesize magnetic semiconducting materials with the following advantages:

1. High quality bulk samples of Mo/W-based semiconducting TMDs can be grown. The systems can be doped to make both n-type and p-type semiconductors available.
2. The materials are cleavable down to a monolayer thickness and readily grown in large-area form. As it is well established in these materials, the bandgap is strongly dependent on the thickness, providing tunability over the semiconductor properties.
3. The chemical potential and electric field in thin films are easily tuned by electrostatic gates, opening the possibility to tune magnetism, as demonstrated in GaAs.

4. Finally, these materials can be easily layered by van der Waals heteroepitaxy, allowing the creation of unique new device concepts. For instance, one can grow the heterostructure of the magnetic TMDs (2H-MoTe$_2$, 2H-MoSe$_2$, 2H-WSe$_2$) and superconducting Weyl semimetal [3] (T$_d$-MoTe$_2$) or the heterostructure of the magnetic TMDs (2H-MoTe$_2$, 2H-MoSe$_2$, 2H-WSe$_2$) and topological insulator (TI) [59,60] (Bi$_2$Se$_3$) and study exotic magnetic/superconducting properties at the interface [61–63]. Even though the SC in the bulk can be topologically trivial, the robust Z$_2$ related topological states at the surface or at the interfaces is expected to feature helical or other exotic Cooper pairing due to spin-momentum locking [64], and the external magnet field induced vortex line can support Majorana fermions. When a topological insulator/metal is in contact with a ferromagnet, both time reversal and inversion symmetries are broken at the interface. An energy gap is formed at the TI surface, and its electrons gain a net magnetic moment through short-range exchange interactions. Magnetic/superconducting proximity effects at the interface between magnet and Weyl semimetal (Figure 7a) or magnet and topological insulators (Figure 7b) is considered to have great potential in spintronics as, in principle, it allows realizing the quantum anomalous Hall and topological magneto-electric effects. However, detailed experimental investigations of induced magnetism at the interfaces has remained a challenge. Low-energy μSR [63,65] experiments, which allows to probe the magnetism and superconductivity as a function of distance from the free surface to the interface, will be crucial to address the interesting magnetic/superconducting aspects of the interfaces.

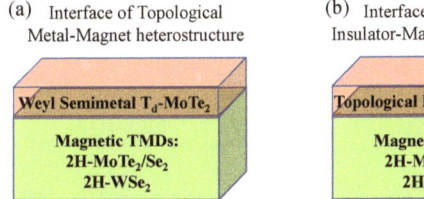

Figure 7. (a) Schematics of the heterostructure of the magnetic transition metal dichalcogenide (TMD) (2H-MoTe$_2$, 2H-MoSe$_2$, 2H-WSe$_2$) and superconducting Weyl semimetal (T$_d$-MoTe$_2$). (b) Schematics of the heterostructure of the magnetic TMD (2H-MoTe$_2$, 2H-MoSe$_2$, 2H-WSe$_2$) and topological insulator (Bi$_2$Se$_3$).

Funding: Zurab Guguchia gratefully acknowledges the financial support by the Swiss National Science Foundation (SNSF-fellowships P2ZHP2-161980 and P300P2-177832).

Acknowledgments: The author started to investigate magnetism in 2H-MoTe$_2$ with μSR, during his postdoctoral research in the Laboratory for Muon Spin Spectroscopy at the Paul Scherrer Institute, Switzerland, working with Rustem Khasanov, Alex Amato and Elvezio Morenzoni. Later on, he joined the group of Yasutomo Uemura at the Columbia University, where he spent two years and carried out extended studies of transition metal dichalcogenides in collaboration with the groups of Abhay Pasupathy and Simon Billinge at the Columbia University and the group of Elton Santos at the Queens University of Belfast. This work benefited from a combination of various experimental techniques and from DFT calculations. Samples were provided by Fabian von Rohr from the University of Zürich and by Daniel Rhodes from the Columbia University. The μSR experiments were carried out at the Swiss Muon Source (SμS) of the Paul Scherrer Institute using low background GPS (πM3 beamline) [37] and high pressure GPD (μE1 beamline) [42] instruments. The μSR time spectra were analyzed using the free software package MUSRFIT [66]. STM and SQUID experiments were carried out at the Columbia University. DFT calculations were performed at the Queens University of Belfast. X-ray PDF experiments were carried out at the Brookhaven National Laboratory. The author thank all the people involved in this project.

Conflicts of Interest: The authors declare no conflict of interest.

References

1. Lombardi, G.C.; Bianchi, G.E. *Spintronics: Materials, Applications and Devices*; Nova Science Pub Inc.: Hauppauge, NY, USA, 2009.
2. Jungwirth, T.; Marti, X.; Wadley, P.; Wunderlich, J. Antiferromagnetic spintronics. *Nat. Nanotechnol.* **2016**, *11*, 231–241. [CrossRef] [PubMed]
3. Soluyanov, A.A.; Gresch, D.; Wang, Z.; Wu, Q.; Troyer, M.; Dai, X.; Bernevig, B.A. Type-II Weyl semimetals. *Nature* **2015**, *527*, 495–498. [CrossRef] [PubMed]
4. Xu, X.; Yao, W.; Xiao, D.; Heinz, T.F. Spin and pseudospins in layered transition metal dichalcogenides. *Nat. Phys.* **2014**, *10*, 343–350. [CrossRef]
5. Ali, M.N.; Xiong, J.; Flynn, S.; Tao, J.; Gibson, Q.D.; Schoop, L.M.; Liang, T.; Haldolaarachchige, N.; Hirschberger, M.; Ong, N.P.; et al. Large, non-saturating magnetoresistance in WTe_2. *Nature* **2014**, *514*, 205–208. [CrossRef] [PubMed]
6. Qian, X.; Liu, J.; Fu, L.; Li, J. Quantum spin Hall effect in two-dimensional transition metal dichalcogenides. *Science* **2014**, *346*, 1344–1347. [CrossRef]
7. Costanzo, D.; Jo, S.; Berger, H.; Morpurgo, A.F. Gate-induced superconductivity in atomically thin MoS_2 crystals. *Nat. Nanotechnol.* **2016**, *11*, 339. [CrossRef]
8. Guguchia, Z.; von Rohr, F.; Shermadini, Z.; Lee, A.T.; Banerjee, S.; Wieteska, A.R.; Marianetti, C.A.; Frandsen, B.A.; Luetkens, H.; Gong, Z.; et al. Signatures of topologically non-trivial s^{+-} superconducting order parameter in type-II Weyl semimetal T_d-$MoTe_2$. *Nat. Commun.* **2017**, *8*, 1082. [CrossRef]
9. Guguchia, Z.; Kerelsky, A.; Edelberg, D.; Banerjee, S.; von Rohr, F.; Scullion, D.; Augustin, M.; Scully, M.; Rhodes, D.A.; Shermadini, Z.; et al. Magnetism in Semiconducting Molybdenum Dichalcogenides. *Sci. Adv.* **2018**, *4*, eaat3672. [CrossRef]
10. Tongay, S.; Varnoosfaderani, S.S.; Appleton, B.R.; Wu, J.; Hebard, A.F. Magnetic properties of MoS_2: Existence of ferromagnetism. *Appl. Phys. Lett.* **2012**, *101*, 123105. [CrossRef]
11. Ataca, C.; Sahin, H.; Akturk, E.; Ciraci, S. Mechanical and Electronic Properties of MoS_2 Nanoribbons and Their Defects. *J. Phys. Chem. C* **2011**, *115*, 3934–3941. [CrossRef]
12. Li, Y.; Zhou, Z.; Zhang, S.; Chen, Z. MoS_2 nanoribbons: high stability and unusual electronic and magnetic properties. *J. Am. Chem. Soc.* **2008**, *130*, 16739–16744. [CrossRef] [PubMed]
13. Krasheninnikov, A.V.; Lehtinen, P.O.; Foster, A.S.; Pyykko, P.; Nieminen, R.M. Embedding Transition-Metal Atoms in Graphene: Structure, Bonding, and Magnetism. *Phys. Rev. Lett.* **2009**, *102*, 126807. [CrossRef] [PubMed]
14. Santos, E.J.G.; Sanchez-Portal, D.; Ayuela, A. Magnetism of substitutional Co impurities in graphene: Realization of single π-vacancies. *Phys. Rev. B* **2010**, *81*, 125433. [CrossRef]
15. Santos, E.J.G.; Ayuela, A.; Sanchez-Portal, D. Universal magnetic properties of sp^3-type defects in covalently functionalized graphene. *New J. Phys.* **2012**, *14*, 043022. [CrossRef]
16. Zhang, Z.; Zou, X.; Crespi, V.H.; Yakobson, B.I. Intrinsic Magnetism of Grain Boundaries in Two-Dimensional Metal Dichalcogenides. *ACS Nano* **2013**, *7*, 10475. [CrossRef] [PubMed]
17. Cai, L.; He, J.; Liu, Q.; Yao, T.; Chen, L.; Yan, W.; Hu, F.; Jiang, Y.; Zhao, Y.; Hu, T.; et al. Vacancy-Induced Ferromagnetism of MoS_2 Nanosheets. *J. Am. Chem Soc.* **2015**, *137*, 2622. [CrossRef]
18. Pümpin, B.; Keller, H.; Kündig, W.; Odermatt, W.; Patterson, B.D.; Schneider, J.W.; Simmler, H.; Connell, S.; Müller, K.A.; Bednorz, J.G.; et al. Internal magnetic fields in the high-temperature superconductor $YBa_2Cu_3O_{7-\delta}$ from muon spin rotation experiments. *Z. Phys. B - Condensed Matter* **1988**, *72*, 175–180. [CrossRef]
19. Keller. H. Muon spin rotation experiments in high-T_c superconductors. In *Earlier and Recent Aspects of Superconductivity*; Springer: Berlin, Heidelberg, Germany, 1990; pp. 222–239.
20. Pümpin, B.; Keller, H.; Kündig, W.; Odermatt, W.; Savić, I.M.; Schneider, J.W.; Simmler, H.; Zimmermann, P.; Kaldis, E.; Rusiecki, S.; et al. Muon-spin-rotation measurements of the London penetration depths in $YBa_2Cu_3O_{6.97}$. *Phys. Rev. B* **1990**, *42*, 8019–8029. [CrossRef]
21. Morenzoni, E.; Wojek, B.M.; Suter, A.; Prokscha, T.; Logvenov, G.; Bozovic, I. The Meissner effect in a strongly underdoped cuprate above its critical temperature. *Nat. Commun.* **2011**, *2*, 272. [CrossRef]

22. Uemura, Y.J.; Luke, G.M.; Sternlieb, B.J.; Brewer, J.H.; Carolan, J.F.; Hardy, W.N.; Kadono, R.; Kempton, J.R.; Kiefl, R.F.; Kreitzman, S.R.; et al. Universal Correlations between T_c and n_s/m^* (Carrier Density over Effective Mass) in High-T_c Cuprate Superconductors. *Phys. Rev. Lett.* **1989**, *62*, 2317. [CrossRef]
23. Sonier, J.E.; Brewer, J.H.; Kiefl, R.F. μSR studies of the vortex state in type-II superconductors. *Rev. Mod. Phys.* **2000**, *72*, 769. [CrossRef]
24. Khasanov, R.; Shengelaya, A.; Maisuradze, A.; La Mattina, F.; Bussmann-Holder, A.; Keller, H.; Müller, K.A. Experimental Evidence for Two Gaps in the HighTemperature $La_{1.83}Sr_{0.17}CuO_4$ Superconductor. *Phys. Rev. Lett.* **2007**, *98*, 057007. [CrossRef]
25. Keller, H.; Bussmann-Holder, A.; Muller, K.A. Jahn-Teller physics and high-T_c superconductivity. *Mater. Today* **2008**, *11*, 38. [CrossRef]
26. Uemura, Y.J. Condensation, excitation, pairing, and superfluid density in high-T_c superconductors: Magnetic resonance mode as a roton analogue and a possible spin-mediated pairing. *J. Phys. Condens. Matter* **2004**, *16*, S4515–S4540. [CrossRef]
27. Bernhard, C.; Drew, A.J.; Schulz, L.; Malik, V.K.; Rossle, M.; Niedermayer, C.; Wolf, T.; Varma, G.D.; Mu, G.; Wen, H.H.; et al. Muon spin rotation study of magnetism and superconductivity in $BaFe_{2-x}Co_xAs_2$ and $Pr_{1-x}Sr_xFeAsO$. *New J. Phys.* **2009**, *11*, 055050. [CrossRef]
28. Luetkens, H.; Klauss, H.-H.; Kraken, M.; Litterst, F.J.; Dellmann, T.; Klingeler, R.; Hess, C.A.; Khasanov, R.; Amato, A.; Baines, C.; et al. The electronic phase diagram of the $LaO_{1-x}F_xFeAs$ superconductor. *Nature Mater.* **2009**, *8*, 305. [CrossRef] [PubMed]
29. Luke, G.M.; Fudamoto, Y.; Kojima, K.M.; Larkin, M.I.; Merrin, J.; Nachumi, B.; Uemura, Y.J.; Maeno, Y.; Mao, Z.Q.; Mori, Y.; et al. Time-Reversal Symmetry Breaking Superconductivity in Sr_2RuO_4. *Nature* **1998**, *394*, 558–561. [CrossRef]
30. Dalmas de Reotier, P.; Yaouanc, A. Muon spin rotation and relaxation in magnetic materials. *J. Phys. Condens. Matter* **1997**, *9*, 9113. [CrossRef]
31. Amato, A. Heavy-fermion systems studied by μSR technique. *Rev. Modern Phys.* **1997**, *69*, 1119. [CrossRef]
32. Uemura, Y.J.; Goko, T.; Gat-Malureanu, I.M.; Carlo, J.P.; Russo, P.L.; Savici, A.T.; Aczel, A.; MacDougall, G.J.; Rodriguez, J.A.; Luke, G.M.; et al. Phase separation and suppression of critical dynamics at quantum phase transitions of MnSi and $(Sr_{1-x}Ca_x)RuO_3$. *Nat. Phys.* **2007**, *3*, 29–35. [CrossRef]
33. Blundell, S.J. Spin-polarized muons in condensed matter physics. *Contemporary Phys.* **1999**, *40*, 175. [CrossRef]
34. Schenk, A. *Muon Spin Rotation Spectroscopy: Principles and Applications in Solid State*; Physics, Adam Hilger: Bristol, UK, 1985.
35. Yaouanc, A.; Dalmas de Reotier, P. *Muon Spin Rotation, Relaxation, and Resonance: Applications to Condensed Matter*; Oxford University Press: Oxford, UK, 2011.
36. Amato, A. Physics with Muons: from Atomic Physics to Condensed Matter Physics. Available online: https://www.psi.ch/en/lmu/lectures (accessed on 15 June 2020).
37. Amato, A.; Luetkens, H.; Sedlak, K.; Stoykov, A.; Scheuermann, R.; Elender, M.; Raselli, A.; Graf, D. The new versatile general purpose surface-muon instrument (GPS) based on silicon photomultipliers for μSR measurements on a continuous-wave beam. *Rev. Sci. Instrum.* **2017**, *88*, 093301. [CrossRef] [PubMed]
38. Kubo, R.; Toyabe, T. A stochastic model for low field resonance and relaxation. In *Magnetic Resonance and Relaxation*; North-Holland: Amsterdam, The Netherlands, 1967; pp. 810–823.
39. Guguchia, Z. Investigations of Superconductivity and Magnetism in Iron-Based and Cuprate High-Temperature Superconductors. Ph.D. Thesis, University of Zurich, Zürich, Switzerland, 2013.
40. Guguchia, Z.; Amato, A. SPG Mitteilungen Nr. 60, January 2020, p.12 ff. Available online: https://www.sps.ch/fileadmin/doc/Mitteilungen/Mitteilungen.60.pdf (accessed on 15 June 2020).
41. Salman, Z.; Prokscha, T.; Amato, A.; Morenzoni, E.; Scheuermann, R.; Sedlak, K.; Suter, A. Direct spectroscopic observation of a shallow hydrogen-like donor state in insulating $SrTiO_3$. *Phys. Rev. Lett.* **2014**, *113*, 156801. [CrossRef] [PubMed]
42. Khasanov, R.; Guguchia, Z.; Maisuradze, A.; Andreica, D.; Elender, M.; Raselli, A.; Shermadini, Z.; Goko, T.; Knecht, F.; Morenzoni, E.; et al. High pressure research using muons at the Paul Scherrer Institute. *High Press. Res.* **2016**, *36*, 140–166. [CrossRef]

43. Guguchia, Z.; Amato, A.; Kang, J.; Luetkens, H.; Biswas, P.K.; Prando, G.; von Rohr, F.; Bukowski, Z.; Shengelaya, A.; Keller, H.; et al. Direct evidence for a pressure-induced nodal superconducting gap in the $Ba_{0.65}Rb_{0.35}Fe_2As_2$ superconductor. *Nat. Commun.* **2015**, *6*, 8863. [CrossRef]
44. Kerelsky, A.; Nipane, A.; Edelberg, D.; Wang, D.; Zhou, X.; Motmaendadgar, A.; Gao, H.; Xie, S.; Kang, K.; Park, J.; et al. Absence of a Band Gap at the Interface of a Metal and Highly Doped Monolayer MoS_2. *Nano Lett.* **2017**, *17*, 5962–5968. [CrossRef] [PubMed]
45. Santos, E.J.G.; Ayuela, A.; Sanchez-Portal, D. First-principles study of substitutional metal impurities in graphene: Structural, electronic and magnetic properties. *New J. Phys.* **2010**, *12*, 053012. [CrossRef]
46. Oliva, R.; Wozniak, T.; Dybala, F.; Kopaczek, J.; Scharoch, P.; Kudrawiec, R. Hidden spin-polarized bands in semiconducting 2H-$MoTe_2$. *Mat. Res. Lett.* **2020**, *8*, 75–81. [CrossRef]
47. Avsar, A.; Ciarrocchi, A.; Pizzochero, M.; Unuchek, D.; Yazyev, O.V.; Kis, A. Defect induced, layer-modulated magnetism in ultrathin metallic $PtSe_2$. *Nat. Nanotechnol.* **2019**, *14*, 674–678. [CrossRef]
48. Bonilla, M.; Kolekar, S.; Ma, Y.; Coy Diaz, H.; Kalappattil, V.; Das, R.; Eggers, T.; Gutierrez, H.R.; Phan, M.-H.; Batzill, M. Strong room-temperature ferromagnetism in VSe_2 monolayers on van der Waals substrates. *Nat. Nanotechnol.* **2018**, *13*, 289. [CrossRef]
49. Zhang, F.; Zheng, B.; Sebastian, A; Olson, H.; Liu, M.; Fujisawa, K.; Pham, Y.T.H.; Jimenez, V.O.; Kalappattil, V.; Miao, L.; et al. Monolayer Vanadium-doped Tungsten Disulfide: A Room-Temperature Dilute Magnetic Semiconductor. *arXiv* **2020**, arXiv:2005.01965.
50. Pham, Y.T.H.; Liu, M.; Jimenez, V.O.; Zhang, F.; Kalappattil, V.; Yu, Z.; Wang, K.; Williams, T.; Terrones, M.; Phan, M.-H. Tunable Ferromagnetism and Thermally Induced Spin Flip in Vanadium-doped Tungsten Diselenide Monolayers at Room Temperature. *arXiv* **2020**, arXiv:2005.00493.
51. Duong, D.L.; Yun, S.J.; Kim, Y.; Kim, S.-G.; Lee, Y.H. Long-range ferromagnetic ordering in vanadium-doped WSe_2 semiconductor. *Appl. Phys. Lett.* **2019**, *115*, 242406. [CrossRef]
52. Gibertini,M.; Koperski, M.; Morpurgo, A.F.; Novoselov, K.S. Magnetic 2D materials and heterostructures. *Nat. Nanotechnol.* **2019**, *14*, 408–419. [CrossRef]
53. McGuire, M.A.; Dixit, H.; Cooper, V.R.; Sales, B.C. Coupling of crystal structure and magnetism in the layered, ferromagnetic insulator CrI_3. *Chem. Mat.* **2015**, *27*, 612–620. [CrossRef]
54. Prokscha, T.; Chow, K.H.; Stilp, E.; Suter, A.; Luetkens, H.; Morenzoni, E.; Nieuwenhuys, G.J.; Salman, Z.; Scheuermann, R. Photo-induced persistent inversion of germanium in a 200-nm-deep surface region. *Sci. Rep.* **2013**, *3*, 2569. [CrossRef]
55. Ding, C.; Man, H.; Qin, C.; Lu, J.; Sun, Y.; Wang, Q.; Yu, B.; Feng, C.; Goko, T.; Arguello, C.J.; et al. $(La_{1-x}Ba_x)(Zn_{1-x}Mn_x)AsO$: A two-dimensional 1111-type diluted magnetic semiconductor in bulk form. *Phys. Rev. B* **2013**, *88*, 041102(R). [CrossRef]
56. Dietl, T. A ten-year perspective on dilute magnetic semiconductors and oxides. *Nat. Mater* **2010**, *9*, 965–974. [CrossRef]
57. Masek, J.; Kudrnovsky, J.; Maca, F.; Gallagher, B.L.; Campion, R.P.; Gregory, D.H.; Jungwirth, T. Dilute Moment n-Type Ferromagnetic Semiconductor Li(Zn,Mn)As. *Phys. Rev. Lett.* **2007**, *98*, 067202. [CrossRef]
58. Glasbrenner, J.K.; Zutic, I.; Mazin, I.I. Theory of Mn-doped I-II-V Semiconductors. *Phys. Rev. B* **2014**, *90*, 140403(R). [CrossRef]
59. Hasan, M.Z.; Kane, C.L. Colloquium: Topological insulators. *Rev. Mod. Phys.* **2010**, *82*, 3045–3067. [CrossRef]
60. Ando, Y.; Fu, L. Topological Crystalline Insulators and Topological Superconductors: From Concepts to Materials. *Annu. Rev. Condens. Matter Phys.* **2015**, *6*, 361. [CrossRef]
61. Eremeev, S.V.; Otrokov, M.M.; Chulkov, E.V. New Universal Type of Interface in the Magnetic Insulator/Topological Insulator Heterostructures. *Nano Lett.* **2018**, *18*, 6521–6529. [CrossRef]
62. Chen, J.; Wang, L.; Zhang, M.; Zhou, L.; Zhang, R.; Jin, L.; Wang, X.; Qin, H.; Qiu, Y.; Mei, J.; et al. Evidence for Magnetic Skyrmions at the Interface of Ferromagnet/Topological-Insulator Heterostructures. *Nano Lett.* **2019**, *19*, 6144–6151. [CrossRef]
63. Krieger, J.A.; Pertsova, A.; Giblin, S.R.; Dobeli, M.; Prokscha, T.; Schneider, C.W.; Suter, A.; Hesjedal, T.; Balatsky, A.V.; Salman, Z. Proximity-induced odd-frequency superconductivity in a topological insulator. *arXiv* **2020**, arXiv:2003.12104.
64. Xu, S.-Y.; Alidoust, N.; Belopolski, I.; Richardella, A.; Liu, C.; Neupane, M.; Bian, G.; Huang, S.-H.; Sankar, R.; Fang, C.; et al. Momentum-space imaging of Cooper pairing in a half-Dirac-gas topological superconductor. *Nat. Phys.* **2014**, *10*, 943–950. [CrossRef]

65. Jackson, T.J.; Riseman, T.M.; Forgan, E.M.; Gluckler, H.; Prokscha, T.; Morenzoni, E.; Pleines, M.; Niedermayer, C.; Schatz, G.; Luetkens, H.; et al. Depth-Resolved Profile of the Magnetic Field beneath the Surface of a Superconductor with a Few nm Resolution. *Phys. Rev. Lett.* **2000**, *84*, 4958. [CrossRef]
66. Suter, A.; Wojek, B. Musrfit: A free platform-independent framework for μSR data analysis. *Phys. Procedia* **2012**, *30*, 69. [CrossRef]

© 2020 by the author. Licensee MDPI, Basel, Switzerland. This article is an open access article distributed under the terms and conditions of the Creative Commons Attribution (CC BY) license (http://creativecommons.org/licenses/by/4.0/).

Article

First-Principles Calculation of Copper Oxide Superconductors That Supports the Kamimura-Suwa Model

Hiroshi Kamimura [1,*], Masaaki Araidai [2], Kunio Ishida [3], Shunichi Matsuno [4], Hideaki Sakata [5], Kenji Shiraishi [2], Osamu Sugino [6] and Jaw-Shen Tsai [5,7]

1. Tokyo University of Science,1-3 Kagurazaka, Shinjuku-ku, Tokyo 162-8601, Japan
2. Institute of Materials and Systems for Sustainability, Nagoya University, Furo-cho, Chikusa-ku, Nagoya 464-8601, Japan; araidai@nagoya-u.jp (M.A.); shiraishi@cse.nagoya-u.ac.jp (K.S.)
3. Graduate School of Engineering, Utsunomiya University, Yoto, Utsunomiya, Tochigi 321-8585, Japan; ishd_kn@cc.utsunomiya-u.ac.jp
4. School Marine Science and Technology, Tokai University, Shimizu 424-8610, Japan; smatsuno@scc.u-tokai.ac.jp
5. Department of Physics, Faculty of Science, Tokyo University of Science,1-3 Kagurazaka, Shinjuku-ku, Tokyo 162-8601, Japan; sakata@rs.kagu.tus.ac.jp (H.S.); tsai@riken.jp (J.-S.T.)
6. Institute for Solid State Physics, The University of Tokyo, 5-1-5 Kashiwanoha, Kashiwa, Chiba 277-8581, Japan; sugino@issp.u-tokyo.ac.jp
7. RIKEN Center for Emergent Matter Science (CEMS), Wako, Saitama 351-0198, Japan
* Correspondence: kamimura@rs.kagu.tus.ac.jp

Received: 27 September 2020; Accepted: 22 October 2020; Published: 2 November 2020

Abstract: In 1986 Bednorz and Műller discovered high temperature superconductivity in copper oxides by chemically doping holes into La_2CuO_4 (LCO), the antiferromagnetic insulator. Despite intense experimental and theoretical research during the past 34 years, no general consensus on the electronic-spin structures and the origin of pseudogap has been obtained. In this circumstance, we performed a first-principles calculation of underdoped cuprate superconductors $La_{2-x}Sr_xCuO_4$ (LSCO) within the meta-generalized gradient approximation of the density functional theory. Our calculations clarify first the important role of the anti Jahn-Teller (JT) effect, the backward deformation against the JT distortion in La_2CuO_4 by doping extra holes. The resulting electronic structure agrees with the two-component theory provided by the tight-binding model of Kamimura and Suwa (K–S), which has been also used to elucidate the d-wave superconductivity. Our first-principles calculation thus justifies the K–S model and demonstrates advanced understanding of cuprates. For example, the remarkable feature of our calculations is the appearance of the spin-polarized band with a nearly flat-band character, showing the peaky nature in the density of states at the Fermi level.

Keywords: cuprates; LSCO; anti-Jahn-Teller effect; first-principles calculation; Kamimura-Suwa model; spin-polarized band; Hund's coupling spin-triplet and spin-singlet multiplets

1. Prologue by Hiroshi Kamimura

In 1987, Georg Bednorz and Alex Műller received the Nobel Prize in Physics for "their discovery of new superconducting materials". In 1988, the Steering Committee of NEC Symposium on Fundamental Approaches to the New Material Phases decided to choose "Mechanism of High Temperature Superconductivity" as the subject of the second NEC Symposium. This Symposium was held in Hakone, near Mt. Fuji, Japan on 24–27 October 1988. Alex was chosen as a plenary speaker, and I was the Symposium chairperson. The Symposium's group photo is shown in Figure 1.

Figure 1. Symposium's group photo in the front of the Hakone Kanko Hotel.

In 2005, we organized the International Workshop on Electronic Structure and Lattice Effects in Cuprates on October 27 and 28 in the National Institute of Advanced Industrial Science and Technology (AIST) in Tsukuba. Alex was a plenary speaker, and a number of Alex's friends, companions, colleagues, and former students attended this Workshop (see Figure 2).

Figure 2. International Workshop's group photo in the front of the Conference Hall in AIST.

On 27–28 March 2006, the International Symposium in Honor of J.G. Bednorz and K.A. Műller was held in University of Zurich, celebrating their discovery of cuprate superconductors 20 years ago. From Japan, six HTSC researchers, Yoichi Ando, Masatoshi Arai, Yoshiteru Maeno, Hiroyuki Oyanagi, Masaki Takashige, and Hiroshi Kamimura were invited. The group photo is shown below in Figure 3.

Figure 3. International Symposium' group photo. The front row: From the left, Prof. Hugo Keller, Prof. Alex Műller, Dr. Georg Bednorz, Prof. Hiroshi Kamimura.

2. Introduction

By the discovery of high-temperature superconductivity (HTSC) in cuprates by Bednorz and Műller [1,2], it is well known that cuprate superconductors are different from ordinary metallic superconductors in the production process of carriers. Cuprates are ionic crystals. The carriers in La_2CuO_4, an antiferromagnetic insulator [3], are produced from doping chemical elements or making oxygen deficiencies, for example, by replacing 3+ cations (La^{3+}) by 2+ ones (Sr^{2+}) in $La_{2-x}Sr_x CuO_4$ (LSCO). Thus, the electronic structures of cuprate superconductors may vary drastically upon doping holes [4].

In this circumstance, we pay attention to the experimental facts that a CuO_6 octahedron in a parent material La_2CuO_4 is deformed into an elongated shape (tetragonal) due to the Jahn–Teller (JT) theorem for an orbitally doubly degenerate state of a Cu^{2+} ion [1,5]. When extra holes are doped, the apical O^{2-} ions in elongated CuO_6 octahedrons in LCO tend to approach toward Cu^{2+} ions to gain the attractive electrostatic energy [6]. This backward deformation against the JT deformations is called the anti-JT effect [7,8]. As a result, the energy separation between the two JT-split levels becomes smaller, i.e., pseudo-degenerate. This anti-JT effect in cuprates was supported experimentally by the neutron scattering [9] and the polarization-dependent spectroscopy [10,11].

Under these circumstances, there are two views concerning the electronic structures of LSCO. One view is based on the single-component theory in which only orbitals extended in a CuO_2 plane are considered. Since a doped hole moves within a CuO_2 plane, this model is irrelevant to the anti-JT effect. A typical model to take this view is the t-J model [12,13].

An alternative view is based on the two-component theory initiated by the Kamimura–Suwa (K–S) model [14,15], in which the two kinds of molecular orbitals (MOs) of each CuO_6 octahedron parallel and perpendicular to a CuO_2 plane are considered. Those MOs are $|a^*_{1g}>$ antibonding and $|b_{1g}>$ bonding orbitals, as shown in Figure 4. In this figure, the schematic MO levels, b_{1g} bonding, a^*_{1g} antibonding, and b^*_{1g} antibonding orbitals, and the hole configurations in Hund's coupling multiplet $^3B_{1g}$ and spin-singlet multiplet $^1A_{1g}$ are shown. On the left side of the figure, a hole occupying a^*_{1g} orbital (red arrow) aligns its spin in parallel to the up spin of a localized hole occupying the b^*_{1g} orbital (green arrow) to form $^3B_{1g}$, owing to the Hund's coupling exchange interaction with the coupling constant, K_{a1g} (−2 eV) [16].

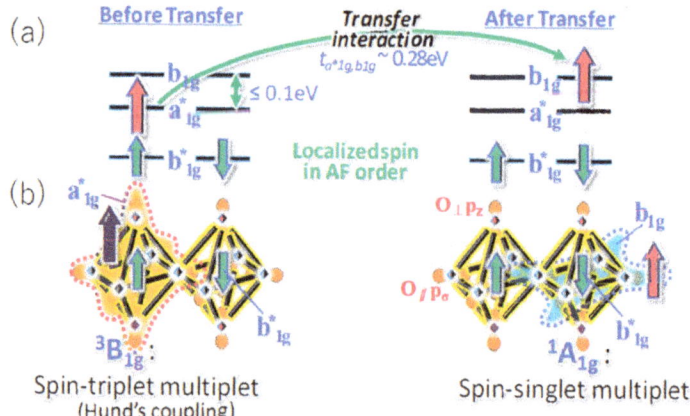

Figure 4. The K–S model: the coexistence of a metallic state and the AF order. In this figure, (a) the transfer (*a long curved arrow*) of a doped hole (red arrows) from $^3B_{1g}$ to $^1A_{1g}$ without destroying the AF order is shown, and (b) the spatial extension of a^*_{1g} MO (left side) and b_{1g} MO (right side) are shown, where $O_{/\!/}$ and $O_\perp$ represent in-plane and apical oxygen, respectively.

On the other hand, on the right side of the figure, a hole occupying the b_{1g} level (up red arrow) forms a spin-singlet (SS) multiplet with the down spin of a localized hole occupying the b^*_{1g} level (green arrow), $^1A_{1g}$, owing to the larger exchange coupling K_{b1g} (= 4 eV) [16]. The key-feature of the K–S model is the coexistence of a metallic state and the AF order.

The above two types of exchange interaction, the Hund's coupling spin-triplet and the spin-singlet (SS), can be described by the scalar product of the spin of the localized $|b^*_{1g}>$ hole, S_i, and that of the itinerant $|a^*_{1g}>$ or $|b_{1g}>$ hole, $s_{i,a^*_{1g}}$ or $s_{i,b_{1g}}$, at each CuO_6 site, i, yielding the Hamiltonian interaction

$$H_{ex} = \sum_i \left(K_{a^*_{1g}} s_{i,a^*_{1g}} \cdot S_i + K_{b_{1g}} s_{i,b_{1g}} \cdot S_i \right). \tag{1}$$

This Hamiltonian interaction constitutes the K–S Hamiltonian to describe the K–S model, in which the localized spins form an AF order [14,15]. The whole K–S Hamiltonian is shown below.

$$H = H_{orbital} + H_{tr} + H_{AF} + H_{ex}$$
$$= \sum_{im\sigma} \varepsilon_m c^+_{im\sigma} c_{im\sigma} + \sum_{(i,j)mn} t_{mn}(c^+_{im\sigma} c_{jn\sigma} + h.c.) + J \sum_{(i,j)} S_i \bullet S_j + \sum_{im} K_m s_{im} \bullet S_i, \tag{2}$$

where ε_m ($m = a_{1g}{}^*$, or b_{1g}) represents the one-electron energy of the a^*_{1g} or b_{1g} MO states, $C^\dagger_{im\sigma}$ and $C_{im\sigma}$ are the creation and annihilation operators of a doped hole in the m-type MO with spin σ in the i-th CuO_6 octahedron, respectively, t_{mn} is the transfer integral of a doped hole between the m- and n- type MOs of neighboring CuO_6 octahedrons, J the superexchange interacrion (as regards the details of the K–S Hamiltonian, see Ref. [17]).

3. First-Principles Calculations

In order to decide whether the one-component t–J model or the two-component K–S model is suitable to describe the electronic-spin state of cuprates, Kamimura et al. [18] recently performed the first-principles non-empirical calculation. In their calculations, the first-principle method of the constrained-and-appropriately normed (SCAN) density functional was adopted to calculate a non-rigid electronic-spin energy bands of LCO and LSCO, following Furness et al. [19]. Kohn-Sham density functional theory (DFT) is a widely-used electronic structure theory for materials as well as condensed matters. This SCAN method is more useful for the electronic-spin structures of diversely-bonded materials (including covalent, metallic, ionic, Jahn-Teller deformed bonds, cuprates consisting of CuO_6 octahedrons and CuO_5 pyramids).

The purpose of this contribution is to show how the first-principles SCAN calculations agree with the K–S model in the tight binding form. In the first-principles calculations, a unique method was adopted—that is, to investigate the electronic states of LSCO by calculating both the energy bands in k-space and the wavefunctions in r-space simultaneously. For this purpose, we calculated not only the non-rigid energy band and the density of states (DOS) in k-space but also the partial DOS (PDOS) which gives information on MOs in a CuO_6 octahedron in r-space. Here PDOS means the density of states projected to the wavefunctions, or orbitals.

4. Calculated Results

4.1. La_2CuO_4 (LCO): Antiferromagnetic (AF) Insulator

The La_2CuO_4 (LCO) transforms from the high temperature tetragonal phase (HTT) into the low temperature orthorhombic phase (LTO) when temperature decreases. The calculated results on the crystal structure explains the Jahn–Teller distortion of an elongated octahedron. The calculated Cu magnetic moment, 0.510 μ_B, is in agreement with the observed values of 0.50 μ_B [20,21]. This means that the DFT-SCAN method [18,19] treats the spin-fluctuation effect properly. The calculated result shows that LCO is the antiferromagnetic insulator.

4.2. $La_{2-x}Sr_x CuO_4$ (LSCO)

In order to calculate the non-rigid energy bands for LSCO with AF order [22], the $2\sqrt{2} \times 2\sqrt{2}$ supercell is adopted, in which there are two CuO_2 planes and each CuO_2 plane includes eight CuO_6 octahedrons. The Brillouin zone (BZ) for this $2\sqrt{2} \times 2\sqrt{2}$ supercell is shown in Figure 5. In this supercell, we cannot discuss the stability of stripe phases which were recently reported by Zhang et al. for YBCO systems [23].

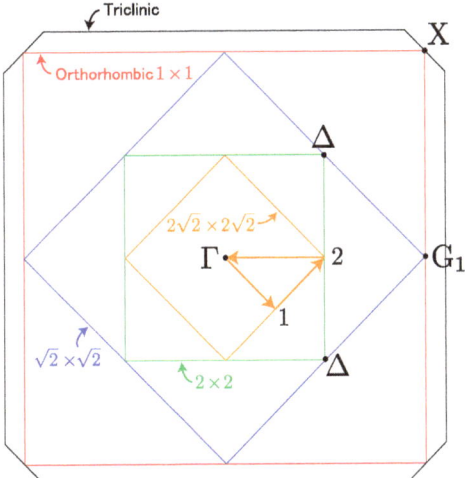

Figure 5. The Brillouin zone (BZ) for the $2\sqrt{2} \times 2\sqrt{2}$ supercell. Inside the BZ, a special triangle root starting form Γ is shown, where the special points 1 and 2 are used later in Figure 10.

Here we show that the key points obtained from the first-principles calculations for LSCO agree with those of the K–S model shown in Figure 4. In particular, it is shown that the Hund's coupling spin-triplet in the K–S model agrees with the appearance of a spin-polarized band in cuprates.

4.3. A Metallic Phase with x = 0.125 (the Underdoped Regime)

In this case, Sr is substituted for one La in each CuO_2 plane in the supercell, leading to $x = 0.125$. The calculated Cu and apical O shortest distance in a CuO_6 octahedron in LSCO is 2.216 Å, which is considerably smaller than 2.448 Å in LCO. This shrinkage of the Cu and apical O distance by doping holes indicates that the anti-JT effect really occurs in LSCO. This result also supports the K–S model.

Although the calculated magnetization at each Cu site suggests the existence of AF order, the calculated value at each site takes a different value. For example, among the calculated values of the magnetization at each Cu site in Figure 6, the absolute value at Cu_4 site shows the maximum value of 0.643 μ_B, which is considerably larger than the magnetization at a Cu site in LCO, 0.510 μ_B. Here we discuss the origin of the high value of magnetization, 0.643 μ_B in LSCO with $x = 0.125$.

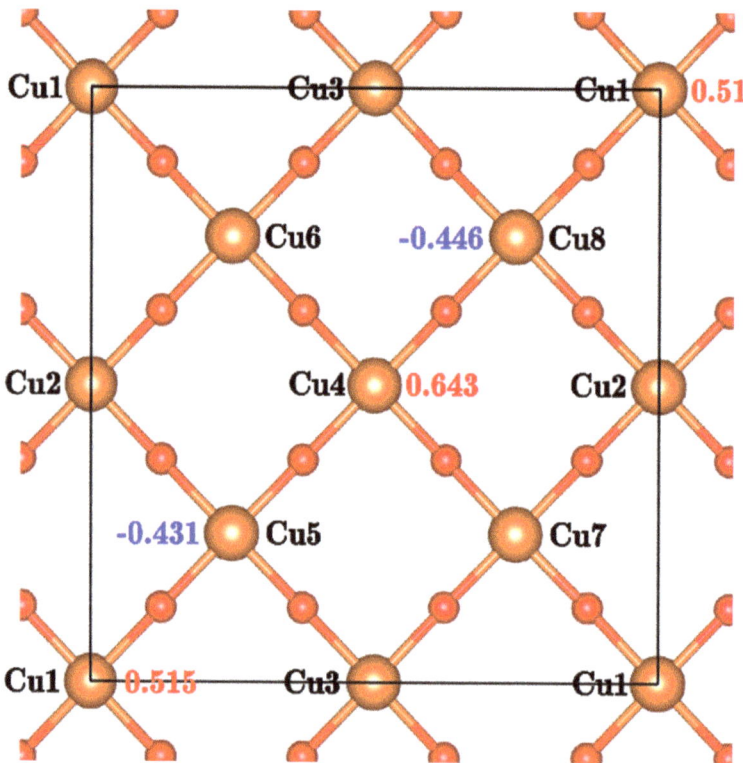

Figure 6. The calculated magnetization at each Cu site in $La_{2-x}Sr_x CuO_4$ (LSCO) with $x = 0.125$.

In order to obtain information on the wavefunctions in a CuO_6 octahedron, the PDOS (partial density of states) of dz^2 orbital and its ligand O p_z orbitals in a CuO_6 octahedron at the Cu_4 site is calculated. The calculated results are shown in Figure 7. Using these orbitals in a CuO_6 octahedron at the Cu_4 site, an antibonding $|a^*_{1g}\rangle$ MO is constructed, in order to investigate a relation between the $|a^*_{1g}\rangle$ orbital in the energy band and the antibonding $|a^*_{1g}\rangle$ MO in the K–S model in Figure 4, where Cu in CuO_6 octahedron is assigned to Cu_4.

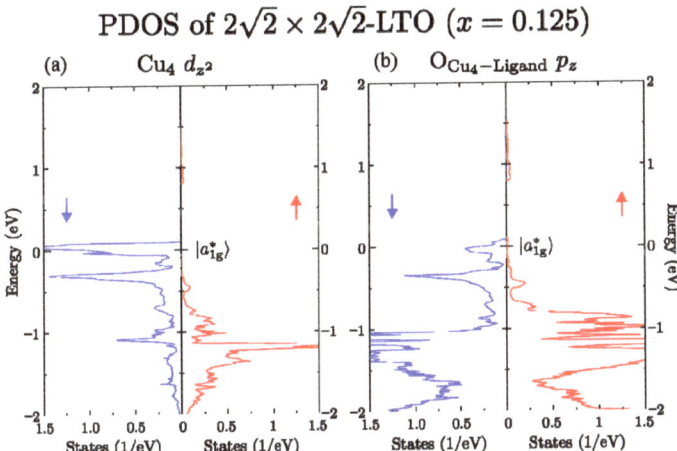

Figure 7. Calculated partial density of states (PDOS) of orbitals in CuO$_6$ at Cu$_4$ site: energy 0 is E_F. (a) PDOS of Cu-dz^2 orbital; (b) PDOS of ligand apical O-p$_z$ orbital.

4.4. The Close Relation between a Spin-Polarized Band in the First Principles Calculations and Hund's Coupling Spin Triplet in the K–S Model

4.4.1. The Appearance of a Spin-Polarized Band in Cuprates

Another remarkable feature of the highest hole band is the spin polarization at the Cu$_4$ site (Figure 7). The difference in the highest states between up-spin and down-spin bands is considered as the "ferromagnetic band splitting" discussed by Slater [24], and Connolly [25]. Indeed, according to the criterion for the occurrence of a spin-polarized band in LSCO derived in the paper [18], the Hund's coupling exchange interaction $K_{a_{1g}^*}(x)\vec{s}_{i,a_{1g}^*} \cdot \vec{S}_i$ in Equation (1) may cause a spin-polarized band when $K(x)D(E_F) > 1$, where $K(x) = |K_{a*1g}(x)|$ and $D(E_F)$ is the density of states at E_F. The magnitude of $K(x)$ is 2.0 eV for LCO ($x = 0$) and is larger in LSCO because of the anti-JT effect. Figure 7 shows that $D(E_F)$ is much larger than 1.5 states/eV. Thus, the condition for the appearance of the spin-polarized band is satisfied. We also notice that the PDOS of the highest down-spin Cu$_4$ dz^2 hole band in Figure 7a, i.e., the spin-polarized band, shows a "peaky" nature due to the existence of a nearly-flat highest hole band, which originates from the JT and the anti-JT deformations. In fact, the effective mass of a hole carrier in this band is six times heavier than the free electron mass. Thus, we consider that this hole carrier is a "Jahn-Teller polaron", proposed by Alex [26–28]. We expect that the peaky nature may be observed for underdoped LSCO by performing an STM or STS measurement

4.4.2. Relation between the Cu$_4$ dz^2 Band and the Hund's Coupling Spin-Triplet in the K–S Model

Now, in order to investigate whether a doped hole in the Cu$_4$ dz^2 band($|a^*_{1g}>$ MO) forms the Hund's coupling spin-triplet $^3B_{1g}$ with a localized spin at the same Cu$_4$ site, we calculated the PDOS of Cu$_4$ dx^2-y^2 and ligand O-p$_x$, p$_y$ orbitals in x-y plane, which form the antibonding $|b^*_{1g}>$ MO and the bonding $|b_{1g}>$ MO in a CuO$_6$ octahedron. The calculated results are shown in Figure 8. The Cu$_4$ dx^2-y^2 orbital and the ligand O-p$_x$, p$_y$ orbitals that appear in the energy region of 1 to 2eV above E_F in Figure 8 form the antibonding $|b^*_{1g}>$ MO. This MO is occupied by the localized holes with down spins. Since both spins of the localized holes in $|b^*_{1g}>$ MO and of the doped holes in the Cu$_4$ dz^2 band ($|a^*_{1g}>$ MO) take the same direction, down spins, the total magnetization at the Cu$_4$ site becomes the maximum value of 0.643μ_B. Thus, one can conclude that the Hund's coupling between a localized hole in $|b^*_{1g}>$ MO and a doped hole in the Cu$_4$ dz^2 band ($|a^*_{1g}>$ MO) plays an important role in LSCO with x = 0.125 in the first-principles calculations.

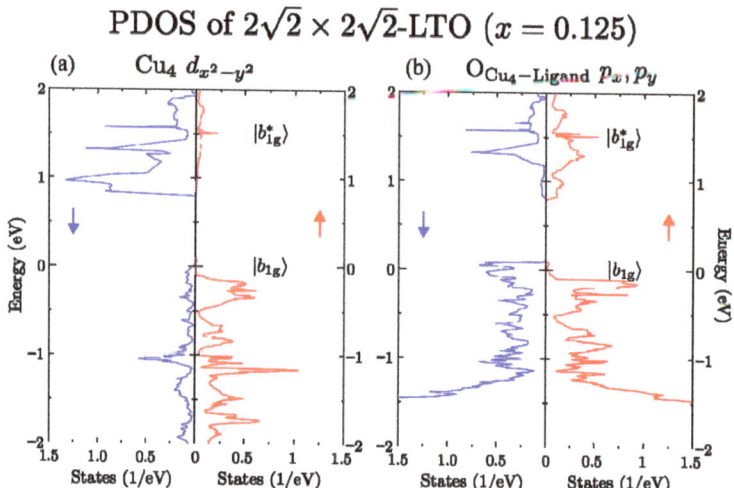

Figure 8. Calculated PDOS of orbitals in CuO$_6$ at Cu$_4$ site: Energy 0 is E_F. (**a**) PDOS of Cu-dx^2-dy^2 orbital; (**b**) PDOS of ligand in-plane-O p$_x$, p$_y$ orbitals.

4.4.3. Itinerancy of a Hole in the Cu$_4$ dz^2 Band

The present first-principle calculations also show how a doped hole in the Cu$_4$ dz^2 band (|a$^*_{1g}$> MO) itinerates in LSCO in the underdoped regime.

Let us pay attention to the hopping of a doped hole with down spin from a Cu$_4$ site to a neighboring Cu$_8$ site in the AF order, whose magnetization, $-0.446\mu_B$, is opposite to that at the Cu$_4$ site, $+0.643$ μ_B. We calculated the PDOS of Cu-dx^2-y^2 orbital and ligand O-p$_x$, p$_y$ orbitals in the x-y plane at the neighboring Cu$_8$ site, as shown in Figure 9. In this figure, we find that these orbitals form bonding b$_{1g}$ and antibonding b$^*_{1g}$ MOs in a CuO$_6$ octahedron at the Cu$_8$ site. The |b$^*_{1g}$> MO with Cu-dx^2-y^2 (a localized hole) appears in the energy region of 1 to 2 eV above E_F (red color) in Figure 9a while the |b$_{1g}$> MO (a doped hole) appears just above E_F. By comparing the calculated results of the spin-direction of a localized hole in Figure 9a (red color) with that in Figure 8a (blue color), we notice that the localized spins at the Cu$_8$ site are directed upwards, while those at the Cu$_4$ site are opposite in the spin directions. Thus, we conclude that the AF order is not destroyed during the itinerancy of a doped hole in LSCO.

Furthermore, when a hole with down spin in the Cu$_4$ dz^2 band (|a$^*_{1g}$ > MO) at the Cu$_4$ site hops into a neighboring Cu$_8$ site in the AF order, the first-principles calculation shows that this hole enters an empty |b$_{1g}$ > MO just above E_F at the Cu$_8$ site, taking a spin antiparallel configuration (the spin-singlet) with the localized hole in |b$^*_{1g}$ > MO at the same site (see Figure 9).

4.4.4. Good Agreement between the First-Principles SCAN Calculations and the K–S Model in the Tight-Binding Form

Summarizing the itinerancy of a doped hole from Cu$_4$ dz^2 orbital to Cu$_8$ dx^2-y^2 orbital, the spin-state of the doped hole changes from the Hund' coupling spin-triplet with the localized spin at Cu$_4$ site to the spin-singlet with the localized spin at Cu$_8$ site.

This itinerancy of a doped hole in the first-principles SCAN calculations is strikingly consistent with the K–S model. In the K–S model (see Figure 4), when a doped hole with up spin enters a Cu site with a up localized spin in b$^*_{1g}$ MO, it occupies an a$^*_{1g}$ MO to align its spin in parallel with the up spin of a localized hole in the b$^*_{1g}$ MO to form ^{3}B$_{1g}$, owing to the Hund's coupling exchange interaction, and owing to the transfer interactions, hops to the b$_{1g}$ MO of a neighboring Cu site in the AF order taking antiparallel spin configuration with the localized spin. By repeating this transfer process, a doped hole itinerates in LSCO without destroying the AF order. We call this doped hole a "KS particle".

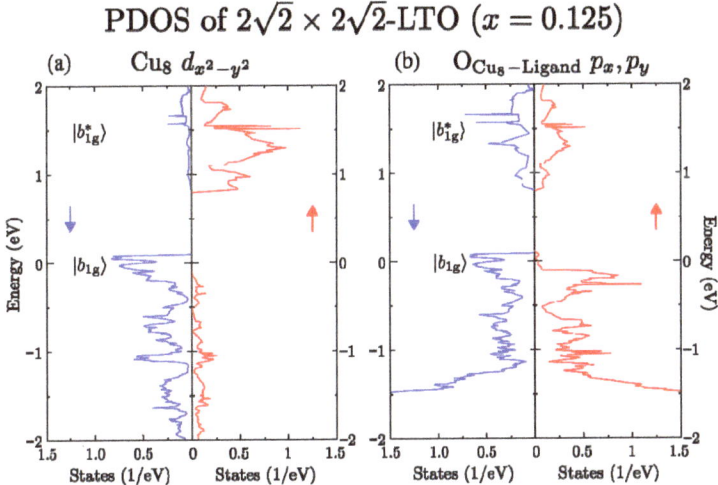

Figure 9. Calculated PDOS of orbitals in CuO_6 at Cu_8 site: (a) PDOS of Cu-dx^2-dy^2 orbital; (b) PDOS of ligand in-plane-O p_x, p_y orbitals. Energy 0 is E_F.

Thus, we understand that the first-principle SCAN calculation is in good agreement with the K–S model in the tight-binding form [14,15]. Namely, in LSCO a doped hole itinerates by taking the Hund's coupling and the spin-singlet states alternately without destroying the AF order.

4.5. Construction of Fermi Surfaces of LSCO with x = 0.125

Based on the present first-principle SCAN calculated results for $x = 0.125$, we construct a Fermi surface (FS) in the underdoped LSCO. In doing so, we would like to point out one problem in our supercell model. In the supercell, a size of a unit cell is wider than that of an ordinal LSCO crystal with the LTO phase. As seen in BZs for various supercell-models in Figure 5, the BZ of a supercell is obtained by folding that of a regular crystal, as shown in Figure 5. As a result, in the case of $2\sqrt{2} \times 2\sqrt{2}$ model, the symmetry points $\Gamma(0, 0)$, $X(\pi,\pi)$, $G_1(\pi, 0)$, and $\Delta(\pi/2, \pi/2)$ gather at Γ. Thus, one may have a difficulty in assigning each calculated energy band to either of symmetry points Γ, X, G_1, and Δ.

As an example, we show the calculated energy bands and DOS of LSCO with $x = 0.125$ in Figure 10. In this figure one may notice that there are two hole-bands above E_F, and that the DOS of the second highest hole-band is very large and peaky shape. Since the peaky shape of DOS is originated from the spin-polarized band $|a^*_{1g}>$ in Figure 7 and the spin-polarized band appears at $G_1(\pi, 0)$ in k space, we assign this hole band to a symmetry point $G_1(\pi, 0)$. On the other hand, we also pay attention to a doped hole in the highest hole-band ($|b_{1g}>$) and a localized hole in the band ($|b^*_{1g}>$) in the energy 1 to 2 eV above E_F in Figure 8b. These holes form an SS (spin-singlet) particle (see Figure 4), and contribute to the formation of Fermi arcs at the Δ point in BZ.

Based on the above-mentioned results of the first-principles calculations, we construct the Fermi surface of LSCO with $x = 0.125$. The feature of the thus-obtained Fermi surface (FS) of LSCO with $x = 0.125$ in the underdoped regime is the coexistence of Fermi pockets (G_1 points) and Fermi arcs (Δ points), and the KS particles occupy the Fermi pockets while the SS particles occupy the Fermi arcs, as shown in Figure 11. This feature seems consistent with the experimental results of angle-resolved photoemission spectroscopy (ARPES) for LSCO with $x = 0.15$ by Yoshida et al. [29].

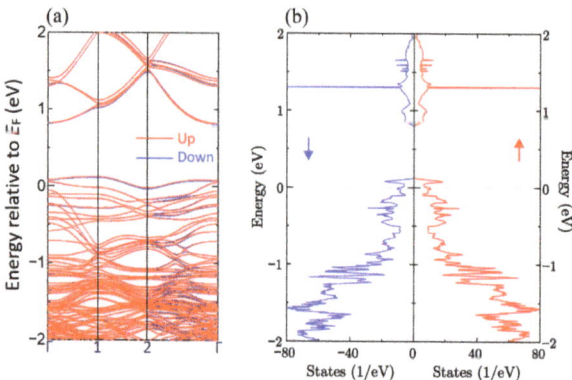

Figure 10. Spin-dependent Non-rigid energy bands and DOS of LSCO with $x = 0.125$. E_F is energy zero. (a) Energy bands. (b) DOS.

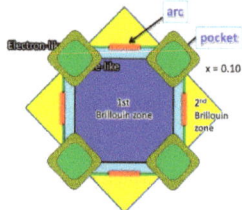

Figure 11. Fermi surface of LSCO with $x = 0.10$ (the underdoped regime).

5. Final Remarks

The feature of the obtained Fermi surface (FS) of LSCO with $x = 0.125$ in the underdoped regime mentioned in Section 4.5 is the coexistence of Fermi pockets (G_1 points) and Fermi arcs (Δ points), and the KS particles occupy the Fermi pockets while the SS particles occupy the Fermi arcs. We find that the FS for the underdoped LSCO (Figure 10) is similar to that of the K–S model [14,15], which is shown in Figure 11. This feature is likely to be consistent with the experimental results of angle-resolved photoemission spectroscopy (ARPES) for LSCO with $x = 0.15$ by Yoshida et al. [27]. Here, the wavefunctions of a KS particle with up and down spins, $\Psi_{k,\uparrow}(r)$ and $\Psi_{k,\downarrow}(r)$, have the unique phase relation [30],

$$\Psi_{k,\downarrow}(r) = \exp(i\,k\cdot a)\,\Psi_{k,\uparrow}(r), \qquad (3)$$

where a is a vector connecting Cu atoms along the Cu-O-Cu line.

Thus, the KS particles on the Fermi pockets around $G_1(\pi, 0)$ points contribute to the d-wave superconductivity in the phonon mechanism below Tc [30–33] (see also [4]).

6. Conclusions

In the present paper, we have shown that the K–S model is consistent with the first-principle method of the constrained-and-appropriately normed (SCAN) density functional. Furthermore, the Fermi-surfaces of LSCO in the underdoped region obtained from the first-principle SCAN density functional method coincide with those derived from the K–S model semi-empirically, so that one may consider that the theory of d-wave superconductivity developed by the K–S model [30] is justified by the first-principles theory.

Author Contributions: All authors contributed extensively to the work presented in this paper. M.A. carried out the DFT-SCAN computations. H.K., M.A. K.S., O.S., K.I. developed the non-rigid energy-band theory of correlated electrons in cuprates, especially the spin-dependent band theory, and compared with the K–S model in the tight binding form. S.M. developed the *d*-wave superconductivity based on the K–S model, H.S. and J.-S.T. compared theoretical results with experiments on STM, STS, interlayer tunneling experiments in cuprates. H.K. mainly wrote the manuscript. All authors discussed the results and commented on the manuscript. All authors have read and agreed to the published version of the manuscript.

Funding: This research received no external funding.

Acknowledgments: At this happy occasion of dedicating the present paper to Alex, I, Hiroshi kamimura, sincerely thank Alex for guiding me to our research on cuprates of the Jahn-Teller materials. We would like to thank Atsushi Fujimori, Kazuyoshi Yamada, Setsuo Mitsuda for their valuable discussion on experimental results. This work was supported by Tokyo University of Science. The computation was carried out in part using the computer resource offered under the category of General Projects by Research Institute for Information Technology, Kyushu University.

Conflicts of Interest: The authors declare no conflict of interest.

References

1. Bednorz, J.G.; Müller, K.A. Possible high Tc superconductivity in the Ba-La-Cu-O system. *Z. Phys. B* **1986**, *64*, 189–193.
2. Bednorz, J.G.; Müller, K.A. Perovskite-type oxides-the new approach to high- T_c superconductivity. *Rev. Mod. Phys.* **1988**, *60*, 585. [CrossRef]
3. Anderson, P.W. Resonating valence bond state in La_2CuO_4 and superconductivity. *Science* **1987**, *235*, 1196–1198. [CrossRef] [PubMed]
4. Kamimura, H.; Sugino, O.; Tsai, J.S.; Ushio, H. Jahn-Teller-effect induced superconductivity in copper oxides: Theoretical developments. In *High-T_c Copper Oxide Superconductors and Related Novel Materials, Dedicated to Prof. K.A. Müller on the Occasion of his 90th Birthday*; Bussmann-Holder, A., Keller, H., Bianconi, A., Eds.; Springer: Berlin/Heidelberg, Germany, 2017.
5. Sugano, S.; Tanabe, Y.; Kamimura, H. *Multiplets of Transition Metal U., and Ions in Crystals*; Chapter X; Academic Press: Cambridge, MA, USA, 1970.
6. Shima, N.; Shiraishi, K.; Nakayama, T.; Oshiyama, A.; Kamimura, H. Electronic structures of doped $(La_{(1-x)}Sr_x)_2CuO_4$.in tetragonal phase. In Proceedings of the 1st International Conference on Electronic Materials Tokyo, June 1988: New Materials and New Physical Phenomena for Electronics of the 21st Century, Shigaku-Kaikan, Tokyo, Japan, 13–15 June 1988; Sugano, T., Chang, R.P.H., Kamimura, H., Hayashi, I., Kamiya, T., Eds.; Material Research Society: Pittsburgh, PA, USA, 1989; pp. 51–54.
7. Anisimov, V.L.; Ezhov, S.Y.; Rice, T.H. Singlet and triplet hole-doped configuration in $La_2Cu_{0.5}Li_{0.5}O_4$. *Phys. Rev. B* **1997**, *55*, 12829–12832. [CrossRef]
8. Kamimura, H.; Matsuno, S.; Mizokawa, T.; Sasaoka, K.; Shiraishi, K.; Ushio, H. On the important role of the anti-Jahn-Teller effect in underdoped cuprate superconductors. *J. Phys. Conf. Ser.* **2013**, *428*, 012043. [CrossRef]
9. Egami, T.; Toby, B.H.; Billinge, S.J.L.; Janot, C.; Jorgensen, J.D.; Hinks, D.G.; Subramanian, M.A.; Crawford, M.K.; Farneth, W.E.; McCarron, E.M. Local structural anomaly at T_c observed by neutron scattering. 389. In *High Temperature Superconductivity*; Ashkenazi, J., Barnes, J., Vezzoli, G.C., Klein, B.M., Eds.; Springer Science + Business Media, LLC.: New York, NY, USA, 1991.
10. Chen, C.T.; Tjeng, L.H.; Kwo, H.; Kao, L.; Rudolf, P.; Sette, P.; Fleming, R.M. Out-of-plane orbital characteristics of intrinsic and doped holes in $La_{2-x}Sr_xCuO_4$. *Phys. Rev. Lett.* **1992**, *68*, 2543–2546. [CrossRef] [PubMed]
11. Pellegrin, E.; Nücker, N.; Fink, J.; Molodtsov, S.L.; Gutierre, A.; Navas, E.; Sterebel, O.; Hu, Z.; Domke, M.; Kaindl, G.; et al. Greene, Orbital character of states at the Fermi level in $La_{2-x}Sr_xCuO_4$ and $R_{2-x}Ce_xCuO_4$ (R = Nd_2Sm). *Phys. Rev. B* **1993**, *47*, 3354–3367. [CrossRef]
12. Zhang, F.C.; Rice, T.M. Effective Hamiltonian for the superconducting Cu oxides. *Phys. Rev. B* **1988**, *37*, 3759–3761. [CrossRef]
13. Norman, M.R.; Kanigel, A.; Randeria, M.; Chtterjee, M.; Campuzano, J.C. Modeling the Fermi arc in underdoped cuprates. *Phys. Rev. B* **2007**, *76*, 174501. [CrossRef]
14. Kamimura, H.; Suwa, Y. New theoretical view for high temperature superconductivity. *J. Phys. Soc. Jpn.* **1993**, *62*, 3368–3371. [CrossRef]

15. Kamimura, H.; Hamada, T.; Ushio, H. Theoretical exploration of electronic structure in cuprates from electronic entropy. *Phys. Rev. B* **2002**, *66*, 054504. [CrossRef]
16. Kamimura, H.; Eto, M. $^1A_{1g}$ to $^3B_{1g}$ Conversion at the onset of superconductivity in $La_{2-x}Sr_xCuO_4$ due to the apical oxygen effect. *J. Phys. Soc. Jpn.* **1990**, *59*, 3053–3056. [CrossRef]
17. Kamimura, H.; Ushio, H.; Matusno, S.; Hamada, T. *Theory of Copper Oxide Superconductors*; Chapter 8.3; Springer: Berlin/Heidelberg, Germany, 2005.
18. Kamimura, H.; Sugino, O.; Shiraishi, K.; Araidai, M.; Tsai, J.S.; Sakata, H.; Ishida, K.; Matsuno, S. A First-Principles Non-Rigid Band Theory of Correlated Electrons in Copper Oxide Superconductors. 2020; In Preparation.
19. Furness, J.W.; Zhang, Y.; Lane, C.; Buda, I.G.; Barbielini, B.; Markiewicz, R.S.; Bansil, A.; Sun, J. An accurate first-principles treatment of doping-dependent electronic structure of high-temperature cuprate superconductors. *Commun. Phys.* **2018**, *1*, 11. [CrossRef]
20. Vaknin, D.; Sinha, S.K.; Moncton, D.E.; Johnson, D.C.; Newsam, J.; Safinya, C.R.; King, H. Antiferromagnetism in La_2CuO_{4-y}. *Phys. Rev. Lett.* **1996**, *77*, 723–726.
21. Freltoft, T.; Shirane, G.; Mitsuda, S.; Remeika, J.P.; Cooper, A.S. Magnetic form factor of Cu in La_2CuO_4. *Phys. Rev. B* **1988**, *37*, 137. [CrossRef] [PubMed]
22. Yamada, K.; Lee, C.H.; Wada, J.; Kurahashi, K.; Kimura, H.; Endoh, Y.; Hosoya, S.; Shirane, G.; Birgeneau, R.J.; Kastner, M.A. Spatial modulation of low-frequency spin-fluctuation in hole-doped La_2CuO_4. *J. Supercond.* **1997**, *10*, 343. [CrossRef]
23. Zhang, Y.; Lane, C.; Furness, J.W.; Barbiellini, B.; Perdew, J.P.; Markiewicz, R.S.; Bansil, A.; Suna, J. Competing stripe and magnetic phases in the cuprates from first principles. *Proc. Nat. Acad. Sci. USA* **2020**, *117*, 68. [CrossRef]
24. Slater, J.C. The Ferromagnetism of Nickel. *Phys. Rev.* **1936**, *49*, 537. [CrossRef]
25. Connolly, J.W. Energy bands in ferromagnetic nickel. *Phys. Rev.* **1967**, *159*, 415. [CrossRef]
26. Műller, K.A. Large, small, and especially jahn-teller polarons. *J. Supercond.* **1999**, *12*, 3. [CrossRef]
27. Shengelaya, A.; Zhao, G.M.; Keller, H.; Műller, K.A. EPR evidence of Jahn-Teller polaron formation in $La_{1-x}Ca_xMnO_{3+y}$. *Phys. Rev. Lett.* **1996**, *77*, 5296. [CrossRef]
28. Lanzara, A.; Saini, N.L.; Brunelnelli, M.; Natali, F.; Bianconi, A.; Radaelli, P.G.; Cheong, S.W. Crossover from large to small polarons across the metal-insulator transition in manganites. *Phys. Rev. Lett.* **1998**, *81*, 878. [CrossRef]
29. Yoshida, T.; Zhou, X.J.; Tanaka, K.; Yang, W.L.; Hussain, Z.; Shen, Z.-X.; Fujimori, A.; Sahrakorpi, S.; Lindroos, M.; Markiewicz, R.S.; et al. Systematic doping evolution off underlying Fermi surface of $La_{2-x}Sr_xCuO_4$. *Phys. Rev. B* **2006**, *74*, 224510. [CrossRef]
30. Kamimura, H.; Matsuno, S.; Suwa, Y.; Ushio, H. Occurrence of d-wave pairing in the phonon-mediated mechanism of high temperature superconductivity in cuprates. *Phys. Rev. Lett.* **1996**, *77*, 723. [CrossRef] [PubMed]
31. Tsuei, C.C.; Kirtley, J.R. Pairing symmetry in cuprate superconductors. *Rev. Mod. Phys.* **2000**, *72*, 969–1016. [CrossRef]
32. Takagi, H.; Ido, T.; Ishibashi, S.; Uota, M.; Uchida, S.; Tokura, Y. Superconductor-to-nonsuperconductor transition in $(La_{1-x}Sr_x)_2CuO_4$ as investigated by transport and magnetic measurements. *Phys. Rev. B* **1989**, *40*, 2254. [CrossRef] [PubMed]
33. Kamimura, H.; Ushio, H. On the Interplay of Jahn-Teller Physics and Mott Physics leading to the occurrence of Fermi pockets without pseudogap hypothesis and d-wave high T_c superconductivity in underdoped cuprate superconductivity. *J. Supercond. Nov. Magn.* **2012**, *25*, 677. [CrossRef]

Publisher's Note: MDPI stays neutral with regard to jurisdictional claims in published maps and institutional affiliations.

© 2020 by the authors. Licensee MDPI, Basel, Switzerland. This article is an open access article distributed under the terms and conditions of the Creative Commons Attribution (CC BY) license (http://creativecommons.org/licenses/by/4.0/).

Article

Suppression of the *s*-Wave Order Parameter Near the Surface of the Infinite-Layer Electron-Doped Cuprate Superconductor Sr$_{0.9}$La$_{0.1}$CuO$_2$

Rustem Khasanov [1,*], Alexander Shengelaya [2], Roland Brütsch [3] and Hugo Keller [4]

1. Laboratory for Muon Spin Spectroscopy, Paul Scherrer Institut, CH-5232 Villigen PSI, Switzerland
2. Department of Physics, Tbilisi State University, Chavchavadze 3, GE-0128 Tbilisi, Georgia; alexander.shengelaya@tsu.ge
3. Laboratory for Material Behaviour, Paul Scherrer Institut, CH-5232 Villigen PSI, Switzerland; roland.bruetsch@psi.ch
4. Physik-Institut der Universität Zürich, Winterthurerstrasse 190, CH-8057 Zürich, Switzerland; keller@physik.uzh.ch
* Correspondence: rustem.khasanov@psi.ch

Received: 29 June 2020; Accepted: 24 July 2020; Published: 3 August 2020

Abstract: The temperature dependencies of the in-plane (λ_{ab}) and out-of-plane (λ_c) components of the magnetic field penetration depth were investigated near the surface and in the bulk of the electron-doped superconductor Sr$_{0.9}$La$_{0.1}$CuO$_2$ by means of magnetization measurements. The measured $\lambda_{ab}(T)$ and $\lambda_c(T)$ were analyzed in terms of a two-gap model with mixed $s+d$-wave symmetry of the order parameter. $\lambda_{ab}(T)$ is well described by an almost pure anisotropic d-wave symmetry component ($\simeq 96\%$), mainly reflecting the surface properties of the sample. In contrast, $\lambda_c(T)$ exhibits a mixed $s+d$-wave order parameter with a substantial s-wave component of more than 50%. The comparison of $\lambda_{ab}^{-2}(T)$ measured near the surface with that determined in the bulk by means of the muon-spin rotation/relaxation technique demonstrates that the suppression of the s-wave component of the order parameter near the surface is associated with a reduction of the superfluid density by more than a factor of two.

Keywords: superconductivity; cuprates; magnetic penetration depth; order parameter; superconducting gap structure

---

Preface

Three authors of this paper, namely Rustem Khasanov, Alexander Shengelaya, and Hugo Keller, had the opportunity to work with Alex Müller during his stay at the University of Zürich. By being specialized in measurements of the magnetic penetration depth (λ) by means of muon-spin rotation/relaxation and magnetization techniques, we try to test the prediction of Alex Müller that the symmetry of the superconducting order parameter in cuprate high-temperature superconductors (HTSs) changes "from purely d at the surface to more s inside" (Müller, K.A. *Phil. Mag. Lett.* **2002**, *82*, 279–288). As a result of our studies, complex order parameters were detected in $\lambda(T)$ measurements for hole-doped HTSs such as La$_{1.83}$Sr$_{0.17}$CuO$_4$, YBa$_2$Cu$_3$O$_{7-\delta}$, and YBa$_2$Cu$_4$O$_8$. Here, we present evidence that the mixed order parameter symmetry is realized in Sr$_{0.9}$La$_{0.1}$CuO$_2$, i.e., in a superconductor belonging to the family of electron-doped cuprate HTSs.

1. Introduction

The order parameter in cuprate high-temperature superconductors (HTSs) is generally considered to be of pure d-wave symmetry. Pertinent evidence for this assumption for both the electron- and the hole-doped classes of HTSs stems from experiments where mainly surface phenomena are probed (e.g., angular resolved photoemission [1–4] or tricrystal experiments [5–8]). On the other hand, experimental data obtained by using techniques that probe the bulk of the material, such as nuclear magnetic resonance [9], Raman scattering [10,11], neutron crystal-field spectroscopy [12–14], and muon-spin rotation/relaxation (μSR) [15–17] provide strong evidence for the presence of a substantial s-wave component in the order parameter. Based on these experimental findings and on an earlier idea that in HTSs, two superconducting condensates with different order parameter symmetries (s- and d-wave) coexist [18–20], Müller [21–24] proposed a scenario in which the order parameter symmetry in HTSs changes from primarily d-wave at the surface to more $d + s$-wave in the bulk. At first glance, this scenario seems to contradict the accepted possible symmetries of the order parameter in HTSs [7]. However, by applying an interacting boson-model used in nuclear physics theory to the D_{4h} symmetry of HTSs, Iachello [25,26] demonstrated that a crossover from a d-wave order parameter symmetry at the surface to a $d + s$-wave symmetry in the bulk is indeed possible from a group theoretical point of view.

This scenario [21,22] can be directly tested by investigating the temperature dependence of the magnetic penetration depth λ near the surface and in the bulk of an HTS, since the behavior of $\lambda^{-2}(T)$ for a d-wave and a s-wave superconductor differs considerably. An isotropic s-wave pairing state leads to an almost constant value of the superfluid density $\rho_s \propto \lambda^{-2}$ for $T \leq 0.3 T_c$ [27–30], while the presence of nodes in the gap gives rise to a continuum of low lying excitations, resulting in a linear temperature dependence of $\lambda^{-2}(T)$ at low temperatures [30–33].

Here, we report on a magnetization study of the magnetic penetration depth near the surface and in the bulk of the electron-doped HTS $Sr_{0.9}La_{0.1}CuO_2$. The in-plane (λ_{ab}) and the out-of-plane (λ_c) components of the magnetic penetration depth were extracted from measurements of the AC susceptibility on a c-axis oriented powder sample in the Meissner state and analyzed within a two-gap $s + d$-wave scenario. The present results are compared with previous muon-spin rotation (μSR) experiments on $Sr_{0.9}La_{0.1}CuO_2$ [34], which probe the magnetic penetration depth in the bulk of the sample.

2. Experimental Details

Details on the sample preparation for $Sr_{0.9}La_{0.1}CuO_2$ can be found elsewhere [35]. The $Sr_{0.9}La_{0.1}CuO_2$ sample used in the present study was the same as measured by means of μSR in [34].

The sintered $Sr_{0.9}La_{0.1}CuO_2$ sample was grounded for about 20 min in order to obtain very small grains needed for the determination of the magnetic penetration depth from magnetic susceptibility measurements. In order to perform measurements of the in-plane and the out-of plane components of the magnetic penetration depth, samples were c-axis oriented in a static 9T magnetic field and cured at elevated temperatures in epoxy resin. In total, four c-axis oriented samples were prepared and measured.

The grain size distribution of the powder was determined by analyzing scanning electron microscope photographs (see Figure 1a). The measured particle size distribution $N(R)$ (R is the grain radius) is shown in Figure 1b.

The AC magnetization (M_{AC}) experiments were performed by using a commercial PPMS Quantum Design magnetometer ($\mu_0 H_{AC} = 0.3$ mT, $\nu = 333$ Hz) at temperatures between 1.75 K and 50 K. In two sets of experiments, the AC field was applied parallel and perpendicular to the c-axis. The absence of weak links between grains was confirmed for both field orientations by the linear magnetic field dependence of M_{AC} at $T = 10$ K for AC fields ranging from 0.1 to 1.0 mT and frequencies between 49 and 599 Hz.

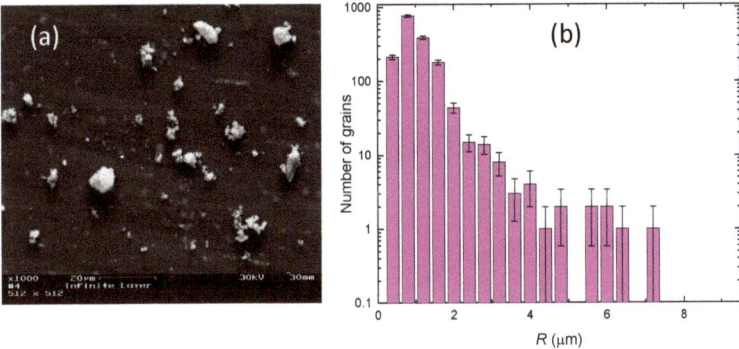

Figure 1. (a) An example of the scanning electron microscope photograph of the powdered $Sr_{0.9}La_{0.1}CuO_2$ sample. (b) The grain size distribution $N(R)$ determined from scanning electron microscope photographs. The thin vertical lines represent the statistical errors ($\pm\sqrt{N(R)}$).

3. Results and Discussion

The temperature dependence of the magnetic penetration depth was extracted from the measured M_{AC} by using the Shoenberg formula [36], modified for the known grain size distribution $N(R)$ [37]:

$$\chi(T)_{\parallel,\perp} = -\frac{3}{2}\int_0^\infty \left(1 - \frac{3\lambda^*_{\parallel,\perp}(T)}{R}\coth\frac{R}{\lambda^*_{\parallel,\perp}(T)} + \frac{3[\lambda^*_{\parallel,\perp}(T)]^2}{R^2}\right) N(R)R^3 dR / \int_0^\infty N(R)R^3 dR. \quad (1)$$

Here, $\chi_{\parallel,\perp} = M_{AC}\cdot\rho_X/m$ (m is the sample mass, and ρ_X is the X-ray density of $Sr_{0.9}La_{0.1}CuO_2$) is the volume susceptibility, and $\lambda^*_{\parallel,\perp}$ is the effective magnetic penetration depth for the magnetic field applied parallel ($\parallel$) and perpendicular ($\perp$) to the c-axis. In order to determine λ_{ab} and λ_c from the measured $\lambda^*_{\parallel,\perp}$, we followed the procedure of Porch et al. [37]:

(i) When a small magnetic field is applied along the c-axis, the screening currents flow in the ab-plane, decaying at a distance λ_{ab} from the grain surface, so that $\lambda^*_\parallel = \lambda_{ab}$.
(ii) With the magnetic field applied perpendicular to the c-axis, the screening currents flow within the ab-plane and along the c-axis, thus implying that both components (λ_{ab} and λ_c) enter the measured AC magnetization. For $\lambda_c \gg \lambda_{ab}$ (which is generally the case for highly anisotropic HTSs), the effective penetration depth $\lambda^*_\perp$ is mainly determined by the out-of-plane component, and for grains of arbitrary size, the relation $\lambda^*_\perp \simeq 0.7\lambda_c$ holds [37].

Bearing this in mind, the temperature dependencies of the in-plane and the out-of plane components of the magnetic penetration depth are determined as:

$$\lambda_{ab}(T) = \lambda^*_\parallel \quad \text{and} \quad \lambda_c = 1.43\lambda^*_\perp,$$

respectively.

The resulting temperature dependencies of λ_{ab}^{-2} and λ_c^{-2} are shown in Figure 2.

The zero-temperature values of λ_{ab} and λ_c were obtained by a linear extrapolation of $\lambda_{ab,c}^{-2}(T)$ for $T < 5$ K to $T = 0$, yielding $\lambda_{ab}(0) \simeq 157$ nm and $\lambda_c(0) \simeq 1140$ nm. An uncertainty in the absolute values of $\lambda_{ab,c}$ was considered by taking into account the statistical nature of the grain size distribution ($N(R) \pm \sqrt{N(R)}$; see Figure 1b), which resulted in a relative error of about ~7% for both λ_{ab} and λ_c.

An additional source of uncertainty stemmed from the deviation of the grain shapes from the spherical one. Assuming a small deviation of the demagnetization factor ($1/3 \pm 10\%$), the relative error of $\lambda_{ab,c}(0)$ was of the order of 3%. Taking both sources of errors into account yielded: $\lambda_{ab}(0) = 157(15)$ nm and $\lambda_c(0) = 1140(100)$ nm. The value of $\lambda_{ab}(0)$ obtained here was in a good agreement with the $\lambda_{ab}(0) = 147(7)$ nm reported by Kim et al. [38] based on the analysis of reversible magnetization data.

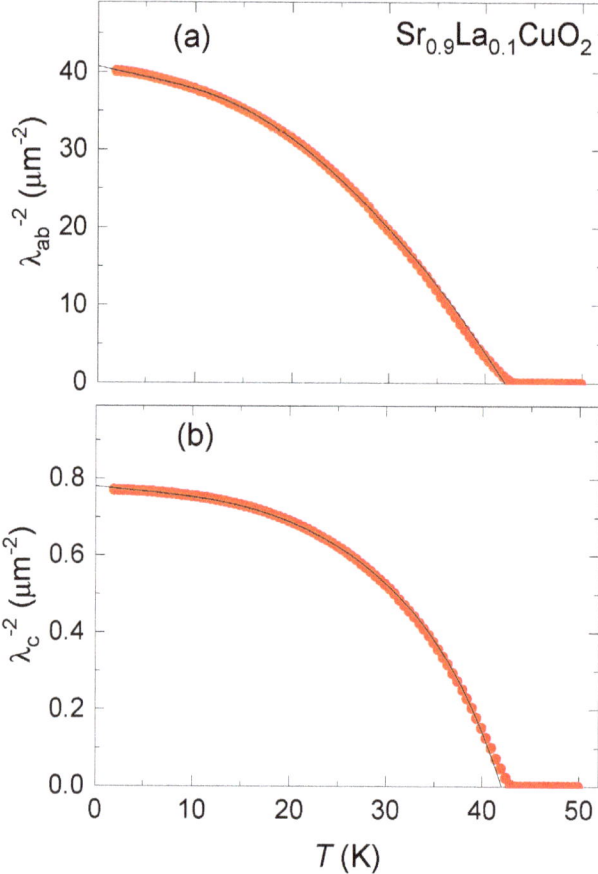

Figure 2. Temperature dependencies of λ_{ab}^{-2} (a) and λ_c^{-2} (b) for $Sr_{0.9}La_{0.1}CuO_2$ extracted from the measured $M_{AC}(T)$ by using Equation (1). Solid lines represent fits with the two-gap $s + d$-wave model. $\lambda_{ab}^{-2}(T)$ and $\lambda_c^{-2}(T)$ were analyzed simultaneously by means of Equation (2) with ω_{ab}, ω_c, $\lambda_{ab}(0)$, and $\lambda_c(0)$ as the individual fitting parameters and common s-wave and anisotropic d-wave gap functions as described by Equations (4)–(6). See the text for details.

In order to test the predictions of [18–22] and in analogy to our previous results on cuprate HTSs [15–17,34], the experimental data presented in Figure 2 were analyzed by decomposing $\lambda_{ab}^{-2}(T)$ and $\lambda_c^{-2}(T)$ into two contributions with s-wave and d-wave symmetry [39]:

$$\frac{\lambda_{ab,c}^{-2}(T)}{\lambda_{ab,c}^{-2}(0)} = \omega_{ab,c} \cdot \frac{\lambda_{ab,c}^{-2}(T, \Delta^s)}{\lambda_{ab,c}^{-2}(0, \Delta^s)} + (1 - \omega_{ab,c}) \cdot \frac{\lambda_{ab,c}^{-2}(T, \Delta^d)}{\lambda_{ab,c}^{-2}(0, \Delta^d)}. \quad (2)$$

Here, Δ^s and Δ^d denote the s-wave and the d-wave gap, respectively, and $\omega_{ab,c}$ is the weighting factor ($0 \leq \omega_{ab,c} \leq 1$), representing the relative contribution of the s-wave gap to $\lambda_{ab,c}^{-2}$. Both (s- and d-wave) components can be expressed by [16]:

$$\frac{\lambda_{ab,c}^{-2}(T, \Delta^{s,d})}{\lambda_{ab,c}^{-2}(0, \Delta^{s,d})} = 1 + \frac{1}{\pi} \int_0^{2\pi} \int_{\Delta^{s,d}(T,\varphi)}^{\infty} \left(\frac{\partial f}{\partial E}\right) \frac{E}{\sqrt{E^2 - \Delta^{s,d}(T,\varphi)^2}} dEd\varphi. \quad (3)$$

Here, $f = [1 + \exp(E/k_B T)]^{-1}$ is the Fermi function, φ is the angle along the Fermi surface ($\varphi = \pi/4$ corresponds to a zone diagonal), and:

$$\Delta^{s,d}(T, \varphi) = \Delta_0^{s,d} \delta(T/T_c) g^{s,d}(\varphi). \quad (4)$$

Here, $\Delta_0^{s,d}$ is the maximum value of the gap at $T = 0$. The temperature dependence of the gap is approximated by $\delta(T/T_c) = \tanh\{1.82[1.018(T_c/T - 1)]^{0.51}\}$ [40,41]. The function $g^{s,d}(\varphi)$ describes the angular dependence of the gap and is given by:

$$g^s(\varphi) = 1 \quad (5)$$

for the s-wave gap and:

$$g^{dAn}(\varphi) = \frac{3\sqrt{3}a}{2} \frac{\cos 2\varphi}{(1 + a\cos^2 2\varphi)^{3/2}} \quad (6)$$

for the anisotropic d-wave gap [42] (a is a constant). We want to stress that series of experimental [3,43] and theoretical works [42] suggest that the angular dependence of the gap in the electron-doped HTSs differs significantly from the simple functional form $\Delta_0 \cos 2\varphi$, observed for various hole-doped HTSs, and has the so-called anisotropic d-wave symmetry (with the gap maximum in between the nodal and the antinodal points on the Fermi surface).

The measured $\lambda_{ab}^{-2}(T)$ and $\lambda_c^{-2}(T)$ displayed in Figure 2 were analyzed simultaneously by means of Equation (2) with ω_{ab}, ω_c, $\lambda_{ab}(0)$, and $\lambda_c(0)$ as the individual fitting parameters and common s-wave and anisotropic d-wave gap functions, as described by Equations (4)–(6). The results of the analysis are summarized in Figure 2 and Table 1.

Panels (a) and (b) of Figure 3 represent the angular dependencies of the s and the d_{An} superconducting energy gaps at $T = 0$. The solid blue and the red lines in Figure 3c correspond to the individual s-wave and d-wave contributions, respectively.

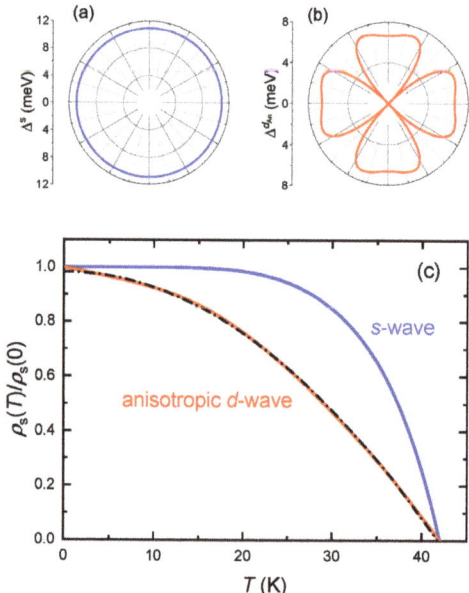

Figure 3. (a) The angular dependence of the s-wave gap at $T = 0$ ($\Delta_0^s \cdot g^s(\varphi)$; see Equation (5) and Table 1). (b) The angular dependence of the anisotropic d-wave gap ($\Delta_0^d \cdot g^d(\varphi)$; Equation (6) and Table 1). (c) Contributions of the s-wave (blue line) and the anisotropic d-wave (red line) gaps to the superfluid density ($\rho_s \propto \lambda^{-2}$) obtained by means of Equation (3). The dashed-dotted line represents the T^2 behavior, which is generally observed in various electron-doped HTSs (see [30] and the references therein).

From the results presented in Figures 2 and 3 and Table 1, the following important points emerge:

(i) The absolute value of the s-wave gap is larger than the maximum value of the anisotropic d-wave gap (Table 1 and Panels (a) and (b) of Figure 3) with $2\Delta_0^s/k_B T_c = 6.02(6)$ and $2\Delta_0^{d_{An}}/k_B T_c = 3.89(3)$, respectively ($T_c \simeq 42$ K). This implies that in electron-doped $Sr_{0.9}La_{0.1}CuO_2$, the s-wave component of the order parameter is the dominant one. This agrees with the results of small-angle neutron scattering experiments revealing that at fields higher than 1.5 T, the superfluid density of $Sr_{0.9}La_{0.1}CuO_2$ is determined entirely by the s-wave component of the order parameter [44].

(ii) Point (i) is in contrast to hole-doped cuprate HTSs where Δ_0^s is always smaller than Δ_0^d [10,15–17,45].

(iii) The temperature dependence of the anisotropic d-wave contribution to the superfluid density (solid red line in Figure 3c) is very close to the quadratic (T^2) dependence (dash-dotted line in Figure 3c), which is often observed in various electron-doped HTSs (see [30] and the references therein). Generally, the T^2 behavior is attributed to a "dirty" d-wave scenario and is explained by impurity scattering of the carriers. However, it is difficult to explain how an order parameter that changes sign persists in the dirty limit, since any scattering centers would act as pair breakers [46]. Therefore, we believe that the anisotropic d-wave approach is more appropriate for electron-doped HTSs.

(iv) For $\lambda_{ab}^{-2}(T)$, the s-wave contribution to the superfluid density is almost negligible ($\omega_{ab} = 0.04$), whereas for $\lambda_c^{-2}(T)$, it is substantial ($\omega_c = 0.54$) (see Table 1). Bearing in mind that our experiments were performed in the Meissner state, the different behavior of $\lambda_{ab}^{-2}(T)$ and $\lambda_c^{-2}(T)$ can be explained within the scenario proposed by Müller [21,22]. Since $\lambda_{ab}(0)$ is rather small (see Table 1), one can assume that its temperature dependence is mainly determined by surface properties and therefore follows the one expected for a d-wave superconductor. In contrast, $\lambda_c(0)$ is almost a factor 10 larger than $\lambda_{ab}(0)$, and thus, $\lambda_c^{-2}(T)$ contains contributions from both the surface and the bulk (mixed $s+d$-wave order parameter).

Table 1. Summary of the analysis of $\lambda_{ab}^{-2}(T)$ and $\lambda_c^{-2}(T)$ for $Sr_{0.9}La_{0.1}CuO_2$ by means of Equation (2). The absolute errors of $\lambda_{ab,c}(0)$ account for the uncertainties in the grain size distribution $N(R) \pm \sqrt{N(R)}$ and that of the demagnetization factor $1/3 \pm 10\%$; see the text for details. TF-μSR, denotes the transverse-field muon-spin rotation/relaxation (TF-μSR) experiments.

Method	Quantity	Δ_0^s (meV)	$\Delta_0^{d_{An}}$ (meV)	a	$\omega_{ab,c}$	$\lambda_{ab,c}(0)$ (nm)
ACsusc.	$\lambda_{ab}^{-2}(T)$ $\lambda_c^{-2}(T)$	10.9(1)	7.03(6)	0.90(2)	0.04(2) 0.54(2)	157(15) 1140(100)
TF-μSR [34]	$\lambda_{ab}^{-2}(T)$	10.9 ^{a}	7.03 ^{a}	0.90 ^{a}	0.72(4)	93(2)

^{a} From the AC susceptibility data.

Additional arguments pointing to the validity of this scenario [21,22] come from the comparison of the in-plane magnetic penetration depth $\lambda_{ab}(T)$ measured near the surface in this work with that determined in μSR experiments on a similar $Sr_{0.9}La_{0.1}CuO_2$ sample [34] (see Figure 4). Note that μSR is a powerful technique to probe the magnetic penetration depth in the bulk of a superconductor in the vortex state [31,47]. It is evident that $\lambda_{ab}^{-2}(T)$ measured near the surface decreases more strongly with increasing temperature than that obtained in the bulk. On the other hand, $\lambda_{ab}^{-2}(T)/\lambda_{ab}^{-2}(0)$ (μSR) is relatively close to $\lambda_c^{-2}(T)/\lambda_c^{-2}(0)$ determined by AC magnetization experiments, thus indicating that $\lambda_c^{-2}(T)$ is mainly governed by bulk properties. Another interesting issue stems from the comparison of the absolute λ_{ab} values obtained in the bulk and the surface sensitive experiments. The analysis of $\lambda_{ab}^{-2}(T)$ (μSR) within the above described scheme, with Δ_0^s and $\Delta_0^{d_{An}}$ fixed to the values determined from the AC susceptibility experiments, results in $\lambda_{ab}(0) = 93(2)$ nm (see Table 1), which is more than 50% shorter than $\lambda_{ab}(0) = 157(15)$ nm obtained near the surface. We may assume, therefore, that the difference in the absolute values and the temperature dependencies of the bulk and the surface λ_{ab} can be explained within the scenario proposed in [21,22] and is caused by a substantial reduction of the s-wave contribution near the surface ($\omega_{ab} \simeq 4\%$) in comparison with that in the bulk ($\omega_{ab} \simeq 72\%$; see Table 1).

To the best of our knowledge, the above presented results give a first example of different order parameter symmetries near the surface and in the bulk in electron-doped high-temperature cuprate superconductors. As for the hole-doped representatives of HTSs, the comprehensive analysis was made in a series of works of K.A. Müller [18–24]. By comparing the results of the "surface" and the "bulk" sensitive experiments, it was concluded that the superconducting order parameter in the hole-doped cuprate HTSs changes from purely d at the surface to the mixture of s and d in the bulk. The theory explanation was given by Iachello [25,26], based on purely symmetry considerations and in analogy with atomic nuclei. The model consists of s and d pairs (approximated as bosons) in a two-dimensional Fermi system with a surface. The transition takes place between the phase where only one type of boson condensate to the phase consisting of a mixture of two type of bosons.

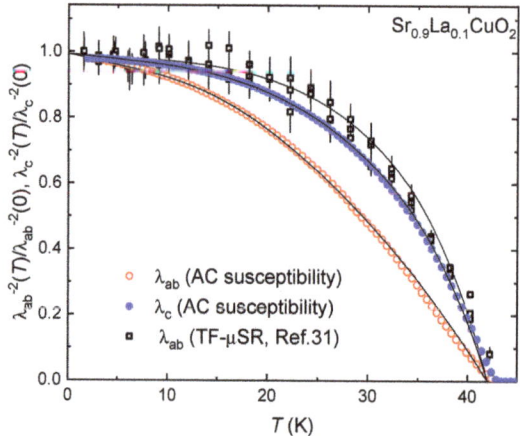

Figure 4. The normalized superfluid density $\lambda_{ab}^{-2}(T)/\lambda_{ab}^{-2}(0)$ (open circles and squares) and $\lambda_c^{-2}(T)/\lambda_c^{-2}(0)$ (closed circles) obtained in the present study (closed and open circles) and by the transverse-field μSR experiments (open squares) in [34]. The solid lines correspond to fits by means of Equation (2) with the parameters summarized in Table 1.

4. Conclusions

In conclusion, the temperature dependence of the in-plane (λ_{ab}) and the out-of-plane (λ_c) components of the magnetic penetration depth of $Sr_{0.9}La_{0.1}CuO_2$ were determined in the Meissner state from AC susceptibility measurements. The temperature dependence of λ_{ab}^{-2} is well described by assuming that the superconducting order parameter is mainly of d-wave symmetry ($\simeq$ 96%) with $\lambda_{ab}(0) = 157(15)$ nm. The out-of-plane component was found to be much longer, $\lambda_c(0) \simeq 1140(100)$ nm. The temperature dependence of λ_c^{-2} is in accordance with a mixed s + d-wave order parameter with a substantial s-wave component of more than 50%. A comparison of $\lambda_{ab}^{-2}(T)$ reported in this work with that obtained in the bulk by μSR [34] reveals that the s-wave component of the order parameter is strongly suppressed near the surface of the superconductor, associated with a substantial reduction of the superfluid density by more than a factor of two. The results presented here are consistent with the scenario of a complex mixed s + d-wave symmetry order parameter proposed by Müller [18–24]. In particular, the prediction of a strongly suppressed s-wave component near the surface was confirmed experimentally. This study clearly demonstrates that special care must be taken when experimental results obtained by surface sensitive and bulk sensitive techniques are compared, since they do not necessarily probe the same properties of high-temperature superconductors.

Author Contributions: R.K., A.S., and H.K. specified the topic of the studies. R.K. performed the experiment, analyzed the data, and wrote the manuscript. R.B. performed the electron-microscopy experiments. A.S. and H.K. contributed to finalizing the manuscript. All authors have read and agreed to the published version of the manuscript.

Funding: This work was supported by the Swiss National Science Foundation, by the K. Alex Müller Foundation, and in part by SCOPES Grant No. IB7420-110784.

Acknowledgments: The authors are grateful to A. Bussmann-Holder and K.A. Müller for stimulating discussions. D.J. Jang and S.-I. Lee are acknowledged for providing the samples used for the experiments.

Conflicts of Interest: The authors declare no conflict of interest.

References and Note

1. Shen, Z.-X.; Dessau, D.S.; Wells, B.O.; King, D.M.; Spicer, W.E.; Arko, A.J.; Marshall, D.; Lombardo, L.W.; Kapitulnik, A.; Dickinson, P.; et al. Anomalously large gap anisotropy in the $a - b$ plane of $Bi_2Sr_2CaCu_2O_{8+\delta}$. *Phys. Rev. Lett.* **1993**, *70*, 1553–1556. [CrossRef] [PubMed]
2. Ding, H.; Norman, M.R.; Campuzano, J.C.; Randeria, M.; Bellman, A.F.; Yokoya, T.; Takahashi, T.; Mochiku, T.; Kadowaki, K. Angle-resolved photoemission spectroscopy study of the superconducting gap anisotropy in $Bi_2Sr_2CaCu_2O_{8+x}$. *Phys. Rev. B* **1996**, *54*, R9678–R9681. [CrossRef] [PubMed]
3. Matsui, H.; Terashima, K.; Sato, T.; Takahashi, T.; Fujita, M.; Yamada, K. Direct observation of a nonmonotonic $d_{x^2-y^2}$-wave superconducting gap in the electron-doped high-T_c superconductor $Pr_{0.89}LaCe_{0.11}CuO_4$. *Phys. Rev. Lett.* **2005**, *95* 017003. [CrossRef]
4. Harter, J.W.; Maritato, L.; Shai, D.E.; Monkman, E.J.; Nie, Y.; Schlom, D.G.; Shen, K.M. Nodeless superconducting phase arising from a strong (π,π) antiferromagnetic phase in the infinite-layer electron-doped $Sr_{1-x}La_xCuO_2$ compound. *Phys. Rev. Lett.* **2012**, *109*, 267001. [CrossRef] [PubMed]
5. Tsuei, C.C.; Kirtley, J.R.; Chi, C.C.; Yu-Jahnes, L.S.; Gupta, A.; Shaw, T.; Sun, J.Z.; Ketchen, M.B. Pairing symmetry and flux quantization in a tricrystal superconducting ring of $YBa_2Cu_3O_{7-\delta}$. *Phys. Rev. Lett.* **1994**, *73*, 593–596. [CrossRef] [PubMed]
6. Tsuei, C.C.; Kirtley, J.R. Phase-sensitive evidence for d-wave pairing symmetry in electron-doped cuprate superconductors. *Phys. Rev. Lett.* **2000**, *85*, 182–185. [CrossRef]
7. Tsuei, C.C.; Kirtley, J.R. Pairing symmetry in cuprate superconductors. *Rev. Mod. Phys.* **2000**, *72*, 969–1016. [CrossRef]
8. Tomaschko, J.; Scharinger, S.; Leca, V.; Nagel, J.; Kemmler, M.; Selistrovski, T.; Koelle, D.; Kleiner, D. Phase-sensitive evidence for $d_{x^2-y^2}$-pairing symmetry in the parent-structure high-T_c cuprate superconductor $Sr_{1-x}La_xCuO_2$. *Phys. Rev. B* **2012**, *86*, 094509. [CrossRef]
9. Martindale, J.A.; Hammel, P.C.; Hults, W.L.; Smith, J.L. Temperature dependence of the anisotropy of the planar oxygen nuclear spin-lattice relaxation rate in $YBa_2Cu_3O_y$. *Phys. Rev. B* **1998**, *57*, 11769–11774. [CrossRef]
10. Masui, T.; Limonov, M.; Uchiyama, H.; Lee, S.; Tajima, S.; Yamanaka, A. Raman study of carrier-overdoping effects on the gap in high-T_c superconducting cuprates. *Phys. Rev. B* **2003**, *68*, 060506(R). [CrossRef]
11. Friedl, B.; Thomsen, C.; Cardona, M. Determination of the superconducting gap in $RBa_2Cu_3O_{7-\delta}$. *Phys. Rev. Lett.* **1990**, *65*, 915–918. [CrossRef] [PubMed]
12. Furrer, A. Admixture of an s-wave component to the d-wave gap symmetry in high-temperature superconductors. In *High-T_c Superconductors and Related Transition Metal Compounds*; Bussmann-Holder, A., Keller, H., Eds.; Springer: Berlin/Heidelberg, Germany, 2007; pp. 135–141.
13. Rubio Temprano, D.; Mesot, J.; Janssen, S.; Conder, K.; Furrer, A.; Mutka, H.; Müller, K.A. Large isotope effect on the pseudogap in the high-temperature superconductor $HoBa_2Cu_4O_8$. *Phys. Rev. Lett.* **2000**, *84*, 1990. [CrossRef] [PubMed]
14. Rubio Temprano, D.; Conder, K.; Furrer, A.; Mutka, H.; Trounov, V.; Müller, K.A. Oxygen and copper isotope effects on the pseudogap in the high-temperature superconductor $La_{1.81}Ho_{0.04}Sr_{0.15}CuO_4$ studied by neutron crystal-field spectroscopy. *Phys. Rev. B* **2002**, *66*, 184506. [CrossRef]
15. Khasanov, R.; Shengelaya, A.; Maisuradze, A.; La Mattina, F.; Bussmann-Holder, A.; Keller, H.; Müller, K.A. Experimental evidence for two gaps in the high-temperature $La_{1.83}Sr_{0.17}CuO_4$ superconductor. *Phys. Rev. Lett.* **2007**, *98*, 057007. [CrossRef]
16. Khasanov, R.; Shengelaya, A.; Bussmann-Holder, A.; Karpinski, J.; Keller, H.; Müller, K.A. s-wave symmetry along the c-axis and $s + d$ in-plane superconductivity in bulk $YBa_2Cu_4O_8$. *J. Supercond. Nov. Magn.* **2008**, *21*, 81–85. [CrossRef]
17. Khasanov, R.; Strässle, S.; Di Castro, D.; Masui, T.; Miyasaka, S.; Tajima, S.; Bussmann-Holder, A.; Keller, H. Multiple gap symmetries for the order parameter of cuprate superconductors from penetration depth measurements. *Phys. Rev. Lett.* **2007**, *99*, 237601. [CrossRef]

18. Müller, K.A. Possible coexistence of s-and d-wave condensates in copper oxide superconductors. *Nature (London)* **1995**, *377*, 133–135. [CrossRef]
19. Müller, K.A. Two gap behavior observed in YBCO (100) *a*-axis, (110) *c*-axis tunnel junctions. *J. Phys. Soc. Jpn.* **1996**, *65*, 3090–3091. [CrossRef]
20. Müller, K.A.; Keller, H. "s" and "d" wave symmetry components in high-temperature cuprate superconductors. In *High-T_c Superconductivity 1996: Ten Years after Discovery*; Kaldis, E., Liarokapis, E., Müller, K.A., Eds.; Springer: Dordrecht, The Netherland, 1997; pp. 7–29.
21. Müller, K.A. On the macroscopic s- and d-wave symmetry in cuprate superconductors. *Philos. Mag. Lett.* **2002**, *82*, 279–288. [CrossRef]
22. Müller, K.A. Estimation of the surface-*d* to bulk-*s* crossover in the macroscopic superconducting wavefunction in cuprates. *J. Supercond. Nov. Magn.* **2004**, *17*, 3–6. [CrossRef]
23. Müller, K.A. The unique properties of superconductivity in cuprates. *J. Supercond. Nov. Magn.* **2014**, *27*, 2163–2179. [CrossRef]
24. Müller, K.A. The polaronic basis for high-temperature superconductivity. *J. Supercond. Nov. Magn.* **2017**, *30*, 3007–3018. [CrossRef]
25. Iachello, F. A model of cuprate superconductors based on the analogy with atomic nuclei. *Philos. Mag. Lett.* **2002**, *82*, 289–295. [CrossRef]
26. Iachello, F. Symmetry of high-T_c superconductors. In *Symmetry and Heterogeneity in High Temperature Superconductors*; Bianconi, A., Ed.; NATO science Series II: Mathematics, Physics and Chemistry; Springer: Dordrecht, The Netherlands, 2006; Volume 214, pp. 165–180.
27. Khasanov, R.; Eshchenko, D.G.; Di Castro, D.; Shengelaya, A.; La Mattina, F.; Maisuradze, A.; Baines, C.; Luetkens, H.; Karpinski, J.; Kazakov, S.M.; et al. Magnetic penetration depth in $RbOs_2O_6$ studied by muon spin rotation. *Phys. Rev. B* **2005**, *72*, 104504. [CrossRef]
28. Khasanov, R.; Klamut, P.W.; Shengelaya, A.; Bukowski, Z.; Savić, I.M.; Baines, C.; Keller, H. Muon-spin rotation measurements of the penetration depth of the Mo_3Sb_7 superconductor. *Phys. Rev. B* **2008**, *78*, 014502. [CrossRef]
29. Khasanov, R.; Zhou, H.; Amato, A.; Guguchia, Z.; Morenzoni, E.; Dong, X.; Zhang, G.; Zhao, Z. Proximity-induced superconductivity within the insulating $(Li_{0.84}Fe_{0.16})OH$ layers in $(Li_{0.84}Fe_{0.16})OHFe_{0.98}Se$. *Phys. Rev. B* **2016**, *93*, 224512. [CrossRef]
30. Prozorov, R.; Giannetta, R.W. Magnetic penetration depth in unconventional superconductors. *Supercond. Sci. Technol.* **2006**, *19*, R41–R67. [CrossRef]
31. Sonier, J.E.; Brewer, J.H.; Kiefl, R.F. μSR studies of the vortex state in type-II superconductors. *Rev. Mod. Phys.* **2000**, *72*, 769–811. [CrossRef]
32. Khasanov, R.; Kondo, T.; Strässle, S.; Heron, D.O.G.; Kaminski, A.; Keller, H.; Lee, S.L.; Takeuchi, T. Evidence for a competition between the superconducting state and the pseudogap state of $(BiPb)_2(SrLa)_2CuO_{6+\delta}$ from muon spin rotation experiments. *Phys. Rev. Lett.* **2008**, *101*, 227002. [CrossRef]
33. Khasanov, R.; Kondo, T.; Bendele, M.; Hamaya, Y.; Kaminski, A.; Lee, S.L.; Ray, S.J.; Takeuchi, T. Suppression of the antinodal coherence of superconducting $(Bi,Pb)_2(Sr,La)_2CuO_{6+\delta}$ as revealed by muon spin rotation and angle-resolved photoemission. *Phys. Rev. B* **2010**, *82*, 020511(R). [CrossRef]
34. Khasanov, R.; Shengelaya, A.; Maisuradze, A.; Di Castro, D.; Savić, I.M.; Weyeneth, S.; Park, I.M.; Jang, I.M.; Lee, S.-I.; Keller, H. Nodeless superconductivity in the infinite-layer electron-doped cuprate superconductor $Sr_{0.9}La_{0.1}CuO_2$. *Phys. Rev. B* **2008**, *77*, 184512. [CrossRef]
35. Jung, S.I.; Kim, J.Y.; Kim, M.S.; Park, M.S.; Kim, H.J.; Yao, Y.; Lee, S.Y.; Lee, S.-I. Synthesis and pinning properties of the infinite-layer superconductor $Sr_{0.9}La_{0.1}CuO_2$. *Physica C* **2002**, *366*, 299–305. [CrossRef]
36. Shoenberg, D. Properties of superconducting colloids and emulsions. *Proc. R. Soc. A* **1940**, *175*, 49–70.
37. Porch, A.; Cooper, J.R.; Zheng, D.N.; Waldram, J.R.; Campbell, A.M.; Freeman, P.A. Temperature dependent magnetic penetration depth of Co and Zn doped $YBa_2Cu_3O_7$ obtained from the AC susceptibility of magnetically aligned powders. *Phycica C* **1993**, *214*, 350–358. [CrossRef]
38. Kim, M.S.; Lemberger, T.R.; Jung, C.U.; Choi, J.H.; Kim, J.Y.; Kim, H.J.; Lee, S.-I. Anisotropy and reversible magnetization of the infinite-layer superconductor $Sr_{0.9}La_{0.1}CuO_2$. *Phys. Rev. B* **2002**, *66*, 214509. [CrossRef]

39. $\lambda_{ab}^{-2}(T)$ and $\lambda_c^{-2}(T)$ can be equally well analyzed within the framework of s + s-wave and s+ anisotropic s-wave models.
40. Carrington, A.; Manzano, F. Magnetic penetration depth of MgB$_2$. *Physica C* **2003**, *385*, 205–214. [CrossRef]
41. Khasanov, R.; Gupta, R.; Das, D.; Leithe-Jasper, A.; Svanidze, E. Single-gap versus two-gap scenario: Specific heat and thermodynamic critical field of the noncentrosimmetric superconductor BeAu. *Phys. Rev. B* **2020**, *102*, 014514. [CrossRef]
42. Eremin, I.; Tsoncheva, E.; Chubukov, A.V. Signature of the nonmonotonic *d*-wave gap in electron-doped cuprates. *Phys. Rev. B* **2008**, *77*, 024508. [CrossRef]
43. Blumberg, G.; Koitzsch, A.; Gozar, A.; Dennis, B.S.; Kendziora, C.A.; Fournier, P.; Greene, R.L. Nonmonotonic $d_{x^2-y^2}$ superconducting order parameter in Nd$_{2-x}$Ce$_x$CuO$_4$. *Phys. Rev. Lett.* **2002**, *88*, 107002. [CrossRef]
44. White, J.S.; Forgan, E.M.; Laver, M.; Häfliger, P.S.; Khasanov, R.; Cubitt, R.; Dewhurst, C.D.; Park, M.S.; Jang, D.J.; Lee, S.-I. Finite gap behaviour in the superconductivity of the 'infinite layer' n-doped high-T_c superconductor Sr$_{0.9}$La$_{0.1}$CuO$_2$. *J. Phys. Condens. Matter.* **2008**, *20*, 104237 [CrossRef]
45. Kohen, A.; Leibovitch, G.; Deutscher, G. Andreev reflections on Y$_{1-x}$Ca$_x$Ba$_2$Cu$_3$O$_{7-\delta}$: Evidence for an unusual proximity effect. *Phys. Rev. Lett.* **2003**, *90*, 207005. [CrossRef] [PubMed]
46. Millis, A.J.; Sachdev, S.; Varma, C.M. Inelastic scattering and pair breaking in anisotropic and isotropic superconductors. *Phys. Rev. B* **1988**, *37*, 4975–4986. [CrossRef] [PubMed]
47. Pümpin, B.; Keller, H.; Kündig, W.; Odermatt, W.; Savić, I.M.; Schneider, J.W.; Simmler, H.; Zimmermann, P.; Kaldis, E.; Rusiecki, S.; et al. Muon-spin-rotation measurements of the London penetration depths in YBa$_2$Cu$_3$O$_{6.97}$. *Phys. Rev. B* **1990**, *42*, 8019–8029. [CrossRef] [PubMed]

© 2020 by the authors. Licensee MDPI, Basel, Switzerland. This article is an open access article distributed under the terms and conditions of the Creative Commons Attribution (CC BY) license (http://creativecommons.org/licenses/by/4.0/).

Review

SrTiO₃—Glimpses of an Inexhaustible Source of Novel Solid State Phenomena

Wolfgang Kleemann [1,*], Jan Dec [2], Alexander Tkach [3] and Paula M. Vilarinho [3]

1. Applied Physics, University Duisburg-Essen, D-47048 Duisburg, Germany
2. Institute of Physics, University of Silesia, PL-40-007 Katowice, Poland; jan.dec@us.edu.pl
3. Department of Materials and Ceramic Engineering, CICECO—Aveiro Institute of Materials, University of Aveiro, P-3810-193 Aveiro, Portugal; atkach@ua.pt (A.T.); paula.vilarinho@ua.pt (P.M.V.)
* Correspondence: wolfgang.kleemann@uni-due.de; Tel.: +49-1575-226-3908

Received: 6 August 2020; Accepted: 30 September 2020; Published: 4 October 2020

Abstract: The purpose of this selective review is primarily to demonstrate the large versatility of the insulating quantum paraelectric perovskite SrTiO₃ explained in "Introduction" part, and "Routes of SrTiO₃ toward ferroelectricity and other collective states" part. Apart from ferroelectricity under various boundary conditions, it exhibits regular electronic and superconductivity via doping or external fields and is capable of displaying diverse coupled states. "Magnetoelectric multiglass (Sr,Mn)TiO₃" part, deals with mesoscopic physics of the solid solution SrTiO₃:Mn²⁺. It is at the origin of both polar and spin cluster glass forming and is altogether a novel multiferroic system. Independent transitions at different glass temperatures, power law dynamic criticality, divergent third-order susceptibilities, and higher order magneto-electric interactions are convincing fingerprints.

Keywords: strontium titanate; quantum paraelectricity; quantum fluctuations; ferroelectricity; isotope exchange; external stress; polar metal; superconductivity; phase coexistence; magnetoelectric multiglass

---

1. Introduction

In this review, we focus onto two research lines of strontium titanate, SrTiO₃ (STO):

(1) the low-temperature phases around the quantum critical point of pure STO, and (2) the disordered electric and magnetic dipolar glassy phases in the solid solution STO: Mn. It is not intended to describe the full extent of all phenomena observed and to detail all of their properties from abundantly issued publications. We merely try to give an impression of some actual fields, which are partially related to our earlier cooperation with K. A. Müller and J. G. Bednorz.

STO is probably the most versatile perovskite-type oxide and one of the richest materials in terms of functionalities. In 1979, Müller and Burkard [1] reported that the planar permittivity of STO strongly increased upon cooling from ≈ 300 at room temperature and saturated with $\varepsilon'_{\langle 110 \rangle} \approx 2.5 \cdot 10^4$ as $T \to 0$ (comparable to $\varepsilon'_{\langle 100 \rangle}$ vs. T, cf. Figure 1, curve 1 [2]). They conjectured a "quantum paraelectric" ground state in close proximity to a ferroelectric (FE) one, where the centrosymmetric tetragonal lattice structure of STO becomes stabilized by quantum fluctuations of the nearly softening in-plane F_{1u} lattice mode.

Quantum corrections to the temperature were proposed to describe the critical behavior of STO from the beginning [1]. In order to account for the obvious deviations from the mean-field Curie–Weiss behavior, some of us proposed a generalized modified "quantum Curie–Weiss law" [3].

$$\varepsilon' = C/\left(T^Q - T_0^Q\right)^\gamma \tag{1}$$

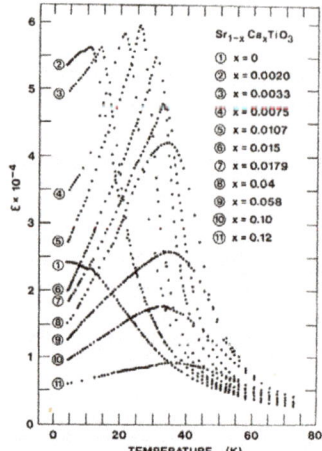

Figure 1. Temperature dependence of the dielectric permittivity $\varepsilon'_{[100]}$ of $Sr_{1-x}Ca_xTiO_3$ crystals with $0 \leq x \leq 0.12$ [2].

where C stands for the Curie constant, γ for the critical exponent, and $T_0^Q = T_S \coth(T_S/T_0)$ for the quantum critical temperature with classic critical temperature T_0 and saturation temperature T_S being related to the ground state energy of the quantum oscillator, $E_0 = k_B T_S$. Crucial novel ingredients are the free parameter γ and the quantum temperature scale, $T^Q = T_S \coth(T_S/T)$, which replaces T. The best fit to the STO data [3] within $2 \leq T \leq 110$ K yields $C = (3 \pm 2)\,10^6$, $T_S = (17 \pm 1)$ K, $T_0^Q \approx 0$, and the highly non-classic exponent $\gamma = 1.7 \pm 0.2$. Interestingly, a very similar value, $\gamma \approx 2$, was recently obtained on pure STO from a conventional power-law fit for $4 \leq T \leq 50$ K [4], where saturation effects deviated below ≈ 4 K. While our approach matches with this extreme quantum regime without any extra conditions, the Φ^4 model of [4] requires corrections due to long-range dipolar interactions and coupling of the electric polarization field to acoustic phonons.

2. Routes of SrTiO$_3$ toward Ferroelectricity and Other Collective States

Overcoming quantum paraelectricity and reaching stable long-range ordered states by proper treatments has remained a major challenge for ongoing research on STO.

(i) In 1976, Uwe and Sakudo [5] succeeded in stabilizing uniaxial ferroelectricity in STO at liquid He temperature by intraplanar symmetry breaking with uniaxial stress perpendicularly to the c axis, σ_{110}. This was the first experience to overcome quantum paraelectricity of STO by an external perturbation.

(ii) About 20 years later, in 1999, Itoh et al. [6] discovered a very efficient internal perturbation via isotopic exchange of ^{16}O by ^{18}O in order to establish the FE state at $T_0 \approx 25$–50 K.

(iii) Again, 20 years later, in 2019, Nova et al. [7] realized transient FE in STO being metastable up to T > 290 K under optical strain due to intense femtosecond laser pulses, while Li et al. [8] recorded similar events after photoexcitation of quantum paraelectric STO with THz laser pulses in resonance with the FE soft mode at $T < 36$ K.

(iv) Metallic behavior of n-type STO has been achieved by substituting transition-metal dopants, e.g., La^{3+} for Sr^{2+} [9], Nb^{5+} for Ti^{4+}, or by reducing pure STO into $SrTiO_{3-\delta}$, $0 < \delta < 1$, where each oxygen vacancy generates two "doped" electrons [10]. The insulator-to-metal transformation occurs at a relatively low critical electron density $n \approx 10^{18}$ cm^{-3} [11], i.e., two orders of magnitude less than in the analogous case of barium titanate [12]. Here the word "metal" is not used in its common material meaning but merely stands for featuring metallic electronic conduction.

A combined method of substitution and reduction was utilized in the case of Cr-doped STO [11]. A significant spatial correlation between oxygen vacancies and Cr^{3+} ions in bulk was established in thermally reduced Cr-doped STO. In the presence of electron donors, the Cr atoms change their valence from 4+ to 3+. Consequently, this reduction drives a symmetry change of the crystal field experienced by the Cr ions from cubic to axial [11], which may be controlled by means of thermal annealing and/or doping with electron donors. This capability of controlling oxygen vacancies or transition-metal dopants has been essential for the development of semiconducting electronic devices [13].

(v) Theoretical predictions of superconductivity in degenerate semiconductors motivated research on reduced n-type STO, which revealed the critical temperature $T_c \approx 0.28$ K as early as 1964 [14]. However, 32 years later, perovskite-like cuprates were to open the door to modern high-T_c superconductivity with $T_c \approx 30$ K [15] and to the physics Nobel Prize [16]. On the other hand, gated n-type STO has reached at most only $T_c \approx 0.6$ K [17].

Meanwhile numerous other processes have made STO a nearly inexhaustible source of activating novel solid state phenomena that suggest future applications. This has remained an attractive research goal even more than 60 years after the pioneering experiments. Extending the initial idea of breaking the local symmetry by stress [5], Bednorz and Müller [2] introduced an A-site doping route by random replacement of Sr^{2+} ions with smaller Ca^{2+} ions in single crystals of $Sr_{1-x}Ca_xTiO_3$ (SCT). The local decrease of volume creates random strain ("negative stress field"), which has an enormous effect on the dielectric response for doping levels $0.002 \leq x \leq 0.12$, as shown in Figure 1 (curves 2–11). Sharp peaks occur at finite temperatures, $10 < T_m < 40$ K, which clearly hint at polar phase transitions (PTs). Their easy axes are actually lying along [110] and [1$\bar{1}$0] within the basal xy-plane and yield, e.g., $\varepsilon_{max}^{\langle 110 \rangle} = 1.1 \cdot 10^5$ for $x = 0.0107$ [2]. Discussion within a random-field concept of PTs reveals xy-type quantum ferroelectricity above $x_c = 0.0018$ along the a axes of the paraelectric parent phase and a PT into a random phase above $x_r \approx 0.016$ (Figure 1). Quantum corrections to the temperature (see Equation (1)) are essential to describe the critical behavior.

In order to understand more details, some of us measured the optical linear birefringence (LB), $\Delta n_{ac} = n_c - n_a$, where n_c and n_a are the principal refractive indices at light wavelength $\lambda = 589.3$ nm, being linearly polarized along the c and a axes of the SCT crystal, respectively, as functions of temperature, T [18]. It is well-known that the LB is sensitive to both the axial rotation of the TiO_6 octahedra, $\langle \Delta \Phi^2 \rangle$, below the antiferrodistortive phase transition temperature $T_a = 105$ K (for $x = 0$), and to the FE short-range order parameter, $\langle P_x^2 \rangle$, where $x \parallel \langle 110 \rangle_c$ (Figure 1). Indeed, non-zero LB arises in pure STO at the transition temperature $T_a = 105$ K and at 115, 140, and 255 K (arrows) for $x = 0.002$, 0.0107, and 0.058, respectively, as shown in Figure 2. Additional FE anomalies, $\delta(\Delta n_{ac})$, are superposed at low T. Being non-morphic, they start smoothly with fluctuation tails and bend over into steeply rising long-range order parts below inflection points $T_1 \approx 15$, 28, and 50 K, respectively (arrows). These temperatures systematically exceed the ε' vs. T peak temperatures, $T_m = 14$, 26, and 35 K, respectively (Figure 1), where discontinuities of $d(\Delta n_{ac})/dT$ would be expected in case of PTs into long-range order. Absence of anomalies of this type and increasing differences, T_1-T_m, at increasing x hint at continuously growing smearing of the PTs. Simultaneously, as $T \to 0$, the polarization was calculated by use of the ordinary refractive index $n_o = 2.41$ and the electro-optic coefficient difference $g_{11}-g_{31} = 0.14$ m^4/C^2 as $\langle P_x^2 \rangle^{1/2} = \{2\delta(\Delta n_{ac})/[n_o^3(g_{11}-g_{31})]\} = 9.8$, 29.4, and 42.6 mC/m^2 for $x = 0.002$, 0.0107, and 0.058, respectively [18]. Since $\langle P_x^2 \rangle^{1/2}$ varies less than proportionally with x, comparatively incomplete FE order is observed. Further, the low-T polarization saturates, albeit slowly, at increasing electric field, E. This strongly hints at random-field induced nanodomains, whose average size increases with an applied ordering field. The increase of the average order parameter gives credit for disappearing domain walls as known from the domain-state FE $K_{0.974}Li_{0.026}TaO_3$ [19].

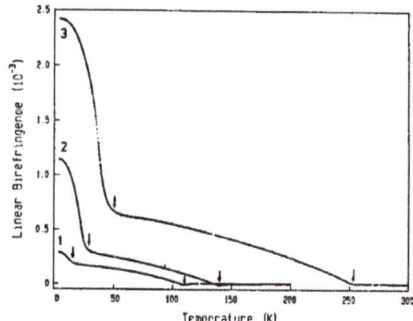

Figure 2. Linear birefringence Δn_{ac} vs. T measured at $\lambda = 589.3$ nm on crystallographic single domains of $Sr_{1-x}Ca_xTiO_3$ with $x = 0.002$ (1), 0.0107 (2), and 0.058 (3), respectively [18].

Further insight into FE SCT is gained from its relaxational behavior. Figure 3 shows the temperature dependence of the real and imaginary parts of the dielectric permittivity of SCT ($x = 0.002$), ε', and ε'' vs. T, at frequencies $10^3 \le f \le 10^4$ Hz [20]. In view of the rounded peaks of $\varepsilon'(T)$, a polydomain state of this FE is conjectured. This has first been interpreted within the concept of "dynamical heterogeneity" [21], which assumes a manifold of mesoscopic "dynamically correlated domains", relaxing exponentially with uniform single relaxation times. Their superposition defines the observed polydispersivity of the sample. It represents aggregates of polar clusters surrounding the quenched off-center Ca^{2+} dopant dipoles.

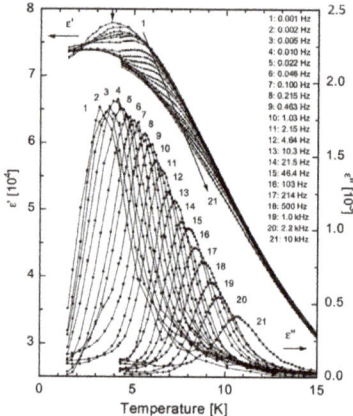

Figure 3. Real and imaginary parts of the permittivity, ε' and ε'' vs. T, of $Sr_{0.998}Ca_{0.002}TiO_3$ measured within $1.5 \le T \le 15$ K at frequencies $10^{-3} \le f \le 10^4$ Hz [20]. $T_g \approx 3.8$ K is indicated by an arrow.

It is noticed that ε' vs. T peaks at a "glass temperature", $T_g \approx 3.8$ K (Figure 3, arrow), in the quasi-static limit, $f = 1$ mHz, although at first glance, no glassy criticality as in spin glass is expected. However, in view of recently ascertained magnetic superspin glasses (SSG) of dipolarly coupled magnetic nanoparticles at low concentration [22], a related electric superdipolar glass (SDG) has become envisaged. It should behave like a relaxor ferroelectric [23] in terms of a superglassy critical power law behavior of the $\varepsilon''(f)$ vs. T peak position T_m.

$$f(T_m) \propto \left(T_m - T_g^e\right)^{zv}. \tag{2}$$

Evaluation over the whole range of frequencies, $10^{-3} \leq f \leq 10^4$ Hz, yields the expected dynamic critical exponent $zv \approx 10$ at $f > 1$Hz, while systematic deviations occur at lower f due to the well-known additional tunneling dynamics. Tests on the expected non-ergodicity of the SDG phase at $T < T_g$ upon zero-field- and field-cooled temperature cycles, respectively (cf. Section 3) are in preparation.

At higher concentration of Ca^{2+}, the polar nanoregions (PNRs) percolate into an FE ground state, as proven by first-order Raman scattering at the softening F_{1u} phonon mode in SCT ($x = 0.007$) at $T < T_0 = 18$ K [24]. SCT thus succeeds in demonstrating stable ferroelectricity. However, systematic research at increasing Ca content showed that T_0 is limited to ≈ 35 K, where the dielectric anomaly becomes increasingly smeared [25]. Better success was achieved by the classic method of stress-induced ferroelectricity in pure STO [5]. To this end Haeni et al. [26] utilized 50 nm thick films of STO, which were epitaxially grown with approximately +1.5% biaxial tensile strain on a (110) DyScO$_3$ substrate, while −0.9% uniform compression due to a $(LaAlO_3)_{0.29}(SrAl_{0.5}Ta_{0.5}O_3)_{0.71}$ (LSAT) substrate was barely active in this respect (Figure 4). The high permittivity in the films on DyScO$_3$, ε' up to 7000 at 10 GHz and room temperature, as well as its sharp dependence on an electric field is promising for device applications [4,26]. The observation of stress-induced ferroelectricity in STO films has confirmed theoretical predictions of Pertsev et al. [27]. While substrate induced tensile strain in epitaxial STO films via lattice parameter mismatch favors in-plane FE, compressive strain provides out-of-plane directed ferroelectricity [27]. This was observed by Fuchs et al. [28] at an STO film epitaxially grown on an STO substrate coated by compressive YBa$_2$Cu$_3$O$_7$.

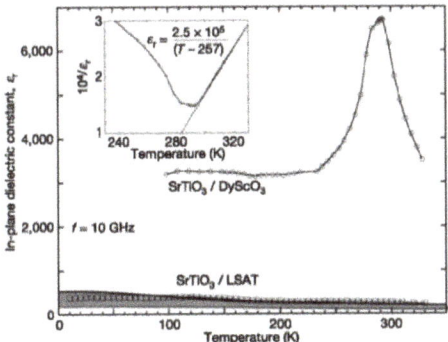

Figure 4. In-plane permittivity ε' vs. T of a strained 50 nm epitaxial STO/(110)DyScO$_3$ film at $f = 10$ GHz as compared to a compressed STO/LSAT film. The inset shows a Curie–Weiss fit to $(\varepsilon_r)^{-1}$ with $T_0 \approx 260$ K [26].

Another realization of room temperature ferroelectricity in STO confirms theoretical predictions of proximity effects at interfaces of metals to oxides containing PNRs such as, for example, STO [29]. Lee et al. [30] reported emergence of room temperature ferroelectricity at reduced dimensions, thus refuting a long-standing contradicting notion. Piezoelectric force microscopy (PFM) was able to evidence room-temperature ferroelectricity in strain-free epitaxial films with 24 unit-cell-thickness of otherwise non-ferroelectric STO (Figure 5). Following arguments from defect engineering in SCT, the authors claimed that electrically induced alignments of PNRs at Sr deficiency related defects are responsible for the appearance of a stable net of ferroelectric polarization in these films. This insight might be useful for the development of low-D materials of emerging nanoelectronic devices.

To systematically control ferroelectricity in thin films of STO at room temperature, Kang et al. [31] selectively engineered elemental vacancies by pulsed laser epitaxy (PLE). Sr^{2+} vacancies play an essential role in inducing the cubic-to-tetragonal transition, since they break the inversion symmetry, which is necessary for switchable electric polarization. The tetragonality turns out to increase with increasing vacancy density, thus strengthening the ferroelectricity, as shown in Figure 6a. This research has optimized tetragonality-induced ferroelectricity in STO with reliable growth control of the behavior. PFM yields stable hysteresis loops at room temperature, as shown in Figure 6b, where low and high laser fluences during PLE clearly demonstrate their key role in creating FE polarization. Similar propositions were made by the Barthélémy–Bibes group, which invoked both an electric field-switchable two-dimensional electron gas emerging in ferroelectric SCT films [32] and the non-volatile electric control of spin–charge conversion in an STO Rashba system [33].

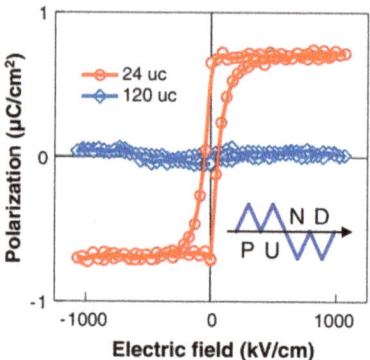

Figure 5. Polarization hysteresis of 24 and 120 unit-cell-thick STO films at room temperature, measured by using the double-wave PUND technique with a triangular *ac* electric field of 10 kHz (see schematic inset). The hysteresis component is obtained by subtracting the non-hysteretic (up (U) and down (D)) from the total (positive (P) and negative (N)) polarization runs [30].

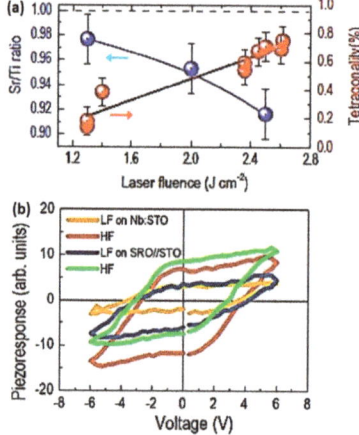

Figure 6. (a) Sr/Ti elemental concentration ratio (blue circles) and tetragonality measured at room temperature (red circles) plotted as functions of the laser fluence during pulsed laser epitaxy (PLE). (b) Ferroelectric hysteresis loops recorded by piezoelectric force microscopy (PFM) at 5-nm-thick $SrTiO_3$ films grown with low and high laser fluences (LF and HF, respectively; see (a)) on different bottom electrodes (STO:Nb and $SrRuO_3$/STO) [33].

Only recently has another insight into the ferroelectric state of compressively strained STO become available from high-angle annular dark-field imaging in scanning transmission electron microscopy. Salmani–Rezaie et al. [34] observed local polar regions in the room-temperature paraelectric phase of (001)-strained STO films, which were grown on (001) faces of LSAT and underwent an FE transition at low T. This unexpected feature was explained by a locally dipolar-ordered, but globally random phase of displaced Ti^{4+} columns, which underwent a disorder–order transition on cooling.

This Section started with different methods to establish long-range order, such as ferroelectricity (FE), metallicity (MT), or superconductivity (SC) in suitably modified STO [2,9,14]. Lately, more demanding procedures have become successful to stabilize the co-existence of apparently contradictory properties, e.g., FE-SC and FE-MT, which appear self-excluding at first glance. In this context, obscure terms such as, for example, "polar metal" and "metallic ferroelectric" or "ferroelectric metal" have been used interchangeably by the research community. Only recently have subtle distinctions of these variants with respect to their electric field switchability been clarified [12], although this topic still remains under debate.

Rischau et al. [35] showed that SC can coexist with an FE-like instability in oxygen-reduced ("n-doped") $Sr_{1-x}Ca_xTiO_{3-\delta}$ (0.002 < x < 0.009, 0 < δ < 0.001), where both long-range orders are intimately linked. The FE transition of insulating SCT was found to survive in this reduced modification. Owing to its metallic conductivity, the latter does not show a bulk reversible electric polarization and hence cannot be a true ferroelectric. However, it shows anomalies in various physical properties at the Curie temperature of the insulator, e.g., Raman scattering evidences that the hardening of the FE soft mode in the dilute metal is identical with what is seen in the insulator. The anomaly in resistivity was found to terminate at a threshold carrier density (n^*), near to which the SC transition temperature is enhanced [35]. This evidences the link between SC pairing and FE dipolar ordering, a subject of current attention [36].

Moreover, it is widely accepted that the low-T phase of STO lies in the vicinity of a quantum critical point, where different phases (i.e., paraelectric, antiferrodistortive, FE, MT, and SC) with similar energies compete, while weak residual interactions may stabilize one or several of these states [37–39]. The coexistence of MT and FE states in STO has been addressed under the keyword charge transport in a polar metal by Wang et al. [40], who studied the low-T electrical resistivity in several $Sr_{1-x}Ca_xTiO_{3-\delta}$ single-crystals at δ > 0 within 0.002 < x < 0.01 (Figure 7). Since both MT and FE are dilute, the distance between mobile MT electrons and fixed FE dipoles can be separately tuned but kept much longer than the interatomic distance. This opens the chance of activating a Ruderman–Kittel–Kasuya–Yosida-like interaction [41] of carriers with local electric moments, which was originally proposed by Glinchuk and Kondakova [42]. They introduced this indirect interaction of FE off-center ions with conduction electrons in order to explain high FE transition temperatures in certain narrow-gap semiconductors with high conductivity, such as $Pb_{1-x}Ge_xTe$. In agreement with this theory, it is expected that the threshold concentration of carriers, n^*, is proportional to x, which indicates that it occurs at a fixed ratio between inter-carrier and inter-dipole distances.

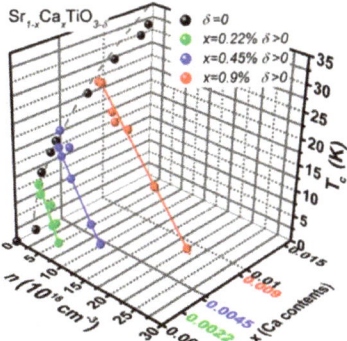

Figure 7. Ferroelectric (FE) phase transition temperatures T_c in insulating $Sr_{1-x}Ca_xTiO_3$ as functions of x (black balls [2]) and in metallic $Sr_{1-x}Ca_xTiO_{3-\delta}$ (δ > 0) as functions of charge carrier density n and x = 0.0022, 0.0045 and 0.009 (green, blue, and red balls, respectively [40]).

Tomioka et al. [17] finally demonstrated the simultaneous occurrence of three states, FE, MT, and SC, by independently controlling two concentrations of electron-doped $Sr_{1-x}La_xTi(^{16}O_{1-z}{}^{18}O_z)_3$ single crystals. They precisely controlled the "dome-like" SC characteristic by n doping via the La^{3+} content, while independently enhancing T_c by substitution of $^{18}O^{2-}$ ions for $^{16}O^{2-}$. At an electron concentration of $n \approx 5 \times 10^{19}$ cm^{-3}, they found the apex of the SC dome at $T_c \approx 0.44$ K, where they subsequently shifted its height to a record-high $T_c \approx 0.6$ K by adjusting $z(^{18}O)$.

Being arbitrarily close to the quantum critical point of non-centrosymmetric SC, experiments have thus come into reach to probe mixed-parity pairing mechanisms with topological aspects to their SC states, such as extremely large and highly anisotropic upper critical fields and topologically protected spin currents. A decisive step toward this aim was done by Schumann et al. [43] using La^{3+} or Sm^{3+} n-doped STO films on (001)-strained LSAT substrates. Being in their polar phase, they reveal enhanced superconducting T_c, while some of them show signatures of an unusual SC state, where the in-plane critical field is higher than both the paramagnetic and orbital pair breaking limits. Moreover, nonreciprocal transport is observed, which reflects the ratio of odd versus even pairing interactions. A similar highlight was observed in a gate-induced 2D SC of interfacial STO [44]. Due to its Rashba-type spin orbit interaction, it reveals nonreciprocal transport, where the inequivalent rightward and leftward currents reflect simultaneous spatial inversion and time-reversal symmetry breaking—an exciting prospect of forthcoming research on STO.

3. The Magnetoelectric Multiglass (Sr,Mn)TiO$_3$

The nature of glassy states in disordered materials has long been controversially discussed. In the magnetic community, generic spin glasses have long been accepted to undergo phase transitions at a static glass temperature T_g, where they exhibit criticality and originate well-defined order parameters [45]. In addition, disordered polar systems are expected to transit into generic "dipolar" or "orientational glass" states [46], which fulfil similar criteria as spin glasses. Hence, it appears quite natural to introduce the term "multiglass" for a new kind of multiferroic material revealing both polar and spin glass properties, which were discovered by some of us in the ceramic solid solution $Sr_{0.98}Mn_{0.02}TiO_3$ [47]. By various experimental methods [48–50] it has been ascertained that the Mn^{2+} ions are randomly substituting Sr^{2+} ions on A-sites in quantum paraelectric STO (Figure 8a), where they become off-centered due to their small ionic size and undergo covalent bonding with one of the twelve nearest neighboring O^{2-} ions. These elementary dipoles readily form polar nanoclusters with frustrated dipolar interactions, as illustrated in Figure 8b. It depicts the local cluster formation of Mn^{2+} ions with antiparallel electric dipole moments and antiferromagnetically correlated spins.

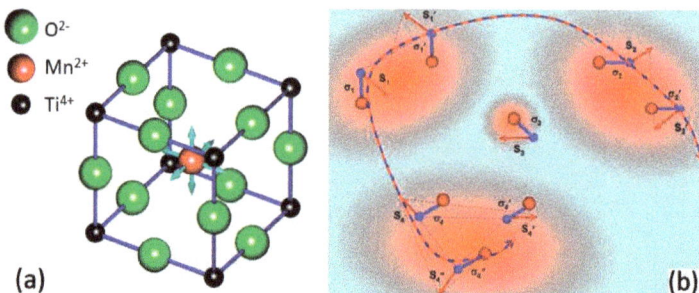

Figure 8. (a) A site substituted Mn^{2+} ion in its cage of 12 nearest neighboring oxygen ions in the ABO$_3$ lattice of STO going off-center along <100> [46]. (b) Schematic structure of SrTiO$_3$: Mn^{2+} highlighting a percolating multiglass path of randomly distributed Mn^{2+} ions (red–blue broken line) carrying dipole moments σ_j (blue lines) and spins S_j (red arrows) with electric dipolar and antiferromagnetic correlations, respectively, within polar STO clusters (red "clouds") [51].

The dipolar glass formation can easily be judged from the asymptotic shift of the dynamic dielectric susceptibility peak, $T_m(f)$, at frequencies within the range $10^{-1} \leq f \leq 10^6$ Hz in Figure 9a. It obeys glassy critical behavior according to Equation (2), where $zv = 8.5$ is the dynamic critical exponent and $T_g^e \approx 38$ K the electric glass temperature [51]. On the other hand, frustrated and random $Mn^{2+}-O^{2-}-Mn^{2+}$ superexchange is at the origin of spin glass formation below the magnetic glass temperature $T_g^m \approx 34$ K. This temperature marks the confluence of three characteristic magnetization curves recorded in $\mu_0 H = 10$ mT after zero-field cooling (ZFC) to $T = 5$ K upon field heating (m^{ZFC}), upon subsequent field cooling (m^{FC}), and thereafter the thermoremanence (m^{TRM}) upon zero-field heating (ZFH) as shown in Figure 9c. It should be noticed that both glassy states have unanimously been confirmed by clear-cut individual aging, rejuvenation, and memory effects in their respective dc susceptibilities [51]. "Holes" burnt into the electric and magnetic susceptibilities by waiting in zero external field for 10.5 h at 32.8 K and for 2.8 h at 33 K, respectively, and subsequent heating with weak electric or magnetic probing fields are shown in Figure 9b,d, respectively. They corroborate the glassy ground states of both polar and magnetic subsystems and their compatibility with spin glass theory [45]. Observation of the biquadratic ME interaction in the free energy [47],

$$F(E,H) = F_0 - (\delta/2)E_i E_j H_k H_l (i,j,k,l = 1,2,3), \tag{3}$$

is compatible with the low symmetry of the compound and is thought to crucially reinforce the spin glass ordering, as schematically depicted in Figure 8b [51]. Similarly to the dielectric anomaly [52], the magnetic anomaly has been found to depend not only on the frequency, but crucially also on the Mn content, confirming its intrinsic origin [53]. Furthermore, apart from ceramics, both glassy states have also been detected in equivalent thin films [54].

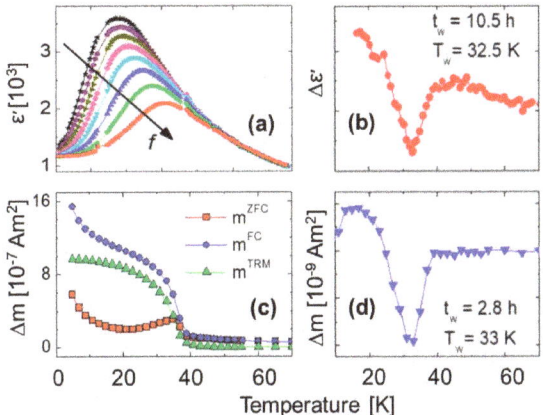

Figure 9. (a) Dielectric susceptibility $\varepsilon'(T)$ of $Sr_{0.98}Mn_{0.02}TiO_3$ ceramics recorded at frequencies $10^{-1} \leq f \leq 10^6$ Hz and (c) magnetization measured in $B = 10$ mT on field heating after ZFC (m^{ZFC}), on FC (m^{FC}), and on ZFH after FC (m^{TRM}). Holes $\Delta\varepsilon(T)$ and $\Delta m(T)$ burnt in zero fields at $T_{wait} = 32.5$ K for 10.5 h (b) and $T_{wait} = 33$ K for 2.8 h (d) confirm memory and rejuvenation of both electric and magnetic glassy subsystems [47].

Starting from a mean-field ansatz within the framework of a transverse Ising model [51], the complete theory of the ME multiglass is still under debate. In particular, the final steps for establishing the spin glass are missing. It is thought to emerge from multipolar interaction of spin clusters (Figure 8b) and probably comes close to the formation of a superspin glass as in systems of magnetic nanoparticles [22]. Since these probably consist of antiferromagnetic $MnTiO_3$ and carry merely surface magnetization [55], special care has to be taken.

In search of other ME multiglasses, we successfully examined also Mn^{2+} doped $KTaO_3$, which in the undoped case is a quantum paraelectric like STO, but nevertheless has slightly different properties on doping [56]. Other research groups have made similar experiments, and all of them reported considerable complexity [57–60]. Moreover, various other ME multiglasses have also been observed in disordered solid solutions such as $CuFe_{0.5}V_{0.5}O_2$ [61], La_2NiMnO_6 [62], Fe_2TiO_5 [63], and $(Ba_3NbFe_3Si_2O_{14})$:Sr [64].

4. Conclusions

STO still enjoys vivid interest in research and technological development. Having overcome the low-T bottleneck by advanced nanotechnologies, STO belongs to the most promising nanoelectronic materials. Unusual properties around the quantum critical point such as the co-existence of regular and superconductivity with ferroelectricity are still the focus of attention. On the other hand, the novel disordered phases of a superglass in $Sr_{0.998}Ca_{0.002}TiO_3$ and a multiglass in $Sr_{0.98}Mn_{0.02}TiO_3$ also still require dedicated activity.

Author Contributions: Conceptualization, W.K.; writing—original draft preparation, W.K.; writing—review and editing, W.K., J.D., A.T. and P.M.V. All authors have read and agreed to the published version of the manuscript.

Funding: This research received no external funding.

Acknowledgments: We are grateful to K.A. Müller and J.G. Bednorz for providing their outstanding single crystal samples of SCT and cooperating within common publications. In addition, we acknowledge valuable cooperation with A. Albertini, S. Bedanta, U. Bianchi, P. Borisov, A. Hochstrat, S. Miga, F.J. Schäfer, and V.V. Shvartsman.

Conflicts of Interest: The authors declare no conflict of interest.

Glossary

FC	field cooling
FE	ferroelectric or ferroelectricity
FH	field heating
LB	linear birefringence
LSAT	$(LaAlO_3)_{0.29}(SrAl_{0.5}Ta_{0.5}O_3)_{0.71}$
MT	metallic or metallicity
PLE	pulsed laser epitaxy
PNR	polar nanoregion
PT	phase transition
SC	superconductive or superconductivity
SCT	$Sr_{1-x}Ca_xTiO_3$
STO	$SrTiO_3$
TRM	thermoremanence
ZFC	zero-field cooling
ZFH	zero-field heating

References

1. Müller, K.A.; Burkard, H. $SrTiO_3$: An intrinsic quantum paraelectric below 4 K. *Phys. Rev. B* **1979**, *19*, 3593–3602. [CrossRef]
2. Bednorz, J.G.; Müller, K.A. $Sr_{1-x}Ca_xTiO_3$: An XY quantum ferroelectric with transition to randomness. *Phys. Rev. Lett.* **1984**, *52*, 2289–2293. [CrossRef]
3. Dec, J.; Kleemann, W. From Barrett to generalized quantum Curie-Weiss law. *Solid State Commun.* **1998**, *106*, 695–699. [CrossRef]
4. Rowley, S.E.; Spalek, L.J.; Smith, R.P.; Dean, M.P.M.; Itoh, M.; Scott, J.F.; Lonzarich, G.G.; Saxena, S.S. Ferroelectric quantum criticality. *Nat. Phys.* **2014**, *10*, 367–372. [CrossRef]
5. Uwe, H.; Sakudo, T. Stress-induced Ferroelectricity and soft phonon modes in $SrTiO_3$. *Phys. Rev. B* **1976**, *13*, 271–286. [CrossRef]

6. Itoh, M.; Wang, R.; Inaguma, Y.; Yamaguchi, T.; Shan, Y.-J.; Nakamura, T. Ferroelectricity induced by oxygen isotope exchange in strontium titanate perovskite. *Phys. Rev. Lett.* **1999**, *82*, 3540–3543. [CrossRef]
7. Nova, T.F.; Disa, A.S.; Fechner, M.; Cavalleri, A. Metastable ferroelectricity in optically strained $SrTiO_3$. *Science* **2019**, *364*, 1075–1079. [CrossRef]
8. Li, X.; Qiu, T.; Zhang, J.; Baldini, E.; Lu, J.; Rappe, A.M.; Nelson, K.A. Terahertz field–induced ferroelec-tricity in quantum paraelectric $SrTiO_3$. *Science* **2019**, *364*, 1079–1082. [CrossRef]
9. Tokura, Y.; Taguchi, Y.; Okada, Y.; Fujishima, Y.; Arima, T.; Kumagai, K.; Iye, Y. Filling dependence of electronic properties on the verge of metal-Mott-insulator. *Phys. Rev. Lett.* **1993**, *70*, 2126–2129. [CrossRef]
10. Spinelli, A.; Torija, M.A.; Liu, C.; Jan, C.; Leighton, C. Electronic transport in doped $SrTiO_3$: Conduction and potential applications. *Phys. Rev. B* **2010**, *81*, 155110. [CrossRef]
11. La Mattina, F.; Bednorz, J.G.; Alvarado, S.F.; Shengelaya, A.; Müller, K.A.; Keller, H. Controlled oxygen vacancies and space correlation with Cr^{3+} in $SrTiO_3$. *Phys. Rev. B* **2009**, *80*, 075122. [CrossRef]
12. Zhou, W.X.; Ariando, A. Review on Ferroelectric/polar metals. *Jpn. J. Appl. Phys.* **2020**, *59*, S10802. [CrossRef]
13. Alvarado, S.P.; La Mattina, P.; Bednorz, J.G. Electroluminescence in $SrTiO_3$: Cr single-crystal nonvolatile memory cells. *Appl. Phys. A* **2007**, *89*, 85–89. [CrossRef]
14. Schooley, J.F.; Hosler, W.R.; Cohen, M.L. Superconductivity in semiconducting $SrTiO_3$. *Phys. Rev. Lett.* **1964**, *12*, 474–475. [CrossRef]
15. Bednorz, J.G.; Müller, K.A. Possible high-T_c superconductivity in the Ba-La-Cu-O system. *Z. Phys. B Cond. Matter* **1986**, *64*, 189–193. [CrossRef]
16. Bednorz, J.G.; Müller, K.A. Perovskite-type oxides—The new approach to high-T_c superconductivity. *Nobel Lect. Dec.* **1987**. Available online: https://www.nobelprize.org/uploads/2018/06/bednorz-muller-lecture.pdf (accessed on 1 April 2020). [CrossRef]
17. Tomioka, Y.; Shirakawa, N.; Shibuya, K.; Inoue, I.H. Enhanced superconductivity close to a nonmagnetic quantum critical point in electron-doped strontium titanate. *Nat. Commun.* **2019**, *10*, 738. [CrossRef]
18. Kleemann, W.; Schäfer, F.J.; Müller, K.A.; Bednorz, J.G. Domain state properties of the random-field xy-model system $Sr_{1-x}Ca_xTiO_3$. *Ferroelectrics* **1988**, *80*, 297–300. [CrossRef]
19. Kleemann, W.; Kütz, S.; Rytz, D. Cluster glass and domain state properties of $K_{1-x}Li_xTaO_3$. *Europhys. Lett.* **1987**, *4*, 239–245. [CrossRef]
20. Kleemann, W.; Albertini, A.; Chamberlin, R.V.; Bednorz, J.G. Relaxational dynamics of polar nano-domains in $Sr_{1-x}Ca_xTiO_3$, $x = 0.002$. *Europhys. Lett.* **1997**, *37*, 145–150. [CrossRef]
21. Chamberlin, R.V.; Haines, D.N. Percolation model for relaxation in random systems. *Phys. Rev. Lett.* **1990**, *65*, 2197–2200. [CrossRef]
22. Bedanta, S.; Kleemann, W. Supermagnetism. *J. Phys. D Appl. Phys.* **2009**, *42*, 013001. [CrossRef]
23. Kleemann, W. Relaxor ferroelectrics: Cluster glass ground state via random fields and random bonds. *Phys. Status Solidi B* **2014**, *251*, 1993–2002. [CrossRef]
24. Bianchi, U.; Kleemann, W.; Bednorz, J.G. Raman scattering of ferroelectric $Sr_{1-x}Ca_xTiO_3$, $x = 0.007$. *J. Phys. Condens. Matter* **1994**, *6*, 1229–1238. [CrossRef]
25. Carpenter, M.A.; Howard, C.J.; Knight, K.S.; Zhang, Z. Structural relationships and a phase diagram for (Ca, Sr)TiO_3 perovskites. *J. Phys. Condens. Matter* **2006**, *18*, 10725–10749. [CrossRef]
26. Haeni, J.H.; Irvin, P.; Chang, W.; Uecker, R.; Reiche, P.; Li, Y.L.; Choudhury, S.; Tian, W.; Hawley, M.E.; Craigo, B.; et al. Room-temperature ferroelectricity in strained $SrTiO_3$. *Nature* **2004**, *430*, 758–761. [CrossRef]
27. Pertsev, N.A.; Tagantsev, A.K.; Setter, N. Phase transitions and strain-induced ferroelectricity in $SrTiO_3$ epitaxial thin films. *Phys. Rev. B* **2000**, *61*, R825–R829. [CrossRef]
28. Fuchs, D.; Schneider, C.W.; Schneider, R.; Rietschel, H. High dielectric constant and tunability of epitaxial thin film capacitors. *J. Appl. Phys.* **1999**, *85*, 7363–7369. [CrossRef]
29. Stengel, M.; Spaldin, N.A. Origin of the dielectric dead layer in nanoscale capacitors. *Nature* **2006**, *443*, 679–682. [CrossRef]
30. Lee, D.; Lu, H.; Gu, Y.; Choi, S.-Y.; Li, S.-D.; Ryu, S.; Paudel, T.R.; Song, K.; Mikheev, E.; Lee, S.; et al. Emergence of room-temperature ferroelectricity at reduced dimensions. *Science* **2015**, *349*, 1314–1317. [CrossRef]
31. Kang, K.T.; Seo, H.I.; Kwon, O.; Lee, K.; Bae, J.-S.; Chu, M.-W.; Chae, S.C.; Kim, Y.; Choi, W.S. Ferroelec-tricity in $SrTiO_3$ epitaxial thin films via Sr-vacancy-induced tetragonality. *Appl. Surf. Sci.* **2020**, *499*, 143930. [CrossRef]

32. Bréhin, J.; Trier, F.; Vicente-Arche, L.M.; Hemme, P.; Noël, P.; Cosset-Chéneau, M.; Attané, J.-P.; Vila, L.; Sander, A.; Gallais, Y.; et al. Switchable two-dimensional electron gas based on ferroelectric Ca: SrTiO$_3$. *Phys. Rev. Mater.* **2020**, *4*, 041002. [CrossRef]
33. Noël, P.; Trier, F.; Vicente-Arche, L.M.; Bréhin, J.; Vaz, D.C.; Garcia, V.; Fusil, S.; Barthélémy, A.; Vila, L.; Bibes, M.; et al. Non-volatile electric control of spin–charge conversion in a SrTiO$_3$ Rashba system. *Nature* **2020**, *580*, 483–486. [CrossRef]
34. Salmani-Rezaie, S.; Ahadi, K.; Strickland, W.M.; Stemmer, S. Order-disorder ferroelectric transition of strained SrTiO$_3$. *Phys. Rev. Lett.* **2020**, *125*, 087601. [CrossRef]
35. Rischau, C.W.; Lin, X.; Grams, C.P.; Finck, D.; Harms, S.; Engelmayer, J.; Lorenz, T.; Gallais, Y.; Fauqué, B.; Hemberger, J.; et al. A ferroelectric quantum phase transition inside the superconducting dome of Sr$_{1-x}$Ca$_x$TiO$_{3-\delta}$. *Nat. Phys.* **2017**, *13*, 643–648. [CrossRef]
36. Wölfle, P.; Balatsky, A.V. Superconductivity at low density near a ferroelectric quantum critical point: Doped SrTiO$_3$. *Phys. Rev. B* **2018**, *98*, 104505. [CrossRef]
37. Takada, Y. Theory of superconductivity in polar semiconductors and its application to *n*-type semicon-ducting SrTiO$_3$. *J. Phys. Soc. Jpn.* **1980**, *49*, 1267–1275. [CrossRef]
38. Gabay, M.; Triscone, J.-M. Superconductivity: Ferroelectricity woos pairing. *Nat. Phys.* **2017**, *13*, 624–625. [CrossRef]
39. Collignon, C.; Lin, X.; Rischau, C.W.; Fauqué, B.; Behnia, K. Metallicity and superconductivity in doped strontium titanate. *Ann. Rev. Cond. Matt. Phys.* **2019**, *10*, 25–44. [CrossRef]
40. Wang, J.L.; Yang, L.W.; Rischau, C.W.; Xu, Z.K.; Ren, Z.; Lorenz, T.; Hemberger, J.; Lin, X.; Behnia, K. Charge transport in a polar metal. *NPJ Quantum Mater.* **2020**, *4*, 61–68. [CrossRef]
41. Available online: https://en.wikipedia.org/wiki/RKKY_interaction (accessed on 15 April 2020).
42. Glinchuk, M.D.; Kondakova, I.V. Ruderman–Kittel–like interaction of electric dipoles in systems with carriers. *Phys. Stat. Sol.* **1992**, *174*, 193–197. [CrossRef]
43. Schumann, T.; Galletti, L.; Jeong, H.; Ahadi, K.; Strickland, W.M.; Salmani-Rezaie, S.; Stemmer, S. Possible signatures of mixed-parity superconductivity in doped polar SrTiO$_3$ films. *Phys. Rev. B* **2020**, *101*, 100503. [CrossRef]
44. Itahashi, Y.M.; Ideue, T.; Saito, Y.; Shimizu, S.; Ouchi, T.; Nojima, T.; Iwasa, Y. Nonreciprocal transport in gate-induced polar superconductor SrTiO$_3$. *Sci. Adv.* **2020**, *6*, eaay9120. [CrossRef] [PubMed]
45. Binder, K.; Young, A.P. Spin glasses: Experimental facts, theoretical concepts, and open questions. *Rev. Mod. Phys.* **1986**, *58*, 801–976. [CrossRef]
46. Binder, K.; Reger, J.D. Theory of orientational glasses: Models, concepts, simulations. *Adv. Phys.* **1992**, *41*, 547–627. [CrossRef]
47. Shvartsman, V.V.; Bedanta, S.; Borisov, P.; Kleemann, W.; Tkach, A.; Vilarinho, P.M. (Sr,Mn)TiO$_3$: A magnetoelectric multiglass. *Phys. Rev. Lett.* **2008**, *101*, 165704. [CrossRef] [PubMed]
48. Laguta, V.V.; Kondakova, I.V.; Bykov, I.P.; Glinchuk, M.D.; Tkach, A.; Vilarinho, P.M.; Jastrabik, L. Electron spin resonance investigation of Mn^{2+} ions and their dynamics in Mn-doped SrTiO$_3$. *Phys. Rev. B* **2007**, *76*, 054104. [CrossRef]
49. Lebedev, A.I.; Sluchinskaya, I.A.; Erko, A.; Kozlovskii, V.F. Direct evidence for off-centering of Mn impurity in SrTiO$_3$. *JETP Lett.* **2009**, *89*, 457–460. [CrossRef]
50. Levin, I.; Krayzman, V.; Woicik, J.C.; Tkach, A.; Vilarinho, P.M. X-ray absorption fine structure studies of Mn coordination in doped perovskite SrTiO$_3$. *Appl. Phys. Lett.* **2010**, *96*, 052904. [CrossRef]
51. Kleemann, W.; Bedanta, S.; Borisov, P.; Shvartsman, V.V.; Miga, S.; Dec, J.; Tkach, A.; Vilarinho, P.M. Multiglass order and magnetoelectricity in Mn2+ doped incipient ferroelectrics. *Eur. Phys. J. B* **2009**, *71*, 407–410. [CrossRef]
52. Tkach, A.; Vilarinho, P.M.; Kholkin, A.L. Polar behavior in Mn-doped SrTiO$_3$ ceramics. *Appl. Phys. Lett.* **2005**, *86*, 172902. [CrossRef]
53. Tkach, A.; Vilarinho, P.M.; Kleemann, W.; Shvartsman, V.V.; Borisov, P.; Bedanta, S. Comment on "The origin of magnetism in Mn-doped SrTiO$_3$". *Adv. Funct. Mater.* **2013**, *23*, 2229–2230. [CrossRef]
54. Tkach, A.; Okhay, O.; Wu, A.; Vilarinho, P.M.; Bedanta, S.; Shvartsman, V.V.; Borisov, P. Magnetic anomaly and dielectric tunability of (Sr,Mn)TiO$_3$ thin films. *Ferroelectrics* **2012**, *426*, 274–281. [CrossRef]
55. Ribeiro, R.A.P.; Andrés, J.; Longo, E.; Lazaro, S.R. Magnetism and multiferroic properties at MnTiO$_3$ surfaces: A DFT study. *Appl. Surf. Sci.* **2018**, *452*, 463–472. [CrossRef]
56. Shvartsman, V.V.; Bedanta, S.; Borisov, P.; Kleemann, W.; Tkach, A.; Vilarinho, P.M. Spin cluster glass and magnetoelectricity in Mn-doped KTaO$_3$. *J. Appl. Phys.* **2010**, *107*, 103926. [CrossRef]

57. Valant, M.; Kolodiazhnyi, T.; Axelsson, A.-K.; Babu, G.S.; Alford, N.M. Spin ordering in Mn-doped KTaO$_3$? *Chem. Mater.* **2010**, *22*, 1952–1954. [CrossRef]
58. Venturini, E.L.; Samara, G.A.; Laguta, V.V.; Glinchuk, M.D.; Kondakova, I.V. Dipolar centers in incipient ferroelectrics: Mn and Fe in KTaO$_3$. *Phys. Rev. B* **2005**, *71*, 094111. [CrossRef]
59. Golovina, I.S.; Shanina, B.D.; Geifman, I.N.; Andriiko, A.A.; Chernenko, L.V. Specific features of the EPR spectra of KTaO$_3$:Mn nanopowders. *Phys. Sol. State* **2012**, *54*, 551–558. [CrossRef]
60. Golovina, I.S.; Lemishko, S.V.; Morozovska, A.N. Percolation magnetism in ferroelectric nanoparticles. *Nanoscale Res. Lett.* **2017**, *12*, 382. [CrossRef]
61. Singh, K.; Maignan, A.; Simon, C.; Hardy, V.; Pachoud, E.; Martin, C. The spin glass Delafossite CuFe$_{0.5}$V$_{0.5}$O$_2$: A dipolar glass? *J. Phys. Condens. Matter* **2011**, *23*, 126005. [CrossRef]
62. Choudhury, D.; Mandal, P.; Mathieu, R.; Hazarika, A.; Rajan, S.; Sundaresan, A.; Waghmare, U.V.; Knut, R.; Karis, O.; Nordblad, P.; et al. Near-room-temperature colossal magnetodielectricity and multiglass properties in partially disordered La$_2$NiMnO$_6$. *Phys. Rev. Lett.* **2012**, *108*, 127201. [CrossRef] [PubMed]
63. Sharma, S.; Basu, T.; Shahee, A.; Singh, K.; Lalla, N.P.; Sampathkumaran, E.V. Multiglass properties and magnetoelectric coupling in the uniaxial anisotropic spin-cluster-glass Fe$_2$TiO$_5$. *Phys. Rev. B* **2014**, *90*, 144426. [CrossRef]
64. Rathore, S.S.; Vitta, S. Effect of divalent Ba cation substitution with Sr on coupled "multiglass" state in the magnetoelectric multiferroic compound Ba$_3$NbFe$_3$Si$_2$O$_{14}$. *Sci. Rep.* **2015**, *5*, 9751. [CrossRef] [PubMed]

© 2020 by the authors. Licensee MDPI, Basel, Switzerland. This article is an open access article distributed under the terms and conditions of the Creative Commons Attribution (CC BY) license (http://creativecommons.org/licenses/by/4.0/).

Communication

Detection of Two Phenomena Opposite to the Expected Ones

Boris I. Kochelaev

Physics Institute, Kazan Federal University, 42008 Kazan, Russia; bkochelaev@gmail.com

Received: 1 September 2020; Accepted: 19 September 2020; Published: 24 September 2020

Abstract: Both phenomena mentioned in the title were revealed by the electron paramagnetic resonance (EPR) method. The first phenomenon was found in superconducting La metal with Er impurities—the spin relaxation rate of the erbium impurities was sharply decreasing after transition into the superconducting state instead of the expected, i.e., the well-known Hebel–Slichter peak. The second unexpected phenomenon was discovered in the YbRh$_2$Si$_2$ compound—an excellent EPR signal from the Yb ions was observed at temperatures below the Kondo temperature determined thermodynamically, while according to the existing belief the EPR signal should not be observed at these temperatures due to the Kondo effect. In this tribute to K. Alex Müller, I describe the nature of the detected phenomena.

Keywords: superconductivity; Kondo effect; spin relaxation rate; magnetic resonance

1. Introduction

It is reasonable to remind readers that the electron paramagnetic resonance (EPR) study by Alex Müller on transition metal ions in the SrTiO$_3$ compound having the perovskite structure was the first step on the way to the discovery of high-temperature superconductivity (HTSC) [1]. His similar EPR study of different properties including the relation to their superconductivity in other perovskites gave him an impulse to search for superconductivity with a higher critical temperature, see [2]. The result is well known (1986). It is remarkable that Alex in his talk in 1969 at the International Conference dedicated to the 25th year of the EPR discovery by E. K. Zavoisky at Kazan University described advantages of the EPR method to study phase transitions, especially for their second type. For me, his opinion was not new due to our meetings before at the AMPERE colloquiums.

Magnetic impurities in transition metals and intermetallic compounds were used a rather long time ago as EPR probes to study the interactions and properties of these materials. It is useful for the following discussion to give some necessary results of these investigations (details can be found in the review by Barnes [3]).

In the case of a low concentration of impurities, their spin relaxation rate is defined mainly by interactions with conduction electrons and phonons. The exchange coupling of impurities with conduction electrons in the case of the axial symmetry can be presented by the following Hamiltonian:

$$H_{ex} = -\sum_i \left\{ J_\perp \left[S_i^x \sigma^x(\mathbf{r}_i) + S_i^y \sigma^y(\mathbf{r}_i) \right] + J_\parallel S_i^z \sigma^z(\mathbf{r}_i) \right\} \tag{1}$$

where $J_\perp, J_\parallel$ are the exchange integrals, $S_i^{x,y,z}$ are the spin components of the impurity, and $\sigma^{x,y,z}(\mathbf{r}_i)$ are the spin density components of the conduction electrons at the $\mathbf{r}_i$ position. The spin–lattice relaxation rate of conduction electrons $\Gamma_{\sigma L}$ is defined usually by their scattering on different defects of the lattice and on phonons due to their spin–orbital and spin–phonon interactions. If this relaxation is effective enough, the spin temperature of conduction electrons remains in the equilibrium state with other degrees of freedom. The EPR line width is defined in this case mainly by the relaxation rate Γ_{SS} (the so-called Korringa relaxation rate) due to the exchange interaction (Equation (1)). The spin–lattice relaxation rate of impurities due to other interactions Γ_{SL} in metals can be usually neglected. In the simplest case where $J_\perp = J_\| = J$, the Korringa relaxation rate is given by the following equation:

$$\Gamma_{SS} = \frac{4\pi}{\hbar}(\rho_F J)^2 k_B T \tag{2}$$

where ρ_F is the density of electronic orbital states at the Fermi energy level, k_B is the Boltzmann constant, and T is the temperature. The vice versa spin relaxation rate of conduction electrons to the equilibrium state of impurities (the Overhauser relaxation rate) $\Gamma_{\sigma\sigma}$ and Γ_{SS} satisfy the following detailed balance equation:

$$\frac{\Gamma_{\sigma\sigma}}{\Gamma_{SS}} = \frac{\chi_S^0 g_\sigma^2}{\chi_\sigma^0 g_S^2}; \quad \chi_S^0 = NS(S+1)\frac{(g_S\mu_B)^2}{3k_B T}, \quad \chi_\sigma^0 = \rho_F\frac{(g_\sigma\mu_B)^2}{2} \tag{3}$$

where χ_S^0, χ_σ^0 are the spin susceptibilities of non-interacting impurities and the conduction electrons, g_S, g_σ are the corresponding g-factors, and N and S are the concentration and spin of impurities, respectively. However, a very different situation appears if the spin systems of impurities and conduction electrons are strongly coupled and their Zeeman frequencies $\hbar\omega_S = g_S\mu_B H_0$, $\hbar\omega_\sigma = g_\sigma\mu_B H_0$ are very close (H_0 is the external magnetic field). This case can be represented by the following relations:

$$\Gamma_{SS}, \Gamma_{\sigma\sigma} \gg \Gamma_{SL}, \Gamma_{\sigma L}, |\omega_S - \omega_\sigma| \tag{4}$$

In this situation, both spin systems cannot be considered in the equilibrium states. The motion equations of magnetic impurities and conduction electrons are coupled by the additional coefficients $\Gamma_{S\sigma}$ and $\Gamma_{\sigma S}$, which coincide in the isotropic case with $\Gamma_{\sigma\sigma}$ and Γ_{SS} correspondingly. The spin relaxation of the whole system is realized then by the two steps as follows: first, the spin systems of magnetic impurities and conduction electrons achieve the common spin temperature and, second, they both relax to the equilibrium state of the lattice (the electron bottleneck regime). The spin dynamics of both spin systems in the bottleneck regime is also sufficiently changed since the collective spin excitations of impurities and conduction electrons appear. We are interested in the EPR signal of the mode, corresponding to the spin oscillations of impurities and conduction electrons in the same phase. In the simplest case $\omega_S = \omega_\sigma$ ($g_S = g_\sigma$) under the relaxation-dominated bottleneck regime, the spin relaxation rate of this mode Γ_{coll} can be presented as follows (neglecting Γ_{SL}):

$$\Gamma_{coll} = \Gamma_{S\sigma} B, \quad B = \Gamma_{\sigma L}/(\Gamma_{\sigma S} + \Gamma_{\sigma L}) \tag{5}$$

where B is the bottleneck factor—the less this factor, the stronger the bottleneck regime; in the case $B = 1$, the bottleneck is absent. If the impurities' concentration is large enough ($\chi_S^0 \gg \chi_\sigma^0$), then Γ_{coll} is strongly reduced and proportional to temperature:

$$\Gamma_{coll} \approx \frac{\Gamma_{S\sigma}}{\Gamma_{\sigma S}}\Gamma_{\sigma L} = \frac{\chi_\sigma^0}{\chi_S^0}\Gamma_{\sigma L} = bT, \quad b = \frac{3k_B\rho_F\Gamma_{\sigma L}}{2NS(S+1)} \tag{6}$$

Since the spin-lattice relaxation of conduction electrons $\Gamma_{\sigma L}$ does not depend on temperature, the behavior according to (Equation (6)) imitates the Korringa relaxation rate (Equation (2)), despite the bottleneck situation. Nevertheless, it should be mentioned that in the strongly anisotropic case ($J_\perp \neq J_\|$) it was found that the EPR line width narrowing is rather weak despite the bottleneck regime [4]. In the case of a parallel orientation of the external magnetic field to the symmetry axis for the spin relaxation rate of the collective mode (instead of Equation (6)), roughly the following result was obtained:

$$\Gamma_{coll} \sim \frac{\chi_\sigma^0}{\chi_S^0}\Gamma_{\sigma L} + \frac{(J_\perp - J_\parallel)^2(g_\perp^2 + g_\sigma^2)}{2J_\perp J_\parallel g_\perp g_\sigma}\Gamma_{S\sigma} \qquad (7)$$

If distances between impurities become relatively short, one should also take into account their magnetic dipole–dipole interactions and the Ruderman–Kittel–Kasuya–Yosida (RKKY) spin–spin interactions via conduction electrons. The latter indirect interaction between two impurities with spins S_1 and S_2 at the distance R between them is as follows:

$$\begin{aligned}H_{ind}(R) &= J_{ind}(\mathbf{S}_1\mathbf{S}_2)f(2k_FR);\\ J_{ind} &= \tfrac{9\pi Z^2 J^2}{2E_F},\ f(x) = \tfrac{\cos x}{x^3} - \tfrac{\sin x}{x^4}.\end{aligned} \qquad (8)$$

where k_F is the electron wave vector at the Fermi surface, Z is the number of conduction electrons per lattice atom, and E_F is the Fermi energy. Although this interaction is rather long range, it should be limited by the free path distance l_p of the conduction electrons: $f(x) \to f(x)\exp(-R/l_p)$. The dipole–dipole interactions give an additional broadening of the EPR signal from impurities, and the RKKY interaction leads to its narrowing of the Anderson–Weiss type. This contribution to the EPR line width is independent of temperature being responsible for the residual line width (at $T = 0$).

2. Peculiarities of the EPR Signal from Impurities in Superconductors

It is well known that NMR gives very important information about properties of superconductors confirming the Bardeen–Cooper–Schrieffer (BCS) theory [5,6]. It was found that the Hebel-Slichter peak appears due to a sharp increase of the nuclear spin relaxation rate to the superconducting electrons via their hyperfine interactions $H_{I\sigma}$:

$$H_{I\sigma} = A\sum_i\left\{I_i^x\sigma^x(\mathbf{r}_i) + I_i^y\sigma^y(\mathbf{r}_i) + I_i^z\sigma^z(\mathbf{r}_i)\right\} \qquad (9)$$

According to the BCS theory, the observed jump of the Korringa-type nuclear spin relaxation rate happens as a consequence of the sharp increase of the electron density states near the edge of the superconducting energy gap and due to the coherence factor of the Cooper pairing. A further lowering of temperature leads to an exponential decrease of the Korringa relaxation rate due to the thermal depopulation of the electronic state above the mentioned gap. The BCS theory explains the NMR result for the nuclear spin relaxation rate rather well (see Figure 1). In the case of superconductors of the second type, an additional broadening of the NMR signal due to the inhomogeneous penetration of the external magnetic field appears.

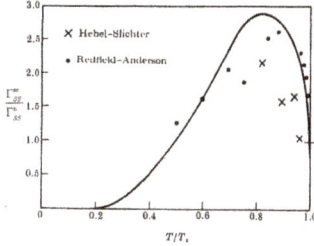

Figure 1. The temperature dependence of the nuclear spin relaxation rates' ratio in the superconducting and normal states of aluminum. The crosses and dots are results obtained in [6,7] accordingly. The solid line was obtained according to the Bardeen–Cooper–Schrieffer (BCS) theory in [8] (after [5], Section 3.3).

Since the structure of the hyperfine interactions (Equation (9)) is identical to the exchange interactions of impurities (Equation (1)), it was quite natural to expect similar behavior of the Korringa relaxation rate of the localized electrons. The only thing to do, for the interpretation of the EPR signal from magnetic impurities, was to substitute the hyperfine parameter A for the exchange integral J. The first observations of the EPR signal from impurities in superconductors were similar to the NMR results [9,10]. It is clearly seen in Figure 2.

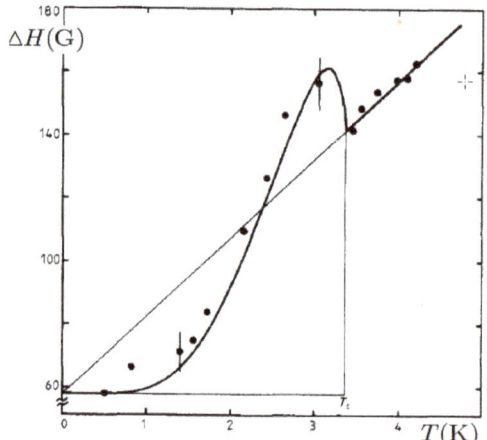

Figure 2. The temperature dependence of the electron paramagnetic resonance (EPR) line width for the Gd^{3+} in the $LaRu_2$ compound. The dots are the experimental results from [9,10], the solid line was calculated in [11] (after [11]).

However, an opposite phenomenon was rather soon observed by the EPR on the Er^{3+} ions in lanthanum (β– phase)—the EPR line width, instead of an expected increase after transition into the superconducting state, was sharply narrowed [12]. Later, a more detailed investigation of such a behavior of the EPR signal was performed for the same system [13]. Experimental results of the EPR line width dependence as functions of the temperature for different concentration of the Er ions are shown in Figure 3.

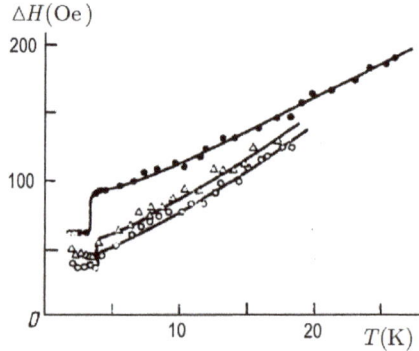

Figure 3. Temperature dependence of the EPR line width for Er^{3+} ions in the superconducting Lanthanum (β–phase) with Er^{3+} ions: black dots—2 at.% Er, triangles—1 at.% Er, open circles—0.5 at.% Er (after [13]).

In order to understand this phenomenon, it was proposed that the superconducting transition leads to a realization of the inequalities (Equation (4)) due to the increase of the spin relaxation rate of impurities to the conduction electrons [14]. As a result, the spin relaxation process of the total spin system transforms into the strong bottleneck regime with narrowing of the EPR signal. The qualitative explanation of this effect can be given in the following way. The effective spin relaxation rate in superconductors can be represented by the equation similar to Equation (5), with B^{sc} as the superconducting bottleneck parameter:

$$\Gamma^{sc}_{coll} \simeq \Gamma^{sc}_{S\sigma} B^{sc}, \quad B^{sc} = \Gamma^{sc}_{\sigma L}/\left(\Gamma^{sc}_{\sigma S} + \Gamma^{sc}_{\sigma L}\right) \tag{10}$$

In the case of $\Gamma^{sc}_{\sigma S} \ll \Gamma^{sc}_{\sigma L}$, the bottleneck regime is absent. The EPR line width in this situation is defined by the Korringa relaxation rate with the Hebel–Slichter peak similar to the NMR results. In the opposite case of $\Gamma^{sc}_{\sigma S} \gg \Gamma^{sc}_{\sigma L}$, we have the strong bottleneck regime and the effective spin relaxation rate is proportional now to the spin–lattice relaxation rate of the superconducting electrons, as follows:

$$\Gamma^{sc}_{coll} \simeq \left(\Gamma^{sc}_{S\sigma}/\Gamma^{sc}_{\sigma S}\right)\Gamma^{sc}_{\sigma L} = \left(\chi^{sc}_{\sigma}/\chi^{0}_{S}\right)\Gamma^{sc}_{\sigma L} \tag{11}$$

It was found earlier that the spin relaxation rate of superconducting electrons $\Gamma^{sc}_{\sigma L}$ becomes temperature dependent and decreases by lowering the temperature [15]. The exchange integral of the localized and the conduction electrons contained in the relaxation rates of the fraction (Equation (11)) is mutually canceled. The spin susceptibility of the superconducting electrons χ^{sc}_{σ} now also depends on the temperature getting much lower due to the Cooper pairing. These qualitative arguments were realized using the Feynman diagram technique, and it was shown that the Hebel—Slichter peak is absent in the case of the strong bottleneck regime [16].

The detailed temperature dependence of the effective spin relaxation rate of magnetic impurities in superconductors was investigated in [11] for different values of the bottleneck parameters B^n in the normal state of superconductors. The result is shown in Figure 4 for the normalized effective spin relaxation rate $R^{sc}_{coll} = \Gamma^{sc}_{coll}/\Gamma^{n}_{coll}$, with Γ^{n}_{coll} as the effective spin relaxation rate in the normal state and for the particular set of the object parameters.

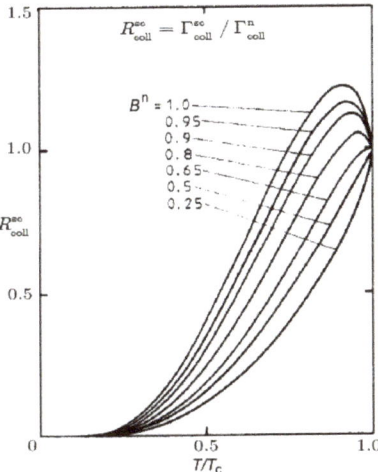

Figure 4. The normalized effective spin relaxation rate R^{sc}_{coll} as a function of the reduced temperature T/T_c for the several values of the normal state bottleneck factor B^n (after [11]).

The important role of the bottleneck regime in the normal state for the temperature dependence of the spin relaxation rate and, therefore, for the EPR signal from impurities in the superconducting state is now clearly seen. It must be mentioned that, in the case of nuclear spins, the bottleneck regime cannot appear because of a great difference in the resonance frequencies of the NMR and the EPR and very small nuclear spin susceptibility.

Another important contribution to the narrowing process of the EPR signal from impurities can arise from the transformation of the RKKY interaction due to the superconducting transition. In the superconducting state, in addition to the usual expression (Equation (8)), a non-oscillating long-range term H_{ind}^{sc} of antiferromagnetic character appears [17]. Since the corresponding expression is rather cumbersome, we mention only that this term slowly varies with distance and has the range of the order of the superconducting coherence length ξ, roughly $f^{sc}(R) \sim (1/R)\exp(-R/\xi)$. The typical value of the coherence length is rather large, i.e., $\xi \gg a_0$, a_0 is the lattice spacing. The detailed theory of the RKKY narrowing of the EPR signal in superconductors was developed in [18]. This type of the EPR line width narrowing has to be taken into account especially in the case of a weak bottleneck regime when interpreting the EPR results in superconductors.

3. The EPR Signal below the Kondo Temperature

The Kondo effect appears in metals with magnetic impurities in the case of their antiferromagnetic exchange interactions with conduction electrons ($J < 0$). To consider scattering of conduction electrons by the impurity at the $r_i = 0$ site, we use the secondary quantization for the conduction electrons in (Equation (1)):

$$H_{ex}^0 = -\frac{1}{2}\sum_{k,k'}[J_\parallel S_z(c_{k\uparrow}^+ c_{k'\uparrow} - c_{k\downarrow}^+ c_{k'\downarrow}) + J_\perp S_-(c_{k\uparrow}^+ c_{k'\downarrow} + S_+ c_{k\downarrow}^+ c_{k'\uparrow})] \quad (12)$$

where $S_\pm = S_x \pm iS_y$ and $c_{k\downarrow}^+$, $c_{k'\uparrow}$ are operators of creation and annihilation of electrons with the corresponding wave vectors and spin orientations. The probability of the electron transition from the state $|k_1 \uparrow\rangle$ into the state $|k_2 \downarrow\rangle$ in first-order perturbation theory is defined by the matrix element $\langle k_2 \downarrow|H_{ex}^0|k_1 \uparrow\rangle = -J_\perp S_+/2$. To obtain the second order for this matrix element, it is convenient to use the perturbation theory in the operator form [19]. We consider the energy band symmetric relative to the Fermi level, the electron energy ξ_k measured from the latter in the range $(-W, W)$. To simplify calculations, it is reasonable to use the following case: $k_B T \ll W$ and $S = 1/2$, see [20]. Taking into account the second-order contribution δH_{ex}^0, one can obtain the following expression for the isotropic symmetry:

$$\langle k_2 \downarrow|H_{ex}^0 + \delta H_{ex}^0|k_1 \uparrow\rangle \approx -(JS_+/2)\{1 - \rho_F J \ln[W/(k_B T)]\} \quad (13)$$

It is clear that in the case of $J < 0$ the matrix element increases with the lowering of temperature. The second term starts to exceed the first one approximately at the following temperature:

$$T_K \sim (W/k_B)\exp[-1/(\rho_F|J|)] \quad (14)$$

which bears the name of Kondo, who obtained these results in order to explain the temperature dependence of the resistivity [21]. The Kondo effect also becomes apparent in the temperature dependences of the specific heat, magnetic susceptibility, and the EPR signal from magnetic impurities (see [22]).

One can see that at the temperature $T \to 0$ the matrix element goes to infinity. It means that the second order of the perturbation theory is not enough. For isotropic symmetry, it has been shown that taking into account all orders of the logarithmic terms leads to very important consequences, i.e., below the Kondo temperature the magnetic moments of impurities are screened by the cloud of the conduction electrons and the EPR line width is of the T_K order (see [23,24]). These arguments created a widely accepted opinion that below the Kondo temperature the EPR signal cannot be observed on the localized electrons in metals.

It was a great surprise for many experts to observe the EPR signal from the Yb^{3+} ions in the $YbRh_2Si_2$ compound at the temperature $T = 1.6$ K and even below, whereas its Kondo temperature is $T_K = 25$ K [25,26], see Figure 5. It was also especially unexpected since this compound is a Kondo lattice system with heavy fermions.

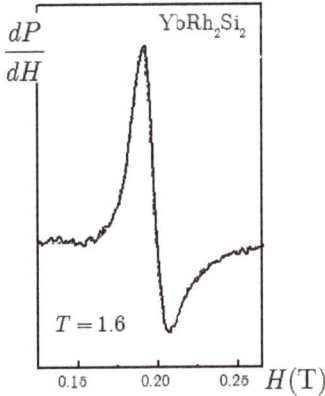

Figure 5. The EPR signal obtained in $YbRh_2Si_2$ below the Kondo temperature ($T_K = 25$ K) (after [26]).

An explanation of this phenomenon can be given on the basis of similar reasons leading to the disappearance of the Hebel–Slichter peak in superconductors, described in the previous Section 2. Since the matrix element (Equation (13)) strongly increases below the Kondo temperature, the same should happen with the Korringa relaxation rate Γ_{SS}. As a result, the bottleneck regime and the collective spin excitation of the Yb^{3+} ions with the conduction electrons should appear. Then, the effective spin relaxation rate can be described by Equation (6), where the exchange integral J is absent, that leads to a sufficient reducing of Γ_{eff} and the corresponding narrowing of the EPR line width.

However, doubts about this scenario can appear due to the extremely anisotropic g-factors of the Yb^{3+} ions ($g_\perp = 3.56$, $g_\| = 0.17$). It is related to the lowest Kramers doublet as a consequence of the tetragonal symmetry of the crystal electric field. One can expect that exchange integrals $J_\perp$ and $J_\|$ have similar anisotropy and that, for the effective spin relaxation rate, one should use Equation (7) instead of Equation (6). It can lead to an increase of the EPR line width instead of its narrowing. In order to understand the real situation, it was necessary to take into account all the orders of the perturbation theory in calculations of the effective spin relaxation rate for the anisotropic case.

To solve this task, it was convenient to use the method known as "Poor Man's Scaling" developed by Anderson [27]. The main idea of this method was to incorporate the total contribution from the high energy levels into the renormalized parameters of the starting Hamiltonian, in our anisotropic case $J_\perp \to J_\perp^R$ and $J_\| \to J_\|^R$. The second-order contribution for the initial anisotropic parameters is the following:

$$\delta J_\| = \rho_F J_\perp^2 \ln\left(\frac{W}{k_B T}\right), \quad J_\perp \delta J_\perp = J_\| \delta J_\| \qquad (15)$$

According to Anderson's idea, we divide the energy band into the low and high energy levels:

$$0 < |\xi_k| < \widetilde{W}, \quad \widetilde{W} < |\xi_k| < W \qquad (16)$$

For the renormalized contributions, we can obtain the following equations, representing the scaling law:

$$\delta \widetilde{J}_\| = \rho_F \widetilde{J}_\perp^2 \left[\ln\left(\frac{W}{k_B T}\right) - \ln\left(\frac{\widetilde{W}}{k_B T}\right) \right] = \rho_F \widetilde{J}_\perp^2 \ln\left(\frac{W}{\widetilde{W}}\right),$$
$$\widetilde{J}_\perp \delta \widetilde{J}_\perp = \widetilde{J}_\| \delta \widetilde{J}_\|.$$
(17)

These equations are valid in the range ($k_B T \leq \widetilde{W} \leq W$) for the running band value $\widetilde{W}$. An integration of the second equation gives the following:

$$\widetilde{J}_\perp^2 - \widetilde{J}_\|^2 = \text{Const} = J_\perp^2 - J_\|^2$$
(18)

The last equality can be accepted since this equation is valid for any values of $\widetilde{W}$ in the range mentioned above, including $\widetilde{W} = W$. To solve the first Equation (17), we consider it for the case of a small difference $W - \widetilde{W} = \delta \widetilde{W}$. Introducing the dimensionless values $U_\perp = \rho_F J_\perp$ and $U_\| = \rho_F J_\|$, we obtain from Equations (17) and (18) the following equation:

$$\delta \widetilde{U}_\| = \widetilde{U}_\perp^2 \ln\left(\frac{\widetilde{W} + \delta \widetilde{W}}{\widetilde{W}}\right) = \widetilde{U}_\perp^2 \frac{\delta \widetilde{W}}{\widetilde{W}} = (\widetilde{U}_\|^2 + U_0^2) \frac{\delta \widetilde{W}}{\widetilde{W}},$$
$$\widetilde{U}_\perp^2 - \widetilde{U}_\|^2 = U_0^2 = \rho_F(J_\perp^2 - J_\|^2).$$
(19)

The integration of this differential equation was performed in the following ranges of the running values: $U_\| \leq \widetilde{U}_\| \leq U_\|^R$ and $k_B T \leq \widetilde{W} \leq W$, where $U_\|^R$ is the final result of the renormalization. The result for renormalized parameters can be represented in a simple form as follows:

$$U_\|^R = U_0 \cot \varphi, \quad U_\perp^R = U_0 / \sin \varphi;$$
$$\varphi = U_0 \ln\left(\frac{T}{T_{GK}}\right), \quad T_{GK} = W \exp\left[-\frac{1}{U_0} \text{arc cot} \frac{U_\|}{U_0}\right].$$
(20)

The parameter T_{GK} is a scaling invariant which characterizes properties of the system at low temperatures. Its index GK reminds us that the ground electronic state of the Yb^{3+} ions is a Kramers doublet. In the case of $J_\perp = J_\|$, this parameter coincides with the Kondo temperature (Equation (14)). The renormalized Hamiltonian of the exchange interaction between impurities and conduction electrons is obtained simply by substituting in Equation (12) of the initial exchange integrals by the renormalized ones: $J_\perp \to J_\perp^R$ and $J_\| \to J_\|^R$.

Due to the extremely small $g_\|$-factor, most of the EPR experiments on $YbRh_2Si_2$ were performed in a geometry where both the static and alternating magnetic fields were perpendicular to the tetragonal axis of the crystal. In this case, the renormalized Korringa and Overhauser relaxation rates are the following:

$$\Gamma_{SS}^R = \frac{\pi}{\hbar} U_0^2 k_B T \left(\cot^2 \varphi + \frac{3}{4} \right), \quad \Gamma_{\sigma\sigma}^R = \Gamma_{SS}^R \frac{g_\perp}{2 g_\sigma \rho_F k_B T}.$$
(21)

Here, we neglected the molecular field appearing due to the RKKY interactions. One can see that both the relaxation rates are logarithmically divergent at temperatures approaching T_{GK} from above as follows: $\Gamma_{SS}^R, \Gamma_{\sigma\sigma}^R \propto 1/\ln^2(T/T_{GK})$. Nevertheless, such a behavior does not confirm the belief mentioned above that an observation of the EPR signal is impossible below the Kondo temperature because of its huge line width. This behavior just points out the strong coupling between the spin systems of ions and conduction electrons, and we cannot consider them separately.

The renormalized coupling coefficients for their equations of motion were found to be as follows:

$$\Gamma_{\sigma S}^R = \frac{\pi}{4\hbar} U_0^2 \frac{k_B T}{\sin^2(\varphi/2)}, \quad \Gamma_{S\sigma}^R = \frac{g_\perp}{2 g_\sigma \rho_F k_B T} \Gamma_{\sigma S}^R$$
(22)

These parameters are also divergent at $T \to T_{GK}$ in a way similar to Equation (21), $\Gamma_{S\sigma}^R, \Gamma_{\sigma S}^R \propto 1/\ln^2(T/T_{GK})$. The spin relaxation rates of both spin systems to the lattice (thermal bath) Γ_{SL} and

$\Gamma_{\sigma L}$ remain the same. The result for the relaxation rate of the resonant collective mode at $T > T_{GK}$ is the following:

$$\Gamma^R_{coll} = \Gamma_{SL} + \Gamma^{eff}_{\sigma L} + \Gamma^{eff}_{SS'}$$
$$\Gamma^{eff}_{\sigma L} = \Gamma_{\sigma L} \frac{\Gamma^R_{S\sigma}\Gamma^R_{\sigma S}}{(\Gamma^R_{\sigma\sigma})^2}, \quad \Gamma^{eff}_{SS} = \frac{(\Gamma^R_{SS})^2 - (\Gamma^R_{\sigma S})^2}{\Gamma^R_{SS}}. \tag{23}$$

The most interesting aspect is the behavior of the relaxation rate at temperatures approaching T_{GK} from above. In this temperature region, we have $U_0 \ln(T/T_{GK}) = \varphi \to 0$. After substitution of the renormalized relaxation rates (Equations (21) and (22)) into Equation (23), we obtain the following result for the effective parameters:

$$\Gamma^{eff}_{\sigma L}(\varphi \ll 1) \approx \left(\frac{2g_\sigma}{g_\perp}\right)\rho_F k_B T\right)\Gamma_{\sigma L}, \quad \Gamma^{eff}_{SS}(\varphi \ll 1) \approx \frac{\pi}{8\hbar}U_0^4 k_B T \ln^2\left(\frac{T}{T_{GK}}\right) \tag{24}$$

It is wonderful that all divergent contributions became mutually canceled even in the case of anisotropic symmetry. Moreover, the renormalized Korringa relaxation rate of the single impurity (Equation (21)) was transformed in the collective mode in such a way that, instead of a divergence, it goes to zero at $T \to T_{GK}$ (Equation (24)). Actually, this result explains the existence of the EPR signal at very low temperatures in the presence of the Kondo effect. As a matter of fact the Kondo effect helps to observe EPR in compounds with a high concentration of magnetic impurities and especially in the Kondo lattices, what is contrary to the common belief [28].

For a detailed comparison of the theory with experimental results, it is necessary to take into account additional interactions of the magnetic impurities and the conduction electrons like spin–phonon, magnetic dipole–dipole, and RKKY interactions and also an inhomogeneity of the static external magnetic field. Some results of these investigations can be found in the review [20].

4. Conclusions

The described very different phenomena appearing in different solids at very different phase transitions have a common reason—these transitions strengthen the electron bottleneck regime and stimulate the formation of collective spin excitations of localized and conduction electrons. As a result, we have rather paradoxical situations depending on the concentration of magnetic impurities in superconductors and normal metals. The reason the Hebel–Slichter peak appeared in the NMR experiments at the transition into the superconducting state turned out to be also the reason for its absence in the EPR experiments in superconductors with impurities in the case of their high enough concentrations. The same paradox happens in the case of the Kondo effect, allowing us to observe the EPR signal of the localized electrons below the Kondo temperature instead of it being blocked.

The EPR study of superconducting cuprates was more difficult because of unusually strong spin–phonon interactions of the Cu^{2+} ions and rather complicated results of the doping process. Nevertheless, Alex Müller was very interested in the EPR investigations of their HTSC properties. To continue this work, he invited an experienced experimenter, A. Shengelaya, to Zürich; I was also later invited for a theoretical support of these investigations. Some of the results are described in the review [29].

Concerning the possibilities of the EPR method, I cannot resist the temptation to quote the concluding words (published already in [29]) of Alex Müller's plenary lecture at the International Conference held in Kazan, dedicated to the 60th year of the EPR discovery:

"60 years after the discovery of EPR in Kazan the method is able to contribute at the forefront in condensed matter physics, such as high temperature superconductivity. This is especially so if properly employed, and the results theoretically interpreted in a scholar way, as well as relate them to other important experiments. Because the EPR spectrometers used are standard, and low cost as are the samples, the research budgets are low; this puts the scientist in a serene mood without stress."

It seems that the revealed two phenomena described in this communication also confirm these words and one can expect a similar role of the EPR method in future investigations.

Funding: This work received no external funding.

Acknowledgments: I am pleased to acknowledge Y.N. Proshin for reading the manuscript, offering useful comments, and helping to improve it.

Conflicts of Interest: The author declares no conflict of interest.

References

1. Müller, K.A. Paramagnetic resonance of Fe^{3+} in single crystals of $SrTiO_3$. *Helv. Phys. Acta* **1958**, *31*, 173–204. (In German)
2. Müller, K.A.; Kool, T.W. (Eds.). *Properties of Perovskites and Other Oxides*; World Scientific Publishing Co. Pte. Ltd.: Singapore, 2010.
3. Barnes, S.E. Theory of electron spin resonance of magnetic ions in metals. *Adv. Phys.* **1981**, *30*, 801–938. [CrossRef]
4. Kochelaev, B.I.; Safina, A.M. Electron-Bottleneck Mode for Magnetic Impurities in Metal in the Case of Anisotropic Exchange Interactions. *Phys. Solid State* **2004**, *46*, 226–230. [CrossRef]
5. Schrieffer, J.R. *Theory of Superconductivity*; W.A. Benjamin: Menlo Park, CA, USA, 1964.
6. Hebel, L.C.; Slichter, C.P. Nuclear Spin Relaxation Rate in Normal and Superconducting Aluminium. *Phys. Rev.* **1959**, *113*, 1504–1519. [CrossRef]
7. Redfield, A.G. Nuclear Spin Relaxation Time in Superconducting Aluminum. *Phys. Rev. Lett.* **1959**, *3*, 85–86. [CrossRef]
8. Hebel, L.C. Theory of Nuclear Spin Relaxation in Superconductors. *Phys. Rev.* **1959**, *116*, 79–81. [CrossRef]
9. Rettori, R.; Davidov, D.; Chaikin, P.; Orbach, R. Magnetic Resonance of a Localized Magnetic Moment in the Superconducting State: LaRu2:Gd. *Phys. Rev. Lett.* **1973**, *30*, 437–440. [CrossRef]
10. Davidov, D.; Rettori, R.; Kim, H.M. Electron-spin resonance of a localized moment in the superconducting state: BRu2:Gd (B=La,Ce,Th). *Phys. Rev. B* **1974**, *9*, 147–153. [CrossRef]
11. Tagirov, L.R.; Trutnev, K.F. Spin kinetics and EPR in superconductors. *J. Phys. F Met. Phys.* **1987**, *17*, 695–713. [CrossRef]
12. Alekseevsky, N.E.; Garifullin, I.A.; Kochelaev, B.I.; Kharakhashyan, E.G. Electron resonance with localized magnetic moments of Er in superconducting La. *Sov. Phys. JETP Lett.* **1973**, *18*, 189–191.
13. Alekseevsky, N.E.; Garifullin, I.A.; Kochelaev, B.I.; Kharakhashyan, E.G. Electron paramagnetic resonance for localized magnetic states in the superconducting La–Er system. *Sov. Phys. JETP* **1977**, *45*, 799–804.
14. Kochelaev, B.I.; Kharakhashyan, E.G.; Garifullin, I.A.; Alekseevsky, N.E. Electron paramagnetic resonance of localized moments in a type-II superconductor. In Proceedings of the 18th AMPERE Congress, Nottingham, UK, 9–14 September 1974; North-Holland Pub. Co.: Amsterdam, The Netherlands, 1975; pp. 23–24.
15. Maki, K. Theory of Electron-Spin Resonance in Gapless Superconductors. *Phys. Rev. B* **1973**, *8*, 191–199. [CrossRef]
16. Kosov, A.A.; Kochelaev, B.I. Electron paramagnetic resonance on the localized magnetic moments in superconductors. *Sov. Phys. JETP* **1978**, *47*, 75–83.
17. Kochelaev, B.I.; Tagirov, L.R.; Khusainov, M.G. Spatial dispersion of the spin susceptibility of conduction electrons in superconductors. *Sov. Phys. JETP* **1979**, *49*, 291–301.
18. Kochelaev, B.I.; Tagirov, L.R. "Exchange-Field-Narrowing" process for the inhomogeneously broadened EPR lines in superconductors. *Solid State Commun.* **1985**, *53*, 961–966. [CrossRef]
19. Bogolyubov, N.N.; Tyablikov, S.V. On an application of perturbation theory to the polar model of a metal. *Zhurnal Ehksperimental'noj i Teoreticheskoj Fiziki* **1949**, *19*, 251–255.
20. Kochelaev, B.I. Magnetic properties and spin kinetics of a heavy-fermion Kondo lattice. *Low Temp. Phys.* **2017**, *43*, 93–103. [CrossRef]
21. Kondo, J. Resistance Minimum in Dilute Magnetic Alloys. *Progr. Theor. Phys.* **1964**, *32*, 37–49. [CrossRef]
22. Baberschke, K.; Tsang, E. ESR Study of the Kondo Effect in Au:Yb. *Phys. Rev. Lett.* **1980**, *45*, 1512. [CrossRef]
23. Abrikosov, A.A. Electron scattering on magnetic impurities in metals and anomalous resistivity effects. *Physics* **1965**, *2*, 5–20. [CrossRef]

24. Suhl, H. Dispersion Theory of the Kondo Effect. *Phys. Rev.* **1965**, *138*, 515–522. [CrossRef]
25. Sichelschmidt, J.; Ivanshin, V.A.; Ferstl, J.; Geibel, C.; Steglich, F. Low Temperature Electron Spin Resonance of the Kondo Ion in a Heavy Fermion Metal: YbRh2Si2. *Phys. Rev. Lett.* **2003**, *91*, 156401. [CrossRef]
26. Sichelschmidt, J.; Wykhoff, J.; Krug von Nidda, H.-A.; Ferstl, J.; Geibel, C.; Steglich, F. Spin dynamics of $YbRh_2Si_2$ observed by electron spin resonance. *J. Phys. Condens. Matter* **2007**, *19*, 116204. [CrossRef]
27. Anderson, P.W. A poor man's derivation of scaling laws for the Kondo problem. *J. Phys. C Solid State Phys.* **1970**, *3*, 2436–2441. [CrossRef]
28. Kochelaev, B.I.; Belov, S.I.; Skvortsova, A.M.; Kutuzov, A.S.; Sichelschmidt, J.; Wykhoff, J.; Geibel, C.; Steglich, F. Why could electron spin resonance be observed in a heavy fermion Kondo lattice? *Eur. Phys. J. B* **2009**, *72*, 485–489. [CrossRef]
29. Kochelaev, B.I. Electron Paramagnetic Resonance in Superconducting Cuprates. In *Springer Series in Materials Science 255: High-Tc Copper Oxide Superconductors and Related Novel Materials Dedicated to Prof. K. A. Müller on the Occasion of His 90th Birthday*; Bussmann-Holder, A., Keller, H., Bianconi, A., Eds.; Springer: New York, NY, USA, 2017; pp. 165–175.

© 2020 by the author. Licensee MDPI, Basel, Switzerland. This article is an open access article distributed under the terms and conditions of the Creative Commons Attribution (CC BY) license (http://creativecommons.org/licenses/by/4.0/).

Creative

Meetings with a Remarkable Man, Alex Müller—The Professor of SrTiO3

Thomas W. Kool

Van 't Hoff Institute for Molecular Sciences, University of Amsterdam, 1098 XH Amsterdam, The Netherlands; tomkool@hotmail.com

Received: 22 June 2020; Accepted: 24 June 2020; Published: 2 July 2020

After my bachelor degree in chemistry with physics and mathematics (in Dutch kandidaatsexamen) at the University of Amsterdam, I chose to study for my master degree (in Dutch doctoraal) a physical chemistry direction. At the laboratory of physical chemistry, I met Ad Lagendijk (later full Professor and Spinoza laureate), who studied with the help of ESR (electron spin resonance) impurities in $SrTiO_3$. Together with Jan Mooij, I did research in hexaurea perchlorate, especially the 300 K phase transition. My master thesis was a theoretical study about the 105 K phase transition in $SrTiO_3$. In this way, I became familiar with the work of Professor K. Alex Müller (KAM), especially the structural phase transition. In 1975, I started a PhD research in $SrTiO_3$ in the group of Professor Max Glasbeek. In the same building, Professor Laurens Jansen (writer of a well-known book about group theory) [1] was working. He was the head of the theoretical chemistry department. Jansen had worked at the Batelle institute (Geneva-Switzerland), which is the same institute where KAM had worked. The PhD thesis of KAM (in German, English translation can be found in Reference [2]) was a study of Fe-doped $SrTiO_3$, especially Fe^{3+} [3]. Prof. Jansen suggested that I contact KAM, and so I did. In the summer of 1976, during my vacation in Italy and Switzerland, I had a break in Rüschlikon near Zürich, the IBM research center, where I met KAM. It was a special meeting. I remember talking about my ESR study of an axial Fe^{5+} center in $SrTiO_3$, telling KAM that I did not know how to explain the rotation angle of this center below the 105K structural phase transition. The rotation angle is different and larger than the intrinsic one. After a few minutes in pin-drop silence, KAM found the solution (see also the first research article below). KAM invited me to do further experiments at the K-band spectrometer at IBM, which I did. There, I met the late Walter Berlinger and performed together with him measurements with Swiss precision. In 1976, KAM visited Amsterdam, giving lectures, and we had discussions about the research we did. We had a wonderful time visiting the Dutch beach, and we also went to classical concerts in Amsterdam. A college of mine, Henk de Jong, studied the $SrTiO_3$ Cr^{5+} center. Lagendijk had the opinion that it was an off-center system [4], but de Jong had the opinion that it was a tetragonal Jahn–Teller system [5]. He also found another Cr^{5+} center and attributed this center to another Jahn–Teller (JT) system, orthorhombic [6]. I had hesitations (see also the second research article below). One evening at a restaurant, I discussed this with KAM, and we found the solution to this second center: it must be an off-center system. He advised me to publish this later after writing my thesis, and so it happened [7]. At that time, KAM was writing a part of a publication about the central peak found in $SrTiO_3$.

Due to personal circumstances, it took a while to write my PhD thesis [8]. Alex was enthusiastic about the thesis, which was completely dedicated to impurity systems in $SrTiO_3$. At a summer holiday in Switzerland, meeting him in his house in the eastern part at Laax-Salums, we decided to write about the $SrTiO_3$ orthorhombic off-center system [7]. He also asked me to translate his thesis from German into English, which I did together with Charlotte Bolliger from the IBM laboratory. The purpose was the publication of a book with the research work of KAM before the Nobel prize and some later work. After a considerable time, Alex invited me to help him to write and edit the book and finish it. In the summer of 2008 and 2009, we were writing the book "Properties of perovskites and other oxides [2]", which was published by World Scientific in 2010. The book contained also work of mine performed in Amsterdam and Osnabrück. The research in Osnabrück was with the late Prof. Ortwin Schirmer, who had worked at the IBM research laboratory before. This research was mainly about impurities in $BaTiO_3$ and explained bosonic bi-polaron behavior, which is important in the cuprate superconductors.

The discussions with Alex were not only about physics but contained a broad spectrum of mutual interest. One of his topics was the number five (in German, Zahl fünf), see reference [9].

In the following part, I will discuss some research in $SrTiO_3$.

The first article is about the 105 K structural phase transition. In this article, the rotation angle of the non-cubic Fe^{5+} is explained as mentioned above [10].

The second article is about the tetragonal off-center Cr^{5+} [11].

1. SrTiO$_3$ with the Non-Cubic Fe^{5+} as Local Probe and a Reinterpretation of Other Fe^{5+} Centers

The 105 K second-order displacive phase transition of $SrTiO_3$ (ST) has been studied with the help of Electron Paramagnetic Resonance. The photochromic non-cubic Fe^{5+} center is used as local probe. Critical phenomena, characterized by an exponent $\beta = 1/3$, are presented. Line broadening effects are interpreted as stemming from time-dependent fluctuations near the phase transition point. The data are consistent with those reported earlier on the Fe^{3+}-V_O pair center, indicating cooperative effects in the crystal. In addition, a model for the non-cubic Fe^{5+} is proposed, i.e., the ion is substitutional for Ti^{4+} with an empty adjacent expanded octahedron. Other Fe^{5+} centers in ST and $BaTiO_3$ (BT) are reviewed and reinterpreted.

1.1. Introduction

Displacive phase transitions were analyzed almost four decades ago with the help of classical theories such as the Landau theory [12] or with microscopic theories using the mean field approximation [13]. More specifically, a typical temperature behavior is predicted for the generalized susceptibility χ, the order parameter and the specific heat c_p of the form $(-\varepsilon)^x$, where $\varepsilon = (T - T_c)/T_c$. Values for the exponent x are:

$x = \beta = \frac{1}{2}$ for the order parameter,
$x = \gamma = -1$ for the susceptibility, and
$x = 0$ for the specific heat, indicating a jump at T_c, where T_c is the critical temperature of the phase transition.

Although classical theories are adequate for temperatures well outside the phase transition point, deviations or critical behavior from these theories occur near T_c. These phenomena arise from correlated fluctuations of the order parameter and become important when the length of the correlated fluctuations exceeds the range of forces. An explanation of these effects and the Electron Spin Resonance (ESR) technique used is reviewed by K. Alex Müller and J.C. Fayet [14]. This article can also be found in Chapter VII of the book: *Properties of Perovskites and other oxides* by K. Alex Müller and Tom W. Kool [2].

Experimentally, the critical behavior of ST near the 105 K phase transition was verified by means of the Fe^{3+} and Fe^{3+}-V_O impurity centers [15]; both are substitutional for Ti^{4+}. In this paper, we present EPR results of the 105 K second-order displacive phase transition in ST, where the non-cubic photochromic Fe^{5+} is used as local probe. This impurity center is a $d^3 (S = 3/2)$ system, substituting for Ti^{4+} and is octahedrally surrounded by a cage of oxygen ions in the presence of a moderate axial field for $T > T_c$. For $T < T_c$, a weak orthorhombic perturbation due to the phase transition is added. This center has been analyzed before by Kool et al. [16].

The study of phase transitions by means of ESR and the use of *different impurities* situated at the *same* site provide more evidence of the cooperative behavior in the crystal. The non-cubic Fe^{5+} center in ST is adequate because of the large anisotropy of the resonance lines, ranging from $g \approx 2$–4 and the very accurate measurements of the rotational order parameter $\varphi(T)$.

1.2. The Non-Cubic Fe^{5+} Centre

For axially distorted (tetragonal or trigonal) octahedrally surrounded d^3 spin systems, the following spin-Hamiltonian is used [17–20]:

$$\mathcal{H} = S \cdot \overline{\overline{D}} \cdot S + \mu_B H \cdot \overline{\overline{g}} \cdot S. \tag{1}$$

The first term represents the zero-field splitting and the latter represents the Zeeman interaction. For systems with $|D| \gg hv$, with v the frequency of a typical ESR experiment, the first term is taken as zero-order Hamiltonian $\mathcal{H}_0$, and the Zeeman splitting is treated as a perturbation $\mathcal{H}_1$. For values $hv/|2D| \geq 0.25$, one has to proceed with exact numerical computer calculations [21]. If the zero-field splitting $|2D|$ is much larger than the Zeeman term, only one ESR transition within the Kramers doublet with $M_S = |\pm 1/2\rangle$ levels is observed, giving a typical g^{eff} ESR spectrum for $S = 3/2$ systems ranging from $g^{eff} \approx 2$ to 4. The ESR spectrum of the non-cubic Fe^{5+} center shows these lines, too [16]. The non-cubic Fe^{5+} center has the following g and D values: $g_\| = 2.0132$, $g_\perp = 2.0116$, and $|2D| = 0.541$ cm^{-1} at 115 K [2,16]. From depopulation measurements at helium temperatures, it could be concluded that the sign of D is negative.

Below the phase transition ($T < T_c$), an extra weak orthorhombic perturbation in the spin-Hamiltonian has to be added:

$$E(S_x^2 - S_y^2). \tag{2}$$

For $|D| \geq hv$ and $|E| \leq hv$, general angular expressions for d^3 ($S = 3/2$) systems were derived [22]. At 77 K, well below T_c, $|2D| = 0.551$ cm^{-1} and $|E| = 0.529 \times 10^{-3}$ cm^{-1}, indicating that D is a little temperature dependent. The rhombic parameter E is temperature dependent and is proportional to φ^2, with φ the intrinsic rotation angle of ST consisting of alternating rotations of neighboring oxygen octahedra below T_c [14,16].

The rotation angle φ^* of the non-cubic Fe^{5+} is larger than the intrinsic one φ. The relative large value for φ^* is interpreted as follows. As can been seen in Figure 1, the octahedron adjacent to the Fe^{5+} ion has been expanded. As a consequence of this expansion, the rotation angle φ^* becomes larger.

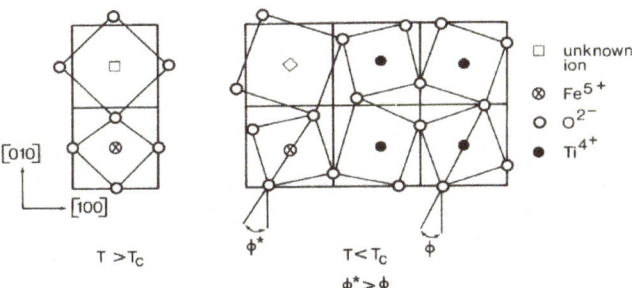

Figure 1. Rotation of the oxygen octahedra around the [001] axis. φ is the intrinsic rotation angle of the crystal, while φ^* is the rotation angle of the non-cubic Fe^{5+} center, which has an expanded adjacent oxygen octahedron due to a Ti^{4+} vacancy.

This expansion can be caused by a Ti^{4+} vacancy or a nearby impurity substitutional for the Ti^{4+} ion. Simple calculations at 77 K showed that with the given intrinsic angle $\varphi = 1.53°$ and $\varphi^* = 1.75°$ of the non-cubic Fe^{5+} and the distance of the lattice constant (above T_c) $a = 3.9$ Å, the expansion must be equal to 0.30 Å. Then, it follows that the unknown ion possesses a radius $r = 0.94$ Å, knowing that $r(Ti^{4+}) = 0.64$ Å [23].

The impurity concentration (in ppm) in the investigated crystal obtained by spectrochemical analyses is as follows (see the Table 1).

Table 1. Impurity concentration.

Fe 18	B < 10
Mo 2	Si 500–2000
Pb 500–1000	Ni 10–50
Sn 200–1000	Al < 10

Impurity concentration in ppm.

None of the above-mentioned impurities and their respective ions fit into the expanded cage [23]. In addition, Ti^{2+}, with $r(Ti^{2+}) = 0.80$ Å, and Sr^{2+}, with $r(Sr^{2+}) = 1.27$ Å, do not fit. Therefore, we assume that the expansion is due to a neighboring Ti^{4+} vacancy and is caused by the repulsion of the O^{2-} ions.

1.3. Critical Effects

1.3.1. Static Critical Exponents

The order parameter (with critical exponent β) corresponds to the displacement parameter, which in ST is represented by the rotation angle φ. The shaping of the crystals was such that after rapid cooling, the crystals became monodomain below the structural phase transition (Figure 2). In monodomain crystals, only the $\pm\varphi^*$ lines are present [24].

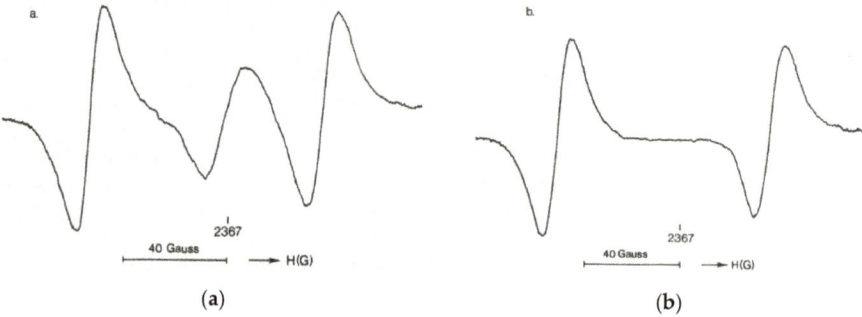

Figure 2. X-band Electron Spin Resonance spectrum of the $-\frac{1}{2} \leftrightarrow +\frac{1}{2}$ transition near $g^{\text{eff}} \approx 2.8$ of a (a) three- and (b) a monodomain crystal of the non-cubic Fe^{5+} in $SrTiO_3$.

In a monodomain crystal, the rotation angle φ^* can be used for the study of static critical exponents and will not be disturbed by extra ESR lines stemming from different domains. In the temperature range 30 K < T < 85 K, this rotation angle was found to be linearly proportional to the intrinsic one. The classical Landau behavior with $\beta = 1/2$ was obtained by plotting $\varphi^{*1/\beta} = \varphi^{*2}$ as a function of the reduced temperature $t = T/T_c$. For $0.7 < t < 0.9$, we found a straight line following a Landau behavior. At $t = 0.9$, the bending down from a straight line becomes noticeable (Figure 3).

It is found that for $0.9 < t < 1$, $\beta = 1/3$, i.e., in this temperature region, the crystal displays critical behavior. In Figure 4, $\varphi^{*1/\beta} = \varphi^{*3}$ is plotted as a function of t. Extrapolation of the plot to $\varphi^{*3} = 0$ yields the phase transition temperature $T_c = 103$ K, which in our sample is lower than the usual value of 105 K. This is due to the presence of impurities, which can alter the phase transition temperature.

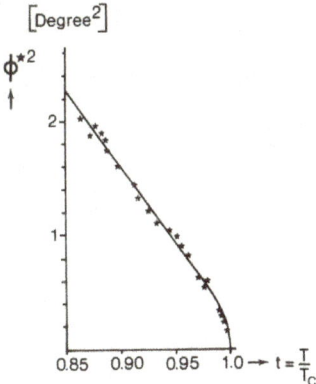

Figure 3. Plot of φ^{*2} of the non-cubic Fe^{5+} center in ST versus reduced temperature $t = T/T_c$, showing the changeover from classical to critical behavior.

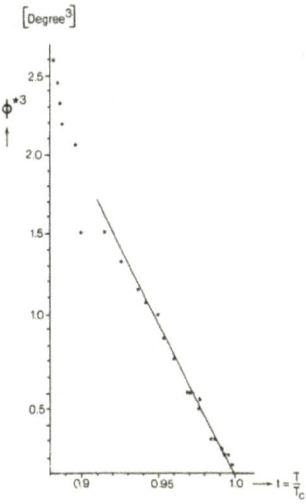

Figure 4. Plot of φ^{*3} of the non-cubic Fe^{5+} center in ST versus reduced temperature $t = T/T_c$.

1.3.2. Asymmetric Line Shapes for $T \rightarrow T_c^-$

Time-dependent fluctuations and line-broadening effects for the non-cubic Fe^{5+} have been published before by Kool et al. [2,16]. In addition, asymmetric line shapes in the critical region were found [2,25,26]. In a monodomain crystal ST, with the magnetic field $H\|[110]$ and the elongated axis of the crystal $c\|[001]$, we found outside the critical region ($T \ll T_c$) a symmetric Lorenzian line shape for each of the $\pm \varphi^*$ lines at $g^{eff} \approx 3.4$ On approaching T_c, the lines have at $T = T_c - 0.8$ K an asymmetric line shape (Figure 5).

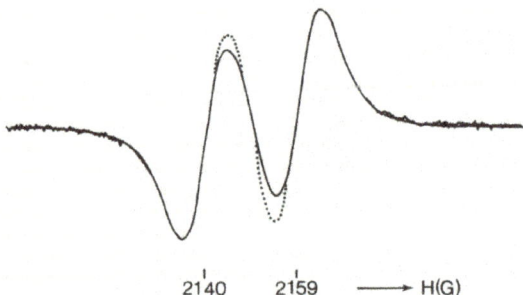

Figure 5. Asymmetric line shape at $g^{eff} \approx 3.4$ of the non-cubic Fe^{5+} center in a monodomain crystal at X-band in the vicinity of T_c. The dotted line is the calculated line composed of a mixture of 50% Gaussian and 50% Lorenzian line shape. The solid curve is the experimental measured one, indicating an asymmetric line form.

The asymmetry is apparent from a comparison of the observed line shape with that of a simulated symmetric line shape. The dotted curve is simulated for a best fit symmetrical line composed of a mixture of 50% Gaussian line shape and a 50% Lorenzian line shape. The discrepancy in the amplitudes of the experimental and simulated curves is a measure of the asymmetry in the experimental line shape. At T_c, the fluctuations of the oxygen octahedra become very slow compared to the ESR measuring time. This means that the local rotation of each non-cubic Fe^{5+} center is seen at rest in the ESR experiment. At T_c, the EPR lines reflect the Gaussian line shape in the case of statistical independence. The origin of the asymmetry is related to the form of the probability distribution for φ^* near T_c in the slow motion limit. Close to T_c, the classical probability distribution $P(\varphi^*)$ of the ensemble $P(\varphi^*,T) = c(T)\exp(-\Delta F(T))$, where $\Delta F(T)$ is the free energy depending on the order parameter. In the Landau theory, ΔF is given by $A(T)\varphi^{*2} + B\varphi^{*4}$, where $A(T) = a(T - T_c)$. $P(\varphi^*,T)$ is a double-peaked function for $T < T_c$ (Figure 6a).

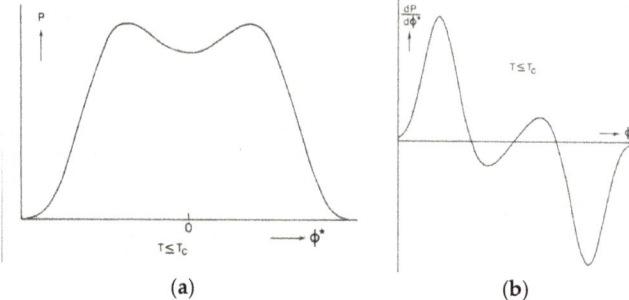

Figure 6. (a) Calculated probability distribution P versus the order parameter φ^* in the vicinity of T_c. (b) First derivative $(dP/d\varphi^*)$ versus the order parameter.

The spread in the value that φ^* may adopt, of course, gives rise to an inhomogeneous broadening (spread in g^{eff}) of the ESR lines. In Figure 6b, it is sketched how the ESR line shape is affected by the distribution function. Additional homogeneous broadening effects result in the observed asymmetrical shaped ESR lines. All the results obtained here are similar to those obtained for the Fe^{3+}-V_O center, reflecting cooperative bulk behavior of the crystal near the 105 K phase transition [15,16].

1.4. Different Fe^{5+} Centers

Different Fe^{5+} centers have been found in ST as well as in $BaTiO_3$ (BT). In ST:Fe^{5+}, ($g_{isotropic}$ = 2.013); only the $-\frac{1}{2} \leftrightarrow +\frac{1}{2}$ transition could be observed, even at helium temperatures [27]. The $\pm 3/2 \leftrightarrow \pm 1/2$ transitions are not observable due to a distribution of strain in the crystal leading to fine structure broadening. As a result of the smaller radius of this center in comparison with that of Ti^{4+}, the ion must be off-centered in one of the <100> directions. In contrast to this, the Fe^{5+} center in BT goes off-centered along one of the <111> directions [28]. In previous unpublished ESR investigations of ST, a vanadium center was found, which was attributed to a V^{5+}-O^- hole-like center (Figure 7) [29]. It is a tetragonal $S = \frac{1}{2}$, $I = 7/2$ light-sensitive center with $g_\parallel$ = 2.017, $g_\perp$ = 2.012, and $A_\parallel$ = 10.3 × 10^{-4} cm^{-1}, $A_\perp$ = 9.9 × 10^{-4} cm^{-1}.

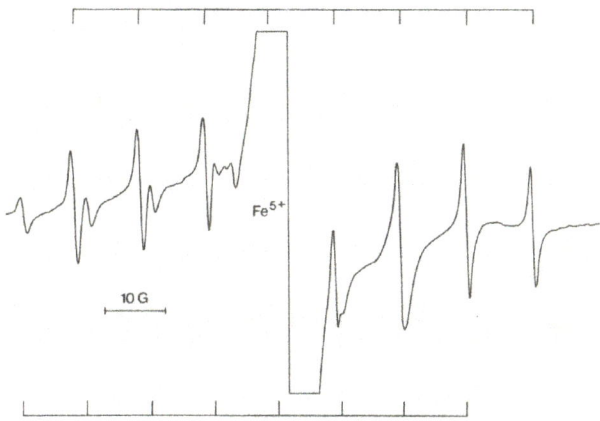

Figure 7. X-band ESR of ST:Fe^{5+}-O^{2-}-V^{5+} (reproduced from the PhD thesis of A.Lagendijk).

The interpretation of this center was largely based on arguments used for a similar center in ST, i.e., the Al^{3+}-O^- hole center [30]. In both centers, the mean g-value is larger than the free electron one, and the hyperfine interaction is relatively small, indicating that the hole is not localized on the central ion. For instance, the localized electron in the V^{4+} (d^1) Jahn–Teller center in ST has much larger hyperfine values of $A_\parallel$ = 147 × 10^{-4} cm^{-1} and $A_\perp$ = 44 × 10^{-4} cm^{-1} [31]. However, Schirmer et al. found a new Al^{3+}-O^- hole center in ST with a different local symmetry [32]. Therefore, they reinterpreted the formerly found hole center to be an Fe^{5+}-O^{2-}-Al^{3+} center, where the Al^{3+} is located in the neighboring oxygen octahedron substitutional for a Ti^{4+} ion. The average g-value of this center, $g_{av} = 1/3(g_\parallel + 2g_\perp)$ = 2.013, is the same as the isotropic g-value found for the Fe^{5+} in ST [2,27,32]. New ESR experiments revealed that by applying [011] uniaxial stress, no change in the ESR line intensity took place, indicating that no reorientation of the axes of this center occurs [33]. In addition, also the vanadium hole center could not be reoriented by applying uniaxial [011] stress [33]. Former stress studies showed that under the influence of uniaxial externally applied stress, the hole-like centers Fe^{2+}-O^- in ST [34] and Na^+-O^- in BT [35] could be reoriented. Therefore, we ascribe this vanadium hole center to an Fe^{5+}-O^{2-}-V^{5+} association with a similar structure as the Fe^{5+}-O^{2-}-Al^{3+} center. The $g_{av} = 1/3(g_\parallel + 2g_\perp) = 2.0137$ is equal to that of the Fe^{5+} center in ST [2,27,32].

All the discussed iron impurity centers are non-centro symmetric systems in contrast to, for instance, the ST:Mo^{3+} [36] and ST:Cr^{3+} [2] centers with Mo^{3+} and Cr^{3+} sitting on center.

1.5. Conclusions

The non-cubic Fe^{5+} center in ST shows the same behavior in the vicinity of the 105 K structural phase transition as the Fe^{3+}-V_O and Fe^{3+} centers, indicating cooperative effects in the ST crystal. Furthermore, the non-cubic Fe^{5+} is probably an Fe^{5+} O^{2-} V_{Ti} center with an expanded adjacent oxygen octahedron due to a Ti^{4+} vacancy. New stress experiments confirmed that the hole center Al^{3+}-O^- must be an Fe^{5+}-O^{2-}-Al^{3+} center, and a previous found vanadium hole center is now attributed to an Fe^{5+}-O^{2-}-V^{5+} association.

2. The Tetragonal Elastic Dipole Cr^{5+} in $SrTiO_3$

We reinvestigate the tetragonal Cr^{5+} impurity center in single crystals of the perovskite ST. After careful analyses, we came to the conclusion that this center is an off-center system and is not a Jahn–Teller impurity system. From previous stress experiments in ESR, we calculated the linear stress coupling tensor as $\beta = 3.56 \times 10^{-30}$ m^3; which is in agreement with other stress coupling tensors of off-center systems in ST as well as BT.

The research of the Cr^{5+} impurity center in ST started with ESR experiments performed by Lagendijk [4]. He concluded that the Cr^{5+} (d^1) ion is substituted for Ti^{4+} sitting off-centered along one of the <100> directions. It was a logical conclusion, because the ion radius of tetragonal Cr^{5+} in octahedral symmetry (0.54 Å) is considerably smaller than that of the Ti^{4+} (0.64 Å) [23]. Furthermore, the experiments performed at 77 K revealed a rotation of the oxygen octahedral of 1.6°, which is larger than one measured by on-center impurities, i.e., 1.53 [2], indicating a shift of the center from the on-center position. The Cr^{5+} center in ST has also been measured by Müller and Berlinger [37].

Later, de Jong et al. performed uniaxial stress experiments on the system [5] with the help of specialized stress equipment [38]. The intensity changes under stress P showed that centers with their tetragonal axes along the stress direction were favorable to centers with their tetragonal axes perpendicular to the stress direction. Therefore, they concluded that the system is a strong Jahn–Teller (JT) ion ($T_{2g} \times e_g$) with a squeezed oxygen octahedron. The research went on with static electric field experiments (E) [39]. Surprisingly, there was an increase of the ESR line intensities due to centers with their main center axes perpendicular to the E-field at the cost of those centers with main axes parallel to E. Since there was a quadratic change of the line intensities with E, they concluded that tetragonal Cr^{5+} is a JT ion in ST.

We make the following objections against their conclusions. First of all, by applying a static electric field, one expects that the centers parallel to E increase in intensities. This was shown for the impurity center ion V^{4+} (d^1) in ST [40]. V^{4+} has the same ion radius as the Ti^{4+} that it replaces, so it must be an on-center system, and therefore it *must* be a JT ion. In this system, the tetragonal axis of the system is directed along the applied E-field in contrast with that of Cr^{5+}. In addition there is an E^2 dependence. The experiments of de Jong et al. can be explained by assuming that the observed quadratic change in intensities is due to a large electro-strictive coupling and must be a P^2 dependence. So, their interpretation that the ST:Cr^{5+} system is a JT system is in our opinion not right, and the interpretation of Lagendijk is the right one.

In the following part, we give more evidence in favor of the off-center system of tetragonal Cr^{5+}. In a later study, Koopmans et al. studied the orthorhombic Cr^{5+} ion in ST. They concluded that this orthorhombic system is a $T_{2g} \times (e_g + t_{2g})$ JT system [6,41]. However, Müller, Blazey and Kool concluded that the center must be an-off center system that is off-centered in one of the <110> directions and tetrahedrally surrounded by four O^{2-} ions [7]. The assignment was based on the g-values of the ion, which was almost the same as in other Cr^{5+} systems. In addition, the ionic radius of the ion (0.34 Å) is considerably smaller than that of Ti^{4+}, which it replaces [23]. Moreover, the local rotation of the octahedra is different from that of the intrinsic one, 1.6° versus 2.0° at 4.2 K. The value of this last rotation we also found for the V^{4+} octahedron at 4.2 K [31]. The off-center assignment was confirmed by static electric field experiments in ESR [42]. From these measurements, an *electric dipole* of $\mu = 2.88 \times 10^{-30}$ Cm and an off-center displacement of about 0.2 Å was determined [43]. In ST, only one type of V^{4+} ion was found in contrast with the two Cr^{5+} centers, which gives further support to the off-center model. The two different Cr^{5+} centres were also found in the perovskite BT [44].

From Figure 2 of the article of de Jong et al. [5], we were able to calculate the *elastic dipole* of tetragonal Cr^{5+} in ST. The concept of an elastic dipole can be found in an article by Nowick and Heller [45]. From the plot of $ln(2_a/l_b)$ versus the stress P and assuming Boltzmann distribution, the elastic dipole moment $\beta_{[100]}$ could be determined.

$ln(2_a/l_b) = \Delta U/kT$, with $\Delta U = -V_0 \lambda^{(p)}$ $\sigma \equiv \beta \cdot \sigma$, with $\lambda^{(p)} = \frac{1}{2}\lambda_1 - \frac{1}{2}\lambda_2$

This results in a value of $\beta = 3.56 \times 10^{-30}$ m^3.

This value is in agreement and in the order of magnitude of other off-center systems in ST and BT (see the Table 2).

Table 2. Elastic dipole values in SrTiO3 and BaTiO3.

SrTiO$_3$	β (10^{-30} m^3)	Reference
Fe^{2+}-O^-	3.56	[34]
Cr^{5+} (orthorhombic)	4.19	[42]
Cr^{5+} (tetragonal–this study)	3.56	
BaTiO$_3$		
Ni^{1+} (substituted for Ba^{2+})	3.8	[46]
Na^+-O^-	1.23	[35]
Fe^{5+}	4.13	[28]

As a last remark, we would like to mention that we do not agree with the conclusion of Yu-Guang Yang et al [47]. First of all, they use for Cr^{5+} in octahedral surroundings an ionic radius of 0.63 Å and for the Ti^{4+} which it replaces an ionic radius of 0.745 Å. It is better to use the values that are given in Reference [23]. Therefore, the ionic radius of Cr^{5+} is much smaller than the intrinsic one 0.54 Å versus 0.64 Å, as mentioned before. Furthermore, they say in their discussion: "Upon cooling, only a continuously increasing resolution is observed, and no specific effects are found in ESR when passing the 105 K phase transition." This is not true. Lagendijk et al. observed a splitting of the ESR lines due to the tetragonal domains of ST and inferred a rotation of the oxygen octahedra of 1.6° which, as already mentioned before, is different from the intrinsic one, indicating an off-center displacement.

Funding: This research received no external funding.

Conflicts of Interest: The author declares no conflict of interest.

References and Notes

1. Jansen, L.; Boon, M. *Theory of Finite Groups. Applications in Physics*; North-Holland Publishing Company: Amsterdam, The Netherlands, 1967.
2. Müller, K.A.; Kool, T.W. *Properties of Perovskites and Other Oxides*; World Scientific: Singapore, 2010.
3. Müller, K.A. Paramagnetische Resonanz von Fe^{3+} in SrTiO$_3$ –Einkristallen. *Helv. Phys. Acta* **1958**, *31*, 173–204.
4. Lagendijk, A.; Morel, R.J.; Glasbeek, M. ESR of Cr^{5+} in chromium-doped SrTiO$_3$ single crystals. *Chem. Phys. Lett.* **1972**, *12*, 518. [CrossRef]
5. De Jong, H.J.; Glasbeek, M. Cr^{5+} in SrTiO$_3$: An example of a static Jahn-Teller effect in a d^1 system. *Solid State Commun.* **1976**, *19*, 1179. [CrossRef]
6. Glasbeek, M.; de Jong, H.J.; Koopmans, W.E. Nonlinear Jahn-Teller coupling and local dynamics of SrTiO$_3$:Cr^{5+} near the structural phase transition point. *Chem. Phys. Lett.* **1979**, *66*, 203. [CrossRef]
7. Müller, K.A.; Blazey, K.W.; Kool, T.W. Tetrahedrally coordinated Cr^{5+} in SrTiO$_3$. *Solid Sate Commun.* **1993**, *85*, 381–384. [CrossRef]

8. Kool, T.W. EPR Studies of Transition Metal Impurity Ions in Strontium Titanate. Ph.D. Thesis, University of Amsterdam, Amsterdam, The Netherlands, 1991.
9. Müller, K.A. Einiges zur Symmetrie und Symbolik der Zahl Fünf. In *Der Pauli-Jung-Dialog und seine Bedeutung für die moderne Wissenschaft*; Springer: Berlin/Heidelberg, Germany, 1995.
10. Kool, T.W. EPR studies of the 105 K phase transition of SrTiO3 with the non-cubic Fe5+ as local probe and a reinterpretation of other Fe5+ centres. *arXiv* 2012, arXiv:1201.3386.
11. Kool, T.W. The tetragonal elastic dipole Cr5+ in SrTiO3. *ChemRxiv* 2018. [CrossRef]
12. Landau, L.D.; Lifschitz, E.M. *Statistical Physics*; Pergamon Press: Oxford, UK, 1970.
13. Pytte, E.; Feder, J. Theory of a structural phase transition in perovskites-types crystals. *Phys. Rev.* 1969, *187*, 1077. [CrossRef]
14. Müller, K.A.; Fayet, J.C. *Structural Phase Transitions II*; Müller, K.A., Thomas, H., Eds.; Springer: Berlin/Heidelberg, Germany, 1991; pp. 1–82.
15. Müller, K.A.; Berlinger, W. Static critical exponents at structural phase transitions. *Phys. Rev. Lett.* 1971, *26*, 13. [CrossRef]
16. Kool, T.W.; Glasbeek, M. Photochromic Fe^{5+} in non-cubic local fields in SrTiO3. *Solid State Commun.* 1977, *22*, 193. [CrossRef]
17. Abragam, A.; Bleaney, B. *Electron Paramagnetic Resonance of Transition Ions*; Clarendon Press: Oxford, UK, 1970.
18. Griffith, J.S. *The Theory of Transition-Metal Ions*; Cambridge University Press: Cambridge, UK, 1971.
19. Carrington, A.; McLachlan, A.D. *Introduction to Magnetic Resonance*; Harper and Row: New York, NY, USA, 1969.
20. Pake, G.E.; Estle, T.L. *The Physical Principles of Electron Paramagnetic Resonance*, 2nd ed.; W.A. Benjamin Inc.: Reading, UK, 1973.
21. Kool, T.W.; Bollegraaf, B. Calculation of the zero-field splitting D and g perpendicular parameters for d^3 spin systems in strong and moderate axial fields. *arXiv* 2010, arXiv:1005.0159.
22. Kool, T.W. Calculation of the orthorhombic E-parameter in EPR for d^3 spin systems. *arXiv* 2011, arXiv:1105.0616.
23. Megaw, H.D. *Crystal Structures: A Working Approach*; W.B. Saunders: London, UK, 1973.
24. Müller, K.A.; Berlinger, W.; Capizzi, M.; Gränicher, H. Monodomain strontium titanate. *Solid State Commun.* 1970, *8*, 549. [CrossRef]
25. Müller, K.A.; Berlinger, W. Critical asymmetry in local fluctuations in SrTiO3 for $T \to T_c^-$. *Phys. Rev. Lett.* 1972, *29*, 715. [CrossRef]
26. Von Waldkirch, T.; Müller, K.A.; Berlinger, W. Fluctuations in SrTiO3 near the 105K phase transition. *Phys. Rev. B* 1973, *7*, 1052. [CrossRef]
27. Müller, K.A.; von Waldkirch, T.; Berlinger, W.; Faughnan, B.W. Photochromic Fe^{5+} ($3d^3$) in SrTiO3 evidence from paramagnetic resonance. *Solid State Commun.* 1971, *9*, 1097.
28. Kool, T.W.; Lenjer, S.; Schirmer, O.F. Jahn-Teller and off-center defects in BaTiO3: Ni^+, Rh^{2+}, Pt^{3+}, and Fe^{5+} as studied by EPR under uniaxial stress. *J. Phys. Condens. Matter* 2007, *19*, 496214. [CrossRef]
29. Lagendijk, A. An ESR study of induced defects in SrTiO3. Ph.D. Thesis, University of Amsterdam, Amsterdam, The Netherlands, 1974.
30. Ensign, T.C.; Stokowski, S.E. Shared holes trapped by charge defects in SrTiO3. *Phys. Rev. B* 1970, *1*, 2799. [CrossRef]
31. Kool, T.W.; Glasbeek, M. V^{4+} inSrTiO3: A Jahn-Teller impurity. *Solid State Commun.* 1979, *32*, 1099. [CrossRef]
32. Schirmer, O.F.; Berlinger, W.; Müller, K.A. Holes trapped near Mg^{2+} and Al^{3+} in SrTiO3. *Solid State Commun.* 1976, *18*, 1505. [CrossRef]
33. Kool, T.W. Stress induced effects on the Fe5+-O2—V5+ and Fe5+-O2—Al3+ centers in SrTiO3. Unpublished results.
34. Kool, T.W.; Glasbeek, M.J. Electron paramagnetic resonance of photochromic Fe^{2+}-O^- in SrTiO3. *Phys. Condens. Matter* 1993, *5*, 361. [CrossRef]
35. Varnhorst, T.; Schirmer, O.F.; Kröse, H.; Scharfschwerdt, R.; Kool, T.W. O^- holes associated with alkali acceptors in BaTiO3. *Phys. Rev. B* 1996, *53*, 116–125. [CrossRef] [PubMed]
36. Kool, T.W. EPR of photochromic Mo^{3+} in SrTiO3. *arXiv* 2010, arXiv:1003.1448.
37. Müller, K.A. Magnetic Resonance and related phenomena. In Proceedings of the 16th Congress Ampere, Bucharest, Romania, 1–5 September 1971; p. 173.
38. Kool, T.W. Description of a specialized stress equipment for EPR X-band measurements. *arXiv* 2017, arXiv:1710.04427.

39. De Jong, H.J.; Glasbeek, M. Electric field effects in EPR of the SrTiO$_3$:Cr^{5+} Jahn-Teller system. *Solid State Commun.* **1978**, *28*, 683. [CrossRef]
40. Kool, T.W.; Glasbeek, M. Electric field effects in EPR of the SrTiO$_3$:V^{4+} Jahn-Teller system. *J. Phys. Condens. Matter* **1991**, *3*, 9747–9755. [CrossRef]
41. Koopmans, W.E. Jahn-Teller interactions of chromium (V) in strontium titanate. Ph.D. Thesis, University of Amsterdam, Amsterdam, The Netherlands, 1985.
42. Kool, T.W.; de Jong, H.J.; Glasbeek, M.J. Electric field effects in the EPR of tetrahedrally coordinated Cr^{5+} in SrTiO$_3$. *Phys. Condens. Matter* **1994**, *6*, 1571–1576. [CrossRef]
43. See the introduction to off-center systems in the book by Müller and Kool, Reference [8].
44. Possenriede, E.; Jacobs, P.; Schirmer, O.F. Paramagnetic defects in BaTiO$_3$ and their role in light-induced charge transport. I ESR studies. *J. Phys. Condens. Matter* **1992**, *4*, 4719. [CrossRef]
45. Nowick, A.S.; Heller, W.R. Anelasticity and stress-induced ordering of point defects. *Adv. Phys.* **1963**, *12*, 251. [CrossRef]
46. Lenjer, S.; Scharfschwerdt, R.; Kool, T.W.; Schirmer, O.F. An off-center ion near a Ba site in BaTiO3 as studied by EPR under uniaxial stress. *Solid State Commun.* **2000**, *116*, 133–136. [CrossRef]
47. Yang, Y.G.; Zheng, W.C.; Su, P.; Yang, W.Q. Investigations of the spin-Hamiltonian parameters and tetragonal compression due to the static Jahn-Teller effect for Cr^{5+} center in cubic SrTiO$_3$ crystal. *Phys. B* **2010**, *405*, 4871–4874. [CrossRef]

© 2020 by the author. Licensee MDPI, Basel, Switzerland. This article is an open access article distributed under the terms and conditions of the Creative Commons Attribution (CC BY) license (http://creativecommons.org/licenses/by/4.0/).

Creative

The Crucial Things in Science often Happen Quite Unexpectedly—Das Entscheidende in der Wissenschaft Geschieht oft Ganz Unerwartet (K. Alex Müller)

Reinhard K. Kremer [1,*], Annette Bussmann-Holder [1], Hugo Keller [2] and Robin Haunschild [1]

1. Max Planck Institute for Solid State Research, Heisenbergstrasse 1, 70569 Stuttgart, Germany; a.bussmann-holder@web.de (A.B.-H.); R.Haunschild@fkf.mpg.de (R.H.)
2. Physik Institut der Universität Zürich, Winterthurerstr. 190, 8057 Zürich, Switzerland; keller@physik.uzh.ch
* Correspondence: rekre@fkf.mpg.de; Tel.: +49-711-689-1688

Received: 10 June 2020; Accepted: 28 June 2020; Published: 1 July 2020

Abstract: We analyzed the publication output of one of the 1987 Nobel Prize awardees, K. Alex Müller, using bibliometric methods. The time-dependent number of publications and citations and the network with respect to the coauthors and their affiliations was studied. Specifically, the citation history of the Nobel Prize awarded 1986 article on "Possible high-temperature superconductivity in the Ba-La-Cu-O system" has been evaluated in terms of the overall number of articles on superconductivity and the corresponding citations of other most frequently referenced articles. Thereby, a publication with "delayed recognition" was identified.

1. Introduction

Rarely in the history of physics has a discovery had such a prompt and enduring impact as the observation of high-temperature superconductivity (high-T_c) in the Ba-La-Cu-O system by J. Georg Bednorz and K. Alex Müller in 1986 [1]. The unexpected disclosure of superconductivity at temperatures about twice as high as had been known before [2,3] triggered immediate (sometimes hectic and often very competitive and controversial) research activity targeted at the synthesis of compounds with even higher T_c's and the still contentious theoretical explanation of high-T_c superconductivity. The bursting reaction in the search for new and deeper insight into already known high-T_c materials is reflected in the instantaneous rise of the number of publications indexed by the Web of Science (WoS) [4,5] before and after 1986 (see Figure 1).

Figure 1 shows that after 1986 about 30% of the publications related to superconductivity covered high-T_c research (see [6–8]). Since 1986, the number of publications dealing with superconductivity essentially remains constant at a high level of about 6500 per year, whereas the number of high-T_c papers decreased gradually after a peak in 1991 with approximately 3100 publications and currently levels off at about 1400 publications per year. A closer inspection of Figure 1 also reveals a couple of historically very interesting aspects: After the discovery of superconductivity in mercury by H. Kamerling Onnes in 1911 [9] there was an extended period lasting until the late 1950s with a continuous but moderate increase of the publication output with about 10 to 50 superconductivity publications per year. Note the plunge in the number of publications in the late years of the Second World War. Certainly triggered by the development of the Bardeen-Cooper-Schrieffer (BCS) theory of superconductivity [10] but very likely also by the first placing of a satellite into Earth's orbit by the Soviet Union in October 1957, the number of publications rose steeply in the late 1950s and early 1960s [11]. This event, perceived in the Western world as the 'Sputnik crisis', initiated massive funding increase into science, especially in the USA [12]. As a consequence, the number of scientists working in the field and the publication output increased steeply after 1957 [13,14]. The data also show that after 1970 the publication output stalled and flattened, because no new discoveries were made, and increased dramatically after 1987.

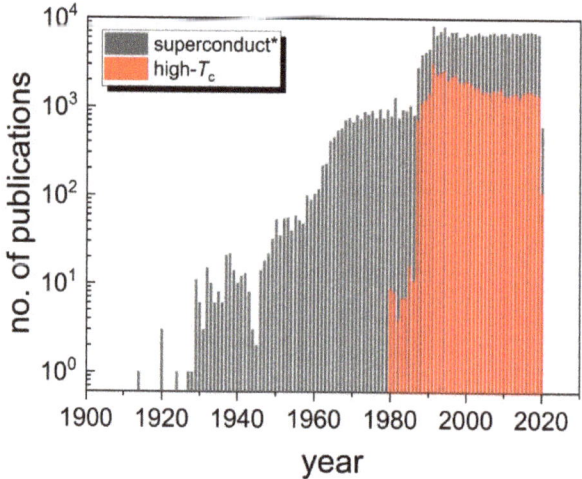

Figure 1. Number of publications indexed by the Web of Science (WoS) either with the keywords 'superconduct*' (grey) or 'high-T_c' (red). [4].

2. The Publication Output of Original Research Publications

K. Alex Müller began his scientific career with his PhD thesis entitled 'Paramagnetische Resonanz von Fe^{3+} in $SrTiO_3$ Einkristallen' submitted to the ETH Zürich in 1958. In his thesis he summarized his microwave resonance experiments on spin $S = 5/2$ ions substituted for Ti in $SrTiO_3$, a compound which fascinated him during all his future scientific activity. The thesis was published in the Helvetica Physica Acta in 1958 [15]. The results of his thesis were summarized in a short single-authored Physical Review Letter (PRL) published in 1959, already mentioning his new affiliation at the Battelle Memorial Institute in Geneva [16]. This PRL referenced a publication, apparently K. Alex Müller's very first one (possibly published more or less simultaneously with his PhD thesis), in the proceedings of the 7th Colloquium Ampère held in 1958 in Geneva (Résonance paramagnétique du Cr^{3+} dans des monocristaux de $SrTiO_3$) [17].

Figure 2a shows the number of papers published by K. Alex Müller per year indexed by the WoS as source items beginning from his first Physical Review Letter, which appeared in 1959. Figure 2b is a 'vintage' diagram, showing the number of citations for papers published in the particular year.

Unfortunately, the 'Archives des Sciences éditées par la Société de Physique et d'Histoire Naturelle de Genève' are not indexed by the WoS and a reliable bibliometric analysis of this publication cannot be carried out [18].

In total, K. Alex Müller published 311 papers in the 61 years of his scientific activity until 2019. Until 1986, he authored between three and eight publications per year (see Figure 2a). Interestingly, this number did not increase significantly over the following years, except in the first two immediate years after the seminal paper on the discovery of high-T_c superconductivity in 1986 [1] when he authored or co-authored up to 30 papers in 1987 and 22 in 1988 (i.e., about one-sixth of Müller's publications appeared in these two years). It is worth mentioning that the number of publications dealing with $SrTiO_3$ (50) is almost as high as those covering cuprates (52). Including 11 additional papers which treated titanates in general (mostly $BaTiO_3$), the number of publications on titanates (61) even exceeds that on cuprates.

Figure 2b displays a 'vintage' diagram showing the number of citations to papers published in a particular year. Besides the citation spikes following immediately after the discovery of high-T_c superconductivity, two more years stand out, namely the years 1979 and 1996. The high citation number in 1979 is due to one particular paper, which also exhibited an unusual citation history (see details below). In 1996, the high citation number is owed to two publications on $La_{1-x}Ca_xMnO_{3+y}$, namely an article in Nature together with Zhao, Conder, and Keller on giant oxygen isotope shifts (622 citations) and a PRL published jointly with Shengelaya, Zhao, and Keller providing electron paramagnetic resonance (EPR) evidence for Jahn-Teller polaron formation (215 citations) (see Table 2 and Ref. [19] for the detailed references).

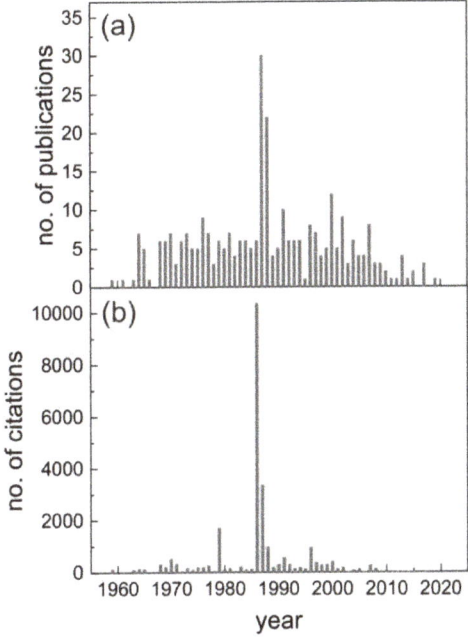

Figure 2. (a) Number of papers published by K. Alex Müller per year indexed by the WoS as source items beginning from his first Physical Review Letter, which appeared in 1959. (b) 'Vintage' diagram showing the number of citations for papers published in the particular year.

K. Alex Müller's publication output compares well, for example, with that of V. L. Ginzburg, another Nobel prize winner for "Pioneering Contributions to the Theory of Superconductors and Superfluids", whose scientific activity covered a similar period of time as of K. Alex Müller's [20].

3. Editorial Activities

Over his long-lasting scientific activity, K. Alex Müller edited and co-edited numerous collections of lectures, conference proceedings, and series volumes, to which he usually contributed an article or a chapter. Table 1 provides a compilation of Müller's editorial activity. Clearly, in the 1970s and 1980s these books focused around phase transition-related phenomena. Later, after the discovery of high-T_c superconductivity, these were in his focus. Noteworthy are the two conference proceedings of international symposia on phase separation problems in cuprate superconductors held in Erice and Cottbus in 1992 and 1993, respectively. These meetings and the contributions summarized in the proceedings reflect Müller's early interest in essential inhomogeneity in high-T_c superconductors.

Table 1. Compilation of K. Alex Müller's editorial activities.

Müller, K.A.; Rigamonti, A. Local Properties at Phase Transitions. Proprietà Locali alle Transizioni di Fase, Societa Italiana di Fisica Bologna, 1976.

Müller, K.A.; Thomas, H. Structural Phase Transitions I. Springer-Verlag, Berlin, Heidelberg, New York, 1981.

Müller, K.A.; Thomas, H. Structural Phase Transitions II. Springer-Verlag, Berlin, Heidelberg, New York, 1991.

Bednorz, J.G.; Müller, K.A. Earlier and Recent Aspects of Superconductivity, Lectures from the International School, Erice, Trapani, Sicily, 4–16 July, 1989. Springer Verlag, 1st Edition 1990, 2nd Edition 1991.

Müller, K.A.; Benedek, G. Proceedings of the Workshop on Phase Separation in Cuprate Superconductors. Erice, Italy, World Scientific Publishing, 1993.

Sigmund, E.; Müller, K.A. Phase Separation in Cuprate Superconductors. Proceedings of the second international workshop on 'Phase Separation in Cuprate Superconductors', 4–10 September, 1993. Cottbus, Germany, Springer-Verlag, 1994.

Müller, K.A.; Bussmann-Holder, A. Superconductivity in Complex Systems, Structure and Bonding 114, Springer-Verlag, 2005.

Müller, K.A.; Kool, T.W. Properties of Perovskites and Other Oxides. World Scientific, 2010.

Müller, K.A. Letters Section in the Journal of Superconductivity and Novel Magnetism, published by Springer Nature.

4. The Citation Record—*Overall*

In his long scientific career, K. Alex Müller's publications were cited about 25,000 times. By far the highest citation count of all of his publications was acquired by the Zeitschrift für Physik article of 1986 on 'Possible High-T_c Superconductivity in the BA-LA-CU-O System' (see Figure 3). Until February 2020, it was cited more than an exceptional 10,000 times. Note that we did not find any publication on superconductivity that was cited more frequently. The citation count for this publication peaks two years after its appearance at a value of about 1250 citations per year. Subsequently, the citations decrease to pass through a broad minimum around the year 2005, whereupon they start to grow again until today. Such a second increase is remarkable and must be attributed to the discovery of new classes of high-T_c superconductors, e.g., MgB_2 or the iron-based superconductors. Forty of K. Alex Müller's 311 scientific papers acquired more than 100 citations. The 10 most frequently cited papers are compiled in Table 2.

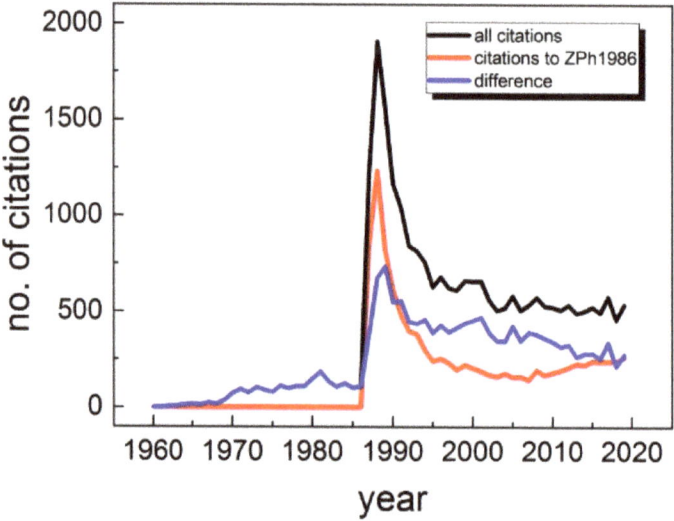

Figure 3. Time evolution of the number of citations to K. Alex Müller's papers. The black solid line gives the number of all citations per year, the red solid line the citations per year of the article in Zeitschrift für Physik in 1986 [1], and the blue solid line the difference between the first and the latter.

Table 2. The 10 most cited publications authored and co-authored by K. Alex Müller (search date February 2020), ordered by their citation count.

1. Bednorz, J.G.; Müller, K.A.
Possible High-Tc Superconductivity in the BA-LA-CU-O System. *Z. Phys. B Condens. Matter* **1986**, *64*, 189–193. Doi: 10.1007/BF01303701.
times cited: 10210

2. Müller, K.A.; Burkard, H.
SrTiO$_3$—Intrinsic Quantum Para-Electric Below 4-K. *Phys. Rev. B* **1979**, *19*, 3593–3602. Doi: 10.1103/PhysRevB.19.3593.
times cited: 1196

3. Müller, K.A.; Takashige, M.; Bednorz, J.G.
Flux Trapping and Superconductive Glass State in La$_2$CuO$_{4-y}$:Ba. *Phys. Rev. Lett.* **1987**, *58*, 1143–1146. Doi: 10.1103/PhysRevLett.58.1143.
times cited: 1094

4. Zhao, G.M.; Conder, K.; Keller, H.; Müller, K.A.
Giant Oxygen Isotope Shift in the Magnetoresistive Perovskite La$_{1-x}$Ca$_x$MnO$_{3+y}$. *Nature* **1996**, *381*, 676–678. Doi: 10.1038/381676a0.
times cited: 622

5. Deutscher, G.; Müller, K.A.
Origin of Superconductive Glassy State and Extrinsic Critical Currents in High-Tc Oxides. *Phys. Rev. Lett.* **1987**, *59*, 1745–1747. Doi: 10.1103/PhysRevLett.59.1745.
times cited: 584

6. Blazey, K.W.; Müller, K.A.; Bednorz, J.G.; Berlinger, W.; Amoretti, G.; Buluggiu, E.; Vera, A.; Matacotta, F.C.
Low-Field Microwave-Absorption in the Superconducting Copper Oxides. *Phys. Rev. B* **1987**, *36*, 7241–7243. DOI: 10.1103/PhysRevB.36.7241.
times cited: 327

7. Bednorz, J.G.; Müller, K.A.
Perovskite-Type Oxides - The New Approach to High-Tc Superconductivity. *Rev. Modern Phys.* **1988**, *60*, 585–600. Doi: 10.1103/RevModPhys.60.585.
times cited: 306

8. Zhao, G.M.; Hunt, M.B.; Keller, H.; Müller, K.A.
Evidence for Polaronic Supercarriers in the Copper Oxide Superconductors La$_{2-x}$Sr$_x$CuO$_4$. *Nature* **1997**, *385*, 236–239. Doi: 10.1038/385236a0.
times cited: 273

9. Müller, K.A.; Berlinger, W.; Waldner, F.
Characteristic Structural Phase Transition in Perovskite-Type Compounds. *Phys. Rev. Lett.* **1968**, *21*, 814–817. Doi: 10.1103/PhysRevLett.21.814.
times cited: 267

10. Šimanek, E.; Müller, K.A.
Covalency and Hyperfine Structure Constant-A of Iron Group Impurities in Crystals. *J. Phys. Chem. Solids* **1970**, *31*, 1027–1040. Doi: 10.1016/0022-3697(70)90313-6.
times cited: 266

5. Citation History of Selected Highly Cited Publications

The citation record as a function of time of a particular article, colloquially called its 'citation history', is specific for every article. However, it often follows a similar pattern [20]. Typically, it takes about one or two years until the citation count markedly starts to rise. Often, the citation count reaches a peak after about three years with a subsequent decrease when newer research results gradually become available. Such exemplary pattern can be observed for three of the four next highest cited publications of Müller, namely those in collaboration with Müller, Takashige, Bednorz, PRL 1987, Zhao, Conder, Keller, Müller, Nature 1996, and Deutscher and Müller, PRL 1987 (see Figure 4). However, the paper by Müller and Burkard, Physical Review B (PRB) 1979 [21] (about 1200 total citations) entitled 'SrTiO$_3$—Intrinsic quantum para-electric below 4 K' exhibits a remarkably different citation history. Published in 1979, it was barely referenced until about 1990. Afterwards, it picked up and kept on increasing until now. It appears that the research reported in this article was well ahead of its time. Publications showing such delayed recognition have sometimes been labeled as 'sleeping beauties'.

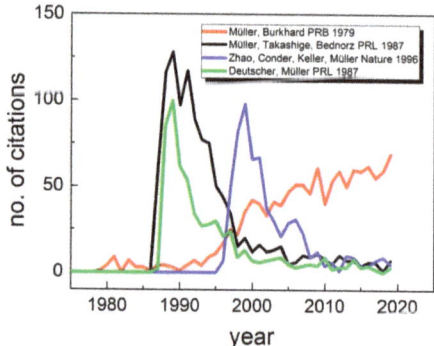

Figure 4. Citation history of the next four highest cited articles of K. Alex Müller besides the 1986 Zeitschrift für Physik paper. Note the unusual citation history characteristic of delayed recognition of the Müller and Burkhard article on quantum para-electric behavior of $SrTiO_3$. For the detailed references see Table 2.

6. The Co-Author Network—*People*

Figure 5 displays an overview (generated using the software VOSviewer [22]) of the co-authors with whom Müller published more than five papers. Above all, ranks J. Georg Bednorz from IBM Rüschlikon (58), with whom he shared the 1987 Nobel Prize. K. Alex Müller continuously emphasized the amount of competent work and inspiration he received from the constant and fruitful assistance of Walter Berlinger. Even nowadays, it is by no means customary to include technical staff as a co-author in a scientific paper. However, K. Alex Müller never ceased to acknowledge Berlinger's work and, consequently, the number of joint publications with Berlinger (55) ranks almost as high as that with Bednorz. K.W. Blazey, also from IBM Rüschlikon, is another regularly named co-author. Other frequently appearing co-authors are H. Keller and his postdoctoral fellow A. Shengelaya and their jointly supervised PhD student G. M. Zhao, all affiliated with the University in Zürich. E. Kaldis, K. Conder, and J. Karpinski from the ETH Zürich were frequent and highly appreciated sources of exquisite samples and crystals. With colleagues from Germany and particularly the Max-Planck-Institute (MPI) for Solid State Research in Stuttgart, K. Alex Müller had a long-lasting collaboration, resulting in several joint publications with A. Bussmann-Holder and A. Simon.

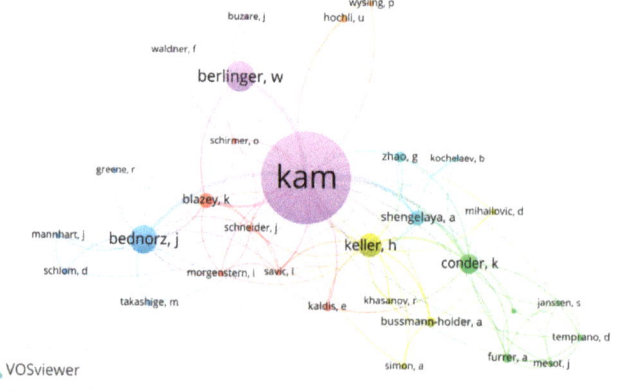

Figure 5. K. Alex Müller's main co-authors. The diameters of the circles correspond to the number of joint publications. Only co-authorships with at least five joint papers are included.

7. Countries

Not surprisingly, K. Alex Müller's main co-authors were affiliated with research institutions in his home country Switzerland (see Figure 6, also generated using the software VOSviewer [22]), which are by far dominating the co-authorship network. Outside Switzerland, the co-author network comprised colleagues from 16 countries (one disconnected country, The Netherlands, was removed). The immediate neighbors, France and Germany (note that Germany includes Western Germany before the reunification), rank second followed by the USA. However, there are also substantial numbers of co-authors from Russia (note that Russia includes the Soviet Union) and smaller countries like Georgia, Slovenia, and Estonia

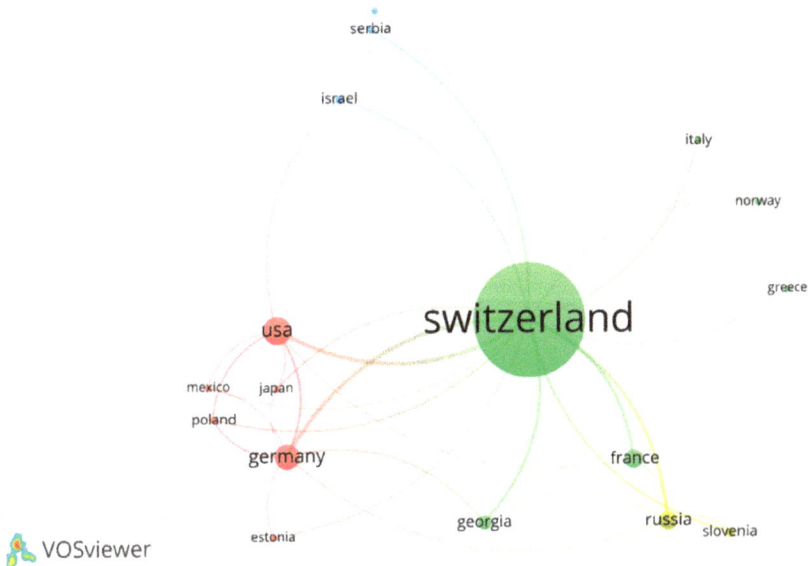

Figure 6. Country of affiliation of K. Alex Müller's main co-authors with whom he published at least one paper. The diameters of the circles correspond to the joint publication record with the named individual country.

8. Conclusions

The publication record and the citation history of K. Alex Müller's publications are exceptional and, from the bibliometric point of view, very interesting: Before J. Georg Bednorz and K. Alex Müller published their seminal paper on possible high-T_c superconductivity in Zeitschrift für Physik in 1986, Müller's scientific work focused on phase transition-related topics with an emphasis on perovskite materials, especially $SrTiO_3$. Naturally, this changed after 1986 and the citation history of all his publications, subtracting out the high-T_c papers, increased by about a factor of four. Since its appearance, the Zeitschrift für Physik article itself has been cited more than 10,000 times. No other article on superconductivity has been cited more frequently. Forty of the 311 articles published by K. Alex Müller received more than 100 citations each. Müller's h-index amounts to 64. Whereas most of his highest cited articles exhibit a citation history with a sharp peak shortly after publication and a steady decay thereafter (typical behavior), one noteworthy publication on $SrTiO_3$, together with H. Burkhard, displays an unusual, delayed recognition-type citation history. Currently, this publication continuous to attract attention and reached approximately 70 citations in 2019.

Funding: This research received no external funding.

Acknowledgments: Very valuable discussions with W. Marx and Th. Scheidsteger as well as a careful reading of manuscript by S. Kremer are gratefully acknowledged.

Conflicts of Interest: The authors declare no conflict of interest.

References and Notes

1. Bednorz, J.G.; Müller, K.A. Possible high Tc superconductivity in the Ba–La–Cu–O system. *Phys. Z. Condens. Matter* **1986**, *64*, 189–193. [CrossRef]
2. Arrhenius, G.; Corenzwit, E.; Fitzgerald, R.; Hull, G.W.J.; Luo, H.L.; Matthias, B.T.; Zachariasen, W.H. Superconductivity of NB(3)(AL, GE) above 20.5 degrees K. *Proc. Natl. Acad. Sci. USA* **1968**, *61*, 621–628. [CrossRef] [PubMed]
3. One of the authors (RKK) recollects a seminar given by B. T. Matthias at the TU Darmstadt in the Elschner/Steglich seminar in the very early eighties of the last century. At the end of his lecture Matthias was asked what he estimates could be the maximum superconducting T_c. Matthias responded with a smile: This question apparently did not come as a great surprise and he must certainly have heard it many times before—and answered that he thinks that T_c will somehow level off between 30 and 35 K.
4. The WoS as Provided by Clarivate Analytics was Used. Available online: https://apps.webofknowledge.com (accessed on 25 February 2020).
5. Birkle, C.; Pendlebury, D.A.; Schnell, J.; Adams, J. Web of Science as a data source for research on scientific and scholarly activity. *Quant. Sci. Stud.* **2020**, *1*, 363–376. [CrossRef]
6. At this occasion we want to point out that such bibliometric searches always are afflicted with some degree of uncertainty since they stick to the search terms. As an example, a publication containing the search term 'high-T_c' but not 'superconduct*' shows up in the high-T_c category only, whereas if it contains both search terms it appears in both categories. How intriguing bibliometric searches can be, may be realized from the fact that the term 'high-temperature superconductors' has already been used for some A15 compounds in 1953/1954 (see e.g., Ref. [7]). Bibliometric searches for authors and citations are comparatively easier and generally more precise. For a detailed treatment of bibliometric analysis and its aspects see e.g., Ref. [8].
7. Gross, R.; Marx, A. *Festkörperphysik*; De Gruyter Mouton: Berlin, Germnay, 2012; p. 754.
8. Marx, W.; Bornmann, L. How to evaluate individual researchers working in the natural and life sciences meaningfully? A proposal of methods based on percentiles of citations. *Scientometrics* **2014**, *89*, 487–509.
9. Kamerling-Onnes, H. Further Experiments with Liquid Helium. C. On the Change of the Electric Resistance of Pure Metals at Very Low Temperature etc. IV. The Resistance of Pure Mercury at Helium Temperature. In *Through Measurement to Knowledge*; Springer: Dordrecht, the Netherlands, 1911; Volume 124.
10. Bardeen, J.; Cooper, L.N.; Schrieffer, J.R. Theory of Superconductivity. *Phys. Rev.* **1957**, *108*, 1175–1204. [CrossRef]
11. Dickson, P. *Sputnik: The Shock of the Century*; Walker Publishing Company, Inc.: New York, NY, USA, 2001.
12. See for Example the Historical Development of the US Federal Spending on Defense and Nondefense Research and Development at the Web-Page. Available online: https://www.aaas.org/programs/r-d-budget-and-policy/historical-trends-federal-rd (accessed on 20 February 2020).
13. Marx, W.; Cardona, M.; Lockwood, D.J. Rutherford's impact on science over the last 110 years: A bibliometric analysis Physics in Canada. *Phys. Can.* **2011**, *67*, 35–40.
14. Marx, W.; Hoffmann, D. Bibliometric analysis of fifty years of physica status solidi. *Phys. Status Solidi* **2011**, *B248*, 2762–2771. [CrossRef]
15. Müller, K.A. Paramagnetische Resonanz von Fe-3+ in SrTiO-3 Einkristallen. *Helv. Phys. Acta* **1958**, *31*, 173–433.
16. Müller, K.A. Electron Paramagnetic Resonance of Manganese IV in SrTiO3. *Phys. Rev. Lett.* **1959**, *2*, 341–343. [CrossRef]
17. Müller, K.A. Resonance paramagnetic du Cr3+ dans des monocristaux de SrTiO3. *Publ. 7ème Colloq. Ampere Arch. Sci.* **1958**, *11*, 150.
18. The Archives des Sciences éditées par la Société de Physique et d'Histoire Naturelle de Genève can be Accessed via. Available online: https://www.unige.ch/sphn/Publications/ArchivesSciences/ (accessed on 1 March 2020).
19. Shengelaya, A.; Zhao, G.; Keller, H.; Müller, K.A. EPR Evidence of Jahn-Teller Polaron Formation in La1−xCaxMnO3+y. *Phys. Rev. Lett.* **1996**, *77*, 5296–5299. [CrossRef] [PubMed]

20. Cardona, M.; Marx, W. Vitaly l. Ginzburg – A bibliometric study. *J. Superconduct. Nov. Magn.* **2006**, *19*, 459–466. [CrossRef]
21. Müller, K.A.; Burkhard, H. SrTiO3: An intrinsic quantum paraelectric below 4 K. *Phys. Rev. B* **1979**, *19*, 3593–3602. [CrossRef]
22. Waltman, L.; van Eck, N.J.; Noyons, E.C.M. A unified approach to mapping and clustering of bibliometric networks. *J. Informetr.* **2010**, *4*, 629–635. [CrossRef]

© 2020 by the authors. Licensee MDPI, Basel, Switzerland. This article is an open access article distributed under the terms and conditions of the Creative Commons Attribution (CC BY) license (http://creativecommons.org/licenses/by/4.0/).

Article

Temperature-Independent Cuprate Pseudogap from Planar Oxygen NMR

Jakob Nachtigal [1], Marija Avramovska [1], Andreas Erb [2], Danica Pavićević [1], Robin Guehne [1] and Jürgen Haase [1,*]

[1] Felix Bloch Institute for Solid State Physics, University of Leipzig, Linnéstr. 5, 04103 Leipzig, Germany; jn43neti@studserv.uni-leipzig.de (J.N.); marija.avramovska@uni-leipzig.de (M.A.); danicas.dp@gmail.com (D.P.); r.guehne@physik.uni-leipzig.de (R.G.)
[2] Walther Meissner Institut, Bayerische Akademie der Wissenschaften, 85748 Garching, Germany; Andreas.Erb@wmi.badw.de
* Correspondence: j.haase@physik.uni-leipzig.de

Received: 23 September 2020; Accepted: 19 October 2020; Published: 21 October 2020

Abstract: Planar oxygen nuclear magnetic resonance (NMR) relaxation and shift data from all cuprate superconductors available in the literature are analyzed. They reveal a temperature-independent pseudogap at the Fermi surface, which increases with decreasing doping in family-specific ways, i.e., for some materials, the pseudogap is substantial at optimal doping while for others it is nearly closed at optimal doping. The states above the pseudogap, or in its absence are similar for all cuprates and doping levels, and Fermi liquid-like. If the pseudogap is assumed exponential it can be as large as about 1500 K for the most underdoped systems, relating it to the exchange coupling. The pseudogap can vary substantially throughout a material, being the cause of cuprate inhomogeneity in terms of charge and spin, so consequences for the NMR analyses are discussed. This pseudogap appears to be in agreement with the specific heat data measured for the YBaCuO family of materials, long ago. Nuclear relaxation and shift show deviations from this scenario near T_c, possibly due to other in-gap states.

Keywords: NMR; cuprates; pseudogap

1. Introduction

Nuclear magnetic resonance (NMR) provides important local information about the electronic properties of materials [1], and it has played a key role in the characterization of cuprate high-temperature superconductors [2,3]. However, different from when NMR proved Bardeen-Cooper-Schrieffer (BCS) theory [4,5], for cuprates a full theoretical understanding is lacking, and thus, it is challenging to decipher NMR data.

In classical metals and superconductors, NMR is known for the local measurement of the electronic spin susceptibility [6–10], including the predicted changes in the density of states at the Fermi surface with a coherence peak in nuclear relaxation [5]. In the normal state, the high density of states near the Fermi surface leads to the distinctive, fast nuclear relaxation ($1/T_1$) that is proportional to temperature ($1/T_1 \propto T$) since temperature increases the available number of electronic states for scattering with nuclear spins. Quite to the contrary, the NMR spin shift that is proportional to the uniform electronic spin susceptibility is temperature-independent, as the increase in temperature also decreases the occupation difference.

These elements of observation were the backdrop against which the cuprate NMR data were discussed, early on. Unfortunately, the cuprates have large unit cells and the important nuclei in the plane, ^{63,65}Cu and ^{17}O, have electric quadrupole moments and thus are affected by the local charges, as well. This leads to multiple resonances that have to be assigned to the chemical structure, and inhomogeneously broadened lines in the non-stoichiometric systems are the rule. This complicates measurement and interpretation. Fortunately, the cuprates are type-II materials and can be investigated in the mixed state below T_c at typical magnetic fields used for NMR, which gives access to the properties of the superfluid, but also complicates shift measurements from residual diamagnetism [11].

Early on, a number of more or less universal magnetic properties of the cuprates were derived, such as spin-singlet pairing, the pseudogap, and special spin fluctuations (for reviews see [2,3]). Here, we will not dwell on a more detailed discussion of previous conclusions, as we believe that while the data are undisputed, the prevailing view needs to be corrected.

In recent years, some of us were involved in special NMR shift experiments that raised suspicions about the description of the magnetic properties based on NMR [12–15]. During the same period of time, a comprehensive picture of the charge distribution in the CuO$_2$ plane was developed [16–18]. It fostered the understanding of charge sharing in electron and hole-doped cuprates, as it was found that $1 + x = n_{Cu} + 2n_O$, i.e., the charges measured with NMR in the planar Cu (n_{Cu}) and O (n_O) bonding orbitals add up to the total charge, inherent plus doped hole ($x > 0$) or electron ($x < 0$) content. An astonishing correlation appeared in this context, as the maximum T_c of a cuprate system ($T_{c,max}$) is nearly proportional to n_O [18,19]. This explains the differences in $T_{c,max}$ between the various families that differ in charge sharing considerably, and it calls into question the usefulness of what one calls the cuprate phase diagram, rather, a phase diagram in terms of n_{Cu} and n_O appears advantageous [20].

These findings suggested that some cuprate properties might be family dependent, and that a broader look at NMR data might be useful, as well. Since planar O NMR requires the exchange of ^{16}O by ^{17}O, which is not easily performed for single crystals and can have consequences for the actual doping and its spatial distribution, the focus was on planar Cu data that appeared more abundant and more reliable.

Immediately, the overview of the Cu shifts across all families [21] demands different shift and hyperfine scenarios, as the changes in the shifts are not proportional to each other (similar to what was found with special NMR experiments before [12,14,15]). Likely, it involves two spin components, one that has a negative uniform response and is located at planar Cu, coupled to a second component (presumably on planar O) with the usual positive response. In the next step, all planar Cu relaxation data were gathered [22,23], and from the associated plots, it became obvious that, surprisingly, the Cu relaxation is quite ubiquitous, very different from what was concluded early on. It turns out that the relaxation rate measured with the magnetic field in the plane ($1/T_{1\perp}$) does neither change significantly between families, nor as a function of doping, with $1/T_{1\perp} T_c \approx 21/\text{Ks}$. Only the relaxation anisotropy changes by about a factor of three across all cuprates. Thus, no enhanced, special spin fluctuations are present in the underdoped systems. This leaves, as an explanation for the failure of the Korringa relaxation (discovered early on [3]), only a suppression of the NMR shifts [22]. This also means that there is no pseudogap effect in planar Cu relaxation, while the Cu shifts do have a temperature dependence above T_c presumably from pseudogap effects. Finally, it was shown that the planar Cu relaxation can be understood in terms of two spin components, as well [24], where a doping dependent correlation of the Cu spin with that of O explains the relaxation anisotropy. Furthermore, the unusual planar Cu shift component that is a function of doping and not necessarily temperature was found to be present in the planar O high temperature data [25], where it causes the hallmark asymmetry of the total quadrupole lineshape, observed long ago [26–28], but not understood.

Here, we present all temperature-dependent shift and relaxation data of planar ^{17}O collected in an intensive literature search (data points from about 80 publications were taken). The main conclusion from the data will be that planar O relaxation, different from Cu, is affected by the pseudogap that also dominates the planar O shifts. Here, the pseudogap represents itself as a loss in the density of states close to the lowest energies (at the Fermi surface) for the underdoped materials, and this gap is temperature-independent, but set by doping, different from what is often assumed [29,30]. This scenario is in agreement with early specific heat data [31] that also discussed such a pseudogap in YBa$_2$Cu$_3$O$_{7-\delta}$. The largest found pseudogap is in agreement with a nodeless suppression of states of the size of the exchange coupling, 1500 K. It rapidly decreases with increasing doping, e.g., it is closed for YBa$_2$Cu$_3$O$_{7-\delta}$ at optimal doping, but not for optimally doped La$_{2-x}$Sr$_x$CuO$_4$.

2. Planar Oxygen Relaxation and Shift for YBa$_2$Cu$_3$O$_{6+y}$ and YBa$_2$Cu$_4$O$_8$

Nuclear relaxation of planar oxygen shows strikingly simple behavior in these most studied materials, and we will find the conclusions to be generic to the cuprates.

2.1. Planar Oxygen Relaxation

In Figure 1, next to a sketch of expected behavior for a Fermi liquid (A) we plot the relaxation rate ($1/T_1$) vs. temperature (T). It is apparent that optimally and overdoped YBa$_2$Cu$_3$O$_{7-\delta}$ (B) are Fermi liquid-like: above T_c, an increase (decrease) in temperature adds (subtracts) additional states for nuclear scattering and even the density of states (DOS) seems to be rather constant up to about 250 K (above that temperature the relaxation appears to begin to lag behind the expected value [32]).

It is important to note that at high temperatures, changes in temperature (ΔT) lead to proportional changes in relaxation ($\Delta(1/T_1)$) with a slope of 0.36 /Ks that intersects the origin. In other words, the proportionality of the rate to temperature is only disturbed by the opening of the superconducting gap at T_c, below which relaxation drops more rapidly as pairing sets in (no Hebel–Slichter peak is observed). Thus, planar O relaxation of optimally and overdoped YBa$_2$Cu$_3$O$_{7-\delta}$ appears determined by Fermi liquid-like electrons, turning into a spin-singlet superconductor.

The underdoped materials behave distinctively different, Figure 1B. Here we observe a rapid change of relaxation with doping at a given temperature, but we find nearly the same high-temperature slope of about 0.36 /Ks, i.e., increasing the temperature adds states at the same rate as for optimally or overdoped systems. However, the shifted slopes signal an offset in temperature below which relaxation must disappear. This means, even at much larger temperatures one is aware of the lost low temperature states. This is exactly what one expects if a temperature-independent, low-energy gap in the DOS develops with doping (a gap that remains open at high temperatures). The same scenario applies to YBa$_2$Cu$_4$O$_8$, cf. Figure 1C, where the intercept of the high-temperature slope with the abscissa is about 70 K.

At lower temperatures, the rates for YBa$_2$Cu$_3$O$_{7-\delta}$ become rather doping-independent, below about 80 K. It appears that the special temperature dependence due to the superconducting gap and pseudogap merge, somewhat different from the behavior with YBa$_2$Cu$_4$O$_8$, but still similar in the sense that the relaxation begins to increase as it departs from the parallel lines.

Note, the relaxation ceases completely at the lowest temperatures for all materials. While electric contributions (electric quadrupole interaction) to the relaxation have been shown to exist and contribute at lower temperatures [33,34] their contribution vanishes, as well. The true magnetic relaxation dependencies might be systematically shifted to lower rates at lower temperatures compared to what is seen in Figure 1. Therefore, the apparent increase in relaxation could signal quadrupolar relaxation, as well. A thorough study of these effects might be in order.

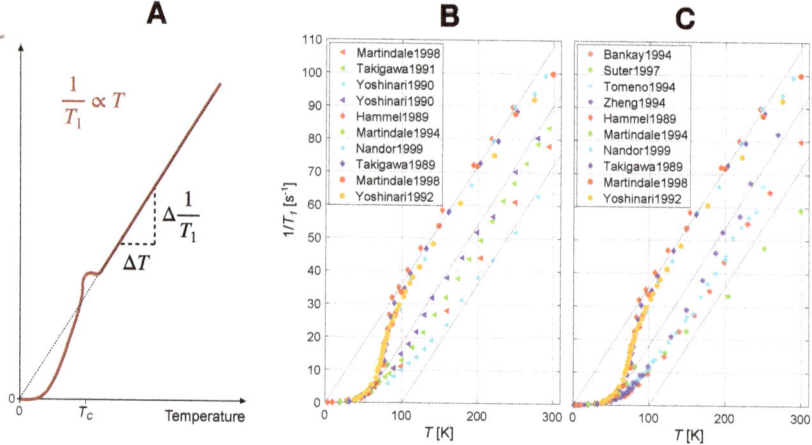

Figure 1. Nuclear Relaxation. (**A**) (sketch), above the critical temperature for superconductivity, T_c, in a Fermi liquid, the relaxation is proportional to temperature, i.e., the slope points to the origin of the plot; only just below T_c, the BCS gap for spin singlet pairing leads to a loss of states and relaxation (after the Hebel–Slichter coherence peak). (**B**), optimally doped $YBa_2Cu_3O_{6.96}$ (full circles) and overdoped $YBa_2Cu_3O_7$ (diamonds) behave Fermi liquid-like above T_c (dotted lines have slope $1/(T_1T) = 0.36/Ks$). Underdoped $YBa_2Cu_3O_{7-\delta}$ (triangles) show identical high-temperature behavior in the sense that as a function of temperature the relaxation increases with the same slope as found for optimally and underdoped systems, i.e., as the Fermi function opens with increasing temperature, it adds states at the same rate. However, the slope does not intersect the origin, which shows that even at high temperatures, low energy states are missing. This is the planar O pseudogap effect that rapidly evolves when the doping is lowered. (**C**), same as (**B**), except the relaxation data for the underdoped materials have been replaced by data for $YBa_2Cu_4O_8$ (starred points); this underdoped, stoichiometric material displays a very similar temperature dependence at higher temperatures. For the references see Appendix A.

2.2. Planar Oxygen Shifts

For planar O the orbital shift is almost negligible [26], making the spin shifts rather reliable with uncertainties arising only from the diamagnetic response below T_c. Shift referencing is simple, as well, as ordinary tap water can be used for ^{17}O NMR referencing (there is significant confusion about Cu shift referencing in the literature [21]). Nevertheless, there appear to be deviations between the shifts measured on similar samples, even for stoichiometric $YBa_2Cu_4O_8$ [35], and it is not always clear if shifts were corrected for the diamagnetic response. We will show the bare shifts without correction, in order to avoid introducing systematic errors. For example, it is possible that the uniform spin response from Cu^{2+} is negative [21,22] leading to a negative term for planar O at low temperatures.

Note that the diamagnetic response of the cuprates was experimentally determined with ^{89}Y NMR, early on [11], by assuming that this nucleus' spin shift is negligible at low temperatures (4.2K). A value of about 0.05% was derived [11]. This value appears to be rather large [36], and as experiments with ^{199}Hg NMR of $HgBa_2CuO_{4+\delta}$ showed [15], the diamagnetic response measured at ^{199}Hg is probably less than 0.01% (note that ^{199}Hg is located far from the plane and should not suffer from large spin shifts, different from ^{89}Y that might be affected by a negative term, as well).

For a Fermi liquid with a fixed DOS near the Fermi surface one expects a temperature-independent spin shift (K) above T_c, since an increase in temperature adds new states from an opening Fermi function, but the occupation decreases at the same rate, cf. Figure 2A. Now, in view of the planar O relaxation, a temperature-independent gap at the Fermi surface should be assumed. Then, qualitatively, we expect a behavior shown in Figure 2A: at the highest temperatures, far above the gap, low temperature states will still be missing, leading to a lower spin shift. As the temperature is lowered, the effect of the gap will be more severe. This is in agreement with data in Figure 2B,C. Below T_c, we note that there is no sudden loss of states as for optimally or overdoped materials, which one might naively expect if the same superconducting gap opens on the states still available. Quite to the opposite, a less rapid decrease of the shifts below T_c is observed (we noted a different low-temperature behavior for relaxation, as well).

Note that the Korringa relation is given by $T_1 T K^2 = (\gamma_e/\gamma_n)^2 \hbar/(4\pi k_B) \equiv S_0$ [8], and with $S_0 = 1.4 \cdot 10^{-5}$ Ks one estimates a spin shift of about $K = 0.23\%$ from the relaxation slope of 0.36 /Ks, not very different from what is observed for optimally or overdoped systems in Figure 2.

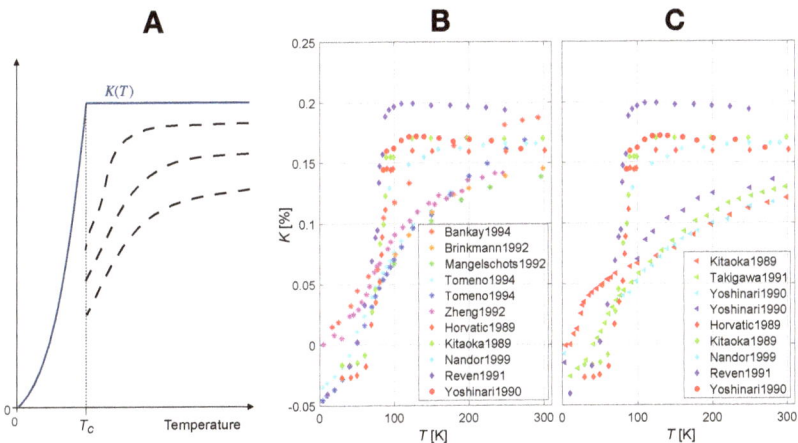

Figure 2. Planar ^{17}O nuclear magnetic resonance (NMR) Shifts. (**A**), (sketch) Fermi liquid behavior with spin singlet pairing at T_c is shown with the full blue line. The dashed lines indicate what one expects based on the relaxation data: above T_c, states are missing increasingly as the doping decreases, and as a function of temperature these lost states become more pronounced. (**B,C**), literature shift data. Optimally doped YBa$_2$Cu$_3$O$_{6.96}$ (circles) and overdoped YBa$_2$Cu$_3$O$_7$ (diamonds) behave Fermi liquid-like, but the underdoped materials YBa$_2$Cu$_3$O$_{7-\delta}$ (triangles), and YBa$_2$Cu$_4$O$_8$ (stars) show the expected high-temperature behavior. Below T_c, the shifts drop less dramatically for the underdoped systems. Some materials appear to show a negative spin shift at the lowest temperatures. For the references see Appendix A.

2.3. Numerical Analysis

The planar O relaxation data point to a pseudogap that is simply caused by missing low energy states. This gap is not temperature dependent, but rapidly increases with decreasing doping. In a very simple picture (that is very likely *not* to be correct, already in view of the planar Cu shift and relaxation data [21–24]), we use the Fermi function with fixed DOS and calculate the relaxation as being proportional to the sum of the product of occupied states times empty states (the nuclear energy change is negligible for the electrons), i.e., $\sum_E p(E)[1 - p(E)]$, where

$$p(E, \mu) = 1/\left[1 + \exp(E - \mu)/k_B T\right]. \tag{1}$$

As a result one finds the Heitler-Teller dependence [6], $1/T_1 \propto T$, cf. Figure 3.

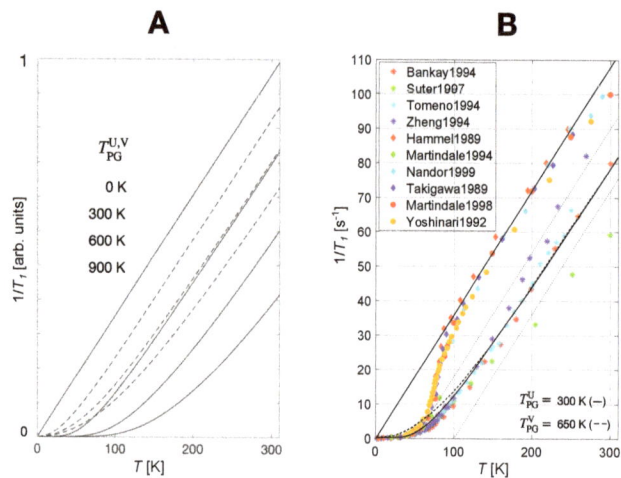

Figure 3. (**A**), model relaxation calculations with a U- and V-shaped gap ($T_{PG}^{U,V}$) in the density of states. (**B**), estimation of the pseudogap temperature by varying the gap size for $YBa_2Cu_4O_8$.

Now, one can remove manually states near the Fermi surface with a width ΔE given in temperature as defined by,

$$T_{PG}^{U,V} = \Delta E^{U,V}/k_B, \tag{2}$$

by assuming a U- or V-shaped gap in the DOS, respectively [31]. For the U-shaped gap all states within ΔE are removed (exponential decrease), for a V-shaped gap a linear decrease in DOS is assumed, vanishing at $E = \mu$. This simple scenario leads to the found behavior, i.e., we obtain nearly parallel high-temperature lines for different sizes of this pseudogap, cf. Figure 3A. For a given offset, the cutoff temperature is different for both gaps, cf. Figure 3B. With such an approach we find for $YBa_2Cu_4O_8$ a gap of about $T_{PG}^{U} \approx 300$ K ($T_{PG}^{V} \approx 650$ K). Obviously, one cannot decide on the shape of the gap. Note that the BCS gap is not included in the fit and that there are uncertainties from quadrupolar relaxation at lower temperatures.

Since the action of the gap is to cause a near parallel shift of the high-temperature dependence, any spatial inhomogeneity of the gap will lead to similar lines, as well, very different from how it affects the shifts that we will discuss now.

One can estimate what such a pseudogap will do for the NMR shifts (by assuming a slightly different μ for spin up and down). Examples are shown in Figure 4 for various T_{PG}^{U} (**A**), and T_{PG}^{V} (**B**). Clearly, for small gap sizes the shift will approach the Fermi liquid value (normalized to 1). The V-shaped gap has more total DOS and the action of the gap is weaker.

Above T_c, one should be able to fit the experimental shifts, and by comparing Figures 2 and 4 one finds qualitative agreement. However, a more quantitative determination of the gap appears difficult since

(i) there is a large spread in shifts already for similar samples, and (ii) at lower temperatures the shifts for the underdoped systems appear larger, cf. Figure 2, pointing to gap inhomogeneity. Note that the dashed lines in Figure 4 are the simple mean shifts of the shown temperature dependences. Thus, any spatial distribution of the pseudogap will change the actual temperature dependence as smaller gaps will lift the apparent shift at lower temperatures. We estimate gap sizes of $T_{PG}^U \approx 200$ K, $T_{PG}^V \approx 400$ K for $YBa_2Cu_4O_8$. These values are less than what relaxation shows, but sufficiently close for the assumed simple scenario and perhaps inhomogeneous samples (see below).

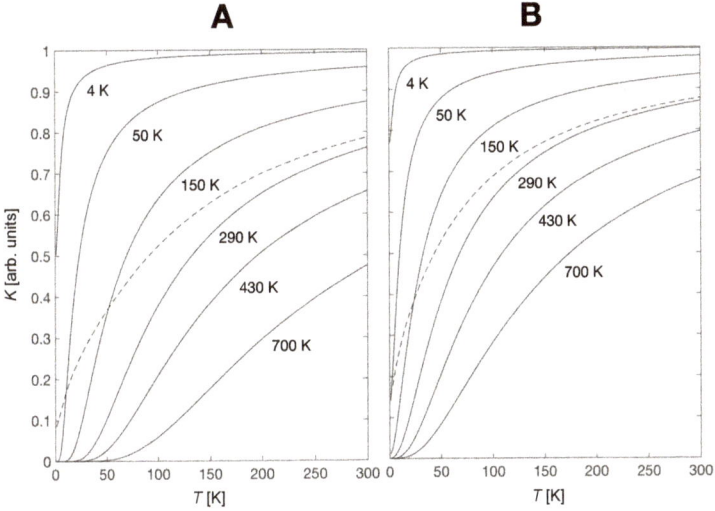

Figure 4. Model calculations of temperature-dependent shifts from a pseudogap at the Fermi surface with the indicated gap temperatures. (**A**), for a U-shaped gap, and, (**B**) for a V-shaped gap. The simple mean of the shifts is indicated by a dashed line, emphasizing that a gap inhomogeneity can cause a different temperature dependence of the apparent magnetic shift. The magnetic linewidths will also behave differently (the linewidths will grow as the temperature decreases, before it finally decreases).

An important feature of this pseudogap is a high-temperature shift offset. It arises from the fact that even far above the pseudogap energy one still misses the low energy states. Even if the shifts are temperature-independent, they can carry a doping dependence (as the pseudogap depends on doping), i.e., two variables are needed to describe the shifts ($K(x, T)$).

3. Planar Oxygen Relaxation in Other Cuprates

In Figure 5 we plot relaxation data from the literature for all other cuprates. Note that only the temperature axis is different (up to 600 K) compared to that in Figure 1B,C.

We note that the slope for optimally and overdoped $YBa_2Cu_3O_{7-\delta}$ (left dashed line) is similar to the dependencies found for the other overdoped cuprates. Thus, the CuO_2 plane appears to have this upper bound on the DOS. However, if we look at optimally doped $La_{2-x}Sr_xCuO_4$, it appears to still have a sizable pseudogap, in fact, similar to that of $YBa_2Cu_4O_8$. The largest gap is observed for the very underdoped $La_{2-x}Sr_xCuO_4$ ($x = 0.025$) with $T_{PG}^U \approx 1450$ K, the size of the exchange coupling in the cuprates. A V-shaped gap appears to better fit the low-temperature behavior. It could be the states near the gap edge that are special (coherence peaks), also in-gap states could play a role in enhancing the relaxation at low temperature. Again, the loss of parts of the inhomogeneous sample with a large gap favors states from lower gap areas with increased relaxation. Quadrupolar relaxation plays some role, as well. Thus, the shape of the gap cannot be deduced from the low-temperature behavior. The gap rapidly closes with doping, as widely assumed.

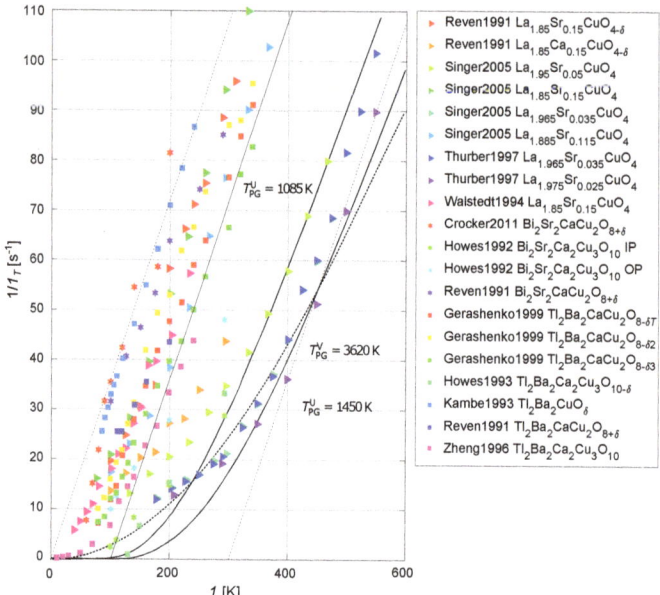

Figure 5. Planar O relaxation rates ($c \parallel B_0$) as a function of temperature for other cuprates. The slopes are rather similar to those observed for YBa$_2$Cu$_3$O$_{7-\delta}$ and YBa$_2$Cu$_4$O$_8$ in Figure 1 as the dotted lines show. The U-shaped gap closes rapidly with increasing doping where all low energy states are recovered. The maximum slope (DOS) appears to be a property of the CuO$_2$ plane, as well as the maximum size of the gap. For the references see Appendix B.

Note that the high-temperature behavior is similar for all materials, which does support the idea of a temperature-independent gap set by doping, and, importantly, very similar high-temperature Fermi liquid-like states.

To conclude, planar O NMR relaxation appears ubiquitous to the cuprates, and it defines and measures the pseudogap in a rather simple way (which is not the case for planar Cu relaxation and shift [22–24]).

4. Planar Oxygen Shifts in Other Cuprates

Shift data from all other materials are presented in Figure 6. The overall qualitative phenomenology is similar to what was found for YBa$_2$Cu$_3$O$_{7-\delta}$ and YBa$_2$Cu$_4$O$_8$. Except for a couple of overdoped materials, the shifts increase monotonously with temperature. Overdoped systems have nearly temperature-independent shifts, as for a Fermi liquid, and drop rapidly near T_c. In the pseudogap regime the shifts begin to show a temperature dependence above T_c, however, a temperature-independent shift as for $La_{1.85}Sr_{0.15}CuO_4$ at high temperatures does not mean there is no pseudogap. Again, Fermi liquid-like shifts can be suppressed in the cuprates due to lost, low-energy states [22].

The superconducting gap is hardly noticeable, as there are no rapid changes in the shifts near T_c. Despite the scarcity of data below T_c, it appears that a number of materials could show a negative shift at the lowest temperatures.

The maximum observed shifts for overdoped materials are expected from the Korringa ratio by using the dominant slope in the relaxation plots ($1/T_1T \approx 0.36/Ks$). Samples with the largest pseudogap ($La_{1.965}Sr_{0.035}CuO_4$) also have the lowest high-temperature shifts. Obviously, the pseudogap can lead to doping-dependent, but not necessarily temperature-dependent spin shift ($K(x,T)$) since the low-energy states are still missing for small pseudogaps at high temperatures.

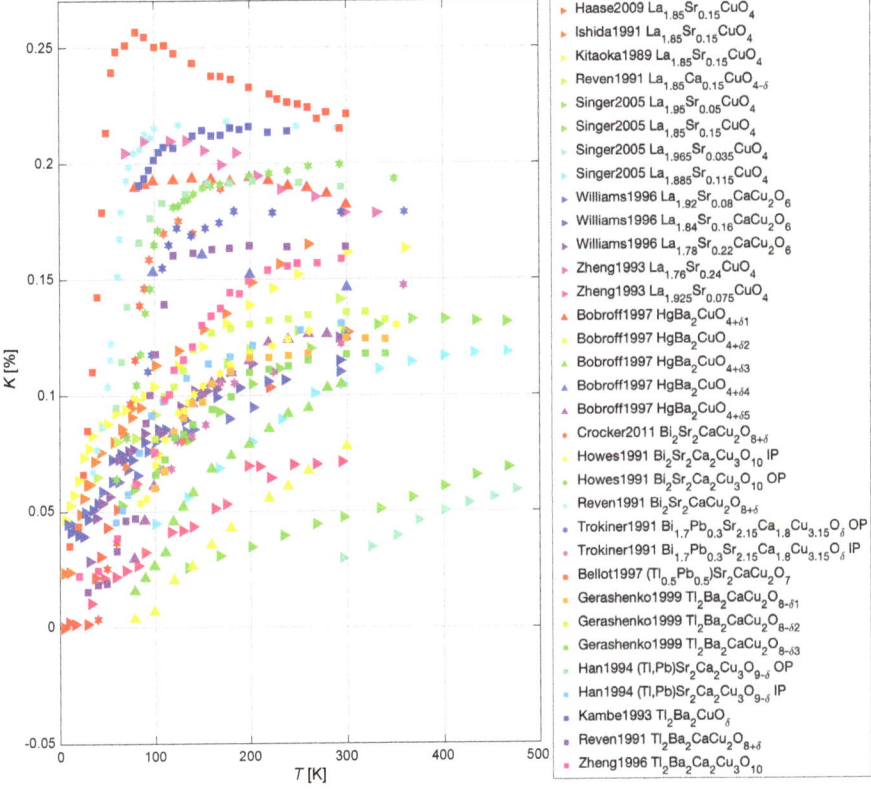

Figure 6. Planar ^{17}O NMR shifts for $c \parallel B_0$ for the other cuprates. Note that the temperature axis extends to 500 K. For more detailed plots see Figures 7 and 8. Note that a high-temperature-independent shift may still show lost states, as for optimally doped $La_{2-x}Sr_xCuO_4$. For references see Appendix B.

The true temperature dependence of the shifts in the pseudogap region is difficult to assess as sample inhomogeneity leads to a loss of the shift from areas that show a larger pseudogap as the temperature is lowered, cf. dashed lines in Figure 4.

It is also clear that optimally doped materials may have almost no pseudogap as for YBa$_2$Cu$_3$O$_{7-\delta}$, but it can be sizable as for La$_{2-x}$Sr$_x$CuO$_4$.

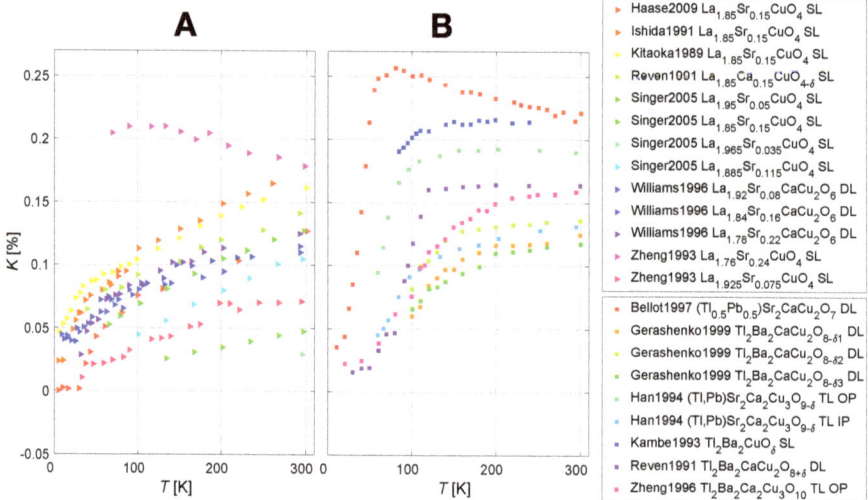

Figure 7. Planar ^{17}O NMR shifts for $c \parallel B_0$ for the other cuprates from Figure 6, separated for clarity. (**A**), La based cuprates, single and double layered. The doping ranges from x = 0.035, highly underdoped (lowest point near 300 K), to x = 0.24, highly overdoped. The shifts cover the range from 0.01% to 0.2%. The highly overdoped sample has the highest shift (there is some discrepancy between optimally doped data from different sources, probably due to inhomogeneity). (**B**), Tl based compounds. The overdoped samples have the highest and Fermi liquid-like shifts and also show an abrupt decrease near T_c. As doping is lowered the shifts become more suppressed. In the triple layer compound the inner plane (IP) has a larger pseudogap than the outer plane (OP).

5. Discussion and Conclusions

Planar O relaxation and spin shift data were collected and simple plots reveal that they demand a temperature-independent pseudogap at the Fermi surface with a size set by doping. The pseudogap rapidly opens, coming from the overdoped side by decreasing doping, and it approaches the size of the exchange coupling, J, for strongly underdoped systems. The states above the pseudogap, no matter what its size is, appear to be the same for all cuprates and carry even a more or less constant density, as perhaps expected from a two-dimensional surface. In fact, in the absence of this pseudogap, shift and relaxation for planar O are Fermi liquid-like and the Korringa relation holds. This supports the view that even in the presence of the pseudogap, the available states above it are the same Fermi liquid-like states. The doping level at which the pseudogap disappears can be different for different materials. For example, at optimal doping, there is a substantial pseudogap already present for La$_{2-x}$Sr$_x$CuO$_4$, while the pseudogap has vanished for optimally doped YBa$_2$Cu$_3$O$_{7-\delta}$. For triple-layer materials, the pseudogap is much larger for the inner layer. A plot of the pseudogap temperature for a U-shaped gap (T_{PG}^U) is shown in Figure 9.

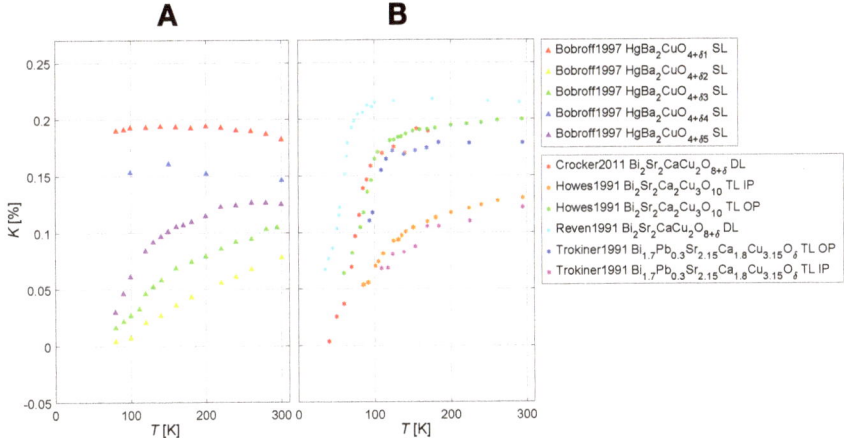

Figure 8. Planar ^{17}O NMR shifts for $c \parallel B_0$ for the other cuprates from Figure 6, separated for clarity. (**A**), single layer mercury based cuprates; the two overdoped samples have temperature-independent shifts. The pseudogap becomes apparent at optimal doping (purple triangles). (**B**), Bi based cuprates; the two double layered and overdoped samples have the highest and temperature-independent shifts, with an abrupt drop near T_c. The outer plane shifts from the two triple layered compounds show Fermi liquid-like behavior, whereas the inner plane (yellow and pink stars) show a large pseudogap.

An important consequence of the temperature-independent pseudogap is a doping dependent spin shift. At high temperatures where the shifts can be nearly temperature-independent (Fermi liquid-like), states can still be missing and thus the magnitude of shift can be suppressed. Consequently, the cuprate planar O spin shifts must carry at least two independent variables, one related to doping and the other to temperature, $K(x,T)$. This is effectively a two-component description. Whether this two-component description is sufficient is not clear (for planar Cu it is not [21]).

At lower energies, there are deviations from the simple behavior, but it is difficult to analyze given the possible influence of inhomogeneity and quadrupolar relaxation. Likely, states in the gap or near the gap edge are responsible for special behavior.

Very recently, it was shown from plots of literature shift data of planar Cu [21] that there is a doping-dependent spin shift at high temperatures, and comparison with planar Cu relaxation data [22,23]—that do not show a pseudogap—led to the conclusion of suppressed planar Cu spin shifts [22,24], as well. Thereafter, it was shown that this doping-dependent planar Cu spin shift explains the conundrum of the correlation of high-temperature spin shifts with the local charge [25], resulting in the hallmark asymmetric total planar O lineshapes (that include the quadrupolar satellites) of the cuprates [25,28].

Here, we argue that it is the doping dependence of the pseudogap that plays the dominant role in these effects. Then it follows that it is the pseudogap that can be spatially very inhomogeneous [25]. This distinction could not be made earlier [28], but it is in agreement with STM data [37]. With a large distribution of the pseudogap, shift and relaxation can be affected. An inhomogeneous broadening changes the apparent temperature dependence of the shift, cf. Figure 4, as small pseudogap areas contribute more to the shift at lower temperatures than those with large pseudogaps. For relaxation, the faster-relaxing regions, i.e., those with a smaller pseudogap, may dominate throughout the whole temperature range, if spin diffusion is possible. Thus, one has to be very careful in analyzing shift and relaxation quantitatively [38].

The inhomogeneity of the pseudogap affects the apparent temperature dependence of the average shift, as discussed with the dashed lines in Figure 4, but also the observed linewidths depend on it. In view of Figure 4 one concludes that in case of inhomogeneity of the pseudogap the NMR linewidths grow towards lower temperatures before they finally decrease again, while the shift is decreasing monotonously. This is exactly what was found experimentally (for YBa$_2$Cu$_3$O$_{7-\delta}$ and La$_{2-x}$Sr$_x$CuO$_4$ [28]), and what was interpreted as proof for two different spin components [25].

The relation of this pseudogap to the intra-unit cell charge variation that was first proposed from NMR data [39] and very recently shown to exist in the bulk of the material [40] is not clear. However, the response of the local charge symmetry to an external magnetic field and pressure found with NMR [40,41], must bear similarities to the discussed charge ordering phenomena and special susceptibilities associated with the pseudogap [29,30], recently. The total charge involved in the ordering is small (1-2% of the total planar O hole content) and may come from states within the pseudogap.

Note that the superconducting transition temperature T_c appears to be not affected by this inhomogeneity, as it is nearly proportional to the average planar oxygen hole density of the parent compounds [18,19]. Then, with the size and distribution of the pseudogap set by doping, there appears no simple relation to the maximum T_c.

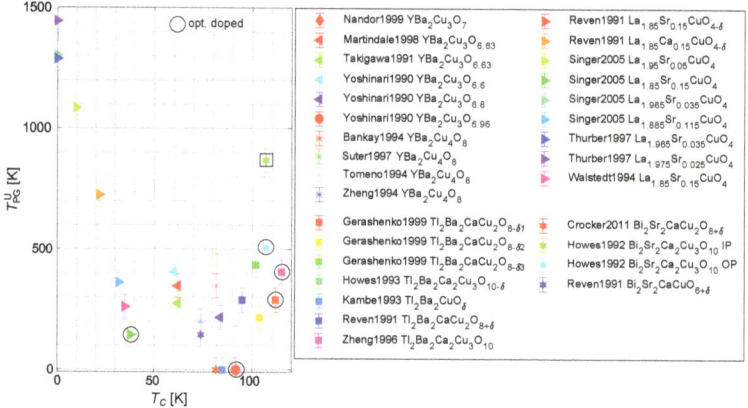

Figure 9. Values for a U-shaped gap, T_{PG}^U, as determined from the relaxation data, vs. the critical temperature, T_c. Optimally doped materials (denoted with circles) can have a vanishing pseudogap as for YBa$_2$Cu$_3$O$_{6.96}$ despite a rather high T_c, but it appears that materials with the highest T_c all have a substantial pseudogap, and their T_c increases with the pseudogap temperature. The inner layer of the triple layer system (denoted with a square) has a significantly larger pseudogap than the outer layers. These findings are in qualitative agreement with the shift data. The data can also be found in the tables in the appendices.

The pseudogap behavior was first reported with measurements above T_c for ^{89}Y NMR of YBa$_2$Cu$_3$O$_{7-\delta}$ [42], and these data show a high-temperature offset in the shifts, as well. So we believe that ^{89}Y NMR data are in agreement with what we found for planar O here.

A U-shaped gap in our simulation means that all states contributing to planar O relaxation vanish suddenly within the gap. With such an assumption the largest pseudogap appears to be set by the exchange coupling. Then, effectively, doping decreases the energy gap that needs to be overcome for electrons to flip the nuclear spin for relaxation. Of course, the true shape of the gap and the nature of the states within the gap are not known.

If the above scenario describes the essential electronic states involved in cuprate conductivity and superconductivity, it should leave its typical signature in electronic specific heat. Indeed, the $YBa_2Cu_3O_{7-\delta}$ family of materials appears to fit the specific heat data by Loram et al. [31] rather well [43]. Loram et al. [31] argue similarly in their specific heat investigations, as the specific heat is linear in temperature in the pseudogap range. Additional states are added by the temperature at the same rate as for overdoped systems where there is no gap. Thus, the specific heat of other materials should be similar in view of all analyzed planar O data.

Planar Cu relaxation was shown not to be affected by the pseudogap, at all [22,23], its relaxation is rather ubiquitous across all cuprates ($1/^{63}T_1T_c \approx 21/Ks$), independent on doping (the relaxation anisotropy changes with doping [24]). With the cuprate specific heat being in agreement with planar O relaxation, the heat involved with the states that relax planar Cu must be small (perhaps nodal particles). Not surprisingly, the planar Cu shifts, as a uniform response, do see the pseudogap. The maximum shift $^{63}K \approx 0.8\%$ is also similar to what follows from the Korringa relation. The details of a comparison between planar Cu and O NMR will be investigated in a forthcoming publication.

Unfortunately, we feel that it is difficult to conclude on the superconducting gap from the planar O data. An inhomogeneous pseudogap dominates the shifts and the relaxation may be partly electric [44] in the vicinity of T_c. The latter clearly points to the involvement of charge fluctuations [45,46], very different from the relaxation of planar Cu [23], which is also rather ubiquitous at low temperatures in the cuprates, when normalized by T_c [23]. Naively, one might assume that the states not already lost to the pseudogap disappear rapidly below T_c, further slowing down relaxation, but the opposite behavior is found, i.e., the rate appears to increase at lower temperature before it finally decreases. This could be due to additional quadrupolar relaxation, alternatively, the magnetic relaxation could show a special increase, but perhaps the inhomogeneity of the pseudogap is most important as regions with fast relaxation (small pseudogap) will dominate. Details of the spin shift, including the behavior below T_c, are difficult to evaluate, as well, not only due to the inhomogeneity, but also because of the uncertainty of the low-temperature data (loss of signal etc.). A small negative spin shift appears to be observed for a number of materials, which would be expected from the suggested shift scenario [21,22].

To conclude, the planar O data in their entirety reveal a simple temperature-independent pseudogap scenario. The gap can be as large as the exchange coupling and vanishes with increasing doping in a family-specific way. The states above the pseudogap are unique and Fermi liquid-like for all cuprates and have even constant density. This leads to a relaxation that increases at the same rate with temperature for all cuprates above the pseudogap, and to shifts that become temperature-independent. However, depending on the size of the pseudogap (located at lower energies), relaxation and shift can still be suppressed at these higher temperatures. This leads to the otherwise unexpected behavior of shift and relaxation found in NMR. The inhomogeneity of the pseudogap becomes apparent from comparison with the total planar O lineshapes and the planar Cu shifts. No simple relation of the pseudogap to the superconducting transition temperature is found. Note, however, that the planar Cu data do not fit this simple scenario with doping-independent relaxation and a two-component shift [21–24], while similarities exist and need to be explored.

Author Contributions: J.H. introduced the main concepts and had the project leadership; J.N. led the final literature data collection and its presentation in the manuscript, M.A., D.P., and A.E. were involved in the earlier stage of discussions; R.G., J.N., J.H. worked mainly on the preparation of the manuscript. All authors have read and agreed to the published version of the manuscript.

Funding: We acknowledge support from Leipzig University, and financial support by the German Science Foundation (HA1893-18-1).

Acknowledgments: We acknowledge the communication with Boris Fine (Moscow), who turned our attention to the specific heat data.

Conflicts of Interest: The authors declare no conflict of interest.

Appendix A

List of all references for $YBa_2Cu_3O_{7-\delta}$ and $YBa_2Cu_4O_8$. We found about 36 publications on these materials, out of a total of about 80 papers on all cuprates. If the same data set appears in multiple papers, typically from the same group, we only show the last published account.

Table A1. References for YBCO literature accounts with critical temperature T_c, label as shown in figures, reference link, external magnetic field during measurement and the size of U-shaped gap from the numerical analysis. All samples were aligned powders, if not stated otherwise *.

Compound	T_c	Label	Ref.	Field	T_{PG}^U
$YBa_2Cu_4O_8$	82 K	Bankay1994	[47]	9.03 T	350 K
$YBa_2Cu_4O_8$	82 K	Brinkmann1992	[48]		
$YBa_2Cu_4O_8$	82 K	Mangelschots1992	[49]	9.129 T	
$YBa_2Cu_4O_8$	81 K	Suter1997	[33]	8.9945 T	490 K
$YBa_2Cu_4O_8$	81 K	Tomeno1994	[50]	5.71 T	290 K
$YBa_2Cu_4O_8$	74 K	Zheng1992	[51]	11 T	
$YBa_2Cu_4O_8$	74 K	Zheng1993	[52]	11 T	
$YBa_2Cu_4O_8$	74 K	Zheng1994	[53]	4.3/11 T	200 K
$YBa_2Cu_3O_7$	93 K	Hammel1989	[54]	7.0 T	
$YBa_2Cu_3O_7$	92 K	Horvatic1989	[55]	5.75 T	
$YBa_2Cu_3O_{6.65}$	61 K	Kitaoka1989	[56]	5.75 T	
$YBa_2Cu_3O_7$	92 K	Kitaoka1989	[56]	5.75T	
$YBa_2Cu_3O_7$	91.2 K	Martindale1993	[57]	0.67 T	
$YBa_2Cu_3O_7$	91.2 K	Martindale1993	[57]	8.30 T	
$YBa_2Cu_3O_7$	93 K	Martindale1994	[58]	0.67 T	
$YBa_2Cu_3O_7$	93 K	Martindale1994	[58]	8.30 T	0 K
$YBa_2Cu_3O_{6.63}$	62 K	Martindale1998	[59]	high field	350 K
$YBa_2Cu_3O_{6.96}$	92.2 K	Martindale1998	[59]	high field	
$YBa_2Cu_3O_7$ *	92 K	Nandor1999	[32]	9.05 T	0 K
$YBa_2Cu_3O_7$ *	92 K	Reven1991	[60]	8.45 T	
$YBa_2Cu_3O_7$	93 K	Takigawa1989	[26]		0K
$YBa_2Cu_3O_{6.63}$	62 K	Takigawa1991	[34]	6/7 T	280 K
$YBa_2Cu_3O_{6.60}$	60 K	Yoshinari1990	[61]	10 T	410 K
$YBa_2Cu_3O_{6.80}$	84 K	Yoshinari1990	[61]	10 T	22 K
$YBa_2Cu_3O_{6.96}$	92 K	Yoshinari1990	[61]	10 T	0 K
$YBa_2Cu_3O_{6.96}$	87 K	Yoshinari1992	[62]	8.97 T	0 K

Appendix B

Here we list the references for other cuprates, about 44 publications with relevant data. If a data set appeared in multiple papers, typically from the same group, we only show the last published account.

Table A2. References to literature accounts of data, with critical temperature T_c, label as shown in figures, reference link, sample type (a.p.(c.)—aligned powder (crystal); r.p.—randomly orientated powder; s.c.—single crystal), the external magnetic field for measurement, and the U-shaped gap size from the numerical analysis (i.p.—inner plain; o.p.—outer plane in case of the triple-layer compound).

Compound	T_c	Label	Ref.	Sample	Field	T_{PG}^U
$La_{1.85}Sr_{0.15}CuO_4$	38 K	Haase2009	[12]	a.p.	9 T	
$La_{1.85}Sr_{0.15}CuO_4$	38 K	Ishida1991	[63]	a.p.	11 T	
$La_{1.85}Sr_{0.15}CuO_4$		Kitaoka1989	[56]	a.p.	5.75 T	
$La_{1.85}Ca_{0.15}CuO_{4+\delta}$	22 K	Reven1991	[60]	a.c	8.45 T	720 K
$La_{1.85}Sr_{0.15}CuO_{4+\delta}$	38 K	Reven1991	[60]	a.c	8.45 T	140 K
$La_{1.95}Sr_{0.05}CuO_4$	~10 K	Singer2005	[64]	a.c.	9 T	1085 K
$La_{1.85}Sr_{0.15}CuO_4$	38 K	Singer2005	[64]	a.c.	9 T	140 K
$La_{1.965}Sr_{0.035}CuO_4$	0 K	Singer2005	[64]	a.c.	9 T	1300 K
$La_{1.885}Sr_{0.115}CuO_4$	~32 K	Singer2005	[64]	a.c.	9 T	360 K
$La_{1.965}Sr_{0.035}CuO_4$	0 K	Thurber1997	[65]	s.c.	9 T	1290 K
$La_{1.975}Sr_{0.025}CuO_4$	0 K	Thurber1997	[65]	s.c.	9 T	1450 K
$La_{1.85}Sr_{0.15}CuO_{4+\delta}$	35 K	Walstedt1994	[66]	a.p.		260 K
$La_{1.92}Sr_{0.08}CaCu_2O_6$	17.7 K	Williams1996	[67]	r.p.	8.45 T	
$La_{1.84}Sr_{0.16}CaCu_2O_6$	31.5 K	Williams1996	[67]	r.p.	8.45 T	
$La_{1.78}Sr_{0.22}CaCu_2O_6$	47 K	Williams1996	[67]	r.p.	8.45 T	
$La_{1.76}Sr_{0.24}CuO_4$	25 K	Zheng1993	[16]	a.c.		
$La_{1.925}Sr_{0.075}CuO_4$	20 K	Zheng1993	[16]	a.c.		
$HgBa_2CuO_{4+\delta 1}$	61 K	Bobroff1997	[68]	a.c.	7.5 T	
$HgBa_2CuO_{4+\delta 2}$	75 K	Bobroff1997	[68]	a.c.	7.5 T	
$HgBa_2CuO_{4+\delta 3}$	87.8 K	Bobroff1997	[68]	a.p.	7.5 T	
$HgBa_2CuO_{4+\delta 4}$	89 K	Bobroff1997	[68]	a.p.	7.5 T	
$HgBa_2CuO_{4+\delta 5}$	95.7 K	Bobroff1997	[68]	a.p.	7.5 T	
$Bi_2Sr_2CaCu_2O_{8+\delta}$	82 K	Crocker2011	[69]	a.p.	9 T	0 K
$Bi_2Sr_2Ca_2Cu_3O_{10}$	107 K	Howes1991	[70]	a.p.	8.45 T	
$Bi_2Sr_2Ca_2Cu_3O_{10}$	107 K	Howes1992	[71]	r.p.	8.45 T	i.p. 870 K
$Bi_2Sr_2Ca_2Cu_3O_{10}$	107 K	Howes1992	[71]	r.p.	8.45 T	o.p. 510 K
$Bi_2Sr_2CaCu_2O_{8+\delta}$	74 K	Reven1991	[60]	r.p.	8.45 T	140 K
$Bi_2Sr_2CaCuO_{6+\delta}$	5.6 K	Reven1991	[60]	r.p.	8.45 T	
$Bi_{1.7}Pb_{0.3}Sr_{2.15}Ca_{1.8}Cu_{3.15}O_\delta$	110 K	Trokiner1991	[72]	r.p.		
$(Tl_{0.5}Pb_{0.5})Sr_2CaCu_2O_7$	65 K	Bellot1997	[73]	r.p.	7 T	
$Tl_2Ba_2CaCu_2O_{8-\delta 1}$	112 K	Gerashenko1999	[74]	a.c.		290 K
$Tl_2Ba_2CaCu_2O_{8-\delta 2}$	104 K	Gerashenko1999	[74]	a.c.		220 K
$Tl_2Ba_2CaCu_2O_{8-\delta 3}$	102 K	Gerashenko1999	[74]	a.c.		430 K
$(Tl,Pb)Sr_2Ca_2Cu_3O_{9-\delta}$	124 K	Han1994	[75]	r.p.	8.45 T	
$Tl_2Ba_2CuO_7$	<4.2 K	Kambe1991	[76]	a.c.		
$Tl_2Ba_2CuO_\delta$	85 K	Kambe1993	[27]	a.c.	12 T	0 K
$Tl_2Ba_2Ca_2Cu_3O_{10-\delta}$	125 K	Howes1993	[77]	s.c.	8.45 T	870 K
$Tl_2Ba_2CaCu_2O_{8+\delta}$	95 K	Reven1991	[60]	r.p.	8.45 T	290 K
$Tl_2Ba_2Ca_2Cu_3O_{10}$	125 K	Zheng1995	[78]	a.p.	11 T	
$Tl_2Ba_2Ca_2Cu_3O_{10}$	125 K	Zheng1996	[79]	a.p.	11 T	410 K

References

1. Slichter, C.P. *Principles of Magnetic Resonance*, 3rd ed.; Springer: Berlin/Heidelberg, Germany, 1990.
2. Slichter, C.P. Magnetic Resonance Studies of High Temperature Superconductors. In *Handbook of High-Temperature Superconductivity*; Schrieffer, J.R., Brooks, J.S., Eds.; Springer: New York, USA, 2007; pp. 215–256. [CrossRef]
3. Walstedt, R.E. *The NMR Probe of High-T_c Materials*, 1st ed.; Springer: Berlin/Heidelberg, Germany, 2008. [CrossRef]

4. Bardeen, J.; Cooper, L.N.; Schrieffer, J.R. Microscopic theory of superconductivity. *Phys. Rev.* **1957**, *106*, 162–164. [CrossRef]
5. Hebel, L.C.; Slichter, C.P. Nuclear Spin Relaxation in Normal and Superconducting Aluminum. *Phys. Rev.* **1959**, *113*, 1504–1519. [CrossRef]
6. Heitler, W.; Teller, E. Time Effects in the Magnetic Cooling Method-I. *Proc. R. Soc. A Math. Phys. Eng. Sci.* **1936**, *155*, 629–639. [CrossRef]
7. Knight, W. Nuclear Magnetic Resonance Shift in Metals. *Phys. Rev.* **1949**, *76*, 1259–1260. [CrossRef]
8. Korringa, J. Nuclear magnetic relaxation and resonnance line shift in metals. *Physica* **1950**, *16*, 601–610. [CrossRef]
9. Schumacher, R.T.; Slichter, C.P. Electron Spin Paramagnetism of Lithium and Sodium. *Phys. Rev.* **1956**, *101*, 58–65. [CrossRef]
10. Yosida, K. Paramagnetic Susceptibility in Superconductors. *Phys. Rev.* **1958**, *110*, 769–770. [CrossRef]
11. Barrett, S.E.; Durand, D.J.; Pennington, C.H.; Slichter, C.P.; Friedmann, T.A.; Rice, J.P.; Ginsberg, D.M. ^{63}Cu Knight shifts in the superconducting state of YBa$_2$Cu$_3$O$_{7-\delta}$ (T$_c$ = 90 K). *Phys. Rev. B* **1990**, *41*, 6283–6296. [CrossRef] [PubMed]
12. Haase, J.; Slichter, C.P.; Williams, G.V.M. Evidence for two electronic components in high-temperature superconductivity from NMR. *J. Phys. Condens. Matter* **2009**, *21*, 455702. [CrossRef] [PubMed]
13. Meissner, T.; Goh, S.K.; Haase, J.; Williams, G.V.M.; Littlewood, P.B. High-pressure spin shifts in the pseudogap regime of superconducting YBa$_2$Cu$_4$O$_8$ as revealed by ^{17}O NMR. *Phys. Rev. B* **2011**, *83*, 220517. [CrossRef]
14. Haase, J.; Rybicki, D.; Slichter, C.P.; Greven, M.; Yu, G.; Li, Y.; Zhao, X. Two-component uniform spin susceptibility of superconducting HgBa$_2$CuO$_{4+\delta}$ single crystals measured using ^{63}Cu and ^{199}Hg nuclear magnetic resonance. *Phys. Rev. B* **2012**, *85*, 104517. [CrossRef]
15. Rybicki, D.; Kohlrautz, J.; Haase, J.; Greven, M.; Zhao, X.; Chan, M.K.; Dorow, C.J.; Veit, M.J. Electronic spin susceptibilities and superconductivity in HgBa$_2$CuO$_{4+\delta}$ from nuclear magnetic resonance. *Phys. Rev. B* **2015**, *92*, 081115. [CrossRef]
16. Zheng, G.Q.; Kuse, T.; Kitaoka, Y.; Ishida, K.; Ohsugi, S.; Asayama, K.; Yamada, Y. ^{17}O NMR study of La$_{2-x}$Sr$_x$CuO$_4$ in the lightly-and heavily-doped regions. *Phys. C Supercond.* **1993**, *208*, 339–346. [CrossRef]
17. Haase, J.; Sushkov, O.P.; Horsch, P.; Williams, G.V.M. Planar Cu and O hole densities in high-T_c cuprates determined with NMR. *Phys. Rev. B* **2004**, *69*, 94504. [CrossRef]
18. Jurkutat, M.; Rybicki, D.; Sushkov, O.P.; Williams, G.V.M.; Erb, A.; Haase, J. Distribution of electrons and holes in cuprate superconductors as determined from ^{17}O and ^{63}Cu nuclear magnetic resonance. *Phys. Rev. B* **2014**, *90*, 140504. [CrossRef]
19. Rybicki, D.; Jurkutat, M.; Reichardt, S.; Kapusta, C.; Haase, J. Perspective on the phase diagram of cuprate high-temperature superconductors. *Nat. Commun.* **2016**, *7*, 1–6. [CrossRef] [PubMed]
20. Jurkutat, M.; Erb, A.; Haase, J. Tc and Other Cuprate Properties in Relation to Planar Charges as Measured by NMR. *Condens. Matter* **2019**, *4*, 67. [CrossRef]
21. Haase, J.; Jurkutat, M.; Kohlrautz, J. Contrasting Phenomenology of NMR Shifts in Cuprate Superconductors. *Condens. Matter* **2017**, *2*, 16. [CrossRef]
22. Avramovska, M.; Pavićević, D.; Haase, J. Properties of the Electronic Fluid of Superconducting Cuprates from ^{63}Cu NMR Shift and Relaxation. *J. Supercond. Nov. Magn.* **2019**, *32*, 3761–3771. [CrossRef]
23. Jurkutat, M.; Avramovska, M.; Williams, G.V.M.; Dernbach, D.; Pavićević, D.; Haase, J. Phenomenology of ^{63}Cu Nuclear Relaxation in Cuprate Superconductors. *J. Supercond. Nov. Magn.* **2019**, *32*, 3369–3376. [CrossRef]
24. Avramovska, M.; Pavićević, D.; Haase, J. NMR Shift and Relaxation and the Electronic Spin of Superconducting Cuprates. *J. Supercond. Nov. Magn.* **2020**, *33*, 2621–2628. [CrossRef]
25. Pavićević, D.; Avramovska, M.; Haase, J. Unconventional ^{17}O and ^{63}Cu NMR shift components in cuprate superconductors. *Mod. Phys. Lett. B* **2020**, *34*, 2040047. [CrossRef]
26. Takigawa, M.; Hammel, P.; Heffner, R.; Fisk, Z.; Ott, K.; Thompson, J. ^{17}O NMR study of YBa$_2$Cu$_3$O$_{7-\delta}$. *Phys. C Supercond.* **1989**, *162–164*, 853–856. [CrossRef]
27. Kambe, S.; Yasuoka, H.; Hayashi, A.; Ueda, Y. NMR study of the spin dynamics in Tl$_2$Ba$_2$CuO$_y$ (Tc=85 K). *Phys. Rev. B* **1993**, *47*, 2825–2834. [CrossRef]

28. Haase, J.; Slichter, C.P.; Stern, R.; Milling, C.T.; Hinks, D.G. Spatial Modulation of the NMR Properties of the Cuprates. *Phys. C Supercond.* **2000**, *341*, 1727–1730. [CrossRef]
29. Mukhopadhyay, S.; Sharma, R.; Kim, C.K.; Edkins, S.D.; Hamidian, M.H.; Eisaki, H.; Uchida, S.I.; Kim, E.A.; Lawler, M.J.; Mackenzie, A.P.; et al. Evidence for a vestigial nematic state in the cuprate pseudogap phase. *Proc. Nat. Acad. Sci. USA* **2019**, *116*, 13249–13254. [CrossRef]
30. Sato, Y.; Kasahara, S.; Murayama, H.; Kasahara, Y.; Moon, E.G.; Nishizaki, T.; Loew, T.; Porras, J.; Keimer, B.; Shibauchi, T.; Matsuda, Y. Thermodynamic evidence for a nematic phase transition at the onset of the pseudogap in $YBa_2Cu_3O_y$. *Nat. Phys.* **2017**, *13*, 1074–1078. [CrossRef]
31. Loram, J.W.; Mirza, K.A.; Cooper, J.R.; Tallon, J.L. Specific heat evidence on the normal state pseudogap. *J. Phys. Chem. Solids* **1998**, *59*, 2091–2094. [CrossRef]
32. Nandor, V.A.; Martindale, J.A.; Groves, R.W.; Vyaselev, O.M.; Pennington, C.H.; Hults, L.; Smith, J.L. High-temperature ^{17}O and $^{89}YNMR$ of $YBa_2Cu_3O_{7-\delta}$. *Phys. Rev. B* **1999**, *60*, 6907–6915. [CrossRef]
33. Suter, A.; Mali, M.; Roos, J.; Brinkmann, D.; Karpinski, J.; Kaldis, E. Electronic crossover in the normal state of $YBa_2Cu_4O_8$. *Phys. Rev. B* **1997**, *56*, 5542–5551. [CrossRef]
34. Takigawa, M.; Reyes, A.P.; Hammel, P.C.; Thompson, J.D.; Heffner, R.H.; Fisk, Z.; Ott, K.C. Cu and O NMR studies of the magnetic properties of $YBa_2Cu_3O_{6.63}$ ($T_c = 62K$). *Phys. Rev. B* **1991**, *43*, 247–257. [CrossRef]
35. Brinkmann, D. The mysterious spin gap in high-temperature superconductors: New NMR/NQR studies. *Appl. Magn. Reson.* **1998**, *15*, 197–202. [CrossRef]
36. Oldfield, E.; Coretsopoulos, C.; Yang, S.; Reven, L.; Lee, H.C.; Shore, J.; Han, O.H.; Ramli, E.; Hinks, D. ^{17}O nuclear-magnetic-resonance spectroscopic study of high-T_c superconductors. *Phys. Rev. B* **1989**, *40*, 6832–6849. [CrossRef] [PubMed]
37. Pan, S.H.; O'Neal, J.P.; Badzey, R.L.; Chamon, C.; Ding, H.; Engelbrecht, J.R.; Wang, Z.; Eisaki, H.; Uchida, S.; Gupta, A.K.; et al. Microscopic electronic inhomogeneity in the high-Tc superconductor $Bi_2Sr_2CaCu_2O_{8+x}$. *Nature* **2001**, *413*, 282–285. [CrossRef] [PubMed]
38. Bussmann-Holder, A. Evidence for s+d Wave Pairing in Copper Oxide Superconductors from an Analysis of NMR and NQR Data. *J. Supercond. Nov. Magn.* **2011**, *25*, 155–157. [CrossRef]
39. Haase, J. Charge density variation in $YBa_2Cu_3O_{6+y}$. *Phys. Rev. Lett.* **2003**, *91*, 189701. [CrossRef]
40. Reichardt, S.; Jurkutat, M.; Guehne, R.; Kohlrautz, J.; Erb, A.; Haase, J. Bulk Charge Ordering in the CuO_2 Plane of the Cuprate Superconductor $YBa_2Cu_3O_{6.9}$ by High-Pressure NMR. *Condens. Matter* **2018**, *3*, 23. doi:10.3390/condmat3030023. [CrossRef]
41. Reichardt, S.; Jurkutat, M.; Erb, A.; Haase, J. Charge Variations in Cuprate Superconductors from Nuclear Magnetic Resonance. *J. Supercond. Nov. Magn.* **2016**, *29*, 3017–3022. [CrossRef]
42. Alloul, H.; Ohno, T.; Mendels, P. ^{89}Y NMR evidence for a Fermi-liquid behavior in $YBa_2Cu_3O_{6+x}$. *Phys. Rev. Lett.* **1989**, *63*, 1700–1703. [CrossRef]
43. Fine, B. (Institute of Physics and Technology, Moscow, Russia). Private communication with J.H, 2020.
44. Suter, A.; Mali, M.; Roos, J.; Brinkmann, D. Charge degree of freedom and the single-spin fluid model in $YBa_2Cu_4O_8$. *Phys. Rev. Lett.* **2000**, *84*, 4938–4941. [CrossRef]
45. Olson Reichhardt, C.J.; Reichhardt, C.; Bishop, A.R. Fibrillar Templates and Soft Phases in Systems with Short-Range Dipolar and Long-Range Interactions. *Phys. Rev. Lett.* **2004**, *92*, 016801. [CrossRef] [PubMed]
46. Mazumdar, S. Valence transition model of the pseudogap, charge order, and superconductivity in electron-doped and hole-doped copper oxides. *Phys. Rev. B* **2018**, *98*, 205153. [CrossRef]
47. Bankay, M.; Mali, M.; Roos, J.; Brinkmann, D. Single-spin fluid, spin gap, and d-wave pairing in $YBa_2Cu_4O_8$: A NMR and NQR study. *Phys. Rev. B* **1994**, *50*, 6416–6425. [CrossRef] [PubMed]
48. Brinkmann, D. Comparing Y–Ba–Cu–O superconductors by Cu, O and Ba NMR/NQR. *Appl. Magn. Reson.* **1992**, *3*, 483–494. [CrossRef]
49. Mangelschots, I.; Mali, M.; Roos, J.; Brinkmann, D.; Rusiecki, S.; Karpinski, J.; Kaldis, E. ^{17}O NMR study in aligned $YBa_2Cu_4O_8$ powder. *Phys. C Supercond.* **1992**, *194*, 277–286. [CrossRef]

50. Tomeno, I.; Machi, T.; Tai, K.; Koshizuka, N.; Kambe, S.; Hayashi, A.; Ueda, Y.; Yasuoka, H. NMR study of spin dynamics at planar oxygen and copper sites in YBa$_2$Cu$_4$O$_8$. *Phys. Rev. B* **1994**, *49*, 15327–15334. [CrossRef] [PubMed]
51. Zheng, G.Q.; Kitaoka, Y.; Asayama, K.; Kodama, Y.; Yamada, Y. ^{17}O NMR study of local hole density and spin dynamics in YBa$_2$Cu$_4$O$_8$. *Phys. C Supercond.* **1992**, *193*, 154–162. [CrossRef]
52. Zheng, G.Q.; Kitaoka, Y.; Asayama, K.; Kodama, Y. Spin susceptibility in YBa$_2$Cu$_4$O$_8$: ^{17}O and ^{63}Cu NMR study. *Phys. B Condens. Matter* **1993**, *186–188*, 1001–1003. [CrossRef]
53. Zheng, G.Q.; Kitaoka, Y.; Asayama, K.; Kodama, Y. Magnetic-field enhancement of ^{63}Cu and ^{17}O relaxation rates in the mixed state of YBa$_2$Cu$_4$O$_8$ observation of fermi-liquid state in the vortex cores. *Phys. C Supercond.* **1994**, *227*, 169–175. [CrossRef]
54. Hammel, P.C.; Takigawa, M.; Heffner, R.H.; Fisk, Z.; Ott, K.C. Spin dynamics at oxygen sites in YBa$_2$Cu$_3$O$_7$. *Phys. Rev. Lett.* **1989**, *63*, 1992–1995. [CrossRef]
55. Horvatić, M.; Berthier, Y.; Butaud, P.; Kitaoka, Y.; Ségransan, P.; Berthier, C.; Katayama-Yoshida, H.; Okabe, Y.; Takahashi, T. ^{17}O NMR study of YBa$_2$Cu$_3$O$_{7-\delta}$ (T_c = 92 K). *Phys. C Supercond.* **1989**, *159*, 689–696. [CrossRef]
56. Kitaoka, Y.; Berthier, Y.; Butaud, P.; Horvatić, M.; Ségransan, P.; Berthier, C.; Katayama-Yoshida, H.; Okabe, Y.; Takahashi, T. NMR study of ^{17}O in high T_c superconducting oxides. *Phys. C Supercond.* **1989**, *162–164 Pt 1*, 195–196. [CrossRef]
57. Martindale, J.A.; Barrett, S.E.; O'Hara, K.E.; Slichter, C.P.; Lee, W.C.; Ginsberg, D.M. Magnetic-field dependence of planar copper and oxygen spin-lattice relaxation rates in the superconducting state of YBa$_2$Cu$_3$O$_7$. *Phys. Rev. B* **1993**, *47*, 9155. [CrossRef] [PubMed]
58. Martindale, J.A.; Barrett, S.E.; Durand, D.J.; O'Hara, K.E.; Slichter, C.P.; Lee, W.C.; Ginsberg, D.M. Nuclear-spin-lattice relaxation-rate measurements in YBa$_2$Cu$_3$O$_7$. *Phys. Rev. B* **1994**, *50*, 13645–13652. [CrossRef]
59. Martindale, J.A.; Hammel, P.C.; Hults, W.L.; Smith, J.L. Temperature dependence of the anisotropy of the planar oxygen nuclear spin-lattice relaxation rate in YBa$_2$Cu$_3$O$_y$. *Phys. Rev. B* **1998**, *57*, 11769–11774. [CrossRef]
60. Reven, L.; Shore, J.; Yang, S.; Duncan, T.; Schwartz, D.; Chung, J.; Oldfield, E. ^{17}O nuclear-magnetic-resonance spin-lattice relaxation and Knight-shift behavior in bismuthate, plumbate, and cuprate superconductors. *Phys. Rev. B* **1991**, *43*, 10466. [CrossRef]
61. Yoshinari, Y.; Yasuoka, H.; Ueda, Y.; Koga, K.I.; Kosuge, K. NMR Studies of ^{17}O in the Normal State of YBa$_2$Cu$_3$O$_{6+x}$. *J. Phys. Soc. Jpn.* **1990**, *59*, 3698–3711. [CrossRef]
62. Yoshinari, Y.; Yasuoka, H.; Ueda, Y. Nuclear Spin Relaxation at Planar Copper and Oxygen Sites in YBa$_2$Cu$_3$O$_{6.96}$. *J. Phys. Soc. Jpn.* **1992**, *61*, 770–773. [CrossRef]
63. Ishida, K.; Kitaoka, Y.; Zheng, G.Q.; Asayama, K. ^{17}O and ^{63}Cu NMR Investigations of High-T_c Superconductor La$_{1.85}$Sr$_{0.15}$CuO$_4$ with T_c = 38 K. *J. Phys. Soc. Jpn.* **1991**, *60*, 3516–3524. [CrossRef]
64. Singer, P.M.; Imai, T.; Chou, F.C.; Hirota, K.; Takaba, M.; Kakeshita, T.; Eisaki, H.; Uchida, S. ^{17}O NMR study of the inhomogeneous electronic state in La$_{2-x}$Sr$_x$CuO$_4$ crystals. *Phys. Rev. B* **2005**, *72*, 014537. [CrossRef]
65. Thurber, K.R.; Hunt, A.W.; Imai, T.; Chou, F.C.; Lee, Y.S. ^{17}O NMR Study of Undoped and Lightly Hole Doped CuO$_2$ Planes. *Phys. Rev. Lett.* **1997**, *79*, 171–174. [CrossRef]
66. Walstedt, R.E.; Shastry, B.S.; Cheong, S.W. NMR, neutron scattering, and the one-band model of La$_{2-x}$Sr$_x$CuO$_4$. *Phys. Rev. Lett.* **1994**, *72*, 3610–3613. [CrossRef] [PubMed]
67. Williams, G.V.M.; Tallon, J.L.; Michalak, R.; Dupree, R. NMR evidence for common superconducting and pseudogap phase diagrams of YBa$_2$Cu$_3$O$_{7-\delta}$ and La$_{2-x}$Sr$_x$CaCu$_2$O$_6$. *Phys. Rev. B* **1996**, *54*, R6909–R6912. [CrossRef] [PubMed]
68. Bobroff, J.; Alloul, H.; Mendels, P.; Viallet, V.; Marucco, J.F.; Colson, D. ^{17}O NMR Evidence for a Pseudogap in the Monolayer HgBa$_2$CuO$_{4+\delta}$. *Phys. Rev. Lett.* **1997**, *78*, 3757–3760. [CrossRef]
69. Crocker, J.; Dioguardi, A.P.; apRoberts-Warren, N.; Shockley, A.C.; Grafe, H.-J.; Xu, Z.; Wen, J.; Gu, G.; Curro, N.J. NMR studies of pseudogap and electronic inhomogeneity in Bi$_2$Sr$_2$CaCu$_2$O$_{8+\delta}$. *Phys. Rev. B* **2011**, *84*, 224502. [CrossRef]
70. Howes, A.; Dupree, R.; Paul, D.; Male, S. ^{17}O NMR of the Bi$_2$Sr$_2$Ca$_2$Cu$_3$O$_{10}$ high temperature superconductor. *Phys. C Supercond.* **1991**, *185–189 Pt 2*, 1137–1138. [CrossRef]

71. Howes, A.; Durpee, R.; Paul, D.; Male, S. An ^{17}O NMR study of the Cu-O planes of $Bi_2Sr_2Ca_2Cu_3O_{10}$. *Phys. C Supercond.* **1992**, *193*, 189–195. [CrossRef]
72. Trokiner, A.; Le Noc, L.; Schneck, J.; Pougnet, A.; Mellet, R.; Primot, J.; Savary, H.; Gao, Y.; Aubry, S. ^{17}O nuclear-magnetic-resonance evidence for distinct carrier densities in the two types of CuO_2 planes of $(Bi,Pb)_2Sr_2Ca_2Cu_3O_y$. *Phys. Rev. B* **1991**, *44*, 2426–2429. [CrossRef]
73. Bellot, P.V.; Trokiner, A.; Zhdonov, Y.; Yakubovskii, A.; Shustov, L.; Verkhovskii, S.; Zagoulaev, S.; Monod, P. Magnetic properties of $(Tl_{1-x}Pb_x)Sr_2CaCu_2O_7$ High-T_c oxide studied by ^{17}O NMR and SQUID: from overdoped superconductor to strongly overdoped metal. *Phys. C Supercond.* **1997**, *282–287 Pt 3*, 1357–1358. [CrossRef]
74. Gerashenko, A.; Piskunov, Y.; Mikhalev, K.; Ananyev, A.; Okulova, K.; Verkhovskii, S.; Yakubovskii, A.; Shustov, L.; Trokiner, A. The ^{63}Cu and ^{17}O NMR studies of spin susceptibility in differently doped $Tl_2Ba_2CaCu_2O_{8-\delta}$ compounds. *Phys. C Supercond.* **1999**, *328*, 163–176. [CrossRef]
75. Han, Z.; Dupree, R.; Howes, A.; Liu, R.; Edwards, P. Charge distribution in $(Tl,Pb)Sr_2Ca_2Cu_3O_{9-\sigma}$ (T_c = 124K) an ^{17}O NMR study. *Phys. C Supercond.* **1994**, *235–240 Pt 3*, 1709–1710. [CrossRef]
76. Kambe, S.; Yoshinari, Y.; Yasuoka, H.; Hayashi, A.; Ueda, Y. ^{17}O, ^{63}Cu and ^{205}Tl NMR study of over-doped $Tl_2Ba_2CuO_y$. *Phys. C Supercond.* **1991**, *185–189*, 1181–1182. [CrossRef]
77. Howes, A.P.; Dupree, R.; Han, Z.P.; Liu, R.S.; Edwards, P.P. Anomalous temperature dependence of the static spin susceptibility of $Tl_2Ba_2Ca_2Cu_3O_{10-\delta}$ ($T_c \approx$ 125 K) in the normal state. *Phys. Rev. B* **1993**, *47*, 11529–11532. [CrossRef] [PubMed]
78. Zheng, G.Q.; Kitaoka, Y.; Ishida, K.; Asayama, K. Local Hole Distribution in the CuO_2 Plane of High-T_c Cu-Oxides Studied by Cu and Oxygen NQR/NMR. *J. Phys. Soc. Jpn.* **1995**, *64*, 2524–2532. [CrossRef]
79. Zheng, G.Q.; Kitaoka, Y.; Asayama, K.; Hamada, K.; Yamauchi, H.; Tanaka, S. NMR study of local hole distribution, spin fluctuation and superconductivity in $Tl_2Ba_2Ca_2Cu_3O_{10}$. *Phys. C Supercond.* **1996**, *260*, 197–210. [CrossRef]

Publisher's Note: MDPI stays neutral with regard to jurisdictional claims in published maps and institutional affiliations.

© 2020 by the authors. Licensee MDPI, Basel, Switzerland. This article is an open access article distributed under the terms and conditions of the Creative Commons Attribution (CC BY) license (http://creativecommons.org/licenses/by/4.0/).

Communication

Polaronic States and Superconductivity in WO$_{3-x}$

Ekhard K. H. Salje

Department of Earth Sciences, University of Cambridge, Downing Street, Cambridge CB2 3EQ, UK; Ekhard@esc.cam.ac.uk

Received: 7 April 2020; Accepted: 27 April 2020; Published: 1 May 2020

Abstract: Superconducting domain boundaries were found in WO$_{3-x}$ and doped WO$_3$. The charge carriers in WO$_3$-type materials were identified by Schirmer and Salje as bipolarons. Several previous attempts to determine the electronic properties of polarons in WO$_3$ failed until Bousque et al. (2020) reported a full first principle calculation of free polarons in WO$_3$. They confirmed the model of Schirmer and Salje that each single polaron is centred around one tungsten position with surplus charges smeared over the adjacent eight tungsten positions. Small additional charges are distributed further apart. Further calculations to clarify the coupling mechanism between polaron to form bipolarons are not yet available. These calculations would help to identify the carrier distribution in Magneli clusters, which were shown recently to contain high carrier concentrations and may indicate totally localized superconductivity in non-percolating clusters.

Keywords: ferroelastic; WO3; polarons; polaronic superconductivity

---

1. Introduction and Some Historic Background

The discovery of polaronic states in WO$_{3-x}$ is related to Alex Muller. I first met Alex in 1970 during a trip to Zurich to visit my friend Fritz Laves. Fritz was then the most preeminent crystallographer in Europe who was at the heart of rapid developments of diffraction tools and the theory of symmetry and finite groups. I was introduced to him by my mathematics professor Heesch in Hannover, who solved the essentials of the four-colour problem but was then discouraged to publish his work on magnetic space groups by Fritz Laves and Raman because it would serve no good purpose'. A few years later Shubnikov published his famous paper and Heesch took it stoically. I was just starting a PhD and struggled to promote the message that perovskite structures would be a good example of structural instabilities without compromising the macroscopic and chemical integrity of the material. The idea was generally rejected in the solid state community for two opposing reasons: the first argument was that anything beyond Si could not be understood properly because our theoretical framework was insufficient, and, secondly, that the perovskite structure was far too simple to show effects of internal structural instabilities where one would need structures of much greater complexity like garnets, feldspars and boracites.

Fritz Laves had discovered the Laves-phases and I expected to get some support from him for my PhD project. He did so enthusiastically and told me to see Alex Müller in Ruschlikon who knew much about the perovskite structures in ferroelectricity and in particular about SrTiO$_3$. He had published an excellent account of the phonon instability of LaAlO$_3$ [1] and some years later he would publish his famous paper on the quantum paraelectric state of SrTiO$_3$ together with H. Burkhard [2]. I wanted to extend the research of structural phase transitions to include electronic effects and guessed that tungsten would be a good B atom in the perovskite structure for this purpose. A prototypic material that I wanted to research was WO$_3$ and its derivatives like Na$_x$WO$_3$. Alex agreed with these ideas and not only encouraged me but also invited me to give a talk in the IBM lab some months later. This was not bad for an early PhD student who had previously done rather little to sharpen his argument and did not have any real data at that time. Thus, I was obliged to grow crystals, measure the electrochromic effects, solve a crystal structure and measure near-infrared absorption to confirm the existence of polaronic states in WO3 and its doped analogues [3,4]. It took three years to establish a small laboratory for the investigation of perovskite structures in my home university in Hannover in Germany, with results coming out after 1975 [5,6]. When I returned to see Alex, he introduced me to Otwin Schirmer who was an expert in ESR spectroscopy. I worked with him for some years when he moved to Freiburg to the Frauenhofer Institut. We firmly established that the charge carriers in WO3 are polarons [7,8]. Through Alex I met Harry Thomas in Basel with whom I worked on the quantum version of the Landau potentials [9], which proved useful to derive the low-temperature saturation effects and, ultimately, quantum fluctuations in SrTiO3 [10]. I then spent time in the IBM lab and observed how Alex created an outstanding, international laboratory, which combined a great number of excellent people from different backgrounds. His leadership was very firm and helped me to a good start in science (and several visits to the 'Zunfthäuser an der Limmat'). In 2020, Alex returned to the investigation of superconductivity in reduced WO3 [11].

2. Results

While superconductivity in the WO_{3-x} materials is well established, the most intriguing property, which makes WO_3 unique, is its closeness to the metal-insulator transition which allows for an unusual phenomenon, as shown in Figure 1. The material undergoes a large number of phase transitions whereby almost all of them are ferroelastic. The transitions generate characteristic twin-domain networks [12–14]. These twin domains are separated by domain boundaries, which are simple twin walls along well-defined orientations. In WO_3, the domain boundaries persist in the ε phase at low temperatures [15]. When the sample is then cooled to ca. 3K, the domain walls—but not the rest of the sample—become superconducting [16,17].

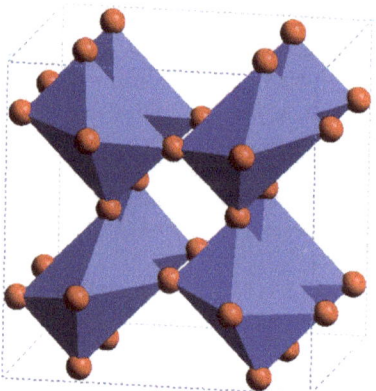

Figure 1. Structure of the tetragonal phase of the superconducting matrix.

The twin domain walls at higher temperatures are already highly conducting—much more so than the surrounding bulk [18]. Similar phenomena were subsequently found in many perovskite materials where the detailed conductivity mechanisms still remain obscure [19]. The effect in WO3 is much clearer by being based on the conduction by bipolarons (the recombination of two polarons into a bipolaron is shown in Figure 2). The domain boundary conductivity is more pronounced than in any other material discussed so far, as displayed in Figure 3.

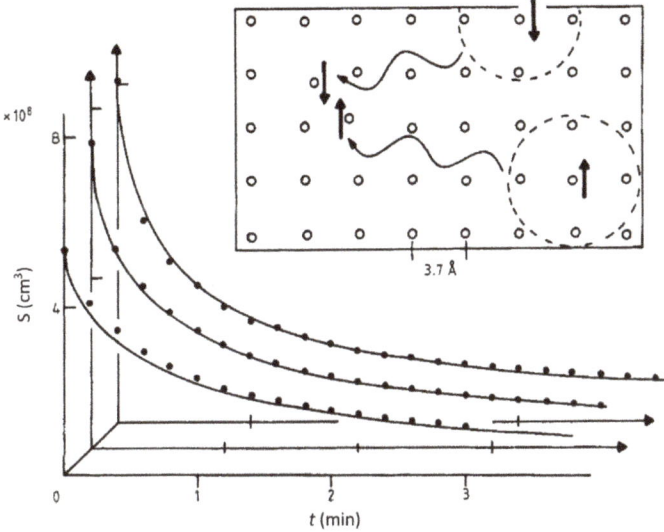

Figure 2. Sketch of the time dependence of the recombination of two polarons to form a bipolaron. Each polaron is marginally stable but does not represent the groundstate. The combination of two polarons form by spin pairing a bipolaron with lower energy. The bipolaron can be split by optical excitation. Once the bipolaron is split, it recombines over time (<60K).

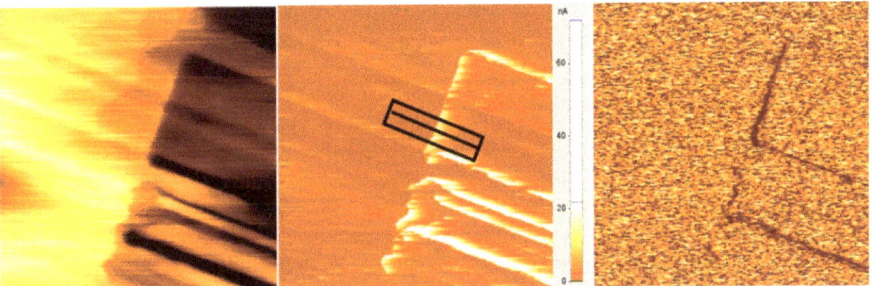

Figure 3. Topology (**left**), conductivity (**middle**) and piezoelectricity (**right**) of WO_3 at room temperature. This tetragonal phase is shown in Figure 1 and is piezoelectric. The piezoelectricity is short-circuited in the highly conducting domain walls, after [18].

First-principles investigations of bulk WO_3 [20] reproduce the essentials of the phase stability of the various crystallographic phases of WO_3. This work constituted a great step forward because it demonstrates that the initially observed deformation amplitudes during phase transitions are indeed related to specific properties of the interatomic bondings and the role of temperature in the form of larger lattice vibrations and increased entropy. It is revealed how a very simple binary compound possesses such a rich phase diagram and accordingly complex microstructures. Local polaronic deformations were calculated in a similar approach. The results are in excellent agreement with the free polaron electron density map of $W5+$ reported by Schirmer and Salje [7,8] from optical absorption and electron spin resonance (ESR) measurements. A single polaron has a 2D disk shape extended on a few neighbours. The possibility to stabilize this $W5+$ state allows us to characterize its electronic and structural properties through real space spin density, density of states analysis and the symmetry adapted mode analysis of the atomic distortion of the crystal. The first principle calculations by

Bousquet et al [21] demonstrated how hard it is to capture polarons as physical objects of material science from first-principles calculations. This makes WO$_3$ a paradigmatic system to study polarons. Bousquet et al. [21] succeeded in calculating the energy and charge distribution of a single self-trapped polaron in WO$_3$ from density functional theory. Their calculations show that the single polaron is at a slightly higher energy than the fully delocalized solution, in agreement with the experiments where a single polaron is an excited state of WO$_3$, as shown in Figure 4. The necessary stabilization energy stems from the spin–spin coupling and the common deformation cloud of the bipolaron. Bipolaron calculations in the first principle approach have still not been successful and are needed for a better understanding of the detailed electronic structure of a bipolaron.

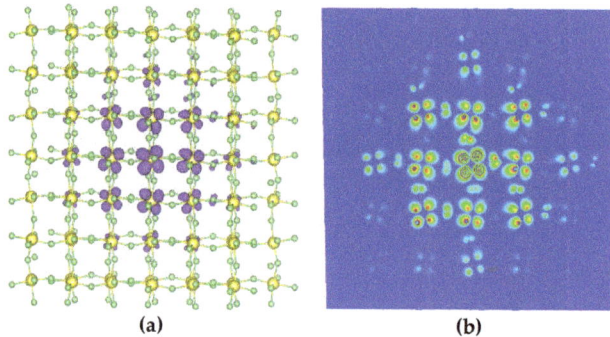

Figure 4. (**a**): 3D visualisation of the calculated density (B1WC functional) of the polaron in the xy plane of a WO3 supercell (W and O atoms are in yellow and green respectively, the charge density is in purple); (**b**): 2D cut plane of the polaron density in the xy and xz planes passing through the central W atom where the charge localizes [21] Note the close similarity with the sketch in Figure 2.

The symmetry-adapted mode decomposition of the polaron distortions shows that, among numerous modes, a polar zone centre mode has the largest contribution and can be at the origin of the observed weak ferroelectricity of WO$_3$.

To analyse the electronic structure of the calculated polaron in WO3, Figure 4 shows the total density of states (DOS) of a supercell with a self-trapped polaron [21]. The localized state appears in the gap of WO3 in the spin up channel. This state in the gap corresponds to the dxy state of W that slightly splits from the conduction bands (by about 0.4 eV), and it spreads over 0.1 eV in the gap. Hence, it is not fully localized, i.e., in agreement with what we observe from the real space density where the electron density of the polaron is spread over a few W atoms. This means that calculations simulate indeed a self-trapped single polaron in the P21/c ground state phase of WO$_3$.

Interestingly, Reich and Tsabba [22] reported superconductivity, possibly related to the sample surface, at much higher temperatures (ca. 80K). Unfortunately, these observations could not be reproduced and remained uncertain and largely ignored. Recently, Muller and collaborators did indeed obtain some similar results. They reported that in samples with the composition WO$_{2.9}$, the signatures of superconductivity with the same transition temperature T$_c$ = 80 K were registered by means of magnetization measurements. By lithium intercalation, the T$_c$ was further increased to 94 K. The observed small superconducting fraction and the absence of clear transition in resistivity measurements indicate that the superconductivity is localized in small regions. No current percolation was found. In contrast, the earlier observation of W^{5+}-W^{5+} electron bipolarons in reduced tungsten oxide samples was confirmed [23].

They proposed that bipolarons form and cluster within crystallographic shear planes which exist in the Magneli phase of WO$_{2.9}$ (W$_{20}$O$_{58}$) and represent charge-carrier rich quasi-1D stripes or puddles [24]. The Magneli structural complexes are common not only in W$_{20}$O$_{58}$ but are locally present in pentagonal tungstate structures [24–26] and many other structures. It can hence be assumed that a much wider class of materials with W-O corner-sharing clusters exist, which may show similar puddle effects. This discovery clearly warrants substantially enhanced research to identify WO_3-derivatives where the coupling between the Magneli complexes in crystallographic shear planes (CS) percolate.

While there is no simple way to assess any superconductivity transition in these compounds, an early attempt where the necessary condition was formulated to obtain metallic conductivity in the normal state [23]. Polaronic systems often show metal–insulator (M–I) transitions as a function of the carrier concentration [27,28]. The M–I transition occurs near the 'overcrowding' point where the packing of polarons with diameters of around 0.5 nm reaches saturation. Geometrically, additional carriers cannot be condensed into polaronic states because the necessary lattice deformation is exhausted and carriers need to transfer to the conduction band. This 'overcrowding' point is possibly the critical point for the formation of carrier into puddles and filaments, which, in turn, promote local superconductivity.

3. Outlook

Superconductivity related to domain boundaries, CS planes, or other structural singularities are not just a simple addition to existing bulk properties, but represent a cornerstone of the emerging field of domain boundary engineering [29] where the domain boundary is the device and the bulk is simply the matrix to hold the domain boundary in space [30]. Novel device designs become possible. An example is a Josephson junction detector where the junction elements are the intersections of superconducting domain boundaries. As such domain boundaries can be designed to appear in high densities [31,32], such devices would have unsurpassed sensitivities and extremely small sizes. Similar effects appear in systems where mesoscopic inhomogeneities on the metal–superconductor transition occur in two-dimensional electron systems, typically at the interface between two perovskite materials. A model with mesoscopic inhomogeneities was considered [33] as a random-resistor network in effective medium theory. Particularly space correlations between the mesoscopic domains were found to dominate over random fluctuations. A typical example is then to utilise the two-dimensional electron gas at the $LaTiO_3/SrTiO_3$ or $LaAlO_3/SrTiO_3$ oxide interfaces, which becomes superconducting when the carrier density is tuned by gating. The measured resistance and superfluid density reveal inhomogeneous superconductivity related to the percolation of filamentary structures of superconducting "puddles" with randomly distributed critical temperatures, embedded in a nonsuperconducting matrix. This scenario is similar to the Magneli phases while the twin walls in WO_3 are, by definition, percolating over the distance of the length of the twin boundaries (which are often spanning from surface to surface). Previously, interfacial superconductivity was modelled using intrapuddle conductivity by a multiband system within a weak coupling BCS scheme [34]. The microscopic parameters, extracted by fitting the transport data with a percolative model, yield a consistent description of the dependence of the average intrapuddle critical temperature and superfluid density on the carrier density [35]. Clearly many of these interfacial features are similar in WO_3 twin boundaries and in $LaAlO_3/SrTiO_3$ oxide interfaces so that one can expect significant progress in either field from the comparison with the other [36,37].

Acknowledgments: E.K.H.S. is indebted to EPSRC for support (grant EP/P024904/1).

Conflicts of Interest: The author declares no conflict of interest.

References

1. Axe, J.D.; Shirane, G.; Mueller, K.A. Zone-boundary phonon instability in cubic $LaAlO_3$. *Bull. Am. Phys. Soc.* **1969**, *183*, 820. [CrossRef]
2. Mueller, K.A.; Burkhard, H. $SrTiO_3$ Intrinsic quantum paraelectric below 4K. *Phys. Rev. B* **1979**, *19*, 3593–3602. [CrossRef]
3. Salje, E. Non-stoichiometric tungsten compound, synthesis and lattice constants of WO_3-$NaWO_3$ mixed crystal compounds. *Z. Fur Allg. Angew. Chem.* **1973**, *396*, 267–270. [CrossRef]

4. Salje, E. New type of electrooptic effect in semiconducting WO_3. *J. Appl. Crystallogr.* **1974**, *7*, 615–617. [CrossRef]
5. Salje, E. Viswanathan K, Physical properties and phase transitions in WO_3. *Acta Crystallogr. Sect. A* **1975**, *31*, 356–359. [CrossRef]
6. Salje, E. Lattice dynamics of WO_3. *Acta Crystallogr. Sect. A* **1975**, *31*, 360–363. [CrossRef]
7. Schirmer, O.F.; Salje, E. Conducting bi-polarons in low-temperature crystalline WO_{3-x}. *J. Phys. C* **1980**, *13*, 1067–1072. [CrossRef]
8. Schirmer, O.F.; Salje, E. W^{5+} polaron in low temperature crystalline WO_3 Electron spin response and optical absorption. *Solid State Commun.* **1980**, *33*, 333–336. [CrossRef]
9. Salje, E.K.H.; Wruck, B.; Thomas, H. Order parameter saturation and low temperature extension of Landau theory. *Z. Phys. Condens. Matter* **1991**, *82*, 399–404. [CrossRef]
10. Kustov, S.; Luibimova, I.; Salje, E.K.H. Domain dynamics in quantum-paraelectric $SrTiO_3$. *Phys. Rev. Lett.* **2020**, *124*, 016801. [CrossRef]
11. Shengelaya, A.; Conder, K.; Mueller, K.A. Signatures of filamentary superconductivity up to 94K in tungsten oxide $WO_{2.9}$. *J. Supercond. Nov. Magn.* **2020**, *33*, 301–306. [CrossRef]
12. Locherer, K.R.; Swainson, I.P.; Salje, E.K.H. Phase transitions in WO_3 at high temperatures—A new look. *J. Phys. Condens. Matter* **2002**, *11*, 6737–6756. [CrossRef]
13. Howard, C.J.; Luca, V.; Knight, K.S. High-temperature phase transitions in tungsten trioxide—The last word? *J. Phys. Condens. Matter* **2002**, *14*, 377–387. [CrossRef]
14. Viehland, D.D.; Salje, E.K.H. Domain boundary dominated systems, adaptive structures and functional twin boundaries. *Adv. Phys.* **2014**, *63*, 267–326. [CrossRef]
15. Salje, E.K.H.; Rehmann, S.; Pobell, F.; Morris, D.; Knight, K.S.; Herrmannsdorfer, T.; Dove, M.T. Crystal structure and paramagnetic behaviour of epsilon-WO_{3-x}. *J. Phys. Condens. Matter* **1997**, *9*, 6563–6577. [CrossRef]
16. Aird, A.; Salje, E.K.H. Sheet superconductivity in twin walls: Experimental evidence of WO_{3-x}. *J. Phys. Condens. Matter* **1998**, *10*, L377–L380. [CrossRef]
17. Aird, A.; Domeneghetti, M.C.; Mazzi, F.; Salje, E.K.H. Sheet superconductivity in WO_{3-x}: Crystal structure of the tetragonal Matrix. *J. Phys. Condens. Matter* **1998**, *33*, L569–L574. [CrossRef]
18. Kim, Y.; Alexe, M.; Salje, E.K.H. Nanoscale properties of twin walls and surface layers in piezoelectric WO_{3-x}. *Appl. Phys. Lett.* **2010**, *96*, 032904. [CrossRef]
19. Seidel, J.; Maksymovych, P.; Batra, Y.; Katan, A.; Yang, S.-Y.; He, Q.; Baddorf, A.P.; Kalinin, S.V.; Yang, C.-H.; Yang, J.-C.; et al. Domain wall conductivity in La doped $BiFeO_3$. *Phys. Rev. Lett.* **2010**, *105*, 197603. [CrossRef]
20. Hamdi, H.; Salje, E.K.H.; Ghosez, P.; Bousquet, P. First-principles investigation of bulk WO_3. *Phys. Rev. B* **2016**, *94*, 245124. [CrossRef]
21. Bousquet, E.; Hamdi, H.; Aguado-Puente, P.; Salje, E.K.H.; Artacho, E.; Ghosez, P. First-principles characterization of single-electron polaron in WO_3. *Phys. Rev. Res.* **2020**, *2*, 012052. [CrossRef]
22. Reich, S.; Tsabba, Y. Possible nucleation of a 2D superconducting phase on WO3 single crystals surface doped with Na^+. *Eur. Phys. J. B* **1999**, *1*, 1–4. [CrossRef]
23. Salje, E.K.H. Polarons and bi-polarons in WO_{3-x}. *Eur. J. Solid State Inorg. Chem.* **1994**, *31*, 805–821.
24. Salje, E.; Gehlig, R.; Viswanathan, K. Structural phase transitions in mixed crystals $W_xMo_{1-x}O_3$. *J. Solid State Chem.* **1978**, *25*, 239–250. [CrossRef]
25. Viswanathan, K.; Salje, E. Crystal-structure and charge carrier behavior of $(W_{12.64}Mo_{1.36})O_{41}$ and its significance to other related compounds. *Acta Cryst.* **1981**, *37*, 4449–4456. [CrossRef]
26. Viswanathan, K.; Brandt, K.; Salje, E. Crystal-structure and charge carrier concentration of $W_{18}O_{49}$. *J. Solid State Chem.* **1981**, *36*, 45–51. [CrossRef]
27. Ruscher, C.; Salje, E.; Hussain, A. The effect of Nb-W distribution of polaronic transport in ternary Nb-W oxides- electrical and optical properties. *J. Phys. C-Solid State* **1988**, *21*, 4465–4480. [CrossRef]
28. Ruscher, C.; Salje, E.; Hussain, A. The effect of polaron concentration on the polaron transport in $NbO_{2.5-x}$ Optical and electric properties. *J. Phys. C-Solid State* **1988**, *21*, 3737–3749. [CrossRef]
29. Salje, E.K.H. Domain boundaries as active memory devices: Trajectories towards domain boundary engineering. *ChemPhysChem* **2010**, *11*, 940–950. [CrossRef]
30. Catalan, G.; Seidel, J.; Ramesh, R.; Scott, J.F. Domain wall nanoelectronics. *Rev. Mod. Phys.* **2012**, *84*, 119–156. [CrossRef]

31. Salje, E.K.H.; Ding, X.; Aktas, O. Domain glass. *Phys. Status Solidi B* **2014**, *251*, 2061–2066. [CrossRef]
32. Ding, X.; Zhao, Z.; Lookman, T.; Saxena, A.; Salje, E.K.H. High junction and twin boundary densities in driven dynamical systems. *Adv. Mater.* **2012**, *24*, 5385–5389. [CrossRef] [PubMed]
33. Caprara, S.; Grilli, M.; Benfatto, L.; Castellani, C. Effective medium theory for superconducting layers: A systematic analysis including space correlation effects. *Phys. Rev. B* **2011**, *84*, 014514. [CrossRef]
34. Caprara, S.; Biscaras, J.; Bergeal, N.; Bucheki, D.; Hurand, S.; Feuillet-Palma, C.; Rastogi, A.; Budhani, R.C.; Lesueur, J.; Grilli, M. Multiband superconductivity and nanoscale inhomogeneity at oxide interfaces. *Phys. Rev. B* **2013**, *88*, 020504. [CrossRef]
35. Scopigno, N.; Bucheli, D.; Caprara, S.; Biscaras, J.; Bergeal, N.; Lesueur, J.; Grilli, M. Phase separation from electron confinement at oxide interfaces. *Phys. Rev. Lett.* **2016**, *116*, 026804. [CrossRef]
36. Dezi, G.; Scopigno, N.; Caprara, S.; Grilli, M. Negative electronic compressibility and nanoscale inhomogeneity in ionic-liquid gated two-dimensional superconductors. *Phys. Rev. B* **2018**, *98*, 214507. [CrossRef]
37. Bovenzi, N.; Caprara, S.; Grilli, M.; Raimondi, R.; Scopigno, N.; Seibold, G. Density inhomogeneities and Rashba spin-orbit coupling interplay in oxide interfaces. *J. Phys. Chem. Solids* **2019**, *128*, 118–129. [CrossRef]

© 2020 by the author. Licensee MDPI, Basel, Switzerland. This article is an open access article distributed under the terms and conditions of the Creative Commons Attribution (CC BY) license (http://creativecommons.org/licenses/by/4.0/).

Article

Ferroelectricity, Superconductivity, and SrTiO$_3$—Passions of K.A. Müller

Gernot Scheerer, Margherita Boselli, Dorota Pulmannova, Carl Willem Rischau, Adrien Waelchli, Stefano Gariglio, Enrico Giannini, Dirk van der Marel and Jean-Marc Triscone *

Department of Quantum Matter Physics, University of Geneva, 24 Quai Ernest-Ansermet, 1211 Geneva 4, Switzerland; gernot.scheerer@unige.ch (G.S.); margherita.boselli@unige.ch (M.B.); Dorota.Pulmannova@unige.ch (D.P.); Willem.Rischau@unige.ch (C.W.R.); Adrien.Waelchli@unige.ch (A.W.); stefano.gariglio@unige.ch (S.G.); enrico.giannini@unige.ch (E.G.); Dirk.VanDerMarel@unige.ch (D.v.d.M.)
* Correspondence: Jean-Marc.Triscone@unige.ch

Received: 9 September 2020; Accepted: 13 October 2020; Published: 15 October 2020

Abstract: SrTiO$_3$ is an insulating material which, using chemical doping, pressure, strain or isotope substitution, can be turned into a ferroelectric material or into a superconductor. The material itself, and the two aforementioned phenomena, have been subjects of intensive research of Karl Alex Müller and have been a source of inspiration, among other things, for his Nobel prize-winning research on high temperature superconductivity. An intriguing outstanding question is whether the occurrence of ferroelectricity and superconductivity in the same material is just a coincidence, or whether a deeper connection exists. In addition there is the empirical question of how these two phenomena interact with each other. Here we show that it is possible to induce superconductivity in a two-dimensional layer at the interface of SrTiO$_3$ and LaAlO$_3$ when we make the SrTiO$_3$ ferroelectric by means of ^{18}O substitution. Our experiments indicate that the ferroelectricity is perfectly compatible with having a superconducting two-dimensional electron system at the interface. This provides a promising avenue for manipulating superconductivity in a non centrosymmetric environment.

Keywords: ferroelectricity; superconductivity; SrTiO$_3$; ^{18}O; isotope substitution; SrTiO$_3$/LaAlO$_3$; interface; heterostructure

1. Introduction

Karl Alex Müller has numerous interests and passions. Most likely quite high in the list are ferroelectricity, superconductivity and SrTiO$_3$—a material that, we believe, he called the drosophila of solid state physics. Known worldwide for their discovery of superconductivity in the cuprates, J.G. Bednorz and K.A. Müller explained in their Nobel lecture that their search for high T$_c$ superconductivity in complex oxides had been partly motivated by SrTiO$_3$, which, once doped, has a maximum T$_c$ of 0.5 K, actually very high when compared to its Fermi energy [1].

Close to ferroelectricity and to superconductivity, SrTiO$_3$ is indeed an amazing material. By itself it is an insulating cubic perovskite at room temperature. Below 105 K, an antiferrodistortive transition makes the system weakly tetragonal. Electronically, SrTiO$_3$ is a quantum paraelectric—a compound often seen as "failed ferroelectric" with its inverse static dielectric constant ϵ versus T revealing a Curie–Weiss behavior. Unlike for ferroelectric materials, however, ϵ never diverges but saturates at low temperatures as shown in 1979 by Müller and Burkard [2]. When doped, SrTiO$_3$ can be turned into a ferroelectric or into a superconductor. To achieve the former, Ca can be partially substituted for Sr [3] or ^{16}O can be replaced by ^{18}O [4]—in thin film form, strain also allows the ferroelectric state to be reached [5]. Superconductivity can be obtained by partially substituting Sr with La, Ti with Nb or by reducing the oxygen content—in all cases, the system is doped with electrons and the maximum T_c is around 500 mK [6,7]. SrTiO$_3$ has, over time, revealed other amazing properties including the emission of blue-light once irradiated with Ar ions [8] or the electrolysis of water [9]. With the discovery in 2004 of conductivity [10] and in 2007 of superconductivity [11] at the interface between LaAlO$_3$ and SrTiO$_3$, this "magic" perovskite was again at the center of worldwide attention. More recently, it is the prediction and discovery of an increase of T_c in electron-doped and Ca or ^{18}O substituted SrTiO$_3$ that triggered a lot of interest, discoveries marrying the passions of K.A. Müller—ferroelectricity, superconductivity and SrTiO$_3$.

In this paper, we aim to discuss how the proximity of a ferroelectric state to the superconducting phase may explain the Cooper pair coupling mechanism. We first review the properties of SrTiO$_3$, presenting a short summary of its phase diagram with the different ground states obtained by the various dopings and substitutions. We then recall the different models proposed since as far back as 1964 that may explain superconductivity in SrTiO$_3$ and we discuss in particular ideas allowing the recent observation of T_c enhancement when SrTiO$_3$ is pushed toward ferroelectricity to be understood. Finally, we briefly introduce the LaAlO$_3$/SrTiO$_3$ system and show some experimental results obtained on these superconducting interfaces for which ^{16}O was partially substituted by ^{18}O in the SrTiO$_3$ single crystal substrate used for the growth of the LaAlO$_3$ layer. We end the paper with a brief conclusion.

2. SrTiO$_3$: Properties, Phase Diagram and Tuning Parameters

The centrosymmetric cubic perovskite structure (tolerance factor $t = 1$) that SrTiO$_3$ adopts at room temperature reflects the perfect balance between the ionic radii of its cations: deviations from $t = 1$ would lead to various types of distortions, the most common ones being the oxygen octahedral rotations occurring for $t < 1$ [12]. As mentioned above, at 105 K SrTiO$_3$ goes through an antiferrodistortive (AFD) transition resulting in a tetragonal structure with oxygen octahedra rotated out of phase about the c-axis ($a^0a^0c^-$ in Glazer notation) [13]. Lowering the temperature further produces a softening of the ferroelectric phonon mode with a strong Curie–Weiss type increase of the static dielectric response, suggesting a transition into a ferroelectric state at 20 K [14]. However, Müller and Burkard discovered that the dielectric constant saturates, reaching a value of 2×10^4 at 4 K [2]: they interpreted this saturation as the signature of an intrinsic quantum paraelectric state, i.e., an avoided ferroelectric state due to the quantum fluctuations of the atoms about their centrosymmetric positions. Monte Carlo calculations have confirmed this scenario and revealed the role of quantum fluctuations on the reduction of the AFD transition temperature [15]. Given such proximity to a ferroelectric state, several groups have explored different approaches to obtain a polar state, applying mechanical [16,17] and epitaxial [5] strain or performing chemical—replacing Sr with Ca [3]—or isotopic substitutions—^{16}O with ^{18}O [4]. These different avenues have induced a ferroelectric ground state with Curie temperatures exceeding, in some cases, room temperature [5]. Figure 1 shows schematically how the ferroelectric state develops beyond the quantum critical point (QCP) for the case of Ca-doping and ^{18}O-substitution.

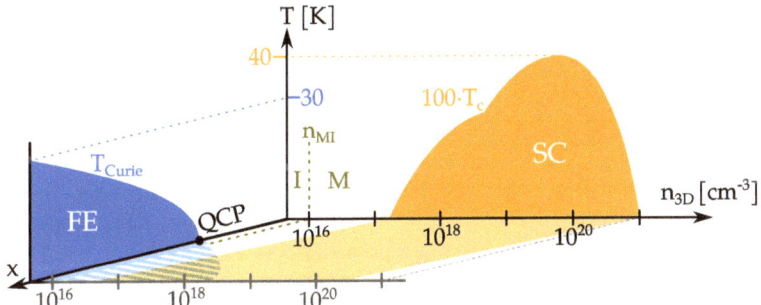

Figure 1. Schematic phase diagram of SrTiO$_3$ showing the ferroelectric (FE) and superconducting (SC) phases. Upon chemical substitution of Ca for Sr, i.e., Sr$_{1-x}$Ca$_x$TiO$_3$ with $0.002 < x < 0.02$ [3], or by oxygen isotope substitution, i.e., SrTi(^{18}O$_y$^{16}O$_{1-y}$)$_3$ with $y > 0.33$ [4], the material develops a FE ground state beyond a quantum critical point (QCP). This FE phase occurs well below the structural transition from cubic to tetragonal [18]. Charge doping turns the material from an insulator (I) into a metal (M) at a critical carrier density (n$_{MI}$) of 10^{16} cm^{-3}, while SC develops in a doping range n$_{3D}$ between 5×10^{17} and 10^{21} cm^{-3}.

The large dielectric susceptibility is thought to be responsible for the fact that the electronic transition from the insulating state to a metallic state occurs at an extremely low carrier density of 10^{16} cm^{-3} [19]. Such doping can be induced by chemical substitution of La for Sr [20], Nb for Ti or by oxygen reduction [21]. At these low dopings the mean free path of the conduction electrons is about 100 times greater than the Fermi wavelength [22]. One of the consequences is that quantum oscillations in the magneto-resistance are observed [23,24], a feature that allows the topology of the Fermi surface to be determined as a function of carrier density. At low temperatures (below 0.4 K), the metallic electron doped system undergoes a phase transition into a superconducting condensate for a carrier density in the range 10^{17}–10^{21} cm^{-3} [24–26]. With 10^{17} cm^{-3}, doped SrTiO$_3$ is the lowest density superconductor and displays a unique broad range of charge concentration over which the superconducting state is observed. The origin of the superconducting state and the dependence of the superconducting temperature on carrier density have been subjects of several studies [26,27]. An appealing proposition is that the two different order parameters may be somehow coupled: according to certain models that apply to perovskite-type structures, the ferroelectric instability is the condition necessary to pair electrons [28]. Such an idea has been explored recently, leading to a clear prediction of the dependence of the superconducting critical temperature upon the proximity to the ferroelectric state [29].

3. Superconductivity in Doped SrTiO$_3$ from 1964 until 2020

Using the linear combination of atomic orbitals method, Kahn and Leyendecker predicted in 1964 that the electronic energy bands in strontium titanate exhibit six conduction band ellipsoids lying along [100] directions of momentum space with minima probably at the edges of the Brillouin zone [30]. In the same year, Marvin Cohen predicted that the attractive electron–electron interaction arising from the exchange of intravalley and intervalley phonons can cause these materials to exhibit superconducting properties [31]. In less than a year, Schooley et al. [6,25] reported superconductivity in electron-doped SrTiO$_3$ with carrier concentrations in the range from 10^{18} to 10^{21} cm^{-3}, and T_c ranging from 50 mK at the lowest doping to about 0.5 K for $n_c = 10^{20}$ cm^{-3}. While these results confirmed Cohen's prediction of superconductivity in electron-doped SrTiO$_3$, it has been demonstrated by later band structure calculations [32] that there is only a single valley, which is located at the Brillouin zone center for each of the three conduction bands.

The three bands at the zone center are non-degenerate due to spin-orbit splitting and (below 105 K) a weak tetragonal crystal field [22], causing the sudden onset of quantum oscillations [23] at the critical dopings where the second and third band become occupied [24]. This also agrees well with the doping dependence of the two superconducting gaps observed by Binnig et al. [33]. While the Fermi surface properties agree well with the ab initio band structure predictions, the experimental values of the effective mass are a factor of two higher than the ab initio predictions [34,35]. From the analysis of the mid-infrared absorption in doped SrTiO$_3$ it has become clear that the factor of two for the mass enhancement observed in the experiments is a consequence of the coupling of the conduction electrons to the longitudinal optical phonons, and that the mid-infrared peaks originate from large polaron formation [36].

In the course of the more than five decades of research on SrTiO$_3$ a variety of models have been proposed for the pairing mechanism: intervalley scattering [27,31], bipolarons [37], two-phonon exchange [38], longitudinal optical phonons [39], full dielectric function for longitudinal phonons and screened Coulomb [40,41], acoustic phonons [42]. A possible role of ferroelectricity was proposed by Bussmann-Holder [28], an idea that has gained momentum in recent years [43]. A detailed theoretical prediction of a giant isotope effect on the superconducting T_c [29] with an opposite sign from the BCS prediction has spurred on a number of isotope-substitution experiments [44,45] and Ca-substitution experiments, which are expected to have a similar effect on T_c [46]. These experiments have confirmed the theoretical predictions. The theory was based on the coupling of electrons to the soft transverse optical phonon (the "TO1" mode). A problem has meanwhile been pointed out, that the coupling to this phonon is far too small to account for T_c on the order of several hundred mK [47]. A possible remedy is to couple the electrons to pairs of transverse optical phonons of opposite momentum [38,48]. A recent analysis of the optical oscillator strength of the TO1 mode has brought to light that this type of bi-phonon exchange is indeed unusually strong in SrTiO$_3$ [49], strong enough in fact to account for superconductivity in this material. In this scenario quantum ferroelectric fluctuations induce the pairing interaction that leads to superconductivity [29,43]; however, the main channel of interaction is mediated by pairs of phonons rather than single phonons as was originally proposed. In this context it is not an accident that superconductivity occurs in proximity to a ferroelectric quantum critical point.

4. Superconductivity in Two Dimensions

Recent research on SrTiO$_3$ has focused on the two-dimensional electron systems that emerge at the crystal surface or in thin films. Mobile electrons can be localized at the surface of undoped SrTiO$_3$ crystals by cleaving in vacuum [50,51], in δ-doped SrTiO$_3$ thin films [52] or in SrTiO$_3$-based heterostructures. The well-known conducting interface between an insulating SrTiO$_3$ substrate and a thin film of LaAlO$_3$ belongs to the last class [10]. This heterostructure hosts a two-dimensional conducting system confined in SrTiO$_3$ within a few nanometers from its interface with LaAlO$_3$. The electrons are transferred to SrTiO$_3$ to compensate for the polar discontinuity occurring between the two materials along the [001] direction [53–55]. Similar to the bulk case, this system undergoes a superconducting transition when cooled down below 300 mK. As-grown samples have a critical temperature of ~200 mK, and a 1D critical current density of 100 μA/cm [11]. The superconductivity in this system has a two-dimensional character. Indeed, the analysis of the critical magnetic field parallel, $H_\parallel^*$, and perpendicular, $H_\perp^*$, to the interface yields a Ginzburg–Landau coherence length, ξ, of 60 nm at T = 0 K, and a thickness of the superconducting slab of 11 nm. As expected for a superconducting thin film (thickness $\ll \xi$) [56], $H_\parallel^*$ is much higher than $H_\perp^*$ as a 2D superconducting layer cannot accommodate a vortex parallel to the plane. Interestingly, the value of $H_\parallel^*$ exceeds the paramagnetic limit set by BCS theory, $\mu_0 \cdot H_\parallel^* = 1.84 \cdot T_c$ (with $\mu_0 \cdot H_\parallel^*$ in T and T_c in K), by a factor of 4–5 [57], and this effect might be linked to the presence of strong spin-orbit coupling in the system [58,59].

In 2008, Caviglia et al. showed that the superconductivity is tunable by the electric field effect [60]. The phase diagram of LaAlO$_3$/SrTiO$_3$ resembles that of bulk SrTiO$_3$, but it extends over a much smaller carrier density range, between 1×10^{19} cm^{-3} and 4×10^{19} cm^{-3} [61]. In the underdoped region of the phase diagram, a quantum critical point separates the superconducting regime from an insulating phase, related to the weak-localization effect [60].

5. ^{18}O Isotope Effect

Following the ferroelectric quantum critical scenario proposed by Rowley et al. [43], Edge et al. considered a specific scenario in which the ferroelectric soft mode is tuned by isotopic ^{18}O-substitution [29]. By tuning the ^{18}O substitution level beyond the QCP, they predicted both an increase of the maximum T_c and a shift of the maximum of the dome to lower carrier densities. Experimentally, the increase of T_c was first observed by Stucky et al. on 35%-isotope substituted samples that were electron-doped by oxygen removal [44]. In the BCS weak-coupling limit, T_c is inversely proportional to the isotope mass M: $T_c \propto M^{-\alpha}$ with an isotope coefficient $\alpha = +0.5$. The experimentally determined increase of T_c of 50% [44] leads, however, to a negative and much larger value of $\alpha \approx -10$, matching the theoretical prediction made by Edge and coworkers, both in sign and order of magnitude [29]. In a later work an enhancement of T_c upon isotope substitution was further confirmed in isotope-substituted samples that were electron-doped by substituting Sr with La [45].

In this context, we measured the electronic properties of LaAlO$_3$/SrTi(^{18}O$^{16}_y$O$_{1-y}$)$_3$ heterostructures: a system where a superconducting two-dimensional electron system is confined at the interface between an insulating thin film and a ferroelectric substrate.

We optimized the isotopic substitution on commercial TiO$_2$ terminated SrTiO$_3$ substrates provided by CrysTec GmbH. Several crystals, $5 \times 2.5 \times 0.5$ mm^3 and $5 \times 2.5 \times 0.25$ mm^3 in size, are put in a standard quartz tube, which is then sealed to fix the internal pressure of ^{18}O$_2$ at 0.4–0.7 bar. The sealed tubes are placed in a tube furnace and heated up at temperatures between 700 and 1100 °C for 20–40 days. Before the LaAlO$_3$ thin film growth, we evaluated the effect of the substitution procedure on the substrate topography using an atomic force microscope (AFM).

As-received TiO$_2$ terminated substrates have an atomically flat surface with a clear step-and-terrace topography (see Figure 2a). AFM imaging revealed that after the thermal treatment the crystal surface is completely reconstructed. Instead of the usual step-and-terrace structure, we found a "block-terrace" structure (see Figure 2b). In order to restore a controlled TiO$_2$ termination, the crystals have been re-polished and then treated in an HF (hydrofluoric acid) bath for 30 s, followed by a rinsing with demineralized water. After this procedure, the substrates recovered the initial step-terrace structure (see Figure 2c), with atomically flat terraces and unit cell-high steps, and were ready for the LaAlO$_3$ deposition. The thin films of LaAlO$_3$ were grown by pulsed laser deposition, following the recipe used for standard LaAlO$_3$/SrTiO$_3$ heterostructures [62]. Their thickness, typically 6–7 unit cells, was monitored during the growth using reflection high energy electron diffraction (see Figure 2d). After the growth, a 20 nm gold layer was sputtered on the back side of the substrate to be used as an electrode for dielectric measurements.

We prepared and analyzed three LaAlO$_3$/SrTi(^{18}O$^{16}_y$O$_{1-y}$)$_3$ heterostructures with nominal ^{18}O contents in the SrTiO$_3$ substrate of 35%, 45%, and 67%, respectively. Compared to pure SrTiO$_3$, the low-temperature dielectric constant, ϵ_r, is strongly enhanced by the presence of ^{18}O (see Figure 2e). At $y = 35\%$ the substrate is on the verge of the ferroelectric transition and ϵ_r saturates at roughly 4.7×10^4 (compared to 2×10^4 for SrTiO$_3$). For $y= 45\%$ and 67%, the dielectric constant has a double peak structure, with the maxima indicating the position of the ferroelectric transition. The first peak occurs approximately at the Curie temperature of 12 K (17 K), which agrees well with the nominal ^{18}O content 45% (67%), as indicated by the black arrow in Figure 2e. The second peak, which occurs at lower temperature (gray arrows), may be due to inhomogeneities in the ^{18}O content. It is worth noting that the substrates are heated up to 800 °C in an ^{16}O atmosphere during the growth of the LaAlO$_3$ film [62] and some ^{16}O may re-substitute part of the ^{18}O present at the SrTiO$_3$ surface. The second peak is visible at 6.9 K (8.1 K) and corresponds to an ^{18}O content of roughly 35% (40%) for a nominal content of 45% (67%).

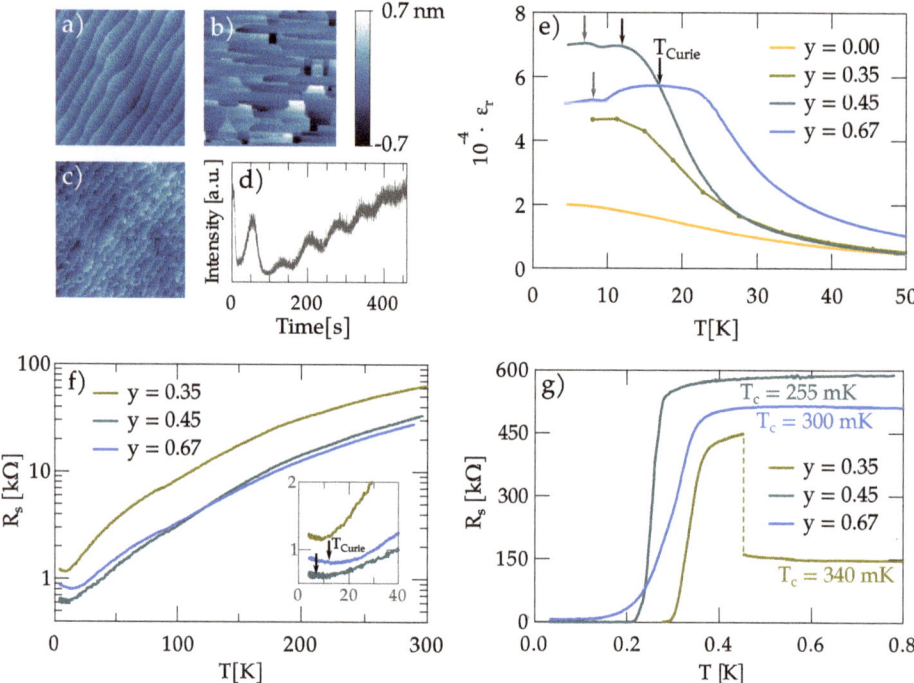

Figure 2. Growth and physical properties of LaAlO$_3$/SrTi(^{18}O$_y^{16}$O$_{1-y}$)$_3$ interfaces. AFM images of the SrTiO$_3$ substrate (**a**) as-received, (**b**) after the ^{18}O$_y$ substitution process, and (**c**) after re-polishing and HF treatment (the size of all AFM images is 4 µm × 4 µm). (**d**) RHEED signal during the growth of the LaAlO$_3$ layer. One oscillation corresponds to the deposition of one unit cell of LaAlO$_3$. (**e**) Dielectric constant versus temperature of a SrTiO$_3$ substrate (y = 0) and of LaAlO$_3$/SrTi(^{18}O$_y^{16}$O$_{1-y}$)$_3$ samples for different values of substitution. The dielectric properties have been measured in a homemade Helium cryostat using the Agilent E4980A Precision LCR Meter. The electric field was applied between the back electrode and the 2DES used as a top-electrode. (**f,g**) Sheet resistance versus temperature of the 2D electron system for ^{18}O substituted samples. The resistance jump visible in the curve for y = 35% at ∼0.45 K is due to an electric spike, which occurred in the measurement system during our study.

Figure 2f,g shows the sheet resistance (R$_s$) as a function of temperature. Between 300 and 1.5 K, the behavior of these samples is similar to that of standard LaAlO$_3$/SrTiO$_3$ heterostructures [63]. The resistance has a slight dip at ∼95 K, which is presumably due to the antiferrodistortive transition at 105 K [64–66]. For all investigated samples, the resistance shows a small upturn below ∼15 K, the origin of which is still under investigation. It should be noted that an anomaly/upturn has been observed in the low-temperature resistivity of bulk Ca-substituted SrTiO$_{3-\delta}$ samples and was associated to a ferroelectric-like state still existing in metallic samples [46,67]. Similarly, a study performed on heterostructures of LaAlO$_3$ grown on top of Ca-substituted SrTiO$_3$ substrates showed the presence of a resistance upturn occurring just below the Curie temperature, possibly linked to the ferroelectricity of the substrate [68]. If the temperature is further decreased, the samples undergo a superconducting transition at T_c = 340, 255, and 300 mK for y = 35, 45, and 67%, respectively. T_c is defined as the temperature at which the resistance is 50% of its value in the normal state (here at 800 mK). The transition temperature observed for the three samples is similar to that reported in the 2D electron system confined in standard SrTiO$_3$ substrates [11,60]. We note that a comparison between the phase diagram shown in Figure 1 and our data for the LaAlO$_3$/ST(^{18}O$_y^{16}$O$_{1-y}$)$_3$ interfaces is difficult due to the uncertainty on the equivalent 3D carrier density of the 2DES, which has an exponential charge profile inside the substrate. This pilot study shows that the presence of a ferroelectric SrTiO$_3$ substrate is compatible with the formation of a conducting—and even superconducting—system at the interface with LaAlO$_3$, and opens the path to the exploration of its effect on the electronic properties.

6. Conclusions

SrTiO$_3$ plays host to a large variety of interesting physical phenomena. In particular, superconductivity can be obtained in the bulk and at a two-dimensional interface. Following the idea that superconductivity can be enhanced by ^{18}O substitution in SrTiO$_3$ we studied the properties of LaAlO$_3$/SrTi(^{18}O$_y^{16}$O$_{1-y}$)$_3$ heterostructures with different ^{18}O concentrations (35%, 45%, and 67%) in the SrTiO$_3$ substrate. The observation of superconductivity at the interface of LaAlO$_3$ and isotope substituted SrTiO$_3$ with T_c on the order of 300 mK demonstrates that it is experimentally possible to induce two-dimensional superconductivity in a ferroelectric-like environment. Further investigations with different levels of doping may reveal higher superconducting critical temperatures in this system that combines ferroelectricity, superconductivity and SrTiO$_3$—the passions of K.A. Müller.

Author Contributions: Data curation, G.S., M.B., C.W.R. and A.W.; Formal analysis, G.S. and M.B.; Investigation, G.S., M.B., D.P., C.W.R., A.W., S.G. and E.G.; Supervision, S.G., E.G., D.v.d.M. and J.-M.T.; Writing—original draft, G.S., M.B., C.W.R., A.W., S.G., D.v.d.M. and J.-M.T.; Writing–review & editing, G.S., M.B., D.P., C.W.R., A.W., S.G., E.G., D.v.d.M. and J.-M.T. All authors have read and agreed to the published version of the manuscript.

Funding: This work was supported by the Swiss National Science Foundation through Division II (projects 200020-179155 and 200020-179157). The research leading to these results has received funding from the European Research Council under the European Union's Seventh Framework Program (FP7/2007-2013)/ERC Grant Agreement 319286 Q-MAC.

Acknowledgments: We would like to thank Jennifer Fowlie for providing useful comments on the manuscript.

Conflicts of Interest: The authors declare no conflict of interest.

References

1. Bednorz, J.G.; Müller, K.A. Perovskite-type oxides—The new approach to high-T_c superconductivity. *Rev. Mod. Phys.* **1988**, *60*, 585–600. [CrossRef]
2. Müller, K.A.; Burkard, H. SrTiO$_3$: An intrinsic quantum paraelectric below 4 K. *Phys. Rev. B* **1979**, *19*, 3593–3602. [CrossRef]
3. Bednorz, J.G.; Müller, K.A. Sr$_{1-x}$Ca$_x$TiO$_3$: An XY Quantum Ferroelectric with Transition to Randomness. *Phys. Rev. Lett.* **1984**, *52*, 2289–2292. [CrossRef]
4. Itoh, M.; Wang, R.; Inaguma, Y.; Yamaguchi, T.; Shan, Y.J.; Nakamura, T. Ferroelectricity Induced by Oxygen Isotope Exchange in Strontium Titanate Perovskite. *Phys. Rev. Lett.* **1999**, *82*, 3540–3543. [CrossRef]
5. Haeni, J.H.; Irvin, P.; Chang, W.; Uecker, R.; Reiche, P.; Li, Y.L.; Choudhury, S.; Tian, W.; Hawley, M.E.; Craigo, B.; et al. Room-temperature ferroelectricity in strained SrTiO$_3$. *Nature* **2004**, *430*, 758–761. [CrossRef]

6. Schooley, J.; Hosler, W.; Ambler, E.; Becker, J.; Cohen, M.; Koonce, C. Dependence of the Superconducting Transition Temperature on Carrier Concentration in Semiconducting SrTiO$_3$. *Phys. Rev. Lett.* **1965**, *14*, 305–307. [CrossRef]
7. Collignon, C.; Lin, X.; Rischau, C.W.; Fauqué, B.; Behnia, K. Metallicity and Superconductivity in Doped Strontium Titanate. *Ann. Rev. Condens. Matter Phys.* **2019**, *10*, 25–44. [CrossRef]
8. Kan, D.; Terashima, T.; Kanda, R.; Masuno, A.; Tanaka, K.; Chu, S.; Kan, H.; Ishizumi, A.; Kanemitsu, Y.; Shimakawa, Y.; et al. Blue-light emission at room temperature from Ar+-irradiated SrTiO$_3$. *Nat. Mater.* **2005**, *4*, 816–819. [CrossRef]
9. Wrighton, M.S.; Ellis, A.B.; Wolczanski, P.T.; Morse, D.L.; Abrahamson, H.B.; Ginley, D.S. Strontium titanate photoelectrodes. Efficient photoassisted electrolysis of water at zero applied potential. *J. Am. Chem. Soc.* **1976**, *98*, 2774–2779. [CrossRef]
10. Ohtomo, A.; Hwang, H.Y. A high-mobility electron gas at the LaAlO$_3$/SrTiO$_3$ heterointerface. *Nature* **2004**, *427*, 423–426. [CrossRef]
11. Reyren, N.; Thiel, S.; Caviglia, A.D.; Kourkoutis, L.F.; Hammerl, G.; Richter, C.; Schneider, C.W.; Kopp, T.; Ruetschi, A.S.; Jaccard, D.; et al. Superconducting Interfaces Between Insulating Oxides. *Science* **2007**, *317*, 1196–1199. [CrossRef]
12. Goodenough, J.B. Electronic and ionic transport properties and other physical aspects of perovskites. *Rep. Prog. Phys.* **2004**, *67*, 1915–1993. [CrossRef]
13. Müller, K.A.; Berlinger, W.; Waldner, F. Characteristic Structural Phase Transition in Perovskite-Type Compounds. *Phys. Rev. Lett.* **1968**, *21*, 814–817. [CrossRef]
14. Cowley, R.A. Lattice Dynamics and Phase Transitions of Strontium Titanate. *Phys. Rev.* **1964**, *134*, A981–A997. [CrossRef]
15. Zhong, W.; Vanderbilt, D. Effect of quantum fluctuations on structural phase transitions in SrTiO$_3$ and BaTiO$_3$. *Phys. Rev. B* **1996**, *53*, 5047–5050. [CrossRef] [PubMed]
16. Burke, W.J.; Pressley, R.J. Stress induced ferroelectricity in SrTiO$_3$. *Solid State Commun.* **1971**, *9*, 191–195. [CrossRef]
17. Uwe, H.; Sakudo, T. Stress-induced ferroelectricity and soft phonon modes in SrTiO$_3$. *Phys. Rev. B* **1976**, *13*, 271–286. [CrossRef]
18. Mishra, S.K.; Ranjan, R.; Pandey, D.; Ranson, P.; Ouillon, R.; Pinan-Lucarre, J.-P.; Pruzan, P. A combined X-ray diffraction and Raman scattering study of the phase transitions in Sr$_{1-x}$Ca$_x$TiO$_3$ (x = 0.04, 0.06, and 0.012). *J. Solid State Chem.* **2005**, *178*, 2846–2857. [CrossRef]
19. Spinelli, A.; Torija, M.A.; Liu, C.; Jan, C.; Leighton, C. Electronic transport in doped SrTiO$_3$: Conduction mechanisms and potential applications. *Phys. Rev. B* **2010**, *81*, 155110. [CrossRef]
20. Ohta, S.; Nomura, T.; Ohta, H.; Koumoto, K. High-temperature carrier transport and thermoelectric properties of heavily La- or Nb-doped SrTiO$_3$ single crystals. *J. Appl. Phys.* **2005**, *97*, 034106. [CrossRef]
21. Tufte, O.; Chapman, P. Electron Mobility in Semiconducting Strontium Titanate. *Phys. Rev.* **1967**, *155*, 796–802. [CrossRef]
22. van der Marel, D.; van Mechelen, J.L.M.; Mazin, I.I. Common Fermi-liquid origin of T^2 resistivity and superconductivity in *n*-type SrTiO$_3$. *Phys. Rev. B* **2011**, *84*, 205111. [CrossRef]
23. Gregory, B.; Arthur, J.; Seidel, G. Measurements of the Fermi surface of SrTiO$_3$:Nb. *Phys. Rev. B* **1979**, *19*, 1039–1048. [CrossRef]
24. Lin, X.; Bridoux, G.; Gourgout, A.; Seyfarth, G.; Krämer, S.; Nardone, M.; Fauqué, B.; Behnia, K. Critical Doping for the Onset of a Two-Band Superconducting Ground State in SrTiO$_3$. *Phys. Rev. Lett.* **2014**, *112*, 207002. [CrossRef]
25. Schooley, J.F.; Hosler, W.R.; Cohen, M.L. Superconductivity in Semiconducting SrTiO$_3$. *Phys. Rev. Lett.* **1964**, *12*, 474–475. [CrossRef]
26. Koonce, C.S.; Cohen, M.L.; Schooley, J.F.; Hosler, W.R.; Pfeiffer, E.R. Superconducting Transition Temperatures of Semiconducting SrTiO$_3$. *Phys. Rev.* **1967**, *163*, 380–390. [CrossRef]
27. Appel, J. Soft-Mode Superconductivity in SrTiO$_{3-x}$. *Phys. Rev.* **1969**, *180*, 508–516. [CrossRef]
28. Bussmann-Holder, A.; Simon, A.; Büttner, H. Possibility of a common origin to ferroelectricity and superconductivity in oxides. *Phys. Rev. B* **1989**, *39*, 207–214. [CrossRef]

29. Edge, J.M.; Kedem, Y.; Aschauer, U.; Spaldin, N.A.; Balatsky, A.V. Quantum Critical Origin of the Superconducting Dome in $SrTiO_3$. *Phys. Rev. Lett.* **2015**, *115*, 247002. [CrossRef]
30. Kahn, A.H.; Leyendecker, A.J. Electronic Energy Bands in Strontium Titanate. *Phys. Rev.* **1964**, *135*, A1321–A1325. [CrossRef]
31. Cohen, M.L. Superconductivity in Many-Valley Semiconductors and in Semimetals. *Phys. Rev.* **1964**, *134*, A511–A521. [CrossRef]
32. Mattheiss, L.F. Energy Bands for $KNiF_3$, $SrTiO_3$, $KMoO_3$, and $KTaO_3$. *Phys. Rev. B* **1972**, *6*, 4718–4740. [CrossRef]
33. Binnig, G.; Baratoff, A.; Hoenig, H.E.; Bednorz, J.G. Two-Band Superconductivity in Nb-Doped $SrTiO_3$. *Phys. Rev. Lett.* **1980**, *45*, 1352–1355. [CrossRef]
34. van Mechelen, J.L.M.; van der Marel, D.; Grimaldi, C.; Kuzmenko, A.B.; Armitage, N.P.; Reyren, N.; Hagemann, H.; Mazin, I.I. Electron-Phonon Interaction and Charge Carrier Mass Enhancement in $SrTiO_3$. *Phys. Rev. Lett.* **2008**, *100*, 226403. [CrossRef]
35. McCalla, E.; Gastiasoro, M.N.; Cassuto, G.; Fernandes, R.M.; Leighton, C. Low-temperature specific heat of doped $SrTiO_3$: Doping dependence of the effective mass and Kadowaki-Woods scaling violation. *Phys. Rev. Mater.* **2019**, *3*, 022001. [CrossRef]
36. Devreese, J.T.; Klimin, S.N.; van Mechelen, J.L.M.; van der Marel, D. Many-body large polaron optical conductivity in $SrTi_{1-x}Nb_xO_3$. *Phys. Rev. B* **2010**, *81*, 125119. [CrossRef]
37. Eagles, D.M. Possible Pairing without Superconductivity at Low Carrier Concentrations in Bulk and Thin-Film Superconducting Semiconductors. *Phys. Rev.* **1969**, *186*, 456–463. [CrossRef]
38. Ngai, K.L. Two-Phonon Deformation Potential and Superconductivity in Degenerate Semiconductors. *Phys. Rev. Lett.* **1974**, *32*, 215–218. [CrossRef]
39. Gor'kov, L.P. Phonon mechanism in the most dilute superconductor n-type $SrTiO_3$. *Proc. Natl. Acad. Sci. USA* **2016**, *113*, 4646–4651. [CrossRef]
40. Klimin, S.N.; Tempere, J.; van der Marel, D.; Devreese, J.T. Microscopic mechanisms for the Fermi-liquid behavior of Nb-doped strontium titanate. *Phys. Rev. B* **2012**, *86*, 045113. [CrossRef]
41. Enderlein, C.; Ferreira de Oliveira, J.; Tompsett, D.A.; Baggio Saitovitch, E.; Saxena, S.S.; Lonzarich, G.G.; Rowley, S.E. Superconductivity mediated by polar modes in ferroelectric metals. *Nat. Commun.* **2020**, *11*, 4852. [CrossRef] [PubMed]
42. Jarlborg, T. Tuning of the electronic screening and electron-phonon coupling in doped $SrTiO_3$ and WO_3. *Phys. Rev. B* **2000**, *61*, 9887–9890. [CrossRef]
43. Rowley, S.E.; Spalek, L.J.; Smith, R.P.; Dean, M.P.M.; Itoh, M.; Scott, J.F.; Lonzarich, G.G.; Saxena, S.S. Ferroelectric quantum criticality. *Nat. Phys.* **2014**, *10*, 367. [CrossRef]
44. Stucky, A.; Scheerer, G.W.; Ren, Z.; Jaccard, D.; Poumirol, J.M.; Barreteau, C.; Giannini, E.; van der Marel, D. Isotope effect in superconducting n-doped $SrTiO_3$. *Sci. Rep.* **2016**, *6*, 37582. [CrossRef]
45. Tomioka, Y.; Shirakawa, N.; Shibuya, K.; Inoue, I.H. Enhanced superconductivity close to a non-magnetic quantum critical point in electron-doped strontium titanate. *Nat. Commun.* **2019**, *10*, 738. [CrossRef]
46. Rischau, C.W.; Lin, X.; Grams, C.P.; Finck, D.; Harms, S.; Engelmayer, J.; Lorenz, T.; Gallais, Y.; Fauqué, B.; Hemberger, J.; et al. A ferroelectric quantum phase transition inside the superconducting dome of $Sr_{1-x}Ca_xTiO_{3-\delta}$. *Nat. Phys.* **2017**, *13*, 643. [CrossRef]
47. Ruhman, J.; Lee, P.A. Superconductivity at very low density: The case of strontium titanate. *Phys. Rev. B* **2016**, *94*, 224515. [CrossRef]
48. Bussmann-Holder, A.; Bishop, A.R.; Simon, A. Enhancement of T_c in BCS theory extended by interband two-phonon exchange. *Z. Phys. B Condens. Matter* **1993**, *90*, 183–186. [CrossRef]
49. van der Marel, D.; Barantani, F.; Rischau, C.W. Possible mechanism for superconductivity in doped $SrTiO_3$. *Phys. Rev. Res.* **2019**, *1*, 013003. [CrossRef]
50. Santander-Syro, A.F.; Copie, O.; Kondo, T.; Fortuna, F.; Pailhes, S.; Weht, R.; Qiu, X.G.; Bertran, F.; Nicolaou, A.; Taleb-Ibrahimi, A.; et al. Two-dimensional electron gas with universal subbands at the surface of $SrTiO_3$. *Nature* **2011**, *469*, 189–193. [CrossRef]

51. Meevasana, W.; King, P.D.C.; He, R.H.; Mo, S.K.; Hashimoto, M.; Tamai, A.; Songsiriritthigul, P.; Baumberger, F.; Shen, Z.X. Creation and control of a two-dimensional electron liquid at the bare SrTiO$_3$ surface. *Nat. Mater.* **2011**, *10*, 114–118. [CrossRef]
52. Kim, M.; Kozuka, Y.; Bell, C.; Hikita, Y.; Hwang, H.Y. Intrinsic spin-orbit coupling in superconducting δ-doped SrTiO$_3$ heterostructures. *Phys. Rev. B* **2012**, *86*, 085121. [CrossRef]
53. Thiel, S.; Hammerl, G.; Schmehl, A.; Schneider, C.W.; Mannhart, J. Tunable Quasi-Two-Dimensional Electron Gases in Oxide Heterostructures. *Science* **2006**, *313*, 1942–1945. [CrossRef]
54. Bristowe, N.C.; Ghosez, P.; Littlewood, P.B.; Artacho, E. The origin of two-dimensional electron gases at oxide interfaces: insights from theory. *J. Phys. Condens. Matter* **2014**, *26*, 143201. [CrossRef]
55. Yu, L.; Zunger, A. A polarity-induced defect mechanism for conductivity and magnetism at polar-nonpolar oxide interfaces. *Nat. Commun.* **2014**, *5*, 5118. [CrossRef]
56. Tinkham, M. Effect of Fluxoid Quantization on Transitions of Superconducting Films. *Phys. Rev.* **1963**, *129*, 2413–2422. [CrossRef]
57. Reyren, N.; Gariglio, S.; Caviglia, A.D.; Jaccard, D.; Schneider, T.; Triscone, J.M. Anisotropy of the superconducting transport properties of the LaAlO$_3$/SrTiO$_3$ interface. *Appl. Phys. Lett.* **2009**, *94*, 112506. [CrossRef]
58. Ben Shalom, M.; Sachs, M.; Rakhmilevitch, D.; Palevski, A.; Dagan, Y. Tuning Spin-Orbit Coupling and Superconductivity at the SrTiO$_3$/LaAlO$_3$ Interface: A Magnetotransport Study. *Phys. Rev. Lett.* **2010**, *104*, 126802. [CrossRef] [PubMed]
59. Gariglio, S.; Gabay, M.; Mannhart, J.; Triscone, J.M. Interface superconductivity. *Phys. C Superconduct. Appl.* **2015**, *514*, 189–198. [CrossRef]
60. Caviglia, A.D.; Gariglio, S.; Reyren, N.; Jaccard, D.; Schneider, T.; Gabay, M.; Thiel, S.; Hammerl, G.; Mannhart, J.; Triscone, J.M. Electric field control of the LaAlO$_3$/SrTiO$_3$ interface ground state. *Nature* **2008**, *456*, 624–627. [CrossRef]
61. Gariglio, S.; Gabay, M.; Triscone, J.M. Research Update: Conductivity and beyond at the LaAlO$_3$/SrTiO$_3$ interface. *APL Mater.* **2016**, *4*, 060701. [CrossRef]
62. Cancellieri, C.; Reyren, N.; Gariglio, S.; Caviglia, A.D.; Fete, A.; Triscone, J.M. Influence of the growth conditions on the LaAlO$_3$/SrTiO$_3$ interface electronic properties. *EPL (Europhys. Lett.)* **2010**, *91*, 17004. [CrossRef]
63. Gariglio, S.; Reyren, N.; Caviglia, A.D.; Triscone, J.M. Superconductivity at the LaAlO$_3$/SrTiO$_3$ interface. *J. Phys. Condens. Matter* **2009**, *21*, 164213. [CrossRef]
64. Unoki, H.; Sakudo, T. Electron Spin Resonance of Fe^{3+} in SrTiO$_3$ with Special Reference to the 110° K Phase Transition. *J. Phys. Soc. Jpn* **1967**, *23*, 546–552. [CrossRef]
65. Shirane, G.; Yamada, Y. Lattice-Dynamical Study of the 110° K Phase Transition in SrTiO$_3$. *Phys. Rev.* **1969**, *177*, 858–863. [CrossRef]
66. Tao, Q.; Loret, B.; Xu, B.; Yang, X.; Rischau, C.W.; Lin, X.; Fauqué, B.; Verstraete, M.J.; Behnia, K. Nonmonotonic anisotropy in charge conduction induced by antiferrodistortive transition in metallic SrTiO$_3$. *Phys. Rev. B* **2016**, *94*, 035111. [CrossRef]
67. Wang, J.; Yang, L.; Rischau, C.W.; Xu, Z.; Ren, Z.; Lorenz, T.; Lin, X.; Behnia, K. Charge transport in a polar metal. *npj Quantum Mater.* **2019**, *4*, 61. [CrossRef]
68. Tuvia, G.; Frenkel, Y.; Rout, P.K.; Silber, I.; Kalisky, B.; Dagan, Y. Ferroelectric Exchange Bias Affects Interfacial Electronic States. *Adv. Mater.* **2020**, *32*, 2000216. [CrossRef] [PubMed]

Publisher's Note: MDPI stays neutral with regard to jurisdictional claims in published maps and institutional affiliations.

© 2020 by the authors. Licensee MDPI, Basel, Switzerland. This article is an open access article distributed under the terms and conditions of the Creative Commons Attribution (CC BY) license (http://creativecommons.org/licenses/by/4.0/).

Article

Unconventional Transport Properties of Reduced Tungsten Oxide $WO_{2.9}$

Alexander Shengelaya [1,2,*], Fabio La Mattina [3] and Kazimierz Conder [4]

1. Department of Physics, Ivane Javakhishvili Tbilisi State University, Chavchavadze 3, GE-0128 Tbilisi, Georgia
2. Andronikashvili Institute of Physics, Ivane Javakhishvili Tbilisi State University, 0177 Tbilisi, Georgia
3. Laboratory for Transport at Nanoscale Interfaces, Empa Swiss Federal Laboratories for Science and Technology, 8600 Dübendorf, Switzerland; Fabio.Lamattina@empa.ch
4. Laboratory for Multiscale Materials Experiments, Paul Scherrer Institute, 5232 Villigen PSI, Switzerland; kazimierz.conder@psi.ch
* Correspondence: alexander.shengelaya@tsu.ge

Received: 12 September 2020; Accepted: 14 October 2020; Published: 16 October 2020

Abstract: The temperature and magnetic field dependence of resistivity in $WO_{2.9}$ was investigated. The variation of resistivity with temperature displayed unusual features, such as a broad maximum around 230 K and a logarithmic increase of resistivity below 16 K. In the temperature range 16–230 K, we observed metallic-like behavior with a positive temperature coefficient. The combined analysis of resistivity and magnetoresistance (MR) data shows that these unusual transport properties of $WO_{2.9}$ can be understood by considering the (bi)polaronic nature of charge carriers. In contrast to magnetization data, superconducting transition below $T_c = 80$ K was not detected in resistivity measurements, indicating that the superconductivity is localized in small regions that do not percolate. We found a strong increase in positive MR below 80 K. This effect is similar to that observed in underdoped cuprates, where the substantial increase of MR is attributed to superconducting fluctuations in small clusters. Therefore, the temperature dependence of MR indicates the presence of non-percolating superconducting clusters in $WO_{2.9}$ below 80 K in agreement with magnetization data.

Keywords: tungsten oxide; polarons; superconductivity

Preface

The authors of this paper have been fortunate and privileged to collaborate with K. Alex Müller over many years. This collaboration still continues and in this Special Issue of the journal dedicated to his life's work, we would like to present the recent results of a project that was initiated and supported by Alex. Since the beginning of his illustrious scientific carrier, he has investigated a remarkable variety of physical properties of oxide perovskites, and especially $SrTiO_3$. This research culminated in the discovery of high-T_c superconductivity in copper oxides. We are grateful to Alex for uncovering the fascinating world of the perovskite oxides, which continue to attract the attention of condensed matter physicists and materials scientists for more than 50 years.

1. Introduction

It is remarkable that reduced $SrTiO_{3-x}$ was the first oxide in which superconductivity was found in 1964 [1]. Only nine months after the $SrTiO_3$ discovery, superconductivity was also reported for another perovskite oxide, sodium-doped WO_3 [2]. Both $SrTiO_{3-x}$ and Na_xWO_3 have a low T_c below 1 K. However, superconducting regions with unusually high critical temperature ($T_c = 90$ K) were observed on the surface of Na-doped WO_3 crystals [3,4]. Unfortunately, the composition of the superconducting regions could not be determined due to their small size. Another possible method to dope charge carriers in a tungsten oxide is to reduce the oxygen content. Filamentary superconductivity below

T_c = 3 K was observed in twin walls of reduced WO$_{3-x}$ single crystals [5]. Despite the filamentary characteristics of superconductivity and low T_c, an unusually large upper critical field $\mu_0 H_{c2}(0)$ above 15 T was found in WO$_{3-x}$. This large $\mu_0 H_{c2}(0)$ violates the Pauli paramagnetic limit by a factor of three. In our opinion, this interesting finding deserves further study.

The parent compound WO$_3$ has a perovskite-like structure with a vacant A site and consists of a three-dimensional array of corner-sharing tungsten-oxygen octahedra. Pure WO$_3$ is insulating since the W^{6+} (5d^0) has an empty d-shell. Creating oxygen deficiency in WO$_3$ leads to a strong increase in conductivity as W^{5+} (5d^1) ions are induced, which are a source of the charge carriers. In the octahedral oxygen coordination, W^{5+} hosts one 5d electron in a triply degenerate orbital and therefore is a Jahn-Teller (JT) ion. A number of experimental studies demonstrated that these charge carriers have a polaronic character and form bipolarons in WO$_{3-x}$ [6]. The discovery of high-T_c superconductivity in copper-oxide materials resulted from the concept of JT polarons [7]. Subsequent experimental and theoretical studies proved the validity of this concept by providing ample evidence of polaron/bipolaron formation and their clustering in cuprates [8–11]. Therefore, high-T_c superconductivity may also exist in oxygen-reduced tungsten oxide. Recently, two authors of this paper, Alexander Shengelaya and Kazimierz Conder, together with Alex Müller, reported that in reduced tungsten oxide with the composition WO$_{2.9}$, the signatures of filamentary superconductivity with T_c = 80 K can be registered by means of magnetization measurements [12]. By lithium intercalation, the T_c was further increased to 94 K. These results indicate that there is a potential for high-T_c superconductivity in oxygen-reduced tungsten oxide, which has not been sufficiently explored.

Until now, not much was known about the low-temperature transport properties of the WO$_{2.9}$ phase. A few reports were published where the temperature dependence of resistivity was measured [13–15]. However, all these measurements were performed above liquid nitrogen temperature. Here, we report results of combined magnetic, transport, and magnetotransport measurements of oxygen reduced tungsten oxide WO$_{2.9}$ in a broad temperature range, including the previously unexplored low temperature (T < 80 K) region.

2. Experimental Details

The polycrystalline WO$_{3-x}$ samples were prepared by the solid-state reaction method starting from WO$_3$ and WO$_2$ reagents in the powder form. Details on the sample preparation can be found in [12].

The DC magnetization measurements were performed on a Quantum Design magnetic property measurement system (MPMS) magnetometer. For the zero-field-cooled (ZFC) magnetization measurements, the samples were first cooled to 5 K in a zero magnetic field, then the magnetic field was applied, and the magnetization was measured with increasing temperature.

The transport measurements were performed on a Quantum Design physical property measurement system (PPMS) via a four-probe technique. Electrical contacts were prepared by indium wires and a silver paste. The magnetic field for magnetoresistance measurements was oriented perpendicular to the current direction.

3. Results and Discussion

Figure 1 shows the temperature dependence of the ZFC magnetizations in WO$_{3-x}$ samples measured in a magnetic field of 100 Oe. Among WO$_{3-x}$ samples with different oxygen contents, only one sample with the composition WO$_{2.9}$ exhibited a clear decrease in magnetization below 80 K. This diamagnetic transition is characteristic of the superconducting state. As was demonstrated in a previous publication, various results obtained from magnetization measurements strongly support the superconducting nature of the diamagnetic transitions below T_c = 80 K in reduced tungsten oxide WO$_{2.9}$ [12].

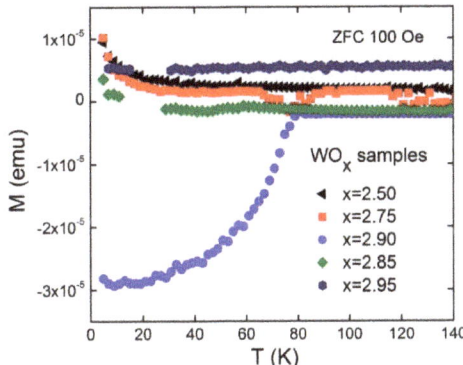

Figure 1. Temperature dependence of the zero-field-cooled (ZFC) magnetizations in a magnetic field of 100 Oe for the WO_{3-x} samples with different oxygen contents.

The magnetization data presented in Figure 1 demonstrate that the reduced tungsten oxide ($WO_{2.9}$) is a special case as it has a high-temperature superconducting phase. To understand why this composition is special, recall that the removal of oxygen from WO_3 induces polaronic charge carriers and also leads to structural changes. The lattice tends to eliminate single oxygen vacancies by the crystallographic shear (CS) process. As the x value in WO_{3-x} increases, the usual corner-sharing arrangement of octahedra is partially replaced by groups of edge-sharing WO_6 octahedra, which form pockets of shear planes [16]. If these shear planes become parallel and equidistant, a crystalline phase with a defined structure arises, as shown by Magnéli [17]. This is the case for the dark blue modification of the tungsten oxide $WO_{2.9}$, which was the subject of the present study. It crystallizes in a large unit cell containing 20 tungsten and 58 oxygens atoms. Therefore, its stoichiometry can be represented as $W_{20}O_{58}$. It belongs to the family of the Magnéli-type oxides with the general formula W_nO_{3n-2} [18]. The $W_{20}O_{58}$ phase is one of the most stable and ordered members of the W_nO_{3n-2} homologous series, where blocks of 2D corner-sharing WO_6 octahedra are mutually connected along CS planes formed by groups of six edge-sharing octahedra [17]. This phase also has a very narrow stability range close to the theoretical stoichiometry value $WO_{2.9}$ with composition limits of $WO_{2.89}$–$WO_{2.92}$ [19]. This well-defined composition of the $W_{20}O_{58}$ phase explains why we observed signatures of superconductivity only within very limited oxygen contents in the series of studied WO_{3-x} samples.

According to Salje and Güttler, $WO_{2.9}$ is also a special compound among reduced tungsten oxides since it is on the verge of the insulator to metal transition [15]. However, they only measured resistivity in this compound down to 80 K. To better understand transport properties of $WO_{2.9}$, we performed resistivity and magnetoresistance measurements in a broad temperature range, including the previously unexplored low temperature region. Figure 2 shows the resistivity as a function of temperature for the $WO_{2.9}$ sample. Metallic behavior ($d\rho/dT > 0$) was observed in the temperature range 16–230 K. In contrast to magnetization measurements, the resistivity had no anomaly at $T_c = 80$ K, indicating that the sample had a low superconducting volume fraction. The estimation of the superconducting volume fraction based upon the magnetization measurement shown in Figure 1 is ~0.01%. Notably, this value is a lower limit valid under the assumption that the size of superconducting regions d is larger than the London penetration depth λ. If $d \ll \lambda$, the volume fraction will be larger. Still, the superconducting volume fraction in $WO_{2.9}$ is small and is apparently below the percolation limit. Therefore, the temperature dependences of magnetization and resistivity could be explained by the formation of filamentary, non-percolative superconducting regions below $T_c = 80$ K in Magnéli-phase tungsten oxide $WO_{2.9}$ ($W_{20}O_{58}$).

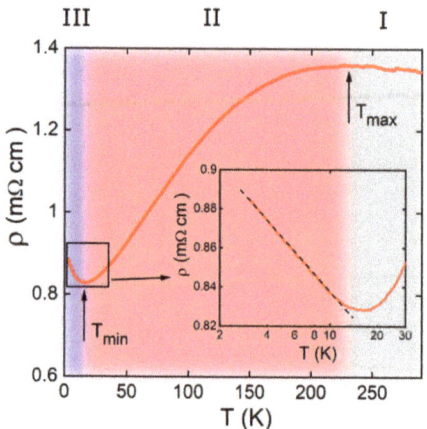

Figure 2. Resistivity as a function of temperature for the WO$_{2.9}$ sample. Colors indicate temperature regions with three distinct transport regimes. Inset: Expanded view of the low-temperature region shown with a logarithmic horizontal scale. Dashed line is a guide for the eye.

It was proposed that WO$_{2.9}$ bipolarons can form and cluster in the edge-sharing WO$_6$ octahedra within CS planes of the Magnéli phase [12]. According to this model, CS planes represent charge-carrier-rich quasi-1D stripes. With decreasing temperature, the superconducting state can be established locally in regions similar to cuprates [20]. However, CS planes are very narrow with a thickness of only six edge-sharing WO$_6$ octahedra in the W$_{20}$O$_{58}$ phase. The nanoscale bipolaron-rich metallic regions in CS planes can be locally superconducting, but because of the small size and limited coupling between them, global phase coherence and superconductivity are difficult to establish. Therefore, in the absence of percolation between the CS planes, the superconductivity has a filamentary character as indicated by magnetization and resistivity measurements.

Now, we discuss the temperature dependence of resistivity in more detail, as shown in Figure 2. There are three distinct transport regimes in the $\rho(T)$ dependence: (I) a semiconducting-like behavior above ~230 K with a negative temperature coefficient, (II) metallic-like behavior in the temperature range of 16–230 K with a positive temperature coefficient, and (III) resistivity upturn below ~16 K.

A broad maximum in the $\rho(T)$ dependence, similar to that shown in Figure 2, was reported previously for WO$_{2.9}$, but its origin was not discussed [13–15]. Notably, a similar crossover from semiconducting-like conductivity at high temperature to metallic-like behavior at a lower temperature was observed in the normal state of quasi-2D organic superconductors (BEDT-TTF)$_2$X (ET$_2$X) with X = Cu[N(CN)$_2$]Cl, Cu[N(CN)$_2$]Br, Cu(NCS)$_2$ [21]. Various explanations have been proposed for the different transport regimes and an anomalous resistance maximum [22]. Some models consider the strongly correlated nature of the electrons in organic superconductors. A smooth crossover from an incoherent "bad-metal" state at high temperatures to a coherent Fermi liquid below the resistivity maximum was obtained within the framework of dynamical mean-field theory [17]. Other explanations involve polaron physics to describe anomalous transport behavior. Optical spectroscopy revealed polaron formation in (ET)$_2$Cu(NCS)$_2$ [23]. In particular, the spectra reveal a crossover from a small-polaron-dominated regime at high T to a large-polaron-dominated regime at low T. At high temperatures, small polarons do not participate in a coherent motion and the conductivity emerges as an activated hopping of self-trapped carriers. At low temperatures, the average size of polarons increases, and a coherent motion of polarons with resulting metallic behavior appears [24]. Other authors also considered an activated hopping of small polarons at high T, but low-temperature metallic behavior was ascribed to tunneling conduction by small polarons according to the prediction of small polaron transport theory [25]. In reduced tungsten oxides, strong electron correlations are not expected, but polaron/bipolaron formation is well documented. Therefore, the observed broad peak in the

T-dependent resistivity for $WO_{2.9}$ can be explained as a result of a crossover from high T activated hopping of small (bi)polarons to low T metallic transport due to their tunneling conduction or transition to large (bi)polaron state.

Another notable feature in the $\rho(T)$ dependence of $WO_{2.9}$ is a resistivity minimum and an upturn toward semiconductor-like behavior, as shown in Figure 2. Interestingly, resistivity increases logarithmically with decreasing temperature, as shown in the inset of Figure 2. Weak localization and the Kondo effect could both lead to this dependence. A logarithmic divergence in the resistivity can be found in Kondo systems, such as dilute magnetic alloys and dense Kondo compounds, where the conduction electrons are scattered by magnetic impurities [26]. However, $WO_{2.9}$ is clearly not a Kondo system as the presence of magnetic impurities is not supported by our low-temperature magnetization measurements. The logarithmic temperature dependence of the resistivity can also be observed due to the quantum corrections to the resistivity, such as weak localization (WL). WL is a quantum effect caused by constructive interference between the closed-loop trajectories of a diffusive electron. In 2D systems, the WL effect results in a logarithmic rise in resistivity at low temperatures [27]. The WL effect is sensitive to an applied magnetic field, which disturbs the wave coherence and the self-interference effects [28]. Therefore, a magnetic field can destroy WL, which is revealed in experiments as a negative magnetoresistance (MR) and the shift of the resistivity minimum T_{min} to lower temperatures.

To obtain more information about the unusual transport properties of $WO_{2.9}$, we performed MR measurements. Figure 3 shows the temperature dependence of resistivity measured in 0 and in 7 T applied field. $WO_{2.9}$ exhibits a significant increase in resistivity in a magnetic field in the whole temperature range including the low-temperature upturn. In addition, the resistivity minimum T_{min} shifts toward higher temperatures from 16 to 23 K by applying a magnetic field of 7 T. These observations are opposite to what is expected for the WL. Therefore, based on the MR measurements, the WL effect can be excluded in our $WO_{2.9}$ sample.

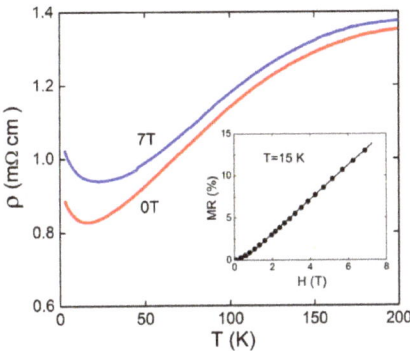

Figure 3. Temperature dependence of resistivity of $WO_{2.9}$ in zero and in 7 T applied field. Inset: Magnetic field dependence of magnetoresistance at $T = 15$ K. Solid line is a guide for the eye.

Notably, a logarithmic low-temperature upturn in the resistivity, similar to that observed here for $WO_{2.9}$ is a well-known feature of underdoped cuprates [29–31]. It was shown that such behavior can be described remarkably well by resonance Wigner scattering of bipolarons by the random potential in underdoped cuprates [32]. Due to the presence of bipolarons in reduced tungsten oxides, this mechanism provides a very plausible explanation of the observed unusual low-temperature transport properties of $WO_{2.9}$.

As MR measurements in reduced tungsten oxides have not been reported before, the large positive MR presented in Figure 3 is a new observation. The inset in Figure 3 shows MR as a function of magnetic field at $T = 15$ K, with the conventional definition MR = $[[\rho(H) - \rho(0)]/\rho(0)] \times 100\%$, where $\rho(H)$ and $\rho(0)$ are the values of resistivity at field H and zero, respectively. Above 1 T, MR increases linearly with the field without any sign of saturation. Such behavior is incompatible with that expected in conventional metals, where a positive MR arises due to the Lorentz force when the electric and magnetic fields are applied transverse to each other. This mechanism leads to the usual quadratic field dependence of the transverse MR [33]. Even more surprisingly, we found that a positive MR with a similar magnitude also exists in a configuration where a magnetic field was applied parallel to the current direction (longitudinal MR). This isotropic MR excludes the possibility of resulting from the orbital motion of charge carriers. Isotropic MR is usually related to spin scattering mechanisms, but in this case, MR is expected to be negative rather than positive [34]. Therefore, we conclude that the unusual, isotropic, and positive MR observed in $WO_{2.9}$ at low temperatures cannot be explained by known spin and orbital mechanisms related to charge carriers in a normal state of metals.

Possibly the most interesting aspect of the MR is its strong temperature dependence, which is presented in Figure 4. It shows that at high temperatures, the MR is small and increases very gradually with decreasing temperature. However, it increases strongly upon cooling below ~80 K and reaches a value of ~15% at low temperatures in a field of 7 T. This temperature coincides with the superconducting transition observed by magnetization measurements (Figure 1). This suggests that the substantial increase in positive MR below 80 K could be related to the onset of superconductivity in $WO_{2.9}$. We note that a considerable positive MR with unusual temperature dependence was reported recently in strongly underdoped $La_{2-x}Sr_xCuO_4$ thin films [35]. For example, in a sample where x = 0.05, a magnetic field of 45 T caused an excess resistivity of 10% at a temperature of 10 K. With increasing temperature, the MR decreased to a constant value of ~2% around 40 K. This temperature corresponds to the maximum T_c observed in optimally doped $La_{2-x}Sr_xCuO_4$. Notably, the sample with x = 0.05 did not show signatures of superconductivity in resistivity measurements down to 1.5 K. The considerable excess positive MR observed at low temperatures, which depends non-quadratically on the magnetic field, was attributed to superconducting fluctuations in underdoped $La_{2-x}Sr_xCuO_4$ [35]. Magnetic field suppresses superconducting fluctuations, leading to an enhanced resistivity. In our opinion, this scenario also applies to the present case and explains the resistivity and MR data obtained in $WO_{2.9}$. It also accounts for the isotropic character of the observed positive MR in this compound at low temperatures.

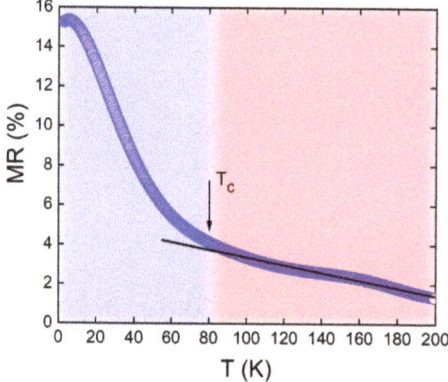

Figure 4. Temperature dependence of magnetoresistance of $WO_{2.9}$ in 7 T applied field. The solid line is a guide for the eye. The arrow marks the superconducting transition observed in magnetization measurements.

As discussed above, there are locally-superconducting bipolaron-rich regions in CS planes of $WO_{2.9}$, but because of the small size and limited coupling between them, global phase coherence and bulk superconductivity are not established. Still, superconducting fluctuations might be present, which are difficult to detect from resistivity measurements. A strong magnetic field would suppress superconducting fluctuations; as a result, resistivity increases. The resulting positive MR will have a strong temperature dependence and should disappear above the T_c of superconducting regions. The positive MR detected in this work for $WO_{2.9}$ and previously in strongly underdoped cuprates seems to be a very sensitive indicator for the presence of small, isolated superconducting clusters embedded in a nonsuperconducting matrix.

4. Summary and Conclusions

To summarize, we performed magnetic, transport, and magnetotransport measurements in Magnéli-phase oxygen-reduced tungsten oxide $WO_{2.9}$ ($W_{20}O_{58}$) in a broad temperature range. We found that the temperature dependence of the resistivity is unusual and three different transport regimes exist in $WO_{2.9}$: a semiconducting-like behavior above ~230 K with a negative temperature coefficient, metallic-like behavior in the temperature range of 16–230 K with a positive temperature coefficient, and resistivity upturn below ~16 K. The broad maximum can be explained as a result of a crossover from high T-activated hopping of small (bi)polarons to low-T metallic transport due to either tunneling conduction of small (bi)polarons or transition to large (bi)polaron state.

We observed for the first time that the resistivity of $WO_{2.9}$ has a minimum with an upturn toward insulator-like behavior below 16 K. We found that the resistivity increases logarithmically with decreasing temperature. MR measurements showed that the observed low-temperature resistivity upturn cannot be explained by known mechanisms, such as weak localization or Kondo effects. One possible explanation of this effect is the resonance Wigner scattering of bipolarons by the random potential in $WO_{2.9}$. This mechanism was suggested previously for the low-temperature upturn of resistivity in underdoped cuprates.

The superconducting transition at $T_c = 80$ K observed in the magnetization measurements was not detected in the temperature dependence of resistivity. This indicates that the superconductivity is localized in small regions that do not percolate. Interestingly, we found a strong increase in positive MR below 80 K. This effect is very similar to that observed in strongly underdoped cuprates, where the considerable excess positive MR was attributed to superconducting fluctuations in small regions. Therefore, we conclude that MR measurements in agreement with magnetization data revealed the presence of superconducting clusters in $WO_{2.9}$ below 80 K. As suggested in a previous publication, such superconducting clusters form within CS planes that exist in the Magnéli phase of $WO_{2.9}$ ($W_{20}O_{58}$) and represent charge-carrier-rich quasi-1D stripes [12]. These nanoscale clusters can host superconducting pairs, but because of the small size and limited coupling between them, global phase coherence and superconductivity cannot be achieved. Generally, the MR effect is a very sensitive tool for detecting the presence of small, isolated superconducting clusters embedded in a nonsuperconducting matrix.

The obtained results demonstrate that the Magnéli phase of the oxygen-reduced tungsten oxide $WO_{2.9}$ ($W_{20}O_{58}$) has interesting normal and superconducting properties that warrant further studies, especially using single crystal samples. We note that, at present, little is known about the electronic band structure of this material. Recent theoretical studies have suggested that there are several flat bands near the Fermi level [36] and that the electronic properties may be governed by the small number of d-bands from the tungsten atoms located in CS planes [37]. So far, experimental studies of the band structure in $WO_{2.9}$ are lacking.

Especially promising, in our opinion, would be to study pressure effects on the normal and superconducting properties of this compound. By applying hydrostatic or uniaxial pressure, the band structure and coupling between CS planes can be tuned. A well-known example is the ladder compound (La, Sr, Ca)$_{14}$Cu$_{24}$O$_{41}$, which becomes superconducting under high pressure when the interactions between the ladders are enhanced [38]. Another interesting and novel approach to achieve bulk high-T_c superconductivity in doped tungstates was proposed recently by Müller [39]. Therefore, by improving the coupling and percolation between CS planes, bulk superconductivity and zero resistance state might be achieved in $WO_{2.9}$.

Author Contributions: Project planning, A.S.; samples synthesis, K.C.; magnetization measurements, A.S.; transport measurements, F.L.M. and A.S.; writing manuscript, A.S.; review and editing manuscript, A.S., F.L.M., and K.C. All authors have read and agreed to the published version of the manuscript.

Funding: This research was funded by the Swiss National Science Foundation SCOPES Grant No. IZ74Z0-160484 and the Shota Rustavely National Science Foundation of Georgia under grant no. STCU 2017_29.

Acknowledgments: The authors are grateful to K.A. Müller and H. Keller for stimulating discussions.

Conflicts of Interest: The authors declare no conflict of interest.

References

1. Schooley, J.F.; Hosler, W.R.; Cohen, M.L. Superconductivity in Semiconducting SrTiO3. *Phys. Rev. Lett.* **1964**, *12*, 474–475. [CrossRef]
2. Raub, C.J.; Sweedler, A.R.; Jensen, M.A.; Broadston, S.; Matthias, B.T. Superconductivity of Sodium Tungsten Bronzes. *Phys. Rev. Lett.* **1964**, *13*, 746–747. [CrossRef]
3. Reich, S.; Tsabba, Y. Possible nucleation of a 2D superconducting phase on WO single crystals surface doped with Na. *Eur. Phys. J. B* **1999**, *9*, 1–4. [CrossRef]
4. Shengelaya, A.; Reich, S.; Tsabba, Y.; Müller, K. Electron spin resonance and magnetic susceptibility suggest superconductivity in Na doped WO samples. *Eur. Phys. J. B* **1999**, *12*, 13–15. [CrossRef]
5. Aird, A.; Salje, E.K.H. Sheet superconductivity in twin walls: Experimental evidence of. *J. Phys. Condens. Matter* **1998**, *10*, L377–L380. [CrossRef]
6. Salje, E.K.H. Polarons and bipolarons in tungsten oxide, WO_{3-x}. *Eur. J. Solid State Inorg. Chem.* **1994**, *31*, 805–821.
7. Bednorz, J.G.; Müller, K.A. Perovskite-Type Oxides-the New Approach to High-TcSuperconductivity. Nobel Lecture. *Angew. Chem. Int. Ed.* **1988**, *27*, 735–748. [CrossRef]
8. Bussmann-Holder, A.; Keller, H.; Müller, K.A. Evidences for Polaron Formation in Cuprates. In *Family Medicine*; Springer Science and Business Media LLC: Berlin/Heidelberg, Germany, 2005; pp. 365–384.
9. Müller, K.A. Essential Heterogeneities in Hole-Doped Cuprate Superconductors. In *Family Medicine*; Springer Science and Business Media LLC: Berlin/Heidelberg, Germany, 2005; pp. 1–11.
10. Innocenti, D.; Ricci, A.; Poccia, N.; Campi, G.; Fratini, M.; Bianconi, A. A Model for Liquid-Striped Liquid Phase Separation in Liquids of Anisotropic Polarons. *J. Supercond. Nov. Magn.* **2009**, *22*, 529–533. [CrossRef]
11. Müller, K.A. The Polaronic Basis for High-Temperature Superconductivity. *J. Supercond. Nov. Magn.* **2017**, *30*, 3007–3018. [CrossRef]
12. Shengelaya, A.; Conder, K.; Müller, K.A. Signatures of Filamentary Superconductivity up to 94 K in Tungsten Oxide $WO_{2.90}$. *J. Supercond. Nov. Magn.* **2019**, *33*, 301–306. [CrossRef]
13. Berak, J.M.; Sienko, M. Effect of oxygen-deficiency on electrical transport properties of tungsten trioxide crystals. *J. Solid State Chem.* **1970**, *2*, 109–133. [CrossRef]
14. Sahle, W.; Nygren, M. Electrical Conductivity and High Resolution Electron Microscopy Studies of WO_{3-x} Crystals with $0 \leq x \leq 0.28$. *J. Solid State Chem.* **1983**, *48*, 154–160. [CrossRef]
15. Salje, E.; Güttler, B. Anderson transition and intermediate polaron formation in WO_{3-x} Transport properties and optical absorption. *Philos. Mag. B* **1984**, *50*, 607–620. [CrossRef]
16. Tilley, R. The crystal chemistry of the higher tungsten oxides. *Int. J. Refract. Met. Hard Mater.* **1995**, *13*, 93–109. [CrossRef]
17. Magnéli, A. Crystal structure studies on beta-tungsten oxide. *Arkiv. Kemi* **1949**, *1*, 513–523.
18. Bursill, L.; Hyde, B. CS families derived from the ReO_3 structure type: An electron microscope study of reduced WO_3 and related pseudobinary systems. *J. Solid State Chem.* **1972**, *4*, 430–446. [CrossRef]
19. Marucco, J.-F.; Gerdanian, P.; Dodé, M. Contribution à l'étude thermodynamique des oxydes de tungstène, à 1000 °C, dans le domaine [math]. *J. Chim. Phys.* **1969**, *66*, 674–684. [CrossRef]
20. Shengelaya, A.; Müller, K.A. The intrinsic heterogeneity of superconductivity in the cuprates. *EPL Europhys. Lett.* **2014**, *109*, 27001. [CrossRef]
21. Ishiguro, T.; Ito, H. *Structure and Phase Diagram of Organic Superconductors*; Springer Science and Business Media LLC: Berlin/Heidelberg, Germany, 1998; pp. 135–146.
22. Lang, M.; Müller, J. *The Physics of Superconductors Vol. II*; Bennemann, K.H., Ketterson, J.B., Eds.; Springer: Berlin/Heidelberg, Germany, 2004; pp. 453–554.

23. Merino, J.; McKenzie, R.H. Transport properties of strongly correlated metals: A dynamical mean-field approach. *Phys. Rev. B* **2000**, *61*, 7996–8008. [CrossRef]
24. Wang, N.L.; Clayman, B.P.; Mori, H.; Tanaka, S. Far-infrared study of the insulator-metal transition in kappa-(BEDT-TTF)$_2$Cu(NCS)$_2$: Evidence for polaron absorption. *J. Phys. Condens. Matter* **2000**, *12*, 2867–2875. [CrossRef]
25. Cariss, C.; Porter, L.; Thorn, R. Temperature dependence of conductivity of κ (BEDT TTF)$_2$Cu(SCN)$_2$; Resolution into two components; Small polaron. *Solid State Commun.* **1990**, *74*, 1269–1273. [CrossRef]
26. Kondo, J. Resistance Minimum in Dilute Magnetic Alloys. *Prog. Theor. Phys.* **1964**, *32*, 37–49. [CrossRef]
27. Abrahams, E.; Anderson, P.W.; Licciardello, D.C.; Ramakrishnan, T.V. Scaling Theory of Localization: Absence of Quantum Diffusion in Two Dimensions. *Phys. Rev. Lett.* **1979**, *42*, 673–676. [CrossRef]
28. Altshuler, B.L.; Khmel'Nitzkii, D.; Larkin, A.I.; Lee, P.A. Magnetoresistance and Hall effect in a disordered two-dimensional electron gas. *Phys. Rev. B* **1980**, *22*, 5142–5153. [CrossRef]
29. Boebinger, G.S.; Ando, Y.; Passner, A.; Kimura, T.; Okuya, M.; Shimoyama, J.; Kishio, K.; Tamasaku, K.; Ichikawa, N.; Uchida, S. nsulator-to-Metal Crossover in the Normal State of La$_{2-x}$Sr$_x$CuO$_4$ Near Optimum Doping. *Phys. Rev. Lett.* **1996**, *77*, 5417–5420. [CrossRef]
30. Ando, Y.; Boebinger, G.S.; Passner, A.; Kimura, T.; Kishio, K. Logarithmic Divergence of both In-Plane and Out-of-Plane Normal-State Resistivities of Superconducting La$_{2-x}$Sr$_x$CuO$_4$ in the Zero-Temperature Limit. *Phys. Rev. Lett.* **1995**, *75*, 4662–4665. [CrossRef]
31. Fournier, P.; Mohanty, P.; Maiser, E.; Darzens, S.; Venkatesan, T.; Lobb, C.J.; Czjzek, G.; Webb, R.A.; Greene, R.L. Insulator-Metal Crossover near Optimal Doping in Pr$_{2-x}$Ce$_x$CuO$_4$: Anomalous Normal-State Low Temperature Resistivity. *Phys. Rev. Lett.* **1998**, *81*, 4720–4723. [CrossRef]
32. Alexandrov, A. Logarithmic normal state resistivity of high-Tc cuprates. *Phys. Lett. A* **1997**, *236*, 132–136. [CrossRef]
33. Pippard, A.B. *Magnetoresistance in Metals*; Cambridge University Press: Cambridge, UK, 1989.
34. Preyer, N.W.; Kastner, M.A.; Chen, C.Y.; Birgeneau, R.J.; Hidaka, Y. Isotropic negative magnetoresistance in La$_{2-x}$Sr$_x$CuO$_{4+y}$. *Phys. Rev. B* **1991**, *44*, 407–410. [CrossRef]
35. Vanacken, J.; Moshchalkov, V.V. Transport properties of high-Tc cuprate thin films as superconductive materials. In *High-Temperature Superconductors*; Qiu, X.G., Ed.; Woodhead Publishing: Cambridge, UK, 2011; pp. 38–102.
36. Sun, L.; Li, Z.; Su, R.; Wang, Y.; Li, Z.; Du, B.; Sun, Y.; Guan, P.; Besenbacher, F.; Yu, M. Phase-Transition Induced Conversion into a Photothermal Material: Quasi-Metallic WO$_{2.9}$ Nanorods for Solar Water Evaporation and Anticancer Photothermal Therapy. *Angew. Chem. Int. Ed.* **2018**, *57*, 10666–10671. [CrossRef]
37. Slobodchikov, A.A.; Nekrasov, I.A.; Pavlov, N.S.; Korshunov, M.M. From the open Heisenberg model to the Landau-Lifshitz equation. *arXiv* **2020**, arXiv:2006.16658v1.
38. Uehara, M.; Nagata, T.; Akimitsu, J.; Takahashi, H.; Mori, N.; Kinoshita, K. Superconductivity in the Ladder Material Sr$_{0.4}$Ca$_{13.6}$Cu$_{24}$O$_{41.84}$. *J. Phys. Soc. Jpn.* **1996**, *65*, 2764–2767. [CrossRef]
39. Müller, K.A. Generation of Bulk HTS with Doped Tungstates. *J. Supercond. Nov. Magn.* **2017**, *30*, 2707–2709. [CrossRef]

Publisher's Note: MDPI stays neutral with regard to jurisdictional claims in published maps and institutional affiliations.

© 2020 by the authors. Licensee MDPI, Basel, Switzerland. This article is an open access article distributed under the terms and conditions of the Creative Commons Attribution (CC BY) license (http://creativecommons.org/licenses/by/4.0/).

MDPI
St. Alban-Anlage 66
4052 Basel
Switzerland
Tel. +41 61 683 77 34
Fax +41 61 302 89 18
www.mdpi.com

Condensed Matter Editorial Office
E-mail: condensedmatter@mdpi.com
www.mdpi.com/journal/condensedmatter

www.ingramcontent.com/pod-product-compliance
Lightning Source LLC
LaVergne TN
LVHW070201100526
838202LV00015B/1980